MONOGRAPHS ON STATISTICS AND APPLIED PROBABILITY

General Editors

D.R. Cox, V. Isham, N. Keiding, T. Louis, N. Reid, R. Tibshirani, and H. Tong

33 Analysis of Infectious Disease Data *N.G. Becker* (1989)
34 Design and Analysis of Cross-Over Trials *B. Jones and M.G. Kenward* (1989)
35 Empirical Bayes Methods, 2nd edition *J.S. Maritz and T. Lwin* (1989)
36 Symmetric Multivariate and Related Distributions
K.T. Fang, S. Kotz and K.W. Ng (1990)
37 Generalized Linear Models, 2nd edition *P. McCullagh and J.A. Nelder* (1989)
38 Cyclic and Computer Generated Designs, 2nd edition
J.A. John and E.R. Williams (1995)
39 Analog Estimation Methods in Econometrics *C.F. Manski* (1988)
40 Subset Selection in Regression *A.J. Miller* (1990)
41 Analysis of Repeated Measures *M.J. Crowder and D.J. Hand* (1990)
42 Statistical Reasoning with Imprecise Probabilities *P. Walley* (1991)
43 Generalized Additive Models *T.J. Hastie and R.J. Tibshirani* (1990)
44 Inspection Errors for Attributes in Quality Control
N.L. Johnson, S. Kotz and X. Wu (1991)
45 The Analysis of Contingency Tables, 2nd edition *B.S. Everitt* (1992)
46 The Analysis of Quantal Response Data *B.J.T. Morgan* (1992)
47 Longitudinal Data with Serial Correlation—A state-space approach
R.H. Jones (1993)
48 Differential Geometry and Statistics *M.K. Murray and J.W. Rice* (1993)
49 Markov Models and Optimization *M.H.A. Davis* (1993)
50 Networks and Chaos—Statistical and probabilistic aspects
O.E. Barndorff-Nielsen, J.L. Jensen and W.S. Kendall (1993)
51 Number-Theoretic Methods in Statistics *K.-T. Fang and Y. Wang* (1994)
52 Inference and Asymptotics *O.E. Barndorff-Nielsen and D.R. Cox* (1994)
53 Practical Risk Theory for Actuaries
C.D. Daykin, T. Pentikäinen and M. Pesonen (1994)
54 Biplots *J.C. Gower and D.J. Hand* (1996)
55 Predictive Inference—An introduction *S. Geisser* (1993)
56 Model-Free Curve Estimation *M.E. Tarter and M.D. Lock* (1993)
57 An Introduction to the Bootstrap *B. Efron and R.J. Tibshirani* (1993)
58 Nonparametric Regression and Generalized Linear Models
P.J. Green and B.W. Silverman (1994)
59 Multidimensional Scaling *T.F. Cox and M.A.A. Cox* (1994)
60 Kernel Smoothing *M.P. Wand and M.C. Jones* (1995)
61 Statistics for Long Memory Processes *J. Beran* (1995)
62 Nonlinear Models for Repeated Measurement Data
M. Davidian and D.M. Giltinan (1995)
63 Measurement Error in Nonlinear Models
R.J. Carroll, D. Rupert and L.A. Stefanski (1995)
64 Analyzing and Modeling Rank Data *J.J. Marden* (1995)
65 Time Series Models—In econometrics, finance and other fields
D.R. Cox, D.V. Hinkley and O.E. Barndorff-Nielsen (1996)

66 Local Polynomial Modeling and its Applications *J. Fan and I. Gijbels* (1996)

67 Multivariate Dependencies—Models, analysis and interpretation *D.R. Cox and N. Wermuth* (1996)

68 Statistical Inference—Based on the likelihood *A. Azzalini* (1996)

69 Bayes and Empirical Bayes Methods for Data Analysis *B.P. Carlin and T.A Louis* (1996)

70 Hidden Markov and Other Models for Discrete-Valued Time Series *I.L. Macdonald and W. Zucchini* (1997)

71 Statistical Evidence—A likelihood paradigm *R. Royall* (1997)

72 Analysis of Incomplete Multivariate Data *J.L. Schafer* (1997)

73 Multivariate Models and Dependence Concepts *H. Joe* (1997)

74 Theory of Sample Surveys *M.E. Thompson* (1997)

75 Retrial Queues *G. Falin and J.G.C. Templeton* (1997)

76 Theory of Dispersion Models *B. Jørgensen* (1997)

77 Mixed Poisson Processes *J. Grandell* (1997)

78 Variance Components Estimation—Mixed models, methodologies and applications *P.S.R.S. Rao* (1997)

79 Bayesian Methods for Finite Population Sampling *G. Meeden and M. Ghosh* (1997)

80 Stochastic Geometry—Likelihood and computation *O.E. Barndorff-Nielsen, W.S. Kendall and M.N.M. van Lieshout* (1998)

81 Computer-Assisted Analysis of Mixtures and Applications—Meta-analysis, disease mapping and others *D. Böhning* (1999)

82 Classification, 2nd edition *A.D. Gordon* (1999)

83 Semimartingales and their Statistical Inference *B.L.S. Prakasa Rao* (1999)

84 Statistical Aspects of BSE and vCJD—Models for Epidemics *C.A. Donnelly and N.M. Ferguson* (1999)

85 Set-Indexed Martingales *G. Ivanoff and E. Merzbach* (2000)

86 The Theory of the Design of Experiments *D.R. Cox and N. Reid* (2000)

87 Complex Stochastic Systems *O.E. Barndorff-Nielsen, D.R. Cox and C. Klüppelberg* (2001)

88 Multidimensional Scaling, 2nd edition *T.F. Cox and M.A.A. Cox* (2001)

89 Algebraic Statistics—Computational Commutative Algebra in Statistics *G. Pistone, E. Riccomagno and H.P. Wynn (2001)*

90 Analysis of Time Series Structure—SSA and Related Techniques *N. Golyandina, V. Nekrutkin and A.A. Zhigljavsky (2001)*

91 Subjective Probability Models for Lifetimes *Fabio Spizzichino (2001)*

92 Empirical Likelihood *Art B. Owen (2001)*

Multivariate Models and Dependence Concepts

Harry Joe

Professor of Statistics
University of British Columbia
Canada

CHAPMAN & HALL/CRC

A CRC Press Company
Boca Raton London New York Washington, D.C.

Library of Congress Cataloging-in-Publication Data

Catalog record is available from the Library of Congress

Direct all inquiries to CRC Press LLC, 2000 N.W. Corporate Blvd., Boca Raton, Florida 33431.

Visit the CRC Press Web site at www.crcpress.com

First edition 1997
First CRC reprint 2001
Originally Published by Chapman & Hall

International Standard Book Number 0-412-07331-5
Printed in the United States of America 1 2 3 4 5 6 7 8 9 0
Printed on acid-free paper

Contents

To the memory of my parents,

CHOU Kum Won and TSE Jip Lin

Preface

This book is devoted to (a) multivariate models for non-normal response, an area of probability and statistics with increasing activity and applications, and (b) dependence concepts that are useful for analysing properties of multivariate models. It also adds to the knowledge of the space of multivariate distributions.

By a multivariate model, I mean a parametric statistical model for a multivariate response, possibly with covariates. Examples are models for multivariate or longitudinal count, binary and ordinal response data. My approach consists of the modelling of the univariate margins followed by adding the appropriate dependence structure, with considerations of positive or negative dependence, and exchangeable, time series or general dependence structure. I find that dependence concepts and dependence analysis are necessary to understand a model and when it might be applicable. This includes analysis of the range of dependence that a model permits and whether the dependence of the model increases as multivariate parameters increase.

This book's special features include:

- methods for constructions of multivariate non-normal distributions and copulas;
- the topics of Fréchet classes and dependence concepts;
- an introduction to new statistical inference theory for multivariate models — the method of *inference functions for margins* is presented in Chapter 10;
- data analysis examples with comparisons of models, diagnostic checking and sensitivity analyses;
- exercises and unsolved problems at the end of chapters;
- some supplementary results in the Appendix, in order to make the book more self-contained.

This book has minimal overlap with earlier books on bivariate and multivariate distributions.

The methods and models of this book extend commonly used univariate models to multivariate models in which parameters of the models can be considered as univariate parameters or dependence parameters, and allow one to make a variety of inferences as well as assess assumptions, do diagnostic checks, make model comparisons and perform sensitivity analyses. These are not all possible with the method of generalized estimating equations (GEEs), which is based on partly specified probability models. There have been many advances in research in multivariate non-normal distributions since researchers proposed methods like the GEE approach partly because of a lack of existing models. The models and methods here are more general and more flexible, and less dependent on assumptions, than are GEEs.

The topics in the book have been largely motivated by applications. Because of space limitations, I cover only the main concepts and ideas that can be used to construct and analyse multivariate distributions and models. There is by no means an exhaustive coverage of what has appeared in the probability and statistics literature in multivariate models and dependence concepts, and there are no comparisons for analysis of multivariate or longitudinal non-normal response data with methods that do not fall within the theme of the 'multivariate' approach. Only the most relevant references are cited and these are mainly in the sections entitled 'Bibliographic notes'.

There is no real linear ordering of the material in this book, so that the more foundational results are given earlier. Different sections are cross-referenced in order that the reader can more easily move around non-linearly. Some features to help the reader are as follows.

- Section 1.3 consists of notation, abbreviations and conventions used throughout.
- Those sections that provide a basic introduction to multivariate models and dependence concepts are indicated with a $^{\circ}$ symbol in the section title and those that are very advanced are indicated with a * symbol.
- There is an appendix at the end to make the book more self-contained.
- The index is arranged so that the first page number listed is usually the definition of a term.

Audience

This book is written for a number of audiences: (i) those who work with and analyse multivariate or longitudinal non-normal response data; (ii) those interested in methods for constructing multivariate non-normal distributions; (iii) researchers (or those who want to become researchers) in multivariate non-normal statistics; and (iv) those who need a reference for multivariate models and dependence concepts.

This book can be read or used in several ways. The reader who is more interested in the theory and foundations can start from the beginning. The reader who is more interested in applications and how the theory applies can start with the examples of data analyses in Chapter 11, and then read the sections with the relevant theory for the multivariate models and inference. This book can also be used as a reference or as a starting point for further research (there are pointers throughout on further research that could be done, for example, in the exercises and unsolved problems).

This book assumes that the reader has some background in mathematical statistics and probability. To implement or to analyse the models, the reader should be able to write out the probability distributions based on stochastic representations. Sometimes the model is given only in terms of stochastic representations because this takes less space and makes properties of the model more evident. In a few places, terminology from measure theory or other areas of mathematics is used; the usage is explained in the Appendix.

This book could be used for a graduate course on multivariate non-normal statistics or as a supplementary book for courses in multivariate statistics, time series, categorical data, and longitudinal data analysis.

Acknowledgments

I have had two opportunities to teach a graduate course on 'Multivariate models for non-normal response, with covariates', first in the Department of Mathematics and Statistics at the University of Pittsburgh from January to April 1994, and second in the Department of Statistics at the University of British Columbia from January to April 1996. I started by writing some notes for the first course, and a first draft of this book was tried out during the second course. I appreciate very much the feedback from the students and audience in these courses that have led to improvements.

I am very grateful to Henry Block and Allan Sampson for hospitable arrangements for my sabbatical leave visit at the University of Pittsburgh. Much difficult research was completed there; without having done that work, this book could not have been considered.

I thank my research collaborators, departmental colleagues, and former M.Sc. and Ph.D. students for their contributions; I am especially grateful for the many contributions of Taizhong Hu and Jianmeng Xu. For inspiration, I acknowledge the publications of Albert Marshall and Ingram Olkin, and Barry Arnold.

As to data sets in the book, I am grateful to Scott Page for permission to use the cardiac surgery data, to Sverre Vedal and John Petkau for permission to use the Port Alberni data, and to Martin Puterman for permission to use the bacteriuria data.

This book was typeset in LaTeX (D.E. Knuth, 1986: *The TeXbook*, Addison-Wesley, Reading, MA; L. Lamport, 1986: LaTeX. *A Document Preparation System*, Addison-Wesley, Reading, MA) with a Chapman & Hall style file. The plots were produced in S-Plus (R.A. Becker, J.M. Chambers and A.R. Wilks, 1988: *The New S Language*, Wadsworth, Pacific Grove, CA; *S-Plus User's Manual*, Statistical Sciences, Inc., Seattle, WA) and MATLAB (*MATLAB User's Guide*, The MathWorks, Inc., Natick, MA).

Finally, I would like to thank the Natural Science and Engineering Research Council of Canada for research grant support, and the Chapman & Hall editors and staff for their efforts and advice.

CHAPTER 1

Introduction

This book is devoted to multivariate models for non-normal response (e.g., binary, ordinal, count, extreme value), an area of probability and statistics with increasing activity and applications. It also contributes to the understanding of the space of multivariate distributions. Important ideas in the book include:

- construction of multivariate models to cover various types of dependence structure;
- development of dependence concepts and their use to analyse multivariate models;
- use of stochastic representations, mixtures and latent variables to construct parametric models with nice dependence properties;
- use of the copula as a summary of the dependence in a multivariate distribution, independent of the univariate margins;
- Fréchet classes with given marginal distributions;
- time series models with given univariate margins;
- the emphasis on properties of multivariates models to decide their applicability;
- the estimation method of inference functions for margins, consisting of parameter estimates from likelihoods of marginal distributions of a multivariate model, together with the jackknife method for standard error estimates;
- comparison of models in examples of data analysis.

In this first chapter, an overview of the book is given in Section 1.1, the style and format are described in Section 1.2, notation and abbreviations are summarized in Section 1.3, some basic results are given in Sections 1.4 to 1.6, and a point of view for statistical modelling is set out in Section 1.7.

1.1 Overview and background

This is a book on multivariate models in which there is a multivariate response vector and possibly a vector of covariates, explanatory variables or factors. For a multivariate response, we will include the cases of

(i) repeated measures, time series or longitudinal data, in which a response variable is measured sequentially at several points in time (with covariates possibly changing with time);

(ii) measurements on m different variables;

(iii) measurements of a variable for each member of a cluster, family or litter;

(iv) combinations of (i), (ii) and (iii).

Concrete examples are the following.

1. For a study of the risk factors of cardiac surgery, binary response variables measured after cardiac surgery are indicators of low-output syndrome and renal, neurological and pulmonary complications (these indirectly measure the quality of life after surgery), and covariates include age, sex, and indicators of chronic obstructive pulmonary disease, prior myocardial infarction, renal disease and diabetes.

2. For an epidemiological study on the effects of air pollution on health, response variables which are measured daily are number of hospital emergency room visits (for respiratory, cardiac and other types of visits), absenteeism and mortality count (for combined, respiratory and other causes of death). Inhalable particulate matter, ozone and organic dust are the principal pollutants of interest, and other covariates are meteorological variables.

3. For extreme value inference concerning air quality in a region with several monitoring stations, measurements are daily maxima of hourly averaged concentrations of several pollutants (e.g., ozone, sulphur dioxide, oxides of nitrogen) at each station.

4. For a study on the psychological effects over time after a disaster, subjects are measured for stress (an ordinal response variable) at several time points; one covariate is the distance from the site of the disaster.

There is a well-developed theory for the case in which the multivariate response vector can be assumed to have a multivariate normal distribution. There is relatively little on multivariate models for non-normal response variables, such as multivariate or longitudinal count, binary and ordinal response data. This is perhaps due to the mathematical intractability of reasonable models and computational problems in statistical inference with the models. The generalized estimating equation (GEE) approach with a partly specified model (some moments and/or marginal distributions but no joint distributions are specified) has been developed since the mid-1980s. However the GEE approach has several disadvantages, including limited types of inferences that can be made, and the lack of a clear accompanying means of diagnostic checking and assessment of implicit assumptions.

This book concentrates on models that can be applied to nonnormal multivariate data, with one chapter devoted to multivariate discrete models, for use with binary, ordinal categorical or count data, another chapter devoted to multivariate extreme value models, etc. Rather than starting with a complex model that can cover many situations, we start by building from the simple cases. The simplest case is that of no covariates, and this reduces to the study of multivariate distributions (with given univariate margins). Then there are several approaches to allow for covariates, including letting parameters in a family of multivariate distributions be functions of the covariates.

The study of multivariate distributions is not easy because one cannot just write down a family of functions and expect it to satisfy the necessary conditions for multivariate cumulative distribution functions (see the conditions in Section 1.4). We will mainly be constructing multivariate distributions through methods such as mixtures, latent variables and stochastic representations, to avoid the need for tedious and perhaps impossible checks on the necessary conditions. Different general methods to obtain families of multivariate distributions are given in Chapter 4, together with their dependence properties, and some parametric families are given in Chapter 5. (Nonparametric multivariate inference requires far more data than parametric multivariate inference; the 'curse of dimensionality' is a problem with the former.) Until recently, little research had been done in the area of multivariate non-normal distributions.

The approach of multivariate models in this book is that of generalizing univariate models or distributions, and obtaining models for which univariate margins belong to a given family. Time series models with univariate margins in a given family are a special case, in which there is a special dependence structure for the response variables. This is different from defining a class of multivariate models or time series models and then asking what are the possible univariate margins (e.g., elliptically contoured distributions, which are discussed briefly in Section 4.9). The construction of multivariate models becomes an existence problem if certain dependence structures are desired.

Methods for obtaining a multivariate family include the following.

1. From a characterizing property of a parametric univariate family, generalize to the multivariate case. Examples are min-stable multivariate exponential distributions and multivariate distributions with univariate margins in a convolution-closed infinitely divisible class.

2. For continuous variables, make use of the probability transformation, so that the multivariate dependence or structure is independent of univariate margins. The copula, which summarizes the dependence structure, is a multivariate distribution with uniform (0,1) margins. It is introduced and some of its properties studied in Section 1.6.

For non-normal random variables, correlation is not the best measure of dependence. More generally useful dependence concepts are introduced in Chapter 2. These are necessary for analysing the type and range of dependence in a parametric family of multivariate models. A parametric family has extra interpretability if some of the parameters can be identified as dependence or multivariate parameters. More specifically, for some general concept of positive and negative dependence, one would like to say that some range of the parameters corresponds to positive dependence and some to negative dependence, and furthermore, it would be desirable to have the amount of dependence increasing as parameters increase.

Chapter 3 contains results on Fréchet classes, including Fréchet bounds, which sometimes are the most dependent multivariate distributions given knowledge of univariate margins and possibly some higher-dimensional margins. For a given parametric family, to know whether it is applicable to given situations, one needs to know the range of dependence that is covered. This can be compared relative to the Fréchet bounds for the magnitude of dependence, and relative to the full range of $\{\delta_{ij} : i < j\}$ over all multivariate distributions, where δ_{ij} is a bivariate dependence measure for the (i, j) bivariate margin. From the latter, one can assess the type of dependence in a parametric family of multivariate distributions.

For example, special types of dependence pattern are: (i) permutation symmetric or exchangeable, in which all bivariate margins are the same (and δ_{ij} does not depend on (i, j)); (ii) partially exchangeable, in which there are only a few distinct bivariate margins among the set of all bivariate margins; (iii) decreasing in dependence with lag, suitable for time series or longitudinal data, in which the amount of dependence in a bivariate margin decreases as $|j - i|$ increases.

This book does not contain an encyclopaedic or exhaustive list of what have been proposed for families of multivariate distributions. The aim is to concentrate on techniques to obtain parametric families of models that (i) cover the types of dependence mentioned above, (ii) have parameters that are all interpretable, (iii) apply to multivariate discrete data (e.g., binary, ordinal, count) or multivariate non-normal continuous data (e.g., extreme value, exponential). The techniques as well as listings of parametric families are given in Chapters 4 to 9. Multivariate survival data are not covered other than the frailty models for survival times associated with members of clusters. Despite the literature on multivariate survival functions as models for lifetimes of components of a system, it is not clear what this means in practice when components are replaced or repaired — perhaps stochastic process models rather than multivariate models are more appropriate.

The hardest part of statistical inference for multivariate non-normal responses has been the multivariate modelling and the relevant dependence concepts. Hence most of this book (Chapters 1 to 9) is devoted to these topics, starting with simpler cases which can serve as building blocks for more complex models. Much of classical inference (e.g., sufficiency, ancillarity, unbiasedness), apart from asymptotic likelihood theory, does not apply to estimation in multivariate models. However, there are some new ideas associated with estimation and data analysis in multivariate models, in particular, parameter estimates from likelihoods of marginal distributions of a multivariate model. Chapters 10 and 11 are devoted to inference, computing and data analysis (and comparisons of models for some real data sets).

1.2 Style and format

Each chapter is divided into sections, some of which have subsections. Most chapters have sections entitled 'Bibliographic notes', 'Exercises' and 'Unsolved problems'. The exercises are roughly ordered by level of difficulty, starting with more basic exercises. Some exercises contain material that supplements results in the text. Generally, background and references for each chapter are given in the sections on 'Bibliographic notes', with the complete citations in the References section at the end of the book. The mention of references and authors in the text is rare. There are results in this book that are definitely new, and other results that may be new; in the latter case, I do not know of any published references. Some results that are conjectured to be true or seem to be true based on some numerical experience are listed in the sections on 'Unsolved problems'.

It will be observed that equations and theorems are numbered by chapter and not by section. Most theorems are combined together into subsections on dependence properties. Key words of definitions are given in bold face.

To make the material in this book easier to read, ideas are presented in simpler cases and then extended to more general cases. This book is mostly self-contained, with some background material in the Appendix, and has minimal overlap with other books on bivariate and multivariate distributions (because of the large void in knowledge about multivariate non-normal distributions).

Short proofs of results are included if they are illuminating, and there are examples throughout the book to illustrate the important concepts. On the other hand, some things that may become repetitive or straightforward to check are left as exercises. The style of the book is partly influenced by the plan to keep it within a certain length. Sometimes, for example, in order to save space, only a stochastic representation and not a probability distribution is given.

There is no real linear ordering of the material in this book, so that results that are more foundational are given earlier. Different sections of the book are cross-referenced in order that the reader can more easily move around non-linearly. Sections that provide a basic introduction to the topic of multivariate models and dependence concepts are indicated with a ° symbol and sections that are very advanced are indicated with a * symbol.

This book can be read or used in several ways. The reader who is more interested in the theory and foundational issues and concepts can start from the beginning. The reader who is more interested in applications and how the theory applies can start with the examples of data analyses in Chapter 11, and then read the sections with the relevant theory for the multivariate models and inference.

This book can also be used as a reference or as a starting point for further research (there are pointers throughout on further research that could be done).

1.3 Notation, abbreviations and conventions

This section consists of notation, abbreviations and conventions that are used throughout the book.

First, **multivariate** as an adjective refers to results that are valid in two or more dimensions. **Bivariate** as an adjective refers to results that are valid in two dimensions, but may not extend to higher dimensions, and similarly for **trivariate**. This usage is not always consistent in the statistical literature; there are papers which use the word *multivariate* but contain only bivariate results that are not extendible.

Second, unless stated otherwise, a **multivariate 'x'** distribution means that all univariate margins are in the class 'x', e.g., multivariate Poisson or multivariate exponential. This is common usage in the statistical literature, but the property does not hold for all existing multivariate distributions that are named.

Other items are enumerated below.

1. Simplifying assumptions, such as existence of derivatives, are used at times for convenience of presentation of ideas and to avoid too much technical detail.
2. All functions involved in integrals (or expectations) are assumed measurable with respect to the appropriate measure.
3. The words 'non-increasing' and 'non-decreasing' are not used; instead 'increasing', 'strictly increasing', 'decreasing' and 'strictly decreasing' are used.
4. cdf is the abbreviation for **cumulative distribution function**, pdf is the abbreviation for **probability density function**, and pmf is the abbreviation for **probability mass function**.
5. rv is the abbreviation for **random variable**.
6. iid is the abbreviation for **independent and identically distributed**.
7. BVN and MVN are the abbreviations for **bivariate normal** and **multivariate normal**; BVSN and MVSN are used when the univariate margins are **standard normal**, i.e., zero mean vector and unit variances.

8. GEV is the abbreviation for the **generalized extreme value** (univariate) distribution.
9. BEV and MEV are the abbreviations for **bivariate extreme value** and **multivariate extreme value**, respectively.
10. MSMVE is the abbreviation for **min-stable multivariate exponential**; it is mainly used in Chapter 6.
11. The abbreviations AR(p) for **autoregressive** of order p, MA(k) for **moving average** of order k, and ARMA for **autoregressive moving average** are used, mainly in Chapter 8.
12. LT is the abbreviation for **Laplace transform**. Some results on Laplace transforms are given in the Appendix. All LTs in this book are assumed to have a limiting value of 0 at ∞ unless otherwise stated.
13. ML and MLE are the abbreviations for **maximum likelihood** and **maximum likelihood estimate** or **estimation**.
14. IFM is the abbreviation for **inference functions for margins**. This is a method for the estimation of parameters in multivariate models that is based on the log-likelihoods of marginal distributions of the model. The theory is given in Chapter 10 and the method is used in Chapter 11.
15. SE is the abbreviation for **standard error**.
16. AIC is the abbreviation for **Akaike information criterion**.
17. $\Re^d$ is the symbol for **Euclidean space** of dimension d, and the real line is denoted by $\Re$.
18. $\sim$ is the symbol for **distributed as**, $\stackrel{d}{=}$ is the symbol for **equality in distribution** or **stochastic equality**, $\rightarrow_d$ is the symbol for **convergence in distribution or law**, $\stackrel{\text{sgn}}{=}$ is the symbol for **equality in sign**, $\stackrel{\text{def}}{=}$ is the symbol for **defined as**, $\uparrow$ is the symbol for **increasing**, $\downarrow$ is the symbol for **decreasing**, $\uparrow_{\text{st}}$ ($\downarrow_{\text{st}}$) is the symbol for **stochastically increasing (decreasing)**.
19. $\prec$, with possibly a subscript or superscript, is used to denote a **partial ordering** or **pre-ordering**.
20. $\prec^{st}$ denotes the **stochastic ordering** for cdfs. For univariate cdfs F, F', $F \prec^{st} F'$ if $F(x) \geq F'(x)$ for all $x \in \Re$; for multivariate cdfs, $F \prec^{st} F'$ if $\int g\, dF \leq \int g\, dF'$ for all increasing functions g for which the expectations exist.
21. $\emptyset$ is used for the **empty set**.

22. The **cardinality** of a finite set S is denoted by $|S|$.
23. The **complement** of a set or event A is denoted by A^c unless indicated otherwise.
24. The **transpose** of a vector or matrix is indicated with a superscript T. Vectors are usually row vectors. Whether a vector is a row or column vector will be clear from the context.
25. m is used for the **dimension** of the multivariate response vector or multivariate distribution.
26. $\mathcal{S} = \mathcal{S}_m$ is used for the set of non-empty subsets of $\{1, \ldots, m\}$.
27. Italic Latin upper-case letters, often X, Y, Z, usually with subscripts, are used for random variables; bold Latin upper-case letters, often $\mathbf{X}, \mathbf{Y}, \mathbf{Z}$, are used for random vectors.
28. Bold Latin lower-case letters are used for vectors, and the components are written in italic form with subscripts, e.g., $\mathbf{a} = (a_1, \ldots, a_m)$.
29. $\mathbf{1}_m$ is used for an m-vector of 1s.
30. $\mathbf{Y} = (Y_1, \ldots, Y_m)$ or $\mathbf{y} = (y_1, \ldots, y_m)$ is used to denote a response vector. A vector of explanatory variables or covariates is usually denoted by $\mathbf{x}$ or $\mathbf{z}$.
31. Script Latin upper-case letters are used for classes of sets or functions, e.g., $\mathcal{L}$, $\mathcal{F}$.
32. Greek lower-case letters, often with subscripts, are used for parameters of families of distributions, e.g., θ, δ. Usually τ is used for **Kendall's tau**, ρ for **Spearman's rho** or **Pearson's correlation**, λ for **tail dependence**. Bold Greek letters are used for parameter vectors, e.g., $\boldsymbol{\theta}, \boldsymbol{\mu}$.
33. ϕ, ψ, often with subscripts, are used mainly for Laplace transforms or strictly decreasing differentiable functions. Classes of such functions that are used are denoted by:

$$\mathcal{L}_m = \{\phi : [0,\infty) \to [0,1] \mid \phi(0) = 1,\ \phi(\infty) = 0,$$
$$(-1)^j \phi^{(j)} \geq 0,\ j = 1, \ldots, m\}, \qquad (1.1)$$

$m = 1, 2, \ldots, \infty$, with $\mathcal{L}_\infty$ being the class of Laplace transforms (with 0 value at ∞). Other classes are:

$$\mathcal{L}_n^* = \{\omega : [0,\infty) \to [0,\infty) \mid \omega(0) = 0, \omega(\infty) = \infty,$$
$$(-1)^{j-1} \omega^{(j)} \geq 0,\ j = 1, \ldots, n\}, \qquad (1.2)$$

$n = 1, 2, \ldots, \infty$. The functions in $\mathcal{L}_n^*$ are usually compositions of the form $\psi^{-1} \circ \phi$ with $\psi, \phi \in \mathcal{L}_1$.

34. The following notation is used for certain special distributions and random variables: $U(a, b)$ for **uniform** on $[a, b]$; $N(\mu, \sigma^2)$ for univariate **normal** with mean μ and variance σ^2; $N_m(\boldsymbol{\mu}, \Sigma)$ for **m-variate normal** with mean vector $\boldsymbol{\mu}$ and covariance matrix Σ; $\text{Gamma}(\alpha, \sigma)$ for **gamma** with shape parameter α and scale parameter σ (and mean $\alpha\sigma$); $\text{NB}(\theta, p)$ for **negative binomial** with probability parameter p and mean $\theta(p^{-1} - 1)$.
35. F, G, H are the common symbols for a (multivariate) cdf; sometimes M is used as the cdf of a mixing variable.
36. For an m-variate cdf F, the set of its **marginal distributions** is denoted by $\{F_S : S \in \mathcal{S}_m\}$; for a specific S, the subscript is written without braces, e.g., F_1, F_{12}, F_{123}, etc.
37. If the **density** of a cdf F and its margins exist, they are denoted by f and f_S, $S \in \mathcal{S}$.
38. **Conditional cdfs and densities** derived from a multivariate cdf are written in the form $F_{S_1|S_2}$, $f_{S_1|S_2}$; the latter is equivalent to $f_{S_1 \cup S_2}/f_{S_2}$.
39. The **survival function** corresponding to a cdf F is denoted by $\overline{F}$; its margins are $\{\overline{F}_S\}$. If $(X_1, \ldots, X_m) \sim F$, then

$$\overline{F}(\mathbf{x}) = \Pr(X_i > x_i,\, i = 1, \ldots, m).$$

For $m = 1$, $\overline{F}(x) = 1 - F(x)$. For $m = 2$, $\overline{F}(x_1, x_2) = 1 - F_1(x_1) - F_2(x_2) + F(x_1, x_2)$; for $m = 3$, $\overline{F}(x_1, x_2, x_3) = 1 - F_1(x_1) - F_2(x_2) - F_3(x_3) + F_{12}(x_1, x_2) + F_{13}(x_1, x_3) + F_{23}(x_2, x_3) - F(x_1, x_2, x_3)$. For general m,

$$\overline{F}(\mathbf{x}) = 1 + \sum_{S \in \mathcal{S}} (-1)^{|S|} F_S(x_j, j \in S). \tag{1.3}$$

A related formula is:

$$F(\mathbf{x}) = 1 + \sum_{S \in \mathcal{S}} (-1)^{|S|} \overline{F}_S(x_j, j \in S). \tag{1.4}$$

40. For a univariate cdf F, F^{-1} denotes the **quantile** or **inverse cdf**. It is defined as usual to be left-continuous, i.e., $F^{-1}(v) = \inf\{x : F(x) \geq v\}$, $0 < v < 1$.
41. The symbol $\mathcal{F}$ is used for **Fréchet classes** given a set of margins, e.g., $\mathcal{F}(F_1, \ldots, F_m)$ denotes the class of multivariate distributions with the given univariate margins $F_1, \ldots, F_m$ and $\mathcal{F}(F_{12}, F_{23})$ denotes the class of trivariate distributions with given (1,2) and (2,3) bivariate margins F_{12}, F_{23}.

1.4 Conditions for multivariate distribution functions

This section consists of the conditions that a function must satisfy in order to be a multivariate cdf. The simpler bivariate case is presented first.

To prove that a function F is a multivariate cdf, it is often necessary to construct rvs $Y_1, \ldots, Y_m$ through latent variables, mixtures and limits, etc., and then show that $\mathbf{Y} \sim F$. In general, it is difficult or impossible to show that a function F is a proper cdf by making use of the conditions given in this section; of course, lower-dimensional cases are easier to handle analytically.

1.4.1 Properties of a bivariate cdf F

Necessary and sufficient conditions for a right-continuous function F on $\Re^2$ to be a bivariate cdf are:

1. $\lim_{x_j \to \infty} F(x_1, x_2) = 0$, $j = 1, 2$;
2. $\lim_{x_j \to \infty \forall j} F(x_1, x_2) = 1$;
3. (rectangle inequality) for all $(a_1, a_2), (b_1, b_2)$ with $a_1 < b_1, a_2 < b_2$,

$$F(b_1, b_2) - F(a_1, b_2) - F(b_1, a_2) + F(a_1, a_2) \geq 0. \qquad (1.5)$$

Note the following observations:

(i) If F has second-order derivatives, then condition 3 is equivalent to $\partial^2 F / \partial x_1 \partial x_2 \geq 0$.

(ii) Conditions 1 and 2 imply that $0 \leq F \leq 1$.

(iii) Let $a_2 \to -\infty$ in (1.5); then $F(b_1, b_2) - F(a_1, b_2) \geq 0$ and F is increasing in the first variable. Similarly, from letting $a_1 \to -\infty$, F is increasing in the second variable.

(iv) The univariate margins F_1, F_2 of $F(x_1, x_2)$ are obtained by letting $x_2 \to \infty$ and $x_1 \to \infty$, respectively.

1.4.2 Properties of a multivariate cdf F

Necessary and sufficient conditions for a right-continuous function F on $\Re^m$ to be a multivariate cdf are:

1. $\lim_{x_j \to \infty} F(x_1, \ldots, x_m) = 0$, $j = 1, \ldots, m$;
2. $\lim_{x_j \to \infty \forall j} F(x_1, \ldots, x_m) = 1$;

3. (rectangle inequality) for all $(a_1, \ldots, a_m)$, $(b_1, \ldots, b_m)$ with $a_i < b_i$, $i = 1, \ldots, m$,

$$\sum_{i_1=1}^{2} \cdots \sum_{i_m=1}^{2} (-1)^{i_1+\cdots+i_m} F(x_{1i_1}, \ldots, x_{mi_m}) \geq 0, \tag{1.6}$$

where $x_{j1} = a_j$, $x_{j2} = b_j$.

Note the following:

(i) If F has mth-order derivatives, then condition 3 is equivalent to $\partial^m F/\partial x_1 \cdots \partial x_m \geq 0$.

(ii) Let $a_2, \ldots, a_m \to -\infty$ in (1.6); then

$$F(b_1, b_2, \ldots, b_m) - F(a_1, b_2, \ldots, b_m) \geq 0$$

and F is increasing in the first variable. Similarly, by symmetry, F is increasing in the remaining variables.

(iii) Let $S \in \mathcal{S}_m$. The margin F_S of $F(\mathbf{x})$ is obtained by letting $x_i \to \infty$ for $i \notin S$.

1.5 Types of dependence

As mentioned earlier, a multivariate model should be analysed for the types of dependence structure that it covers as well as the range of dependence. These dependence properties are important in order for one to know whether a particular model might be suitable for a given application or data set. Types of dependence include: (i) singularities on some curves or surfaces; (ii) positive and negative dependence; (iii) exchangeable dependence or flexible dependence; (iv) dependence decreasing with lag if there is a time index.

Sometimes the type of dependence for a multivariate model and whether the model can be used in a specific instance can be understood from stochastic representations and derivations of the model, so analysis of a model should include the search for one or more stochastic representations.

1.6 Copulas

For continuous multivariate distributions, the univariate marginals and the multivariate or dependence structure can be separated, and the multivariate structure can be represented by a copula.

The **copula** is a multivariate distribution with all univariate margins being $U(0,1)$. Hence if C is a copula, then it is the distribution of a multivariate uniform random vector. For an m-variate distribution $F \in \mathcal{F}(F_1, \ldots, F_m)$, with jth univariate margin F_j, the **copula associated with** F is a distribution function $C : [0,1]^m \to [0,1]$ that satisfies

$$F(\mathbf{x}) = C(F_1(x_1), \ldots, F_m(x_m)), \;\; \mathbf{x} \in \Re^m. \tag{1.7}$$

If F is a continuous m-variate distribution function with univariate margins $F_1, \ldots, F_m$, and quantile functions $F_1^{-1}, \ldots, F_m^{-1}$, then

$$C(\mathbf{u}) = F(F_1^{-1}(u_1), \ldots, F_m^{-1}(u_m))$$

is the unique choice in (1.7). This result essentially follows from two properties: (i) if H is a univariate cdf with inverse cdf H^{-1} and $U \sim U(0,1)$, then $H^{-1}(U) \sim H$; (ii) if H is a *continuous* univariate cdf and $X \sim H$, then $H(X) \sim U(0,1)$. That is, if $\mathbf{X} \sim F$ and F is continuous, then $(F_1(X_1), \ldots, F_m(X_m)) \sim C$, and if $\mathbf{U} \sim C$, then $(F_1^{-1}(U_1), \ldots, F_m^{-1}(U_m)) \sim F$.

The copula can be considered 'independent' of the univariate margins, since if C is a copula, then

$$G(\mathbf{y}) = C(G_1(y_1), \ldots, G_m(y_m))$$

is a distribution (survival) function if $G_1, \ldots, G_m$ are all univariate distribution (survival) functions. If C is parametrized by a (vector) parameter $\boldsymbol{\theta}$, then we call $\boldsymbol{\theta}$ a **multivariate parameter**.

Example 1.1 For $\delta > 0$, the distribution

$$F(x,y) = \exp\{-[e^{-x} + e^{-y} - (e^{\delta x} + e^{\delta y})^{-1/\delta}]\}, \quad -\infty < x, y < \infty,$$

is obtained as a limiting distribution in Section 5.1. By letting $y \to \infty$ and $x \to \infty$ in turn, its univariate margins are $F_1(x) = \exp\{-e^{-x}\}$ and $F_2(y) = \exp\{-e^{-y}\}$. By substituting $u = F_1(x)$ and $v = F_2(y)$, or $x = -\log(-\log u)$ and $y = -\log(-\log v)$, one obtains the copula

$$C(u,v) = uv \exp\{[(-\log u)^{-\delta} + (-\log v)^{-\delta}]^{-1/\delta}\},$$

which is in the family B7 in Section 5.1. A bivariate survival function with exponential margins is $G(s,t) = C(e^{-s}, e^{-t}) = \exp\{-s - t + (s^{-\delta} + t^{-\delta})^{-1/\delta}\}$. □

If F is an m-variate distribution of discrete rvs, then the copula associated with F is not unique. The above argument does not work because if H is a *non-continuous* or discrete univariate cdf and $X \sim H$, then $H(X)$ does not have a $U(0,1)$ distribution. An example of a copula that satisfies (1.7) in the discrete case is given below. The main idea is that the copula is defined over a discrete grid of points and assumed to be conditionally uniform in between; it can be shown that this leads to a multivariate density with $U(0,1)$ margins.

For the purely discrete case, let the points of support for the jth margin be x_{j,i_j}, where i_j is in the ordered index set D_j, $j = 1,\ldots,m$. Suppose that each D_j is a consecutive sequence of integers. Let $F_j(i_j)$ and $f_j(i_j)$ be the univariate cdf and pmf for the jth margin. Let $P(i_1,\ldots,i_m)$ and $p(i_1,\ldots,i_m)$ be the cdf and pmf for the joint m-variate distribution. A copula C associated with P satisfies

$$P(i_1,\ldots,i_m) = C(F_1(i_1),\ldots,F_m(i_m)), \quad \mathbf{i} \in D_1 \times \cdots \times D_m.$$

To define the remaining values of C, suppose that C is uniform in the rectangle $\times_{1\le j\le m}[F_j(i_j-1), F_j(i_j)]$, i.e., the multivariate pdf of C in this rectangle is $p(i_1,\ldots,i_m)/\prod_{j=1}^m f_j(i_j)$. For $F_1(i_1-1) < u_1 \le F_1(i_1)$, integrating over the margins $j = 2,\ldots,m$ leads to

$$\sum_{i_2}\cdots\sum_{i_m} \frac{p(i_1,\ldots,i_m)}{f_1(i_1)\cdots f_m(i_m)} f_2(i_2)\cdots f_m(i_m)$$

$$= [f_1(i_1)]^{-1}\sum_{i_2}\cdots\sum_{i_m} p(i_1,\ldots,i_m) = 1.$$

Hence the first univariate margin is $U(0,1)$; by symmetry, the other univariate margins can also be deduced to be $U(0,1)$.

The non-uniqueness comes from the fact that the copula C satisfying (1.7) need not be uniform over rectangles. The details are left as an exercise.

We now move on to other properties of copulas. Since a copula C is the distribution of a random vector, $\mathbf{U} = (U_1,\ldots,U_m)$, where each $U_j \sim U(0,1)$, C is a continuous function. However C need not be absolutely continuous (there may not be density with respect to Lebesgue measure on $\Re^m$), in which case it has a singular component. Often, in cases where C is not absolutely continuous, the singular component can be identified through a functional relationship. For example, if $C_U(\mathbf{u}) = \min\{u_1,\ldots,u_m\}$, defined as the Fréchet upper bound copula in Section 3.1, then C_U is the distribution of $\mathbf{U}$ such that $U_1 = \ldots = U_m$; this is the functional relationship causing the singularity. (If necessary, please consult the Appendix for background on the concepts referred to here.)

A copula is continuous and increasing, so right derivatives of first order, i.e., $\partial C(\mathbf{u})/\partial u_j$, $j = 1,\ldots,m$, exist. Hence if $\mathbf{U} \sim C$, conditional distributions of the form

$$C_{1,\ldots,j-1,j+1,\ldots,m|j}(u_1,\ldots,u_{j-1},u_{j+1},\ldots,u_m|u_j)$$

exist, and C has a singular component if one or more of these conditional distributions has a jump discontinuity. A similar result holds if C has mixed derivatives of kth order everywhere, $2 \le k < m$. A bivariate result, for identifying the total mass of the singular component, is the following.

Theorem 1.1 *Let F be a continuous bivariate distribution with univariate margins F_1, F_2 and conditional distribution $F_{2|1}$ of the second variable given the first. Suppose that $F_{2|1}(\cdot|x)$ has jump discontinuities totalling a mass of $a(x)$, and $a(\cdot)$ is continuous and positive on an interval. Then F has a singular component and the mass of the singular component is $\int a(x)\,dF_1(x)$. A similar result holds for the conditional distribution $F_{1|2}$ of the first variable given the second.*

Proof. Let $f_{2|1}(\cdot|x)$ be the derivative of $F_{2|1}(\cdot|x)$ where it exists. Because of the jump discontinuities, $\int f_{2|1}(y|x)\,dy = 1 - a(x)$. The conclusion follows. □

Example 1.2 Consider $C(u_1,u_2) = [\min\{u_1,u_2\}]^\theta [u_1u_2]^{1-\theta}$, where $0 < \theta \le 1$ (this is the family B12 in Section 5.1). The conditional distribution is

$$C_{2|1}(u_2|u_1) = \begin{cases} (1-\theta)u_2u_1^{-\theta}, & 0 \le u_2 < u_1, \\ u_2^{1-\theta}, & u_1 \le u_2 \le 1. \end{cases}$$

Therefore $C_{2|1}(\cdot|x)$ has a jump discontinuity at x, $a(x) = x^{1-\theta} - (1-\theta)x^{1-\theta} = \theta x^{1-\theta}$ and the mass of the singular component is $\theta/(2-\theta)$. If $(U_1,U_2) \sim C$, the singular component corresponds to the relationship $U_1 = U_2$ occurring with probability $\theta/(2-\theta)$. □

Next, some results on associated copulas are given. Note that for a given m-variate copula, there are $2^m - 1$ associated copulas. For $m = 2$ and $(U_1,U_2) \sim C$, the **associated copulas** come from the distributions of

$$(1-U_1, 1-U_2), \quad (U_1, 1-U_2) \quad \text{and} \quad (1-U_1, U_2).$$

Hence they are

$$C'(u_1,u_2) = u_1 + u_2 - 1 + C(1-u_1, 1-u_2),$$
$$C''(u_1,u_2) = u_1 - C(u_1, 1-u_2)$$

and

$$C'''(u_1,u_2) = u_2 - C(1-u_1, u_2).$$

So if a bivariate copula C is applied to survival functions $\overline{F}_1, \overline{F}_2$, the copula associated with $C(\overline{F}_1, \overline{F}_2)$ is C'. The multivariate extension is obvious. If C is permutation-symmetric in the m arguments, then there are m distinct associated copulas.

Copulas have many uses. In this book, they are used for construction of models for various types of multivariate data; however, they can also be used for simple examples and counterexamples of dependence properties of rvs. Highlights of applications of copulas are the following.

1. Parametric families of copulas with a logistic univariate margin are used to obtain multivariate logit models for multivariate binary or ordinal data with covariates (Sections 7.1.7, 7.3, 11.1, 11.2).
2. The extreme value limit of copulas is used to construct parametric families of extreme value copulas (Sections 6.2, 6.3); extreme value copulas with generalized extreme value univariate margins are models for multivariate maxima (Section 11.3).
3. Copulas are used to construct Markov chains and k-dependent stationary time series with an arbitrary univariate margin (Sections 8.1, 8.2, 11.5, 11.6).

1.7 View of statistical modelling

Statistical modelling usually means that one comes up with a simple (or mathematically tractable) model without knowledge of the physical aspects of the situation. The statistical model need not be 'real' and is not an end but a means of providing statistical inferences, such as percentiles, exceedance probabilities, predictions and forecasts, etc. The availability of modern computers has been an important factor in the types of multivariate models that can now be used. My view of multivariate modelling, based on experience with multivariate data, is that models should try to capture important characteristics, such as the appropriate density shapes for the univariate margins and the appropriate dependence structure, and otherwise be as simple as possible. The parameters of the model should be in a form most suitable for easy interpretation (e.g., a parameter is interpreted as either a dependence parameter or a univariate parameter but not some mixture); this form of parametrization also helps a lot in the estimation of the parameters, which must typically be done numerically. This and other desirable properties for multivariate models are given in Section 4.1. The properties of a multivariate model are a factor in whether the model is useful in a given situation. For a given data set, I usually like to carry out sensitivity analyses by comparing inferences from several models. If there is much sensitivity, then one must think a lot more about the assumptions in the models. The examples in Chapter 11 all involve comparisons of models for each data set.

1.8 Bibliographic notes

An early reference for copulas is Sklar (1959). The copula is called a uniform representation in Kimeldorf and Sampson (1975) and a dependence function in Galambos (1987) and Deheuvels (1978). A recent historical account of copulas is given in Schweizer (1991). Scarsini (1989) studies copulas for more general probability measures.

Early books on bivariate and multivariate distributions are Mardia (1970) and Johnson and Kotz (1972), and these do not mention copulas. More recent books on bivariate distributions are Hutchinson and Lai (1990) and Kocherlakota and Kocherlakota (1992).

The ideas in Shaked and Shanthikumar (1993) may be useful in the construction of models for multivariate survival in reliability theory. A recent book that includes the generalized estimating equations approach is Diggle, Liang and Zeger (1994).

1.9 Exercises

1.1 For the following bivariate cdfs or survival functions, find the univariate margins and copula:

(a) $F(x,y;\delta) = 1-(e^{-\delta x}+e^{-2\delta y}-e^{-\delta(x+2y)})^{1/\delta}$, $x, y \geq 0$, $\delta \geq 1$.

(b) $F(x,y;\delta) = \exp\{-(e^{-\delta x}+e^{-\delta y})^{1/\delta}\}$, $-\infty < x, y < \infty$, $\delta \geq 1$.

(c) $G(x,y;\theta,\eta) = (1+x^{\eta}+y^{\eta})^{1/\theta}$, $x, y \geq 0$, $\theta > 0$, $\eta > 0$.

(d) $F(x,y) = (1+e^{-x}+e^{-y})^{-1}$, $-\infty < x, y < \infty$.

1.2 Show that the family B10 in Section 5.1 consists of proper bivariate distributions if and only if $|\delta| \leq 1$.

1.3 Write out the associated copulas for the family B10 in Section 5.1.

1.4 Obtain the copula for the multivariate normal distribution. What are its parameters?

1.5 Let S_1, S_2 be two non-empty subsets of $\{1, \dots, m\}$. S_1 and S_2 can have empty or non-empty intersection. If $F_1(x_i : i \in S_1)$ and $F_2(y_j : j \in S_2)$ are cdfs, prove that the product

$$F(z_k : k \in S_1 \cup S_2) = F_1(z_i : i \in S_1)F_2(z_j : j \in S_2)$$

is a cdf. [Hint: in the case of a non-empty intersection, it may be useful to construct a stochastic representation.]

1.6 Suppose $\{F_n\}$ is a sequence of m-variate cdfs such that $F_n \to_d F$. What conditions must be checked to show that F is a proper cdf?

1.7 Find the mass of the singular component for the trivariate copula

$$C(u_1, u_2, u_3; \delta) = [\min\{u_1, u_2, u_3\}]^{\delta}[u_1u_2u_3]^{1-\delta}, \;\; 0 < \delta \leq 1.$$

1.8 Given a bivariate copula C, outline a general approach to simulating (U, V) from C? Extend this approach to a multivariate copula C.

1.9 Find two copulas associated with the bivariate binary pair (Y_1, Y_2) with probabilities $\Pr(Y_1 = Y_2 = 0) = 0.25$, $\Pr(Y_1 = 0, Y_2 = 1) = \Pr(Y_1 = 1, Y_2 = 0) = 0.15$, $\Pr(Y_1 = Y_2 = 1) = 0.45$. For example, can a bivariate normal copula be used?

1.10 Prove that the copula is not unique in the case of a multivariate distribution of discrete rvs.

1.11 What is the most dependence that one can obtain for two dependent Bernoulli rvs, with respective parameters p_1, p_2? What are the maximum and minimum possible correlations for two such rvs?

1.12 Suppose two Bernoulli rvs Y_1, Y_2 depend on a covariate vector $\mathbf{x}$. Consider a model in which Y_1, Y_2 have constant correlation ρ over $\mathbf{x}$. If $(p_1(\mathbf{x}), p_2(\mathbf{x})) = (\Pr(Y_1 = 1|\mathbf{x}), \Pr(Y_2 = 1|\mathbf{x}))$ takes on all values in $(0,1)^2$ as $\mathbf{x}$ varies, what are possible values for ρ? What if $(p_1(\mathbf{x}), p_2(\mathbf{x}))$ lies in a set $\{(\pi_1, \pi_2) : \pi_1^{\alpha} \leq \pi_2 \leq 1 - (1 - \pi_1)^{\alpha}\}$ as $\mathbf{x}$ varies, where $\alpha \in (1, \infty)$?

CHAPTER 2

Basic concepts of dependence

For non-normal random variables, Pearson's correlation and concepts based on linearity are not necessarily the best concepts to work with. More generally useful concepts of positive and negative dependence and measures of monotone dependence are given in Section 2.1. Dependence (partial) orderings which compare the amount of (monotone) dependence in two different random vectors of the same length are studied in Section 2.2. Included are properties that should be satisfied in order for a partial ordering to be considered a dependence ordering.

The dependence concepts that are presented in this chapter are those that are needed and used in analysis of multivariate models in subsequent chapters. There is no attempt to be exhaustive in mentioning all dependence concepts that have ever been proposed in the literature. Highlights of the important use of dependence concepts are the following.

- The concepts of positive quadrant dependence (in Section 2.1.1) and the concordance ordering (in Section 2.2.1) are basic to the parametric families of copulas in Chapter 5 in determining whether a multivariate parameter is a dependence parameter. The concordance ordering is also used in Section 7.1.10 to obtain the most negatively dependent multivariate exchangeable Bernoulli distribution.
- The concept of stochastic increasing positive dependence (in Section 2.1.2) is a key concept in the analysis of the decrease in dependence with lag for stationary Markov chains (Section 8.5).
- The concepts of TP_2 dependence (in Section 2.1.5) and max-infinite divisibility (in Section 2.1.8) are necessary for the method in Section 4.3 of constructing families of closed-form copulas with a wide range of dependence.
- The concept of tail dependence (in Section 2.1.10) is crucial to the construction and analysis of multivariate extreme value

distributions and copulas.

- Kendall's tau and Spearman's rho (in Section 2.1.9) are used as summary measures of dependence for bivariate copulas in Section 5.1; Kendall's tau is also used for compatibility conditions in Sections 3.4 and 3.6.

- The more stochastic increasing ordering (in Section 2.2.4) is useful in the analysis of the Fréchet classes in Sections 3.3 and 3.4, and in the analysis of the range of dependence of the construction method in Section 4.5.

2.1 Dependence properties and measures

Bivariate dependence concepts, properties and measures are easier to define and have appeared more often in the probability and statistics literature than multivariate counterparts. This section consists of a number of dependence concepts; for each, the bivariate version is given first, followed by the intuition behind it, and then a multivariate extension is given if there is one. Examples illustrating the concepts are combined together into a separate subsection.

2.1.1 Positive quadrant and orthant dependence °

Let $\mathbf{X} = (X_1, X_2)$ be a bivariate random vector with cdf F. $\mathbf{X}$ or F is **positive quadrant dependent** (PQD) if

$$\Pr(X_1 > a_1, X_2 > a_2) \geq \Pr(X_1 > a_1)\Pr(X_2 > a_2) \quad \forall a_1, a_2 \in \Re. \tag{2.1}$$

Condition (2.1) is equivalent to

$$\Pr(X_1 \leq a_1, X_2 \leq a_2) \geq \Pr(X_1 \leq a_1)\Pr(X_2 \leq a_2) \quad \forall a_1, a_2 \in \Re. \tag{2.2}$$

The reason why (2.1) or (2.2) is a positive dependence concept is that X_1 and X_2 are more likely to be large together or to be small together compared with X_1' and X_2', where $X_1 \stackrel{d}{=} X_1'$, $X_2 \stackrel{d}{=} X_2'$, and X_1' and X_2' are independent of each other. Reasoning similarly, $\mathbf{X}$ or F is **negative quadrant dependent** (NQD) if the inequalities in (2.1) and (2.2) are reversed.

For the multivariate extension, let $\mathbf{X}$ be a random m-vector ($m \geq 2$) with cdf F. $\mathbf{X}$ or F is **positive upper orthant dependent**

(PUOD) if

$$\Pr(X_i > a_i,\ i = 1, \ldots, m) \geq \prod_{i=1}^{m} \Pr(X_i > a_i) \quad \forall \mathbf{a} \in \Re^m, \qquad (2.3)$$

and $\mathbf{X}$ or F is **positive lower orthant dependent** (PLOD) if

$$\Pr(X_i \leq a_i,\ i = 1, \ldots, m) \geq \prod_{i=1}^{m} \Pr(X_i \leq a_i) \quad \forall \mathbf{a} \in \Re^m. \qquad (2.4)$$

If both (2.3) and (2.4) hold, then $\mathbf{X}$ or F is **positive orthant dependent** (POD). Note that for the multivariate extension, (2.3) and (2.4) are not equivalent.

Intuitively, (2.3) means that $X_1, \ldots, X_m$ are more likely simultaneously to have large values, compared with a vector of independent rvs with the same corresponding univariate margins. If the inequalities in (2.3) and (2.4) are reversed, then the concepts of **negative lower orthant dependence** (NLOD), **negative upper orthant dependence** (NUOD) and **negative orthant dependence** (NOD) result.

2.1.2 Stochastic increasing positive dependence

Let $\mathbf{X} = (X_1, X_2)$ be a bivariate random vector with cdf $F \in \mathcal{F}(F_1, F_2)$. X_2 is **stochastically increasing** (SI) in X_1 or the conditional distribution $F_{2|1}$ is **stochastically increasing** if

$$\Pr(X_2 > x_2 \mid X_1 = x_1) = 1 - F_{2|1}(x_2|x_1) \ \uparrow x_1 \ \forall x_2. \qquad (2.5)$$

By reversing the roles of the indices of 1 and 2, one has X_1 SI in X_2 or $F_{1|2}$ SI. The reason why (2.5) is a positive dependence condition is that X_2 is more likely to take on larger values as X_1 increases. By reversing the direction of monotonicity in (2.5) from $\uparrow$ to $\downarrow$, the **stochastically decreasing** (SD) condition results.

There are two dependence concepts that could be considered as multivariate extensions of SI; they are *positive dependence through the stochastic ordering* and *conditional increasing in sequence.*

Definition. The random vector $(X_1, \ldots, X_m)$ is **positive dependent through the stochastic ordering** (PDS) if $\{X_i : i \neq j\}$ conditional on $X_j = x$ is increasing stochastically as x increases, for all $j = 1, \ldots, m$.

Definition. The random vector $(X_1, \ldots, X_m)$ is **conditional increasing in sequence** (CIS) if X_i is stochastically increasing in $X_1, \ldots, X_{i-1}$ for $i = 2, \ldots, m$, i.e., $\Pr(X_i > x_i \mid X_j = x_j, j = 1, \ldots, i-1)$ is increasing in $x_1, \ldots, x_{i-1}$ for all x_i.

Note that for $m = 2$, PDS is the same as X_2 SI in X_1 and X_1 SI in X_2, and CIS is the same as SI.

2.1.3 Right-tail increasing and left-tail decreasing

Let $\mathbf{X} = (X_1, X_2)$ be a bivariate random vector with cdf $F \in \mathcal{F}(F_1, F_2)$. X_2 is **right-tail increasing** (RTI) in X_1 if

$$\Pr(X_2 > x_2 \mid X_1 > x_1) = \overline{F}(x_1, x_2)/\overline{F}_1(x_1) \uparrow x_1 \; \forall x_2. \tag{2.6}$$

Similarly, X_2 is **left-tail decreasing** (LTD) in X_1 if

$$\Pr(X_2 \le x_2 \mid X_1 \le x_1) = F(x_1, x_2)/F_1(x_1) \downarrow x_1 \; \forall x_2. \tag{2.7}$$

The reason why (2.6) and (2.7) are positive dependence conditions is that, for (2.6), X_2 is more likely to take on larger values as X_1 increases, and, for (2.7), X_2 is more likely to take on smaller values as X_1 decreases. Reversing the directions of the monotonicities lead to negative dependence conditions.

A multivariate extension of RTI for an m-vector $(X_1, \ldots, X_m)$ is: $X_i, i \in A^c$, is RTI in $X_j, j \in A$, if

$$\Pr(X_i > x_i,\, i \in A^c \mid X_j > x_j,\, j \in A) \uparrow x_k,\, k \in A,$$

where A is a non-empty subset of $\{1, \ldots, m\}$. Similarly, there is a multivariate extension of LTD.

2.1.4 Associated random variables

Let $\mathbf{X}$ be a random m-vector. $\mathbf{X}$ is (positively) **associated** if the inequality

$$\mathrm{E}\,[g_1(\mathbf{X})g_2(\mathbf{X})] \ge \mathrm{E}\,[g_1(\mathbf{X})]\mathrm{E}\,[g_2(\mathbf{X})] \tag{2.8}$$

holds for all real-valued functions g_1, g_2 which are increasing (in each component) and are such that the expectations in (2.8) exist. Intuitively, this is a positive dependence condition for $\mathbf{X}$ because it means that two increasing functions of $\mathbf{X}$ have positive covariance whenever the covariance exists.

It may appear impossible to check this condition of association directly given a cdf F for $\mathbf{X}$. Where association of a random vector can be established, it is usually done by making use of a stochastic representation for $\mathbf{X}$. One important consequence of the association condition is that it implies the POD condition; see Section 2.1.7 on relationships between concepts of positive dependence.

For a random variable (with $m = 1$), inequality (2.8) holds whenever the expectations exist.

Lemma 2.1 *For a rv X, $\mathrm{E}\,[g_1(X)g_2(X)] \geq \mathrm{E}\,[g_1(X)]\mathrm{E}\,[g_2(X)]$ for all increasing real-valued functions g_1, g_2 such that the expectations exist.*

Proof. For binary increasing functions $g_j(x) = I_{(a_j,\infty)}(x)$, the left-hand side of (2.8) becomes $\Pr(X > \max\{a_1, a_2\})$ and the right-hand side becomes $\Pr(X > a_1)\Pr(X > a_2)$ so that inequality (2.8) holds and is equivalent to $\mathrm{Cov}\,[g_1(X), g_2(X)] \geq 0$. For general increasing functions g_1, g_2 such that the covariance exist, Hoeffding's identity (see Exercise 2.15) leads to

$$\begin{aligned} &\mathrm{Cov}\,[g_1(X), g_2(X)] \qquad (2.9)\\ &= \int_{-\infty}^{\infty}\int_{-\infty}^{\infty} \mathrm{Cov}\left[I_{(s_1,\infty)}(g_1(X)), I_{(s_2,\infty)}(g_2(X))\right] ds_2 ds_1 \\ &= \int_{-\infty}^{\infty}\int_{-\infty}^{\infty} \mathrm{Cov}\left[I_{(g_1^{-1}(s_1),\infty)}(X), I_{(g_2^{-1}(s_2),\infty)}(X)\right] ds_2 ds_1, \end{aligned}$$

and the integrand is always non-negative from the binary case. □

There exists a definition for negative association but it will not be given here as it is not needed in subsequent chapters.

2.1.5 Total positivity of order 2

A non-negative function b on A^2, where $A \subset \Re$, is **totally positive of order 2** (TP$_2$) if for all $x_1 < y_1$, $x_2 < y_2$, with $x_i, y_j \in A$,

$$b(x_1, x_2)b(y_1, y_2) \geq b(x_1, y_2)b(y_1, x_2). \qquad (2.10)$$

The 'order 2' part of the definition comes from writing the difference $b(x_1, x_2)b(y_1, y_2) - b(x_1, y_2)b(y_1, x_2)$ as the determinant of a square matrix of order 2. Total positivity of higher orders involves the non-negativity of determinants of larger square matrices. If the inequality in (2.10) is reversed then b is **reverse rule of order 2** (RR$_2$).

For a bivariate cdf F with density f, three notions of positive dependence are: (i) f is TP$_2$; (ii) F is TP$_2$; (iii) $\overline{F}$ is TP$_2$. The reasoning behind (i) as a positive dependence condition is that for $x_1 < y_1$, $x_2 < y_2$, $f(x_1, x_2)f(y_1, y_2) \geq f(x_1, y_2)f(y_1, x_2)$ means that it is more likely to have two pairs with components matching high–high and low–low than two pairs with components matching high–low and low–high. Similarly, f RR$_2$ is a negative dependence condition.

It is shown later in Section 2.1.7 that f TP$_2$ implies both F and $\overline{F}$ TP$_2$, and either F TP$_2$ or $\overline{F}$ TP$_2$ implies that F is PQD. Hence both (ii) and (iii) are positive dependence conditions. A direct explanation for (ii) as a positive dependence condition is as follows. The condition of F TP$_2$ is given by:

$$F(x_1,x_2)F(y_1,y_2) - F(x_1,y_2)F(y_1,x_2) \geq 0, \forall x_1 < y_1, x_2 < y_2.$$

This is equivalent to:

$$\begin{aligned} &F(x_1,x_2)[F(y_1,y_2) - F(y_1,x_2) - F(x_1,y_2) + F(x_1,x_2)] \\ &\quad -[F(x_1,y_2) - F(x_1,x_2)][F(y_1,x_2) - F(x_1,x_2)] \geq 0, \quad (2.11) \\ &\qquad \forall\, x_1 < y_1, x_2 < y_2. \end{aligned}$$

If $(X_1, X_2) \sim F$, then the inequality in (2.11) is the same as

$$\begin{aligned} &\Pr(X_1 \leq x_1, X_2 \leq x_2)\Pr(x_1 < X_1 \leq y_1, x_2 < X_2 \leq y_2) \\ &\quad - \Pr(X_1 \leq x_1, x_2 < X_2 \leq y_2)\Pr(x_1 < X_1 \leq y_1, X_2 \leq x_2) \geq 0, \end{aligned}$$

for all $x_1 < y_1$, $x_2 < y_2$. This has an interpretation as before for high–high and low–low pairs versus high–low and low–high. Similarly, the inequality resulting from $\overline{F}$ TP$_2$ can be written in the form of (2.11) with the survival function $\overline{F}$ replacing F.

The conditions of F TP$_2$ and $\overline{F}$ TP$_2$ occur as necessary and sufficient conditions for a bivariate cdf or survival function to be max- or min-infinitely divisible; see Section 2.1.8.

A multivariate extension of TP$_2$ is the following. Let $\mathbf{X}$ be a random m-vector with density f. $\mathbf{X}$ or f is **multivariate totally positive of order 2** (MTP$_2$) if

$$f(\mathbf{x} \vee \mathbf{y})f(\mathbf{x} \wedge \mathbf{y}) \geq f(\mathbf{x})f(\mathbf{y}) \qquad (2.12)$$

for all $\mathbf{x}, \mathbf{y} \in \Re^m$, where

$$\begin{aligned} \mathbf{x} \vee \mathbf{y} &= (\max\{x_1, y_1\}, \max\{x_2, y_2\}, \ldots, \max\{x_m, y_m\}), \\ \mathbf{x} \wedge \mathbf{y} &= (\min\{x_1, y_1\}, \min\{x_2, y_2\}, \ldots, \min\{x_m, y_m\}). \end{aligned}$$

An important property of MTP$_2$ is that if a density is MTP$_2$, then so are all of its marginal densities of order 2 and higher (see the proof in Section 2.1.7).

If the inequality in (2.12) is reversed, then f is **multivariate reverse rule of order 2** (MRR$_2$). This is a weak negative dependence concept because, unlike MTP$_2$, the property of MRR$_2$ is not closed under the taking of margins. (An example of non-closure is given in Section 9.2.1.)

2.1.6 Positive function dependence

Positive function dependence is a concept of dependence for the special case in which all univariate margins are the same, i.e., for $\mathcal{F}(F_1, \ldots, F_m)$, with $F_1 = \cdots = F_m$ ($= F_0$, say). For the bivariate case, let X_1, X_2 be dependent rvs with cdf F_0 and suppose $(X_1, X_2) \sim F$. Then (X_1, X_2) or F is **positive function dependent** (PFD) if

$$\mathrm{Cov}\,[h(X_1), h(X_2)] \geq 0, \ \forall \text{ real-valued } h \tag{2.13}$$

such that the covariance exists. The multivariate extension with $X_1, \ldots, X_m$ being dependent rvs with cdf F_0 and $(X_1, \ldots, X_m) \sim F$ is that $(X_1, \ldots, X_m)$ is **positive function dependent** if

$$\mathrm{E}\left[\prod_{i=1}^{m} h(X_i)\right] \geq \prod_{i=1}^{m} \mathrm{E}\,[h(X_i)], \ \forall \text{real-valued } h \tag{2.14}$$

such that the expectations exist. If m is odd, then there is the further restriction that h be non-negative.

In the statistical literature, this concept has been called 'positive dependence', but here we use the term 'positive function dependence' in order to avoid confusion with the general notion of positive dependence (i.e., many definitions in Section 2.1 are concepts of either positive or negative dependence). In Section 8.5, we show an application of positive function dependence to inference for stationary dependent sequences.

Similar to the definition of association, it looks as if (2.13) and (2.14) would be difficult to establish analytically in general. Again, where PFD can be established, it is usually done by making use of a stochastic representation for $\mathbf{X}$. For example, a condition that implies PFD in the bivariate case is **positive dependent by mixture**, which means that $F(x_1, x_2)$ (or $\overline{F}(x_1, x_2)$) has the representation $\int G(x_1; \alpha) G(x_2; \alpha)\, dM(\alpha)$ (or $\int \overline{G}(x_1; \alpha) \overline{G}(x_2; \alpha)\, dM(\alpha)$), where M is a mixing distribution and $G(\cdot; \alpha)$ is an appropriately chosen family of distributions so that the representation holds. The proof is left as an exercise.

2.1.7 Relationships among dependence properties

In this subsection, invariance results and results on relationships among dependence properties are given. The first theorem is trivial so its proof is omitted.

Theorem 2.2 *All of the dependence properties in Sections 2.1.1 to 2.1.6 are invariant with respect to strictly increasing transformations on the components of the random vector. For example, if* (X_1, X_2) *is PQD then so is* $(a_1(X_1), a_2(X_2))$ *for strictly increasing functions* a_1, a_2.

Theorem 2.3 *Relations in the bivariate case are:*

(a) TP_2 *density* $\Rightarrow$ *SI* $\Rightarrow$ *LTD, RTI;*

(b) LTD or RTI $\Rightarrow$ *association* $\Rightarrow$ *PQD;*

(c) TP_2 *density* $\Rightarrow$ TP_2 *cdf and* TP_2 *survival function;*

(d) TP_2 *cdf* $\Rightarrow$ *LTD, and* TP_2 *survival function* $\Rightarrow$ *RTI.*

Proof. TP_2 density $\Rightarrow$ SI: Let $(X_1, X_2) \sim F$ with density f. We need to show $\Pr(X_2 > y \mid X_1 = x) \le \Pr(X_2 > y \mid X_1 = x')$ for arbitrary $x < x'$. This is equivalent to showing

$$\int_y^\infty f(x,z)\,dz \int_{-\infty}^\infty f(x',w)\,dw \le \int_y^\infty f(x',z)\,dz \int_{-\infty}^\infty f(x,w)\,dw$$

or

$$\int_y^\infty \int_{-\infty}^\infty [f(x',z)f(x,w) - f(x,z)f(x',w)]\,dw\,dz \ge 0.$$

But the left-hand side of the above inequality simplifies to

$$\int_y^\infty \int_{-\infty}^y [f(x',z)f(x,w) - f(x,z)f(x',w)]\,dw\,dz$$

and the integrand is non-negative for all $(z,w) \in (y,\infty) \times (-\infty, y]$ by the TP_2 assumption.

SI $\Rightarrow$ RTI (SI $\Rightarrow$ LTD is similar): Let $x < x'$ and let $(X_1, X_2) \sim F$. Since $\overline{F}_{2|1}(x_2|x_1) = \Pr(X_2 > x_2 \mid X_1 = x_1)$ is increasing in x_1, there is an inequality for the weighted averages when this conditional probability is weighted against the density of X_1, i.e.,

$$\frac{\int_x^\infty \overline{F}_{2|1}(x_2|x_1)\,dF_1(x_1)}{\overline{F}_1(x)} \le \frac{\int_{x'}^\infty \overline{F}_{2|1}(x_2|x_1)\,dF_1(x_1)}{\overline{F}_1(x')}.$$

This inequality is the same as

$$\Pr(X_2 > x_2 \mid X_1 > x) \le \Pr(X_2 > x_2 \mid X_1 > x')$$

and this is the RTI condition.

LTD or RTI $\Rightarrow$ association: The proof of this is lengthy and requires technical details. It is omitted here but is given in Esary and Proschan (1972). We give instead a simple proof of the implication

'SI $\Rightarrow$ association', since this gives an indication of how to work with the concept of associated rvs.

SI $\Rightarrow$ association: Let a and b be increasing functions on $\Re^2$. Assuming that second-order moments exist,

$$\begin{aligned}\mathrm{Cov}\,[a(X_1,X_2),b(X_1,X_2)] &= \mathrm{E}\,\{\mathrm{Cov}\,[a(X_1,X_2),b(X_1,X_2)\mid X_1]\}\\ &+\mathrm{Cov}\,(\mathrm{E}\,[a(X_1,X_2)\mid X_1],\mathrm{E}\,[b(X_1,X_2)\mid X_1]). \qquad (2.15)\end{aligned}$$

Let $a^*(X_1)=\mathrm{E}\,[a(X_1,X_2)|X_1]$, $b^*(X_1)=\mathrm{E}\,[b(X_1,X_2)|X_1]$. Then a^* and b^* are increasing functions since a and b are increasing and X_2 is SI in X_1, so that the second term on the right of (2.15) is non-negative, by Lemma 2.1 in Section 2.1.4. For the first term on the right of (2.15), $a(x_1,X_2)$, $b(x_1,X_2)$ are increasing in X_2 for each fixed x_1 so that the conditional covariance is non-negative for each x_1 (again by Lemma 2.1). Hence unconditionally the expectation in the first term is non-negative.

Association $\Rightarrow$ PQD: The bivariate case is a special case of the multivariate result of 'association $\Rightarrow$ POD', which is proved in the next theorem.

TP_2 density $\Rightarrow$ TP_2 cdf (TP_2 density $\Rightarrow$ TP_2 survival function is similar): Let $(X_1,X_2)\sim F$ with density f. Let $x_1<y_1$, $x_2<y_2$. Then f TP_2 implies that

$$\int_{-\infty}^{x_1}\int_{-\infty}^{x_2}\int_{x_1}^{y_1}\int_{x_2}^{y_2}[f(s_1,s_2)f(t_1,t_2)-f(s_1,t_2)f(t_1,s_2)]dt_2dt_1ds_2ds_1$$

is non-negative or

$$\begin{aligned}F(x_1,x_2)[F(y_1,y_2)-F(y_1,x_2)-F(x_1,y_2)+F(x_1,x_2)]\\ \geq [F(x_1,y_2)-F(x_1,x_2)][F(y_1,x_2)-F(x_1,x_2)].\end{aligned}$$

This is equivalent to the TP_2 condition for the cdf (see inequality (2.11) in Section 2.1.5).

TP_2 cdf $\Rightarrow$ LTD (TP_2 survival function $\Rightarrow$ RTI is similar): Let $y_2\to\infty$ and suppose $x_1<y_1$; then the TP_2 cdf condition implies $F(x_1,x_2)/F(x_1,\infty)\geq F(y_1,x_2)/F(y_1,\infty)$, which is the LTD condition. □

Theorem 2.4 *Relations in the multivariate case are:*

(a) a random subvector of an associated random vector is associated;

(b) association $\Rightarrow$ PUOD and PLOD;

(c) PDS $\Rightarrow$ PUOD and PLOD;

(d) CIS $\Rightarrow$ association.

Proof. (a) Let $(X_1, \ldots, X_m)$ be associated and $\mathbf{X}^s = (X_{i_1}, \ldots, X_{i_k})$ be a subvector, with $1 \le k < m$ and $i_1 < \cdots < i_k$. Let g_1, g_2 be increasing functions on $\Re^k$. Then

$$\mathrm{E}\,[g_1(\mathbf{X}^s)g_2(\mathbf{X}^s)] \ge \mathrm{E}\,[g_1(\mathbf{X}^s)]\mathrm{E}\,[g_2(\mathbf{X}^s)]$$

provided the expectations exist since g_1, g_2 can be considered as functions of $X_1, \ldots, X_m$.

(b) Let $(X_1, \ldots, X_m)$ be associated. Fix real numbers $a_1, \ldots, a_m$. Let $g_1(x_1, \ldots, x_{m-1}) = I_{(a_1,\infty)\times\cdots\times(a_{m-1},\infty)}(x_1, \ldots, x_{m-1})$ and let $g_2(x_m) = I_{(a_m,\infty)}(x_m)$. Then g_1, g_2 are increasing functions. Inequality (2.8) leads to

$$\Pr(X_i > a_i,\, i \le m) \ge \Pr(X_i > a_i,\, i \le m-1)\Pr(X_m > a_m).$$

By (a) and making use of induction,

$$\Pr(X_i > a_i,\, i \le k) \ge \Pr(X_i > a_i,\, i \le k-1)\Pr(X_k > a_k)$$

for $k = m-1, \ldots, 2$. Therefore

$$\Pr(X_i > a_i,\, i = 1, \ldots, m) \ge \prod_{i=1}^{m} \Pr(X_i > a_i)$$

or $\mathbf{X}$ is PUOD.

Similarly, to show the conclusion of PLOD, use the functions $g_1(x_1, \ldots, x_{m-1}) = -I_{(-\infty,a_1]\times\cdots\times(-\infty,a_{m-1}]}(x_1, \ldots, x_{m-1})$ and $g_2(x_m) = -I_{(-\infty,a_m]}(x_m)$.

(c) $(X_1, \ldots, X_m)$ PDS implies

$$\begin{aligned}&\Pr(X_2 > x_2, \ldots, X_m > x_m \mid X_1 = x_1)\\ &\quad\ge \Pr(X_2 > x_2, \ldots, X_m > x_m \mid X_1 = x_1')\end{aligned}$$

for all $x_1 > x_1'$ and for all $x_2, \ldots, x_m$. Then

$$\begin{aligned}&\Pr(X_2 > x_2, \ldots, X_m > x_m \mid X_1 > x_1)\\ &= \int_{x_1}^{\infty} \Pr(X_2 > x_2, \ldots, X_m > x_m | X_1 = z)\, dF_1(z) \Big/ \int_{x_1}^{\infty} dF_1(z)\\ &\ge \int_{x_1'}^{\infty} \Pr(X_2 > x_2, \ldots, X_m > x_m | X_1 = z)\, dF_1(z) \Big/ \int_{x_1'}^{\infty} dF_1(z)\\ &= \Pr(X_2 > x_2, \ldots, X_m > x_m \mid X_1 > x_1'),\end{aligned}$$

for all $x_1 > x_1'$. Letting $x_1' \to -\infty$ yields

$$\Pr(X_j > x_j,\, j \le m) \ge \Pr(X_1 > x_1)\Pr(X_2 > x_2, \ldots, X_m > x_m). \tag{2.16}$$

Since a subset of a vector that is PDS is also PDS, $(X_j, \ldots, X_m)$ is PDS for $j = 2, \ldots, m-1$ and by induction $\Pr(X_j > x_j, j = 1, \ldots, m) \geq \prod_{j=1}^m \Pr(X_j > x_j)$ or $(X_1, \ldots, X_m)$ is PUOD.

The conclusion of PLOD follows similarly since $(X_1, \ldots, X_m)$ PDS implies $\Pr(X_2 \leq x_2, \ldots, X_m \leq x_m \mid X_1 = x_1) \geq \Pr(X_2 \leq x_2, \ldots, X_m \leq x_m \mid X_1 = x_1')$ for all $x_1 < x_1'$. A similar inequality then holds conditional on $X_1 \leq x_1$ and $X_1 \leq x_1'$. An inequality like (2.16) results by letting $x_1' \to \infty$.

(d) The proof is similar to that of 'SI $\Rightarrow$ association' in the preceding theorem.

□

Note that as a consequence of part (d) of Theorem 2.4, independent rvs are associated since they clearly satisfy the CIS condition.

Theorem 2.5 *Let $(X_1, \ldots, X_m)$ have density f which is MTP$_2$. Then all of marginal densities of f of order 2 and higher are also MTP$_2$.*

Proof. This proof is modified from Karlin and Rinott (1980a). Suppose densities for X_j exist relative to the measure ν. Because of symmetry and induction, it suffices to show that the density of $(X_1, \ldots, X_{m-1})$ is MTP$_2$, or that

$$\begin{aligned}\int_{s<t} & [f(\mathbf{x}_{m-1}, s) f(\mathbf{y}_{m-1}, t) + f(\mathbf{x}_{m-1}, t) f(\mathbf{y}_{m-1}, s)]\, d\nu(s)\, d\nu(t) \\ \leq & \int_{s<t} \big[f(\mathbf{x}_{m-1} \vee \mathbf{y}_{m-1}, s) f(\mathbf{x}_{m-1} \wedge \mathbf{y}_{m-1}, t) \\ & + f(\mathbf{x}_{m-1} \vee \mathbf{y}_{m-1}, t) f(\mathbf{x}_{m-1} \wedge \mathbf{y}_{m-1}, s)\big]\, d\nu(s)\, d\nu(t)\end{aligned} \tag{2.17}$$

where $\mathbf{x}_{m-1} = (x_1, \ldots, x_{m-1})$, $\mathbf{y}_{m-1} = (y_1, \ldots, y_{m-1})$. (For the 'discrete' case, the inequality

$$\begin{aligned}\int_{s=t} & [f(\mathbf{x}_{m-1}, s) f(\mathbf{y}_{m-1}, s) + f(\mathbf{x}_{m-1}, s) f(\mathbf{y}_{m-1}, s)]\, d\nu(s)\, d\nu(t) \\ & \leq 2 \int_{s=t} [f(\mathbf{x}_{m-1} \vee \mathbf{y}_{m-1}, s) f(\mathbf{x}_{m-1} \wedge \mathbf{y}_{m-1}, s)]\, d\nu(s)\, d\nu(t),\end{aligned}$$

follows easily from the MTP$_2$ property of f.) In (2.17), let

$$\begin{aligned} a &= f(\mathbf{x}_{m-1}, s) f(\mathbf{y}_{m-1}, t), \quad b = f(\mathbf{x}_{m-1}, t) f(\mathbf{y}_{m-1}, s), \\ c &= f(\mathbf{x}_{m-1} \vee \mathbf{y}_{m-1}, s) f(\mathbf{x}_{m-1} \wedge \mathbf{y}_{m-1}, t), \\ d &= f(\mathbf{x}_{m-1} \vee \mathbf{y}_{m-1}, t) f(\mathbf{x}_{m-1} \wedge \mathbf{y}_{m-1}, s), \end{aligned}$$

with $s < t$. From the MTP$_2$ property for f, $d \geq a, b$ and $ab \leq cd$ (the latter from matching up terms with s and t separately). Then

$(c+d)-(a+b) = d^{-1}[(d-a)(d-b)+(cd-ab)] \geq 0$ and inequality (2.17) holds. □

2.1.8 Max-infinite and min-infinite divisibility

For a univariate cdf F, all positive powers F^γ ($\overline{F}^\gamma$), $\gamma > 0$, are cdfs (survival functions). This need not be the case for multivariate cdfs. In general, for an m-variate cdf F, F^γ ($\overline{F}^\gamma$) is a cdf (survival function) for all $\gamma \geq m-1$. If F^γ is a cdf for all $\gamma > 0$, then F is **max-infinitely divisible** (max-id), and if $\overline{F}^\gamma$ is a survival function for all $\gamma > 0$, then F is **min-infinitely divisible** (min-id).

The explanations for these definitions are as follows. If F is max-id and $\mathbf{X} = (X_1, \ldots, X_m) \sim F$, then for all positive integers n, $F^{1/n}$ is a cdf. If $(X_{i1}^{(n)}, \ldots, X_{im}^{(n)})$, $i = 1, \ldots, n$, are iid with cdf $F^{1/n}$, then

$$\mathbf{X} \stackrel{d}{=} (\max_i X_{i1}^{(n)}, \ldots, \max_i X_{im}^{(n)})$$

where the maxima are over the indices 1 to n. For min-id, replace max by min and cdf by survival function.

The max-id and min-id conditions are equivalent respectively to being a TP_2 cdf and TP_2 survival function in the bivariate case, and hence they are (strong) dependence conditions. These proofs are given next and then conditions are given in the multivariate case.

Theorem 2.6 *Let F be a bivariate cdf.*

(a) F is max-id if and only if F is TP_2.

(b) F is min-id if and only if $\overline{F}$ is TP_2.

Proof. (a) Let $R(x,y) = \log F(x,y)$, so that R is increasing in x and y. Then F TP_2 implies that for $\delta, \epsilon > 0$, $R(x+\delta, y+\epsilon) \geq R(x, y+\epsilon) + R(x+\delta, y) - R(x,y)$. Since e^z is convex and increasing in z, for $\gamma > 0$,

$$\begin{aligned} e^{\gamma R(x,y+\epsilon)} - e^{\gamma R(x,y)} &\leq e^{\gamma[R(x,y+\epsilon)+R(x+\delta,y)-R(x,y)]} - e^{\gamma R(x+\delta,y)} \\ &\leq e^{\gamma R(x+\delta,y+\epsilon)} - e^{\gamma R(x+\delta,y)}. \end{aligned}$$

This is equivalent to $F^\gamma(x+\delta, y+\epsilon) - F^\gamma(x+\delta, y) - F^\gamma(x, y+\epsilon) + F^\gamma(x,y) \geq 0$ for all $\gamma > 0$. Hence, from the rectangle inequality, F^γ is a cdf for all $\gamma > 0$.

For the converse, if F is max-id, then

$$w(\gamma) = F^\gamma(x+\delta, y+\epsilon) - F^\gamma(x+\delta, y) - F^\gamma(x, y+\epsilon) + F^\gamma(x,y) \geq 0$$

for all $\delta, \epsilon, \gamma > 0$. Since w is continuous and differentiable and $w(0) = 0$, the right derivative of $w(\gamma)$ at 0 is non-negative. This leads to

$$\log \left[\frac{F(x+\delta, y+\epsilon)F(x,y)}{F(x+\delta, y)F(x, y+\epsilon)} \right] \geq 0$$

for all $\delta, \epsilon > 0$. Equivalently, F is TP_2.

The proof of (b) is similar and is left as an exercise. □

For a multivariate distribution F to be max-id, a necessary condition is that all bivariate margins are TP_2. Hence max-id is a (strong) positive dependence condition.

A general condition for max-id, which generalizes the above bivariate result to any dimension m, is given next.

Theorem 2.7 *Let $m \geq 2$. Suppose $F(\mathbf{x})$ is an m-variate distribution with a density (with respect to Lebesgue measure) and let $R = \log F$. For a subset S of $\{1, \ldots, m\}$, let R_S denote the partial derivative of R with respect to x_i, $i \in S$. A necessary and sufficient condition for F to be max-id is that $R_S \geq 0$ for all (non-empty) subsets S of $\{1, \ldots, m\}$.*

Proof. We look at the derivatives of $H = F^\gamma = e^{\gamma R}$ with respect to $x_1, \ldots, x_m$, $i = 1, \ldots, m$, and then permute indices. All of the derivatives must be non-negative for all $\gamma > 0$ if F is max-id. The derivatives are:

$\partial H/\partial x_1 = \gamma H R_1$,

$\partial^2 H/\partial x_1 \partial x_2 = \gamma^2 H R_1 R_2 + \gamma H R_{12}$,

$\partial^3 H/\partial x_1 \partial x_2 \partial x_3 = \gamma^3 H R_1 R_2 R_3 + \gamma^2 H[R_1 R_{23} + R_2 R_{13} + R_3 R_{12}] + \gamma H R_{123}$, etc.

For the non-negativity of $\partial^{|S|} H / \prod_{i \in S} \partial x_i$ for $\gamma > 0$ arbitrarily small, a necessary condition is that $R_S \geq 0$. From the form of the derivatives above, it is clear that $R_S \geq 0$ for all S is a sufficient condition. □

For multivariate distributions which have special forms, simpler conditions can be obtained. These are obtained in Section 4.3 where mixtures of powers of a max-id or min-id multivariate distribution are used to obtain families of multivariate distributions.

2.1.9 Kendall's tau and Spearman's rho °

Kendall's tau (denoted by τ) and Spearman's rho (denoted by ρ_S or ρ) are bivariate measures of (monotone) dependence for continuous variables that are (i) invariant with respect to strictly increasing transformations and (ii) equal to 1 for the bivariate Fréchet upper bound (one variable is an increasing transform of the other) and -1 for the Fréchet lower bound (one variable is a decreasing transform of the other). These two properties do not hold for Pearson's correlation, so that τ and ρ_S are more desirable as measures of association for multivariate non-normal distributions. Another property (Exercise 2.10) is that τ and ρ_S are increasing with respect to the concordance ordering of Section 2.2.1.

Definition. Let F be a continuous bivariate cdf and let (X_1,X_2), (X_1',X_2') be independent random pairs with distribution F. Then **Kendall's tau** is

$$\tau = \Pr((X_1 - X_1')(X_2 - X_2') > 0) - \Pr((X_1 - X_1')(X_2 - X_2') < 0)$$
$$= 2\Pr((X_1 - X_1')(X_2 - X_2') > 0) - 1 = 4\int F\,dF - 1.$$

Definition. Let F be a continuous bivariate cdf with univariate margins F_1, F_2 and let $(X_1, X_2) \sim F$; then **Spearman's rho** is the correlation of $F_1(X_1)$ and $F_2(X_2)$. Since $F_1(X_1)$ and $F_2(X_2)$ are $U(0,1)$ rvs (under the assumption of continuity), their expectations are $1/2$, their variances are $1/12$, and Spearman's rho is

$$\rho_S = 12\int\int F_1(x_1)F_2(x_2)\,dF(x_1,x_2) - 3 = 12\int\int \overline{F}\,dF_1dF_2 - 3.$$

The condition $(X_1 - X_1')(X_2 - X_2') > 0$ corresponds to (X_1, X_2), (X_1', X_2') being two concordant pairs in that one of the two pairs has the larger value for both components, and the condition $(X_1 - X_1')(X_2 - X_2') < 0$ corresponds to (X_1, X_2), (X_1', X_2') being two discordant pairs in that for each pair one component is larger than the corresponding component of the other pair and one is smaller. Hence Kendall's tau is the difference of the probability of two random concordant pairs and the probability of two random discordant pairs. If there is an increasing (decreasing) transform from one variable to the other, the probability of a concordant (discordant) pair is 1 and the probability of a discordant (concordant) pair is 0.

Because τ and ρ_S are invariant to strictly increasing transformations, their definitions could be written in terms of the copula C associated with F. That is,

$$\tau = 4\int C\,dC - 1 \quad \text{and}$$
$$\rho_S = 12\int\int uv\,dC(u,v) - 3 = 12\int\int \overline{C}(u,v)\,du\,dv - 3.$$

The relation to the sample version of ρ_S can now be seen. For bivariate data, ρ_S is the rank correlation and the rank transformation is like the probability transform of a rv to $U(0,1)$.

2.1.10 Tail dependence °

The concept of bivariate tail dependence relates to the amount of dependence in the upper-quadrant tail or lower-quadrant tail of a bivariate distribution. It is a concept that is relevant to dependence in extreme values (which depends mainly on the tails) and in the derivation of multivariate extreme value distributions from the taking of limits (see Chapter 6). Because of invariance to increasing transformations, the definition will be given in terms of copulas. The symbol used for a tail dependence parameter is λ.

Definition. If a bivariate copula C is such that

$$\lim_{u\to 1} \overline{C}(u,u)/(1-u) = \lambda_U$$

exists, then C has **upper tail dependence** if $\lambda_U \in (0,1]$ and no upper tail dependence if $\lambda_U = 0$. Similarly, if

$$\lim_{u\to 0} C(u,u)/u = \lambda_L$$

exists, C has **lower tail dependence** if $\lambda_L \in (0,1]$ and no lower tail dependence if $\lambda_L = 0$.

The reasoning behind these definitions is as follows. Suppose $(U_1, U_2) \sim C$. Then

$$\lambda_U = \lim_{u\to 1} \Pr(U_1 > u \mid U_2 > u) = \lim_{u\to 1} \Pr(U_2 > u \mid U_1 > u).$$

A similar expression holds for λ_L. These expressions show that the parameters λ_U, λ_L are bounded between 0 and 1 inclusive. If $\lambda_U > 0$ ($\lambda_L > 0$), there is a positive probability that one of U_1, U_2 takes values greater (less) than u given that the other is greater (less) than u for u arbitrarily close to 1 (0).

2.1.11 Examples

In this subsection, a few examples are used to illustrate the dependence concepts in the preceding subsections.

Example 2.1 Let $f(x_1, x_2; \rho) = (2\pi)^{-1}(1-\rho^2)^{-1/2} \exp\{-\frac{1}{2}(x_1^2 + x_2^2 - 2\rho x_1 x_2)/(1-\rho^2)\}$, $-1 < \rho < 1$, be the BVSN density. Then it is straightforward to show that f is TP_2 (RR_2) if and only if $\rho \ge 0$ ($\rho \le 0$). Also the conditional distribution of the second variable given the first is $F_{2|1}(x_2|x_1) = \Phi((x_2 - \rho x_1)/\sqrt{1-\rho^2})$, and this is decreasing in x_1 for all x_2 if and only if $\rho \geq 0$. Hence this shows directly that $F_{2|1}$ is stochastically increasing (decreasing) if $\rho \geq 0$ ($\rho \leq 0$). □

Example 2.2 For the MVSN distribution with $m \times m$ correlation matrix $R = (\rho_{ij})$, the PDS condition is equivalent to $\rho_{ij} \geq 0$ for all i, j. Also the association condition is equivalent to $\rho_{ij} \geq 0$ for all i, j. Let $A = R^{-1} = (a_{ij})$; then the MTP$_2$ condition is equivalent to $a_{ij} \leq 0$ for all $i \neq j$.

Proof. Let $(X_1, \ldots, X_m)$ be MVSN with correlation matrix $R = (\rho_{ij})$. Note that the mean vector of $(X_2, \ldots, X_m)$ given $X_1 = x_1$ is $(\rho_{12}, \ldots, \rho_{1m})x_1$, so that the stochastic increasing property holds only if $\rho_{1j} \geq 0$ for $j = 2, \ldots, m$. By permuting the indices, all correlations must be non-negative if $(X_1, \ldots, X_m)$ is PDS.

The proof of association is non-trivial; see Joag-dev, Perlman and Pitt (1983).

It is easy to show that the MVN density $\phi_R(\mathbf{x})$, with correlation matrix R, is TP$_2$ in x_i, x_j, for all $i \neq j$, if $a_{ij} \leq 0$ for all $i \neq j$. This implies the MTP$_2$ condition. □

Example 2.3 Consider the family B5 of bivariate copulas in Section 5.1. With $\overline{u} = 1 - u$, $\overline{v} = 1 - v$, the family is

$$C(u, v; \delta) = 1 - (\overline{u}^\delta + \overline{v}^\delta - [\overline{u}\overline{v}]^\delta)^{1/\delta}, \quad 1 \leq \delta < \infty. \tag{2.18}$$

The corresponding family of densities is

$$c(u, v; \delta) = (\overline{u}^\delta + \overline{v}^\delta - [\overline{u}\overline{v}]^\delta)^{-2+1/\delta}[\overline{u}\overline{v}]^{\delta-1}[\delta - 1 + \overline{u}^\delta + \overline{v}^\delta - \overline{u}^\delta\overline{v}^\delta].$$

Note that the case of $\delta = 1$ corresponds to the independence copula $C_I(u, v) = uv$.

The conditional cdf,

$$C_{2|1}(v|u; \delta) = [1 + \overline{v}^\delta \overline{u}^{-\delta} - \overline{v}^\delta]^{-1+1/\delta}[1 - \overline{v}^\delta], \tag{2.19}$$

is decreasing in u for each v, so this proves directly that $C_{2|1}$ is SI for each $\delta \geq 1$.

The demonstration that the density c is TP$_2$ reduces to showing that $h(x, y) = (1 - xy)^{-2+1/\delta}(\delta - xy)$ is TP$_2$ in $0 \leq x, y \leq 1$ or that $h_0(s, t) = (1 - e^{-s-t})^{-2+1/\delta}(\delta - e^{-s-t})$ is TP$_2$ in $s, t \geq 0$. The inequality $h_0(s_1, t_1)h_0(s_2, t_2) \geq h_0(s_1, t_2)h_0(s_2, t_1)$ holds for $0 < s_1 < s_2$, $0 < t_1 < t_2$ if $g(x_1+y_1)+g(x_2+y_2) \geq g(x_1+y_2)+g(x_2+y_1)$ for $0 < x_1 < x_2$, $0 < y_1 < y_2$, where $g(z) = \log(\delta - e^{-z}) + (-2 + 1/\delta)\log(1 - e^{-z})$. But $g(z)$ is convex for $z > 0$ and then the inequality follows from $(x_1 + y_2, x_2 + y_1) \prec_m (x_1 + y_1, x_2 + y_2)$ ($\prec_m$

is the majorization ordering in Marshall and Olkin 1979; see also the Appendix). To show the convexity, $g''(z) = -\delta e^z(\delta e^z - 1)^{-2} + (2-1/\delta)e^z(e^z-1)^{-2} = e^z(\delta e^z-1)^{-2}(e^z-1)^{-2}(\delta-1)w(\delta,z)$, where $w(\delta,z) = 2\delta e^{2z} - 2e^z - 1 + \delta^{-1}$. Note that $w(\delta,0) = \delta^{-1}(2\delta-1)(\delta-1) \geq 0$ and $\partial w/\partial z = 4\delta e^{2z} - 2e^z \geq 0$, so that $g''(z) \geq 0$ for all $\delta \geq 1$.

The upper and lower tail dependence parameters are respectively $2 - 2^{1/\delta}$ and 0, so that (2.18) has upper tail dependence for $\delta > 1$.

The PFD property follows from Exercise 2.4 and results in Sections 4.2 and 5.1. □

Example 2.4 Consider the family B10 of bivariate copulas in Section 5.1:

$$C(u,v;\theta) = uv[1 + \theta(1-u)(1-v)], \quad -1 \leq \theta \leq 1. \tag{2.20}$$

This family is just a perturbation of the independence copula $C_I(u,v) = uv$. The distribution in (2.20) is PQD (NQD) for $0 \leq \theta \leq 1$ ($-1 \leq \theta \leq 0$). It has a limited range of of dependence which is why it is not useful as a model; simple computations show that Kendall's tau is $2\theta/9$ and Spearman's rho is $\theta/3$ so that τ is bounded in absolute value by 2/9 and ρ_S is bounded in absolute value by 1/3. □

Example 2.5 A family of bivariate exponential survival functions due to Gumbel (1960b) is:

$$\overline{F}(x_1,x_2;\theta) = e^{-x_1-x_2-\theta x_1 x_2}, \quad x_1 > 0, x_2 > 0,\ 0 \leq \theta \leq 1.$$

This has negative quadrant dependence and limited range of dependence so that it is not useful as a model. The amount of negative dependence increases as θ increases; for $\theta = 1$, Kendall's tau and Spearman's rho are -0.361 and -0.524, respectively. □

2.2 Dependence orderings

Positive dependence concepts such as PQD, SI and LTD, in the preceding section, result from comparing a bivariate or multivariate random vector with a random vector of independent rvs with the same corresponding univariate distributions. That is, if $F \in \mathcal{F}(F_1,\ldots,F_m)$, the class of m-variate distributions with given univariate margins $F_1,\ldots,F_m$, a positive dependence concept comes from comparing whether F is more positive dependent in some sense than the cdf $\prod_{j=1}^m F_j$. For example, the PUOD concept compares $\Pr(X_i > a_i, i = 1,\ldots,m)$ for $\mathbf{X} \sim F$ with $\mathbf{X} \sim \prod_{j=1}^m F_j$.

However, for a parametric family of multivariate distributions, one would be interested in more information than just positive or negative dependence (or indeterminate type of dependence). A parameter in the family is interpretable as a dependence parameter if the amount of dependence is increasing (or decreasing) as the parameter increases. This is one motivation for comparing whether one multivariate cdf is more dependent than another cdf based on some dependence concept. Comparisons can be made via dependence orderings that are partial orderings within a class $\mathcal{F}(F_1, \ldots, F_m)$. Because of the result (1.7) on copulas, dependence orderings or comparisons should hold the univariate margins fixed, at least in the continuous case.

In this section, we first give the concordance ordering corresponding to the dependence concept of PQD and POD. Then, we discuss and list the types of properties that would be desirable for a dependence ordering; one property is that the bivariate concordance ordering should hold for all sets of corresponding bivariate margins. (There are references to the Fréchet bounds, which are studied in detail in Chapter 3.) Following this, we list some dependence orderings that generalize the concepts in Section 2.1.

2.2.1 Concordance ordering °

We first give the definition of the concordance ordering in the bivariate case.

Definition. Let $F, F' \in \mathcal{F}(F_1, F_2)$ where F_1 and F_2 are univariate cdfs. F' is **more concordant** (or **more PQD**) than F, written $F \prec_c F'$, if

$$F(x_1, x_2) \le F'(x_1, x_2) \quad \forall\, x_1, x_2 \in (-\infty, \infty). \tag{2.21}$$

From the relation between a survival function and a cdf in the bivariate case (see equation (1.3)), (2.21) is equivalent to

$$\overline{F}(x_1, x_2) \le \overline{F}'(x_1, x_2) \quad \forall\, x_1, x_2 \in (-\infty, \infty). \tag{2.22}$$

Note that if $(X_1, X_2) \sim F$ and $(X_1', X_2') \sim F'$, then the concordance ordering means that

$$\Pr(X_1 \le x_1, X_2 \le x_2) \le \Pr(X_1' \le x_1, X_2' \le x_2) \quad \forall\, x_1, x_2$$

and

$$\Pr(X_1 > x_1, X_2 > x_2) \le \Pr(X_1' > x_1, X_2' > x_2) \quad \forall\, x_1, x_2.$$

For random vectors, we may use the notation $(X_1, X_2) \prec_c (X_1', X_2')$ instead of $F \prec_c F'$.

In the multivariate case with dimension $m \geq 3$, the orderings of cdfs and survival functions are not equivalent (i.e., the multivariate extensions of (2.21) and (2.22) are not equivalent). Hence there are various versions that could be considered as simple multivariate dependence orderings.

Definition. Let $F, F' \in \mathcal{F}(F_1, \ldots, F_m)$ where $F_1, \ldots, F_m$ are univariate cdfs. F' is **more PLOD** than F, written $F \prec_{cL} F'$, if

$$F(\mathbf{x}) \leq F'(\mathbf{x}) \quad \forall \mathbf{x} \in \Re^m. \tag{2.23}$$

F' is **more PUOD** than F, written $F \prec_{cU} F'$, if

$$\overline{F}(\mathbf{x}) \leq \overline{F'}(\mathbf{x}) \quad \forall \mathbf{x} \in \Re^m. \tag{2.24}$$

F' is **more concordant** or **more POD** than F, written $F \prec_c F'$, if both (2.23) and (2.24) hold.

The use of the term **concordant** here means that if $\mathbf{X}' \sim F'$ and $\mathbf{X} \sim F$, then the components of $\mathbf{X}'$ are more likely than those of $\mathbf{X}$ to take on small values (or large values) simultaneously.

For the bivariate ordering in $\mathcal{F}(F_1, F_2)$, the most concordant or maximal distribution is the Fréchet upper bound $F_U(x_1, x_2) = \min\{F_1(x_1), F_2(x_2)\}$ and the most discordant or minimal distribution is the Fréchet lower bound $F_L(x_1, x_2) = \max\{0, F_1(x_1) + F_2(x_2) - 1\}$. For the general multivariate ordering in $\mathcal{F}(F_1, \ldots, F_m)$, the maximal distribution is the Fréchet upper bound $F_U(\mathbf{x}) = \min_i F_i(x_i)$.

A nice property of the concordance ordering is that if F, F' are continuous bivariate distributions with Kendall taus $\tau(F), \tau(F')$, Spearman rhos $\rho_S(F), \rho_S(F')$, tail dependence parameters $\lambda(F)$, $\lambda(F')$, and $F \prec_c F'$, then $\tau(F) \leq \tau(F')$, $\rho_S(F) \leq \rho_S(F')$ and $\lambda(F) \leq \lambda(F')$. (The proof is left as an exercise.) The next theorem is a consequence of the $\prec_c$ ordering that is used later and its proof is also left as an exercise.

Theorem 2.8 *Suppose that $s_1, \ldots, s_k$ are all non-negative increasing or all non-negative decreasing functions on the real line and that F, F' are two m-variate cdfs. Let $\phi(x_1, \ldots, x_m) = \prod_{j=1}^m s_j(x_j)$. Then $F \prec_c F'$ implies $\int \phi \, dF \leq \int \phi \, dF'$ provided that the integrals exist.*

Other properties appear in the next subsection as part of an axiomatic approach for defining what properties are needed for an ordering on distributions to be considered a dependence ordering.

2.2.2 Axioms for a bivariate dependence ordering

In this subsection, we list properties or axioms that an ordering of distributions should have in order that higher in the ordering means more positive dependence.

Let $\prec$ be a bivariate dependence ordering (for cdfs in $\mathcal{F}(F_1, F_2)$ or random vectors that have the same corresponding univariate marginal distributions). Desirable properties or axioms for $\prec$ are:

P1. (concordance) $F \prec F'$ implies $F(x_1, x_2) \leq F'(x_1, x_2)$ for all x_1, x_2;

P2. (transitivity) $F \prec F'$ and $F' \prec F''$ imply $F \prec F''$;

P3. (reflexivity) $F \prec F$;

P4. (equivalence) $F \prec F'$ and $F' \prec F$ imply $F = F'$;

P5. (bounds) $F_L \prec F \prec F_U$, where F_U is the Fréchet upper bound and F_L is the Fréchet lower bound;

P6. (invariance to limit in distribution) $F_n \prec F'_n$, $n = 1, 2, \ldots$, and $F_n \to_d F$, $F'_n \to_d F'$ as $n \to \infty$, imply that $F \prec F'$;

P7. (invariance to order of indices) $(X_1, X_2) \prec (X'_1, X'_2)$ implies $(X_2, X_1) \prec (X'_2, X'_1)$;

P8. (invariance to increasing transforms) $(X_1, X_2) \prec (X'_1, X'_2)$ implies $(a(X_1), X_2) \prec (a(X'_1), X'_2)$ for all strictly increasing functions a;

P9. (invariance to decreasing transforms) $(X_1, X_2) \prec (X'_1, X'_2)$ implies $(b(X'_1), X'_2) \prec (b(X_1), X_2)$ for all strictly decreasing functions b.

If property P1 is satisfied, then the bivariate dependence ordering is stronger than the concordance ordering $\prec_c$. Property P5 implies that the Fréchet upper (lower) bound is the most (least) dependent in the ordering. Properties P6 to P9 are fairly natural invariance requirements.

An ordering that satisfies the nine properties is called a **bivariate positive dependence ordering** (BPDO). The concordance ordering is a BPDO and it is the weakest one in that if $F \prec F'$ for any other BPDO $\prec$, then $F \prec_c F'$. Other orderings are given in later subsections.

2.2.3 Axioms for a multivariate dependence ordering

In this subsection, we generalize the properties or axioms of the preceding subsection to the multivariate case. Let $\prec$ be a multivariate dependence ordering (for cdfs in $\mathcal{F}(F_1, \ldots, F_m)$ or random vectors that have the same corresponding univariate marginal distributions) that is defined for all dimensions $m \geq 2$. Desirable properties or axioms for $\prec$ are:

P1. (bivariate concordance) $F \prec F'$ implies that, for all $1 \leq i < j \leq m$, $F_{ij}(x_i, x_j) \leq F'_{ij}(x_i, x_j)$ $\forall x_i, x_j$, where F_{ij}, F'_{ij} are the (i, j) bivariate margins;

P2. (transitivity) $F \prec F'$ and $F' \prec F''$ imply $F \prec F''$;

P3. (reflexivity) $F \prec F$;

P4. (equivalence) $F \prec F'$ and $F' \prec F$ imply $F = F'$;

P5. (bound) $F \prec F_U$, where F_U is the Fréchet upper bound;

P6. (invariance to limit in distribution) $F_n \prec F'_n$, $n = 1, 2, \ldots$, and $F_n \to_d F$, $F'_n \to_d F'$ as $n \to \infty$, imply that $F \prec F'$;

P7. (invariance to order of indices) $(X_1, \ldots, X_m) \prec (X'_1, \ldots, X'_m)$ implies $(X_{i_1}, \ldots, X_{i_m}) \prec (X'_{i_1}, \ldots, X'_{i_m})$ for all permutations $(i_1, \ldots, i_m)$ of $(1, \ldots, m)$;

P8. (invariance to $\uparrow$ transforms) $(X_1, \ldots, X_m) \prec (X'_1, \ldots, X'_m)$ implies $(a(X_1), X_2, \ldots, X_m) \prec (a(X'_1), X'_2, \ldots, X'_m)$ for all strictly increasing functions a;

P9. (closure under marginals) $(X_1, \ldots, X_m) \prec (X'_1, \ldots, X'_m)$ implies $(X_{i_1}, \ldots, X_{i_k}) \prec (X'_{i_1}, \ldots, X'_{i_k})$ for all $i_1 < \cdots < i_k$, $2 \leq k < m$.

Note that bivariate property P5 does not extend completely because there is no Fréchet lower bound in general for dimensions $m \geq 3$. Similarly, the use of a decreasing transformation to reverse the ordering of dependence does not extend to the multivariate case. So property P9 from the bivariate case is replaced by the natural property of closure under marginals.

An ordering that satisfies these properties is called a **multivariate positive dependence ordering** (MPDO). The **pairwise concordance** ordering, which is defined next, satisfies all of the properties except for property P4.

Definition. Let $F, F' \in \mathcal{F}(F_1, \ldots, F_m)$, where $F_1, \ldots, F_m$ are univariate cdfs. F' is **more pairwise concordant** than F, written $F \prec_c^{\text{pw}} F'$, if, for all $1 \leq i < j \leq m$,

$$F_{ij}(x_i, x_j) \leq F'_{ij}(x_i, x_j) \quad \forall\, (x_i, x_j) \in \Re^2,$$

where F_{ij}, F'_{ij} are the (i, j) bivariate margins of F, F', respectively.

It is simple to show that for any MPDO $\prec$, $F \prec F'$ implies $F \prec_c^{\text{pw}} F'$. It is also straightforward to show that $\prec_c$, $\prec_{cU}$ and $\prec_{cL}$

are MPDOs. An MPDO which is stronger than $\prec_c$ is given in Section 2.2.5. It will be seen from the families of multivariate distributions given in Chapters 4 and 5 that the $\prec_c$ ordering is difficult or impossible to show analytically, whereas $\prec_c^{pw}$ and one of $\prec_{cU}$ or $\prec_{cL}$ is not difficult to establish. The reason can be seen in the formulas for obtaining a survival function from a cdf and vice versa (equations (1.3) and (1.4)). If one has a parametric family $F(\mathbf{x};\boldsymbol{\theta})$ in $\mathcal{F}(F_1,\ldots,F_m)$ that is increasing in $\boldsymbol{\theta}$ for all $\mathbf{x}$ (so that the $\prec_{cL}$ ordering holds), then the ordering in $\boldsymbol{\theta}$ holds for all marginal distributions, but this need not imply (analytically) that the survival functions are ordered because of the alternating signs in (1.3). Where the multivariate $\prec_c$ ordering has been established, it is through stochastic representations (e.g., Theorems 2.21 and 4.7 to 4.10).

2.2.4 More SI bivariate ordering *

In this subsection, we define a bivariate ordering $\prec_{SI}$ such that if $F \in \mathcal{F}(F_1, F_2)$, then $F_1F_2\prec_{SI}F$ is equivalent to $F_{2|1}$ SI. This ordering has been called the 'more regression dependent' or 'more monotone regression dependent' ordering in the statistical literature. There are several equivalent versions of the definition. Here we use the forms that will be the most useful in subsequent chapters. Also we impose some conditions, such as F_1, F_2 continuous and differentiable, to avoid technicalities.

Definition. Let $(X_1, X_2) \sim F$, $(X_1', X_2') \sim F'$ with $F, F' \in \mathcal{F}(F_1, F_2)$. Let $G = F_{2|1}$, $G' = F_{2|1}'$ be the respective conditional distributions of the second rv given the first. Suppose that $G(x_2|x_1)$ and $G'(x_2|x_1)$ are continuous in x_2 for all x_1. Then $F_{2|1}'$ is **more SI** than $F_{2|1}$ (written $F\prec_{SI}F'$ or $F_{2|1}\prec_{SI}F_{2|1}'$) if $\psi(x_1, x_2) = G'^{-1}(G(x_2|x_1)|x_1)$ is increasing in x_1. (Note that ψ is increasing in x_2 since, for each fixed x_1, it is a composition of increasing functions.)

We go through a sequence of theorems to establish properties and equivalences for the $\prec_{SI}$ ordering.

Theorem 2.9 *Suppose $X_1 = X_1'$, $X_1 \sim F_1$, $(X_1, X_2) \sim F$ and $(X_1', X_2') \sim F'$. Also suppose $F_{2|1}(x_2|x_1)$ and $F_{2|1}'(x_2|x_1)$ are continuous in x_2 for all x_1. Then a stochastic representation is*

$$(X_1', X_2') \stackrel{d}{=} (X_1, \psi(X_1, X_2))$$

where $\psi(x_1, x_2) = F_{2|1}'^{-1}(F_{2|1}(x_2|x_1)|x_1)$.

Proof. Let $G = F_{2|1}$ and $G' = F'_{2|1}$. Given $X_1 = x_1$, $X_2 \sim G(\cdot|x_1)$. Since $G(x_2|x_1)$ is continuous in x_2, $G(X_2|x_1)$ is uniform conditional on $X_1 = x_1$. Also if $U \sim U(0,1)$, then, conditional on $X_1 = x_1$, $G'^{-1}(U|x_1)$ has distribution $G'(x_2|x_1)$ by the continuity of this function in x_2. Hence $(X_1, G'^{-1}(G(X_2|X_1)|X_1)) \overset{d}{=} (X_1, X'_2)$. □

Note that by symmetry another stochastic relationship is

$$(X_1, X_2) \overset{d}{=} (X'_1, \zeta(X'_1, X'_2)),$$

where $\zeta(x_1, x_2) = F_{2|1}^{-1}(F'_{2|1}(x_2|x_1)|x_1)$.

Theorem 2.10 *With $G = F_{2|1}$, $G' = F'_{2|1}$, such that $G(x_2|x_1)$ and $G'(x_2|x_1)$ are continuous in x_2 for all x_1, equivalent forms for $F_{2|1} \prec_{\mathrm{SI}} F'_{2|1}$ are the following:*

(a) For any $z_1 < z_2$ and u, v in $(0,1)$,

$$G^{-1}(u|z_2) \ge G^{-1}(v|z_1) \quad \Rightarrow \quad G'^{-1}(u|z_2) \ge G'^{-1}(v|z_1).$$

(b) For any $z_1 < z_2$ and any y, y' with y in the support of $G(\cdot|z_1)$ and y' in the support of $G'(\cdot|z_2)$,

$$G(y|z_1) \ge G'(y'|z_1) \quad \Rightarrow \quad G(y|z_2) \ge G'(y'|z_2). \qquad (2.25)$$

(c) $\zeta(x_1, x_2) = F_{2|1}^{-1}(F'_{2|1}(x_2|x_1)|x_1)$ is increasing in x_2 and decreasing in x_1.

Proof. We prove the equivalence of (a) and (b) in the case where $G(x_2|x_1)$ and $G'(x_2|x_1)$ are strictly increasing in x_2. This assumption can be relaxed with some extra technical details.

For (b) ⇒ (a), the proof is by contradiction. Suppose (b) holds and (a) does not. Then there exist $u, v, z_1 < z_2$ such that $G^{-1}(u|z_2) \ge G^{-1}(v|z_1)$ and $G'^{-1}(u|z_2) < G'^{-1}(v|z_1)$. Let $y = G^{-1}(v|z_1)$ and $y' = G'^{-1}(u|z_2)$ so that $v = G(y|z_1)$ and $u = G'(y'|z_2)$. The inequalities become $G^{-1}(u|z_2) \ge y$ or $G(y|z_2) \le G'(y'|z_2)$ and $y' < G'^{-1}(v|z_1)$ or $G'(y'|z_1) < G(y|z_1)$. From the assumption of strictly increasing G, G', there exists $\epsilon > 0$ such that $G'(y'|z_1) \le G(y-\epsilon|z_1)$ and $G(y-\epsilon|z_2) < G'(y'|z_2)$. This contradicts condition (b).

The proof of (a) ⇒ (b) is similar. Suppose (a) holds and (b) does not. Then there exist $y, y', z_1 < z_2$ such that $G(y|z_1) \ge G'(y'|z_1)$ and $G(y|z_2) < G'(y'|z_2)$. Let $v = G(y|z_1)$ and $u = G'(y'|z_2)$ so that $y = G^{-1}(v|z_1)$ and $y' = G'^{-1}(u|z_2)$. The inequalities become $v \ge G'(y'|z_1)$ or $G'^{-1}(v|z_1) \ge G'^{-1}(u|z_2)$, and $G(y|z_2) < u$ or $G^{-1}(v|z_1) < G^{-1}(u|z_2)$. From the assumption of strictly increasing

G, G', there exists $\epsilon > 0$ such that $G^{-1}(v|z_1) \leq G^{-1}(u-\epsilon|z_2)$ and $G'^{-1}(v|z_1) > G'^{-1}(u-\epsilon|z_2)$. This contradicts condition (a).

Next we show the equivalence of condition (b) and the definition of $\prec_{\rm SI}$, again under the assumption of strictly increasing conditional cdfs. (This corrects part of the proof of Theorem 2.2 in Fang and Joe (1992).) Assume that (2.25) holds for $z_1 < z_2$. Let $\psi(z,y) = G'^{-1}(G(y|z)|z)$. It suffices to show that $\psi(z,y)$ is increasing in z. Let $z_1 < z_2$ and fix y. Let y' satisfy $G(y|z_1) = G'(y'|z_1)$. Then (2.25) implies that $\psi(z_2,y) = G'^{-1}(G(y|z_2)|z_2) \geq G'^{-1}(G'(y'|z_2)|z_2) \geq y' = G'^{-1}(G'(y'|z_1)|z_1) = G'^{-1}(G(y|z_1)|z_1) = \psi(z_1,y)$.

For the converse, suppose that (2.25) does not hold for some y, y^* and some $z_1 < z_2$. That is,

$$G(y|z_1) \geq G'(y^*|z_1) \quad \text{and} \quad G(y|z_2) < G'(y^*|z_2).$$

Let y' satisfy $G(y|z_1) = G'(y'|z_1)$ so that $y' \geq y^*$. Then $G(y|z_2) < G'(y'|z_2)$. Furthermore,

$$\psi(z_1,y) = G'^{-1}(G(y|z_1)|z_1) = G'^{-1}(G'(y'|z_1)|z_1) = y',$$

and

$$\psi(z_2,y) = G'^{-1}(G(y|z_2)|z_2) < G'^{-1}(G'(y'|z_2)|z_2) = y',$$

so that $\psi(z_1,y) > \psi(z_2,y)$ and ψ is not increasing in z for all y.

For the equivalence of condition (c) to the definition, we provide a proof in the case where ψ is strictly increasing and differentiable. The general case then follows by a limit of approximations. The transformation from the definition is $(x_1,x_2) \to (x_1',x_2') = (x_1, \psi(x_1,x_2))$ and the inverse transformation is $(x_1',x_2') \to (x_1,x_2) = (x_1', \zeta(x_1',x_2'))$. The Jacobian matrices of the two transformations are inverses of each other, i.e.,

$$\begin{bmatrix} 1 & 0 \\ \psi_1 & \psi_2 \end{bmatrix}^{-1} = (\psi_2)^{-1} \begin{bmatrix} \psi_2 & 0 \\ -\psi_1 & 1 \end{bmatrix} = \begin{bmatrix} 1 & 0 \\ \zeta_1 & \zeta_2 \end{bmatrix},$$

where ψ_j, ζ_j are the partial derivatives with respect to the jth variable, $j = 1,2$. Hence $\zeta_1(x_1',x_2') = -\psi_1(x_1,x_2)/\psi_2(x_1,x_2)$ or ζ_1 and ψ_1 are opposite in sign (i.e., the monotonicities of ζ and ψ in the first variable x_1 are opposite in direction). □

Theorem 2.11 *Let $F \in \mathcal{F}(F_1,F_2)$. Then $F_1F_2 \prec_{\rm SI} F$ if and only if $F_{2|1}$ is SI.*

Proof. $F_1F_2 \prec_{\rm SI} F \iff F_{2|1}^{-1}(F_2(x_2)|x_1) \uparrow x_1 \ \forall x_2 \iff F_{2|1}^{-1}(u|x_1) \uparrow x_1 \ \forall\, 0 < u < 1 \iff F_{2|1}(x_2|x_1) \downarrow x_1 \ \forall x_2 \iff F_{2|1}$ is SI. □

Theorem 2.12 *$F \prec_{SI} F'$ implies $F \prec_c F'$.*

Proof. Let $(X_1, X_2) \sim F$, $(X_1', X_2') \sim F'$, with $X_j \stackrel{d}{=} X_j'$, $j = 1, 2$. From Theorem 2.9, $(X_1', X_2') \stackrel{d}{=} (X_1, \psi(X_1, X_2))$ with $\psi(x_1, x_2) = F_{2|1}'^{-1}(F_{2|1}(x_2|x_1)|x_1)$. The assumption here implies that ψ is increasing in both x_1, x_2. To prove the concordance ordering, we consider two cases.

Case 1. Suppose that x_1, x_2 are such that $\psi(x_1, x_2) \le x_2$. Then

$$\begin{aligned} F'(x_1, x_2) = \Pr(X_1' \le x_1,\, X_2' \le x_2) & \\ = \;\; & \Pr(X_1 \le x_1,\, \psi(X_1, X_2) \le x_2) \\ \ge \;\; & \Pr(X_1 \le x_1,\, \psi(x_1, X_2) \le x_2) \\ \ge \;\; & \Pr(X_1 \le x_1,\, X_2 \le x_2) = F(x_1, x_2), \end{aligned}$$

where the last inequality results since $X_2 \le x_2$ implies $\psi(x_1, X_2) \le \psi(x_1, x_2) \le x_2$ under the starting assumption.

Case 2. Suppose that x_1, x_2 are such that $\psi(x_1, x_2) > x_2$. Then $X_2 > x_2 \Rightarrow \psi(x_1, X_2) \ge \psi(x_1, x_2) > x_2$ or $\psi(x_1, X_2) \le x_2 \Rightarrow X_2 \le x_2$. Therefore, $\Pr(X_1 > x_1, X_2 \le x_2) \ge \Pr(X_1 > x_1, \psi(x_1, X_2) \le x_2) \ge \Pr(X_1 > x_1, \psi(X_1, X_2) \le x_2) = \Pr(X_1' > x_1, X_2' \le x_2)$, so that $F(x_1, x_2) \le F'(x_1, x_2)$. □

We next extend the definitions of $\prec_{SI}$ in order to incorporate the Fréchet upper and lower bounds in the ordering of distributions in $\mathcal{F}(F_1, F_2)$. Then we comment on the properties of a BPDO, and whether they are satisfied for $\prec_{SI}$.

The Fréchet upper bound does not have continuous conditional cdfs but the condition in the definition of $\prec_{SI}$ still holds. Let $F \in \mathcal{F}(F_1, F_2)$ and let $F' = F_U$ be the Fréchet upper bound in $\mathcal{F}(F_1, F_2)$. Then $F_{2|1}'(x_2|x_1) = 1$ if $x_2 \ge F_2^{-1} \circ F_1(x_1)$ and 0 otherwise, and $F_{2|1}'^{-1}(u|x_1) = F_2^{-1} \circ F_1(x_1)$, $0 < u < 1$. Hence $\psi(x_1, x_2) = F_{2|1}'^{-1}(F_{2|1}(x_2|x_1)|x_1) = F_2^{-1} \circ F_1(x_1)$ is increasing in x_1. Furthermore, if $(X_1, X_2) \sim F$ and $(X_1', X_2') \sim F_U$, then $(X_1', X_2') \stackrel{d}{=} (X_1, F_2^{-1} \circ F_1(X_1))$.

Next let $F' \in \mathcal{F}(F_1, F_2)$ and let $F = F_L$ be the Fréchet lower bound in $\mathcal{F}(F_1, F_2)$. If $(X_1, X_2) \sim F_L$ and $(X_1', X_2') \sim F'$, there is no stochastic representation for (X_1', X_2') in terms of (X_1, X_2) because of the relationship $X_2 = F_2^{-1}(1 - F_1(X_1))$. For incorporating the Fréchet lower bound into the $\prec_{SI}$ ordering, we make use of the equivalent condition (c) in Theorem 2.10. Note that $F_{2|1}(x_2|x_1) = 1$ if $x_2 \ge F_2^{-1}(1 - F_1(x_1))$ and 0 otherwise, and

$F_{2|1}^{-1}(u|x_1) = F_2^{-1}(1 - F_1(x_1))$, $0 < u < 1$. Hence

$$\zeta(x_1, x_2) = F_{2|1}^{-1}(F'_{2|1}(x_2|x_1)|x_1) = F_2^{-1}(1 - F_1(x_1)).$$

Also $(X_1, X_2) \stackrel{d}{=} (X'_1, \zeta(X'_1, X'_2))$.

Theorem 2.13 *The (extended) $\prec_{\mathrm{SI}}$ ordering satisfies all properties of a BPDO except for P7. P7 is not satisfied because the definition of $\prec_{\mathrm{SI}}$ is not symmetric in the two variables. P7 is satisfied within the subfamilies of permutation-symmetric distributions.*

An approach that is useful for showing the $\prec_{\mathrm{SI}}$ ordering for a one-parameter family $C(\cdot; \delta)$ of copulas, when $C_{2|1}(v|u; \delta)$ does not have a closed-form inverse, is provided by the following theorem.

Theorem 2.14 *Let $C(u, v; \delta)$ be a family of bivariate copulas. Let $B(u, v, \delta) = g(C_{2|1}(v|u; \delta))$, where g is an arbitrary strictly increasing real-valued function. Assume that B is continuously differentiable in all variables up to second order. The family C is increasing in $\prec_{\mathrm{SI}}$ as δ increases (i.e., $C(u, v; \delta_1) \prec_{\mathrm{SI}} C(u, v; \delta_2)$ for $\delta_1 < \delta_2$) if*

$$\frac{\partial^2 B}{\partial v \partial u} \frac{\partial B}{\partial \delta} - \frac{\partial^2 B}{\partial \delta \partial u} \frac{\partial B}{\partial v} \geq 0.$$

Proof. Let $\delta_1 < \delta_2$. Then $C(u, v; \delta_1) \prec_{\mathrm{SI}} C(u, v; \delta_2)$ if $v^*(u) = v^*(\delta_2) = v^*(u; \delta_1, \delta_2, v)$ is increasing in u with v^* being the root of

$$B(u, v^*, \delta_2) = B(u, v, \delta_1). \tag{2.26}$$

Taking the derivative of (2.26) with respect to u leads to

$$\frac{\partial B}{\partial u}(u, v^*, \delta_2) + \frac{\partial B}{\partial v}(u, v^*, \delta_2) \frac{\partial v^*}{\partial u}(u) = \frac{\partial B}{\partial u}(u, v, \delta_1).$$

Since $\frac{\partial B}{\partial v} \geq 0$, $\frac{\partial v^*}{\partial u} \geq 0$ if $\frac{\partial B}{\partial u}(u, v^*, \delta_2) - \frac{\partial B}{\partial u}(u, v, \delta_1) \leq 0$ or if $\frac{\partial^2 B}{\partial \delta \partial u}(u, v^*(\delta), \delta) \leq 0$. This is equivalent to

$$\frac{\partial^2 B}{\partial v \partial u}(u, v^*, \delta) \frac{\partial v^*}{\partial \delta} + \frac{\partial^2 B}{\partial \delta \partial u}(u, v^*, \delta) \leq 0. \tag{2.27}$$

From (2.26), with $\delta = \delta_2$, $\frac{\partial v^*}{\partial \delta} = -\frac{\partial B}{\partial \delta} / \frac{\partial B}{\partial v}$. Hence (2.27) is equivalent to the condition in the statement of the theorem. □

2.2.5 More TP_2 bivariate orderings *

This subsection is on orderings involving the TP_2 condition. It is mainly included for theoretical interest and completeness. The orderings here are not used subsequently, whereas the more SI ordering is used; they are also difficult to check analytically.

Some notation is needed in order to present the orderings in a simple form. For intervals I_1, I_2 of real numbers, the notation $I_1 < I_2$ means that $x_1 \in I_1$ and $x_2 \in I_2$ imply $x_1 < x_2$. If $I = (a,b)$, $J = (c,d)$ are intervals and F is a bivariate cdf, the notation $F(I,J)$ is shorthand for the rectangle probability $F(b,d) - F(a,d) - F(b,c) + F(a,c)$.

Definition. Let $F, F' \in \mathcal{F}(F_1, F_2)$, where F_1 and F_2 are univariate cdfs. F' is **more TP$_2$ than** F **with respect to rectangles** (written $F \prec_{\text{TPR}} F'$) if, for all intervals I_1, I_2, J_1, J_2, with $I_1 < J_1$, $I_2 < J_2$,

$$\begin{aligned} &F(I_1,I_2)F(J_1,J_2)F'(I_1,J_2)F'(J_1,I_2) \\ &\quad \le F'(I_1,I_2)F'(J_1,J_2)F(I_1,J_2)F(J_1,I_2). \end{aligned} \tag{2.28}$$

F' is **more TP$_2$ than** F **with respect to lower quadrants** (written $F \prec_{\text{TPL}} F'$) if (2.28) holds for all intervals I_1, I_2, J_1, J_2 with $I_1 < J_1$, $I_2 < J_2$ and the extra restriction that I_1, I_2 have lower limits of $-\infty$. Similarly, F' is **more TP$_2$ than** F **with respect to upper quadrants** (written $F \prec_{\text{TPU}} F'$) if (2.28) holds for all intervals I_1, I_2, J_1, J_2 with $I_1 < J_1$, $I_2 < J_2$ and the extra restriction that J_1, J_2 have upper limits of ∞.

For $\prec_{\text{TPL}}$ and $\prec_{\text{TPU}}$, inequality (2.28) could be written respectively as

$$\begin{aligned} & F(x_1,x_2)[F(y_1,y_2) - F(y_1,x_2) - F(x_1,y_2) + F(x_1,x_2)] \\ & \cdot[F'(x_1,y_2) - F'(x_1,x_2)][F'(y_1,x_2) - F'(x_1,x_2)] \\ \le\ & F'(x_1,x_2)[F'(y_1,y_2) - F'(y_1,x_2) - F'(x_1,y_2) + F'(x_1,x_2)] \\ & \cdot[F(x_1,y_2) - F(x_1,x_2)][F(y_1,x_2) - F(x_1,x_2)] \end{aligned}$$

and

$$\begin{aligned} & \overline{F}(x_1,x_2)[\overline{F}(y_1,y_2) - \overline{F}(y_1,x_2) - \overline{F}(x_1,y_2) + \overline{F}(x_1,x_2)] \\ & \cdot[\overline{F}'(x_1,x_2) - \overline{F}'(x_1,y_2)][\overline{F}'(x_1,x_2) - \overline{F}'(y_1,x_2)] \\ \le\ & \overline{F}'(x_1,x_2)[\overline{F}'(y_1,y_2) - \overline{F}'(y_1,x_2) - \overline{F}'(x_1,y_2) + \overline{F}'(x_1,x_2)] \\ & \cdot[\overline{F}(x_1,x_2) - \overline{F}(x_1,y_2)][\overline{F}(x_1,x_2) - \overline{F}(y_1,x_2)] \end{aligned}$$

where $x_1 < y_1$, $x_2 < y_2$.

We go through a sequence of theorems to establish properties of the orderings.

Theorem 2.15 *Let $F \in \mathcal{F}(F_1, F_2)$ and suppose F has density f. Then $F_1F_2 \prec_{\text{TPR}} F$ if and only if f is TP_2.*

Proof. For comparing F_1F_2 with F, (2.28) is equivalent to

$$F(I_1, J_2)F(I_2, J_1) \le F(I_1, J_1)F(I_2, J_2). \tag{2.29}$$

Let $x_1 < y_1$, $x_2 < y_2$ and let $\epsilon > 0$ be sufficiently small. Let $I_1 = (x_1, x_1+\epsilon]$, $I_2 = (y_1, y_1+\epsilon]$, $J_1 = (x_2, x_2+\epsilon]$, $J_2 = (y_2, y_2+\epsilon]$. Divide both sides of (2.29) by ϵ and let $\epsilon \to 0$ to get

$$f(x_1, y_2)f(y_1, x_2) \le f(x_1, x_2)f(y_1, y_2), \tag{2.30}$$

so that $F_1F_2 \prec_{\mathrm{TPR}} F$ implies that f is TP_2. If f is TP_2, so that (2.30) holds for all $x_1 < y_1$, $x_2 < y_2$, then (2.29) holds for all $I_1 < I_2$, $J_1 < J_2$ by integration. □

Theorem 2.16 *$F \prec_{\mathrm{TPR}} F' \Rightarrow F \prec_{\mathrm{TPU}} F'$ and $F \prec_{\mathrm{TPL}} F'$. Both $F \prec_{\mathrm{TPU}} F'$ and $F \prec_{\mathrm{TPL}} F' \Rightarrow F \prec_{\mathrm{c}} F'$.*

Proof. The first statement is obvious. For the second, take $I_1 = (-\infty, x_1]$, $J_1 = (x_1, \infty)$, $I_2 = (-\infty, x_2]$, $J_2 = (x_2, \infty)$. Then (2.28) becomes

$$\begin{aligned} &F(x_1, x_2)\overline{F}(x_1, x_2)(F_1(x_1) - F'(x_1, x_2))(F_2(x_2) - F'(x_1, x_2)) \\ &\quad \le F'(x_1, x_2)\overline{F}'(x_1, x_2)(F_1(x_1) - F(x_1, x_2))(F_2(x_2) - F(x_1, x_2)), \end{aligned}$$

and this implies $F(x_1, x_2) \le F'(x_1, x_2)$ since $h(w) = \log\{[w(1 - F_1 - F_2 + w)]/[(F_1 - w)(F_2 - w)]\}$ is increasing in $w \in [\max\{0, F_1 + F_2 - 1\}, \min\{F_1, F_2\}]$ (its derivative is $w^{-1} + (1 - F_1 - F_2 + w)^{-1} + (F_1 - w)^{-1} + (F_2 - w)^{-1} \ge 0$). □

Remarks. Although a bivariate cdf F with TP_2 density satisfies the SI property, the $\prec_{\mathrm{TPR}}$, $\prec_{\mathrm{TPL}}$ and $\prec_{\mathrm{TPL}}$ orderings have not been shown to imply the $\prec_{\mathrm{SI}}$ ordering. There is no obvious connection between the TP_2 orderings and the $\prec_{\mathrm{SI}}$ ordering.

Theorem 2.17 *$\prec_{\mathrm{TPR}}$, $\prec_{\mathrm{TPL}}$ and $\prec_{\mathrm{TPU}}$ are BPDOs.*

Proof. The proof for $\prec_{\mathrm{TPR}}$ is given in Kimeldorf and Sampson (1987). The proof for the other two orderings is very similar (see also Metry and Sampson 1991). □

*2.2.6 Positive function dependence ordering **

The ordering that generalizes the dependence concept of PFD is given in this subsection.

Let the rvs $X_1, \ldots, X_m, X_1', \ldots, X_m'$ have a common distribution, say F_0, and let $\mathbf{X} \sim F$, $\mathbf{X}' \sim F'$. Then $\mathbf{X}'$ or F' is **more positive function dependent** than $\mathbf{X}$ or F (written $\mathbf{X} \prec_{\mathrm{pfd}} \mathbf{X}'$ or

$F \prec_{\text{pfd}} F'$) if

$$\mathrm{E}\,[h(X_1)\cdots h(X_m)] \le \mathrm{E}\,[h(X_1')\cdots h(X_m')]$$

for all real-valued functions h such that the expectations exist. In the case of m odd, there is the extra constraint of h being non-negative.

This ordering has some applications to multivariate models, but the following results show that it is not a BPDO when $m = 2$. Also a result below shows that two multivariate distributions can be ordered in $\prec_c$ but not in $\prec_{\text{pfd}}$ and vice versa. Generally, the $\prec_{\text{pfd}}$ ordering is useful only for exchangeable and some partially exchangeable multivariate distributions.

Theorem 2.18 *Let F_0 be a given univariate cdf and let $F \in \mathcal{F}(F_0, F_0)$. Furthermore, let $F_U(x_1, x_2) = \min\{F_0(x_1), F_0(x_2)\}$ be the Fréchet upper bound in $\mathcal{F}$. Then $F \prec_{\text{pfd}} F_U$.*

Proof. Let $(X_1, X_2) \sim F$ so that $(X_1, X_1) \sim F_U$. Then $\mathrm{E}\,[h(X_1) - h(X_2)]^2 \ge 0$ implies $2\mathrm{E}\,[h^2(X_1)] \ge 2\mathrm{E}\,[h(X_1)h(X_2)]$. □

Theorem 2.19 *Let F_0 be a given univariate cdf and consider the Fréchet class $\mathcal{F}(F_0, F_0)$; let $F_L(x_1, x_2) = \max\{0, F_0(x_1) + F_0(x_2) - 1\}$ be the Fréchet lower bound. Then it is not true that $F_L \prec_{\text{pfd}} F$ for all $F \in \mathcal{F}(F_0, F_0)$.*

Proof. Let us simplify to the case where F_0 is the cdf of a $U(0,1)$ rv. Let $(U_1, U_2) \sim F$ and then $(U_1, 1 - U_1) \sim F_L$. The $\prec_{\text{pfd}}$ ordering would require $\mathrm{E}\,[h(U_1)h(1 - U_1)] \le \mathrm{E}\,[h(U_1)h(U_2)]$ for all h. However, with $h(x) = x(1 - x)$ on [0,1], $\mathrm{E}\,[U_1^2(1 - U_1)^2] = 1/30$ and, for U_1, U_2 independent, $\{\mathrm{E}\,[U_1(1 - U_1)]\}^2 = 1/36$. □

Theorem 2.20 *The $\prec_{\text{pfd}}$ ordering need not imply the $\prec_c$ ordering, and vice versa.*

Proof. To get a simple example of a family of copulas $C(\cdot;\theta)$ which is ordered by $\prec_c$ but not by $\prec_{\text{pfd}}$, the symmetry in the two variables is eliminated. Let b_1, b_2 be functions on [0,1] which satisfy $\int_0^1 b_j(u)du = 0$, $j = 1, 2$, and $\int_0^x b_1(u)du \int_0^y b_2(v)dv \ge 0$ for all x, y in [0,1]. Then $c(u, v; \theta) = 1 + \theta b_1(u)b_2(v)$ is a proper density on $[0,1]^2$ for θ in a neighbourhood of 0, and the cdfs $C(\cdot;\theta)$ are increasing in $\prec_c$ as θ increases. Now let b_1, b_2, h be piecewise constant with

$$b_1(u) = \begin{cases} 1 & \text{if } 0 \le u < 0.5, \\ -1 & \text{if } 0.5 \le u \le 1, \end{cases}$$

$$b_2(u) = \begin{cases} 1 & \text{if } 0 \le u < 0.25,\ 0.5 \le u < 0.75, \\ -1 & \text{if } 0.25 \le u < 0.5,\ 0.75 \le u \le 1, \end{cases}$$

$$h(u) = \begin{cases} 0 & \text{if } 0 \le u < 0.25,\ 0.75 \le u \le 1, \\ h_2 & \text{if } 0.25 \le u < 0.5, \\ h_3 & \text{if } 0.5 \le u < 0.75, \end{cases}$$

and $h_2 \neq h_3$. Then $\int_0^1 \int_0^1 h(u)h(v)\, c(u,v;\theta)\, du\, dv = [(h_2+h_3)/4]^2 - \theta[(h_2-h_3)/4]^2$ is decreasing in θ.

The above example can be modified to get a family of copulas $C(\cdot;\theta)$ which is ordered by $\prec_{\text{pfd}}$ but not by $\prec_{\text{c}}$. Let b be a function on [0,1] such that $|b| \le 1$, and $\int_0^1 b(s)ds = 0$. Then $c(u,v;\theta) = 1 + \theta b(u)b(v)$ is a proper density for $-1 \le \theta \le 1$. Let h be an integrable function on [0,1]. Note that $\int_0^1 \int_0^1 h(u)h(v)\, c(u,v;\theta)\, du\, dv = [\int_0^1 h(u)du]^2 + \theta[\int_0^1 b(u)h(u)du]^2$ is increasing in θ, so that

$$C(u,v;\theta) = uv + \theta \int_0^u b(s)\, ds \int_0^v b(t)\, dt$$

is increasing in the $\prec_{\text{pfd}}$ ordering as θ increases. Now let

$$b(u) = \begin{cases} -1 & \text{if } 0 < u < 0.25,\ 0.75 < u < 1, \\ 1 & \text{if } 0.25 \le u \le 0.75, \end{cases}$$

so that

$$\int_0^x b(s)\, ds = \begin{cases} -x & \text{if } 0 < x \le 0.25, \\ x - 0.5 & \text{if } 0.25 < x \le 0.75, \\ 1 - x & \text{if } 0.75 < x \le 1. \end{cases}$$

For $u = 1/8$, $v = 5/8$, $C(u,v;\theta) = (5-\theta)/64$ is decreasing in θ, so that $C(\cdot;\theta)$ is not increasing in the $\prec_{\text{c}}$ ordering. □

2.2.7 Examples: bivariate

This subsection consists of bivariate examples that illustrate the dependence orderings.

Example 2.3 (continued). The family B5 of copulas in (2.18) is increasing in concordance as δ increases and the limit is the Fréchet upper bound as $\delta \to \infty$. To show the concordance ordering, one needs $\delta^{-1}\log(u^\delta + v^\delta - u^\delta v^\delta)$ to be decreasing in δ for all $0 < u, v < 1$ or $a \log a + b \log b - ab \log(ab) - (a+b-ab)\log(a+b-ab) \le 0$ for all $0 < a, b < 1$ ($a = u^\delta$, $b = v^\delta$, $\delta \ge 1$). The last inequality follows from the majorization ordering $(a,b) \prec_m (ab, a+b-ab)$ and the convexity of the function $w \log w$ for $w \ge 0$.

The family B5 satisfies the stronger $\prec_{\text{SI}}$ ordering as δ increases. The following is a proof based on Theorem 2.14. Let $B = \log C_{2|1}$, where $C_{2|1}$ is given in (2.19). Let $U = u^\delta$, $V = v^\delta$, $\dot{U} = \partial U/\partial \delta =$

$\delta^{-1}U \log U$, and $\dot{V} = \delta^{-1}V \log V$. Then

$$B(u, v, \delta) = (-1+\delta^{-1}) \log(U+V-UV)+(1-\delta^{-1}) \log U+\log(1-V).$$

Derivatives are:

$$\frac{\partial^2 B}{\partial v \partial u} = \frac{(\delta-1)\delta UV}{(U+V-UV)^2 uv},$$

$$\frac{\partial B}{\partial \delta} = \frac{-1}{\delta^2} \log(U+V-UV)+\frac{\dot{U}}{U}-\frac{\dot{V}}{1-V}+(\delta^{-1}-1)\frac{\dot{U}(1-V)+\dot{V}(1-U)}{U+V-UV},$$

$$\frac{\partial^2 B}{\partial \delta \partial u} = \frac{V(U+V-UV)+(\delta-1)[U\dot{V}-\dot{U}V(1-V)]}{u(U+V-UV)^2},$$

$$\frac{\partial B}{\partial v} = \frac{V[(1-U)(1-V)-\delta]}{v(1-V)(U+V-UV)}.$$

The condition $\frac{\partial^2 B}{\partial v \partial u}\frac{\partial B}{\partial \delta} - \frac{\partial^2 B}{\partial \delta \partial u}\frac{\partial B}{\partial v}$ simplifies to

$$(U+V-UV)^{-3}(uv)^{-1}[A_1+A_2+A_3],$$

where $A_1 = V^2(U+V-UV)[\delta-(1-U)(1-V)]/(1-V) \geq 0$, $A_2 = -\delta^{-1}(\delta-1)UV \log(U+V-UV) \geq 0$ and $A_3 = (\delta-1)\dot{U}(U+V-UV)V(1-V) = \delta^{-1}(\delta-1)UV(U+V-UV)(1-V) \log U \leq 0$. The sum A_2+A_3 can be negative. However $A_1+A_2+A_3 \geq 0$ since $V(U+V-UV)^2/(1-V)+U[-\log(U+V-UV)+(U+V-UV)(1-V) \log U] \geq 0$ for all $0 \leq U, V \leq 1$. This last inequality has been verified by numerical computation and a study at the boundaries. □

Example 2.6 The $\prec_{\mathrm{SI}}$ ordering is shown for the family B3 in Section 5.1, using a direct application of the definition of $\prec_{\mathrm{SI}}$.

Consider the family of copulas

$$C(u, v; \delta) = -\delta^{-1} \log[1-(1-e^{-\delta u})(1-e^{-\delta v})/(1-e^{-\delta})],\ 0 \leq \delta < \infty.$$

Let $0 < \delta_1 < \delta_2$. We show that $C(\cdot; \delta_1) \prec_{\mathrm{SI}} C(\cdot; \delta_2)$. Let $G(v|u; \delta) = C_{2|1}(v|u; \delta) = [1-e^{-\delta}-(1-e^{-\delta u})(1-e^{-\delta v})]^{-1} e^{-\delta u}(1-e^{-\delta v}) = e^{-\delta u}[(1-e^{-\delta})(1-e^{-\delta v})^{-1}-(1-e^{-\delta u})]^{-1}$. Then

$$\begin{aligned}\psi(u, v; \delta_1, \delta_2) &= G^{-1}(G(v|u; \delta_1)|u; \delta_2) \\ &= -\frac{1}{\delta_2} \log\left\{1-\frac{1-\exp\{-\delta_2\}}{(w^{-1}-1)\exp\{-\delta_2 u\}+1}\right\},\end{aligned}$$

where $w = G(v|u; \delta_1)$. Here $\psi(u, v; \delta_1, \delta_2)$ is increasing in u since $(w^{-1}-1)e^{-\delta_2 u} = e^{(\delta_1-\delta_2)u}(1-e^{-\delta_1 v})^{-1}(e^{-\delta_1 v}-e^{-\delta_1})$ is decreasing in u for fixed v. □

Example 2.7 Numerical checks seem to indicate that the $\prec_{\mathrm{TPU}}$ (and $\prec_{\mathrm{TPL}}$) ordering holds for the one-parameter families B1–B8 of copulas in Section 5.1. Note that by numerical checks we mean that conditions are evaluated over a (fine) grid of values of the relevant variables. The $\prec_{\rm TPR}$ ordering has been shown numerically and analytically not to hold for the families B2–B8, by comparing the special case of densities (intervals in the definition collapsed to points). The $\prec_{\rm TPR}$ ordering seems to hold for the BVN family from numerical tests. The $\prec_{\rm SI}$ ordering holds for the families B1–B6; for the families B7 and B8 the $\prec_{\rm SI}$ ordering has not been checkable analytically. □

Example 2.8 Let (X_1, X_2), (X_1', X_2') have BVSN distributions with respective correlations ρ, ρ' such that $0 \le \rho < \rho' \le 1$. Then $(X_1, X_2) \prec_{\rm pfd} (X_1', X_2')$. This result generalizes to equicorrelated MVN random vectors with positive correlations.

Proof. Let $g(u, v, w) = \sqrt{1-\rho'}\, u + \sqrt{\rho' - \rho}\, v + \sqrt{\rho}\, w$. Then we can write

$$\begin{aligned}(X_1, X_2) &= (g(U_1, V_1, W), g(U_2, V_2, W)),\\ (X_1', X_2') &= (g(U_1, V_1, W), g(U_2, V_1, W)),\end{aligned}$$

where U_1, U_2, V_1, V_2, W are iid standard normal rvs. Let $h^*(v, w) = \mathrm{E}\,[h(g(U_1, V_1, W)) \mid V_1 = v, W = w]$. Then

$$\begin{aligned}\mathrm{E}\,[h(X_1)h(X_2)] &= E[h^*(V_1, W)h^*(V_2, W)]\\ &\le E\{[h^*(V_1, W)]^2\} = \mathrm{E}\,[h(X_1')h(X_2')],\end{aligned}$$

by making use of Theorem 2.18 for the inequality. □

2.2.8 Examples: multivariate

This subsection consists of multivariate examples that illustrate the dependence orderings.

Example 2.9 The following is a result on the concordance ordering for elliptically contoured distributions, which include MVN distributions as a special case.

Theorem 2.21 *Let $\mathcal{P}$ be the set of non-negative definite correlation matrices. Let $\mathbf{Z}$ have a spherically symmetric distribution, and let $\mathbf{X}^T = A\mathbf{Z}^T$ where $AA^T = \Sigma = (\sigma_{ij})$ is the Cholesky decomposition of $\Sigma \in \mathcal{P}$, with A lower triangular. Then, for all $\mathbf{b} \in \Re^m$,*

$$\Pr(X_1 \le b_1, \ldots, X_m \le b_m) \tag{2.31}$$

is increasing in σ_{ij} for all $i \ne j$.

Proof. Since a spherically symmetric distribution is a mixture of distributions that are uniform on the surfaces of spheres of different radii, it suffices to prove the result for the uniform distribution on the surface of a sphere with radius 1.

In the bivariate case, the representation from the Cholesky decomposition is $X_1 = Z_1$, $X_2 = \rho Z_1 + (1-\rho^2)^{1/2} Z_2$ as ρ varies from -1 to 1. Then $\Pr(X_2 \le b_2) = \Pr(Z_2 \le (b_2 - \rho Z_1)/(1-\rho^2)^{1/2})$ is a constant; the line $y = (b_2 - \rho x)/(1-\rho^2)^{1/2}$ divides the circle $x^2 + y^2 = 1$ in the same proportions for all ρ between -1 and 1. Hence it is clear from a diagram that $\Pr(X_1 = Z_1 \le b_1, Z_2 \le (b_2 - \rho Z_1)/(1-\rho^2)^{1/2})$ is increasing in ρ because the slope $-\rho/(1-\rho^2)^{1/2}$ is decreasing in ρ. Therefore the case $m = 2$ has been proved.

For $m > 2$, it suffices by symmetry to show that (2.31) is increasing in $\rho = \sigma_{m-1,m}$ with other σ_{ij} held fixed. Let $\mathbf{Z}$ be uniform on the surface of a sphere of radius 1. From the Cholesky decomposition, only $a_{m,m-1}$ and a_{mm} depend on ρ and $a_{m,m-1} = (\rho - \sum_{j=1}^{m-2} a_{m-1,j} a_{mj})/a_{m-1,m-1}$ if $a_{m-1,m-1} > 0$ and $a_{m,m-1} = 0$ if $a_{m-1,m-1} = 0$ (in this case the upper $(m-1) \times (m-1)$ submatrix of Σ is singular and ρ is fixed given the other σ_{ij}). If $a_{m-1,m-1} > 0$, $\Pr(X_1 \le b_1, \ldots, X_m \le b_m)$ becomes a weighted integral over $z_1, \ldots, z_{m-2}$ of

$$\Pr(Z_{m-1} \le c_1(z_1, \ldots, z_{m-2}), \rho^* Z_{m-1} + (1-\rho^{*2})^{1/2} Z_m$$
$$\le c_2(z_1, \ldots, z_{m-2}) \mid (Z_1, \ldots, Z_{m-2}) = (z_1, \ldots, z_{m-2})), \quad (2.32)$$

where $\rho^* = a_{m,m-1}/D$ is increasing in ρ, and

$$c_1(z_1, \ldots, z_{m-2}) = \Big[b_{m-1} - \sum_{j=1}^{m-2} a_{m-1,j} z_j\Big] \Big/ a_{m-1,m-1},$$

$c_2(z_1, \ldots, z_{m-2}) = [b_m - \sum_{j=1}^{m-2} a_{mj} z_j]/D$, $D = [1 - \sum_{j=1}^{m-2} a_{mj}]^{1/2}$. Hence the monotonicity of (2.32) follows from the general $m = 2$ case, since in (2.32), (Z_{m-1}, Z_m) has a density with circular contours. □

Example 2.10 Suppose F_{12}, F_{13}, F_{23} are compatible (1,2), (1,3) and (2,3) bivariate margins. Consider the set of trivariate cdfs $\mathcal{F}(F_{12}, F_{13}, F_{23})$. If $F, F' \in \mathcal{F}(F_{12}, F_{13}, F_{23})$ then $F \prec_{\mathrm{cU}} F'$ implies $F' \prec_{\mathrm{cL}} F$ so that $F \prec_{\mathrm{c}} F'$ implies that $F = F'$. This follows from the relationship between a trivariate cdf and survival function. □

Example 2.11 (Multivariate Fréchet upper bound and $\prec_{\mathrm{pfd}}$.) Suppose $\mathbf{X}$ is such that $X_1 \stackrel{d}{=} \cdots \stackrel{d}{=} X_m$. Then $\mathbf{X} \prec_{\mathrm{pfd}} (X_1, \ldots, X_1)$.

Proof. Consider the case where h is a non-negative function. First let $X_1, \ldots, X_m$ be exchangeable rvs. Then it suffices to prove that

$$\mathrm{E}\,[h(X_1)\cdots h(X_m)] \le \mathrm{E}\,[h^m(X_1)]. \tag{2.33}$$

Note that $h(X_j)$ are exchangeable rvs and then (2.33) follows from Muirhead's theorem (Marshall and Olkin 1979, p. 87). For the non-exchangeable case, note that if $\mathbf{X} \sim F$ and F has density f relative to some measure, then the F can be symmetrized to F^* with density $f^*(\mathbf{x}) = (m!)^{-1} \sum_\pi f(x_{\pi(1)}, \ldots, x_{\pi(m)})$, where the sum is over the permutations of $\{1, \ldots, m\}$. If $\mathbf{X}^* \sim F^*$, then

$$\mathrm{E}\,[h(X_1^*)\cdots h(X_m^*)] = \mathrm{E}\,[h(X_1)\cdots h(X_m)].$$

Hence in general, $\mathbf{X} \prec_{\mathrm{pfd}} (X_1, \ldots, X_1)$.

Next consider the case where m is even and h can have positive and negative values. It is necessary to show (2.33) above for arbitrary h. Similarly to the proof in Theorem 2.18, we start with $\mathrm{E}\,\{[h(X_1)\cdots h(X_{m/2}) - h(X_{m/2+1})\cdots h(X_m)]^2\} \ge 0$. We can assume (by symmetrization) that $X_1, \ldots, X_m$ are exchangeable rvs. Then $\mathrm{E}\,[h(X_1)\cdots h(X_m)] \le \mathrm{E}\,[h^2(X_1)\cdots h^2(X_{m/2})]$. The latter term is dominated by $\mathrm{E}\,[h^m(X_1)]$ from the preceding case. □

2.3 Bibliographic notes

Early references for the concepts of PQD, POD, SI, RTI, LTD, TP_2 density, association and CIS are Barlow and Proschan (1981), Lehmann (1966), Esary and Proschan (1972) and Esary, Proschan and Walkup (1967). Papers that include negative association, not used in this book, are Alam and Saxena (1981) and Joag-dev and Proschan (1983). A reference for multivariate dependence concepts is Block and Ting (1981). A reference for the concept of PDS is Block, Savits and Shaked (1985). The multivariate extension of LTD is from Alzaid and Proschan (1994). For the concept of TP_2 survival functions and generalizations, see Shaked (1977a). Shaked (1977a) shows that the condition of a TP_2 bivariate survival function is the same as an earlier definition of *right corner set increasing* (RCSI) in Harris (1970). The relation between min-id and TP_2 surivival functions is proved in Marshall and Olkin (1990). The concepts of MTP_2 and MRR_2 are from Karlin and Rinott (1980a; 1980b). Results on max-id and min-id are from Joe and Hu (1996). For further results for max-id for bivariate distributions, see Balkema and Resnick (1977).

Original references for Spearman's rho and Kendall's tau are

Spearman (1904) and Kendall (1938); a later reference with connections to copulas is Schweizer and Wolff (1981). The concept of tail dependence is from Joe (1993).

The axioms of a bivariate positive dependence ordering and a framework for positive dependence are given in Kimeldorf and Sampson (1987; 1989). The bivariate concordance ordering is presented in Yanagimoto and Okamoto (1969), Tchen (1980) and Cambanis, Simons and Stout (1976), and the multivariate concordance ordering is presented in Joe (1990c). The more SI ordering, although called the more regression dependent or more monotone regression dependent ordering, is studied in Yanagimoto and Okamoto (1969), Schriever (1986; 1987) and Fang and Joe (1992). Schriever has a more associated ordering and a different equivalent condition for the more SI ordering; these are not given here because they are not needed for the results in this book. The more TP_2 orderings are studied in Kimeldorf and Sampson (1987) and Metry and Sampson (1991); the latter has more versions of more TP_2 orderings than given in this chapter.

Property P9 in Section 2.2.2 is an extension of Kimeldorf and Sampson (1987) from a sign change to a decreasing transformation. Section 2.2.3 consists of new results. However, the usefulness of multivariate positive dependence orderings other than the variations of the concordance ordering seems to be limited because of the difficulty of analytic checking. Theorem 2.14 is new, and the proof of Theorem 2.21 is from Joe (1990c).

The PFD condition and orderings, known as positive dependence and more positive dependent, respectively, are studied in Rinott and Pollak (1980), Gleser and Moore (1983) and Tong (1989). Related ideas, including positive dependence by mixtures, are in Shaked (1977b; 1979).

A comprehensive reference for stochastic orderings is Shaked and Shanthikumar (1994). A reference on total positivity is Karlin (1968). Dependence concepts have many other applications besides those in this book; several dependence concepts arise from reliability (see Barlow and Proschan 1981) and Boland *et al.* (1996) study dependence properties of order statistics.

2.4 Exercises

2.1 Show the equivalence of (2.1) and (2.2).

2.2 Show that (2.3) and (2.4) are not equivalent for $m > 2$.

2.3 Show by means of counterexamples that there are no other implications among the bivariate positive dependence concepts of PQD, SI, LTD, RTI, associated, TP_2 density and TP_2 cdf.

2.4 Show that if $F(x_1, x_2) = \int_0^\infty G(x_1; \alpha) G(x_2; \alpha)\, dM(\alpha)$, then F is PFD.

2.5 For an m-variate cdf F, show that F^γ ($\overline{F}^\gamma$) is a cdf (survival function) for all $\gamma \geq m - 1$.

2.6 Suppose $F \in \mathcal{F}(F_1, F_2)$ has a covariance of 0 but $F \neq F_1 F_2$. Prove that F is neither PQD nor NQD.

2.7 For a bivariate copula C, let $C'(u_1, u_2) = u_1 + u_2 - 1 + C(1 - u_1, 1 - u_2)$, $C''(u_1, u_2) = u_1 - C(u_1, 1 - u_2)$ be two of the associated copulas. Show that if C is PQD, then C' is PQD and C'' is NQD. Show that if C has upper tail dependence, then C' has lower tail dependence.

2.8 For the bivariate normal density with correlation ρ, establish the condition for a TP_2 density and association.

2.9 Let (X_1, X_2) have a BVSN distribution with correlation $\rho < 1$. Show that $\Pr(X_2 > x \mid X_1 > x) \to 0$ as $x \to \infty$ so that the BVN copula does not have tail dependence for $\rho < 1$.

2.10 Let $F \prec_c F'$, where F, F' are continuous cdfs. Show that the Kendall tau, Spearman rho and tail dependence values for F' are respectively greater than or equal to those of F.

2.11 Prove Theorem 2.8.

2.12 Let $C(u, v; \delta) = (u^{-\delta} + v^{-\delta} - 1)^{-1/\delta}$, $0 \leq \delta < \infty$. This is the bivariate family B4 of copulas in Section 5.1. Check whether the dependence concepts in Section 2.1 hold. Also check whether C is increasing with respect to the $\prec_c$ and $\prec_{SI}$ orderings as δ increases.

2.13 For a bivariate cdf F, prove that $\overline{F}$ TP_2 is equivalent to F being min-id. (Marshall and Olkin 1990)

2.14 For the bivariate normal density with correlation ρ, show that Kendall's tau is $\tau = (2/\pi)\arcsin(\rho)$ and Spearman's rho is $\rho_S = (6/\pi)\arcsin(\rho/2)$. (See Kepner, Harper and Keith 1989, for the quadrant probability calculation.)

2.15 Let $(X_1, X_2) \sim F$, where $F \in \mathcal{F}(F_1, F_2)$, and suppose that the covariance of X_1, X_2 exists. Hoeffding's identity, which is

$$\mathrm{Cov}\,(X_1, X_2) = \int_{-\infty}^{\infty}\int_{-\infty}^{\infty} [\overline{F}(x_1, x_2) - \overline{F}_1(x_1)\overline{F}_2(x_2)]\, dx_2 dx_1,$$

is used in (2.9). Verify it. [Hint: use the positive and negative parts of a rv.] (Hoeffding 1940; Shea 1983)

2.16 Find an example of a random vector (X_1, X_2, X_3) which is not PFD but has pairs (X_1, X_2), (X_1, X_3), (X_2, X_3) that are are PFD. Also find an example where (X_1, X_2, X_3) is PFD but (X_1, X_2) is not PFD. [Hint: consider trivariate families that are like the family B10 in Section 5.1.]

2.17 Prove that if an m-variate density f is MTP_2, then its cdf F and survival function $\overline{F}$ are also MTP_2.

2.18 Let X_1, X_2 be continuous rvs such that X_2 is RTI in X_1 and X_2 is LTD in X_1. Prove that the Spearman rho value for X_1, X_2 is larger than the Kendall tau value.

(Capéraà and Genest 1993)

2.19 (Interpretation of Σ^{-1} for a MVN distribution, see Example 2.2.) Let $A = (a_{ij})$ be the inverse of the m-dimensional correlation or covariance matrix Σ. Let $\Sigma^{(uv)}$ be the matrix obtained from Σ by removing the uth, vth rows and columns. Let $\sigma_{ij\cdot r}$ be the partial covariance of the ith and jth variables given the remaining $m-2$ variables. Show that $a_{ij} = -\sigma_{ij\cdot r}|\Sigma^{(ij)}||\Sigma|^{-1}$ for $i \neq j$, so that $\mathbf{X} \sim N(\mathbf{0}, \Sigma)$ has MTP_2 density if and only if, for all $i \neq j$, the partial correlation of X_i, X_j given any subset of the remaining variables is non-negative.

2.20 Show that the orderings $\prec_{\mathrm{c}}$, $\prec_{\mathrm{cU}}$ and $\prec_{\mathrm{cL}}$ are MPDOs.

2.21 Let $F, F' \in \mathcal{F}(F_0, \ldots, F_0)$ be m-variate distributions, with respective continuous densities f, f'. Prove that a necessary condition for $F \prec_{\mathrm{pfd}} F'$ is that $f(x, \ldots, x) \leq f'(x, \ldots, x)$ for all x in the support of F_0.

2.22 Let $\Phi_3(\cdot; \rho_{12}, \rho_{13}, \rho_{23})$ be the family of trivariate standard normal cdfs with means 0 and correlations ρ_{ij}. If $\rho_{ij} \geq 0$, $i < j$, and if the inequality $\rho_{12}\rho_{23} \leq \rho_{13} \leq 1 - |\rho_{12} - \rho_{23}|$ does not hold, then show that $\Phi_3(\cdot; \rho_{12}, \rho_{13}, \rho_{23})$ and $\Phi_3(\cdot; \rho_{12}, \rho_{13} + \epsilon, \rho_{23})$ are not ordered by $\prec_{\mathrm{pfd}}$, where $\epsilon > 0$ is arbitrarily small.

2.23 Let $F, F' \in \mathcal{F}(F_1, F_2)$ be such that $F \prec_{\mathrm{c}} F'$. Let $F_{2|1}, F'_{2|1}$ be the corresponding conditional distributions. For a fixed x_2, show that the number of sign changes of the function $s = F_{2|1}(x_2|\cdot) - F'_{2|1}(x_2|\cdot)$ is odd and $s(x_1)$ is negative (positive) for x_1 near x_L (x_U), the lower (upper) end point of support.

2.24 Show by means of a counterexample that the converse of the result in the preceding exercise is not true.

2.25 Let $F, F' \in \mathcal{F}(F_1, F_2)$ and let $F_{2|1}, F'_{2|1}$ be the (continuous) conditional distributions. Suppose $F \prec_{\mathrm{SI}} F'$. Show that there is a function $b(x_2)$ such that $F_{2|1}(x_2|x_1) - F'_{2|1}(x_2|x_1) \geq 0$ if and only if $b(x_2) \leq x_1$.

2.26 Let $F, F' \in \mathcal{F}(F_1, F_2)$ and let $F_{2|1}, F'_{2|1}$ be the (continuous) conditional distributions. Suppose that there is a function $b(x_2)$ such that $F_{2|1}(x_2|x_1) - F'_{2|1}(x_2|x_1) \geq 0$ if and only if $b(x_2) \leq x_1$. Show that $F \prec_{\mathrm{c}} F'$.

2.27 Consider the ordering for $F, F' \in \mathcal{F}(F_1, F_2)$ defined by $F \prec F'$ if there exists a real-valued function $b(x_2)$ such that $F_{2|1}(x_2|x_1) - F'_{2|1}(x_2|x_1) \geq 0$ if and only if $b(x_2) \leq x_1$. Show that $\prec$ is not transitive.

2.5 Unsolved problems

2.1 Verify or disprove the $\prec_{\mathrm{TPR}}$ ordering for the BVN family.

2.2 Verify or disprove the $\prec_{\mathrm{TPL}}$ and $\prec_{\mathrm{TPU}}$ orderings for the families B1 to B8 in Section 5.1.

2.3 Consider the ordering with definition $F \prec F'$ if $\int \psi\, dF \leq \int \psi\, dF'$ for all L-superadditive functions ψ. A real-valued function ψ on $\Re^m$ is **L-superadditive** or **lattice superadditive** if

$$\psi(\mathbf{x} \vee \mathbf{y}) + \psi(\mathbf{x} \wedge \mathbf{y}) \geq \psi(\mathbf{x}) + \psi(\mathbf{y}), \quad \forall \mathbf{x}, \mathbf{y} \in \Re^m.$$

Is this an MPDO for $m \geq 3$? It is equivalent to the concordance ordering for $m = 2$ (Tchen 1980). For $m \geq 4$, the L-superadditive ordering is strictly stronger, and for $m = 3$ it is unknown if the L-superadditive ordering is strictly stronger.

(Joe 1990c)

CHAPTER 3

Fréchet classes

This chapter is concerned with results on the extremes of and bounds on **Fréchet classes** (or classes of multivariate distributions with some given margins). If the indices of the margins being fixed are overlapping, then one first has to determine whether the margins are compatible. Section 3.1 is devoted to the class of m-variate distributions $\mathcal{F}(F_1, \ldots, F_m)$, in which the univariate margins $F_1, \ldots, F_m$ are given or fixed. Subsequent sections are devoted to other Fréchet classes, starting with classes of trivariate distributions with some fixed bivariate margins. Specifically, $\mathcal{F}(F_{12}, F_3)$, $\mathcal{F}(F_{12}, F_{13})$, $\mathcal{F}(F_{12}, F_{13}, F_{23})$, $\mathcal{F}(F_{ij}, 1 \leq i < j \leq m)$, etc., are studied. For something like $\mathcal{F}(F_{12}, F_{13})$, it is assumed that the first univariate margin of F_{12} and F_{13} are the same. The study of the class $\mathcal{F}(F_{ij}, 1 \leq i < j \leq m)$ is important but not easy; one first has to determine if the set of bivariate margins are compatible and, if so, come up with methods to construct a multivariate distribution with the given margins. Some non-compatibility results for $\mathcal{F}(F_{ij}, 1 \leq i < j \leq m)$ are based on the set $\{\delta_{ij} : i < j\}$, where δ_{ij} is a measure of bivariate association for F_{ij}.

We can also consider Fréchet classes of survival functions given marginal survival functions, e.g., $\mathcal{F}(\overline{F}_1, \ldots, \overline{F}_m)$, $\mathcal{F}(\overline{F}_{12}, \overline{F}_{13})$, $\mathcal{F}(\overline{F}_{12}, \overline{F}_{13}, \overline{F}_{23})$. In some cases, because of the relationship between multivariate cdfs and survival functions, an upper (lower) bound cdf becomes a lower (upper) bound survival function.

For a given Fréchet class $\mathcal{F}$, natural questions to ask are the following.

1. What are the bounds for the multivariate distributions in $\mathcal{F}$?
2. Do the bounds correspond to proper multivariate distributions, and if so, is there a stochastic representation or interpretation of the extremes?
3. Are the bounds sharp if they do not correspond to proper multivariate distributions?

4. What are simple members of $\mathcal{F}$?
5. Can one construct parametric subfamilies in $\mathcal{F}$ with desirable properties?

The results in this chapter mainly concern 1 to 4, and later chapters deal with 5. The discussion of the classes $\mathcal{F}(F_{12}, F_{13})$, $\mathcal{F}(F_{12}, F_{13}, F_{23})$ and $\mathcal{F}(F_{ij}, 1 \le i < j \le m)$ is crucial to the development of the construction methods in Sections 4.5, 4.7 and 4.8.

3.1 $\mathcal{F}(F_1, \dots, F_m)$ °

Let $F_1, \dots, F_m$ be given univariate distribution functions, each of which can be continuous or non-continuous. The Fréchet bounds for $\mathcal{F} = \mathcal{F}(F_1, \dots, F_m)$ are given by the inequalities in the following theorem. They depend on simple inequalities involving probabilities of sets.

Theorem 3.1 *Let $F \in \mathcal{F}(F_1, \dots, F_m)$. Then for all $\mathbf{x} \in \Re^m$,*

$$\max\{0, F_1(x_1) + \dots + F_m(x_m) - (m-1)\} \le F(\mathbf{x}) \le \min_j F_j(x_j).$$

Proof. Let $p_i = \Pr(A_i)$, where $A_i = \{X_i \le x_i\}$ with $X_i \sim F_i$, $i = 1, \dots, m$. Then from Lemma 3.8 below,

$$\max\{0, p_1 + \dots + p_m - (m-1)\} \le \Pr(A_1 \cap \dots \cap A_m) \le \min_j p_j.$$

□

Let the Fréchet upper bound, $\min_j F_j(x_j)$, be denoted by $F_U(\mathbf{x})$ and let the Fréchet lower bound, $\max\{0, F_1(x_1) + \dots + F_m(x_m) - (m-1)\}$, be denoted by $F_L(\mathbf{x})$.

Theorem 3.2 *The Fréchet upper bound F_U is a cdf.*

Proof. Let $X_j \sim F_j$, $j = 1, \dots, m$. If one of the univariate cdfs is continuous, say F_1, then F_U is the joint distribution of $\mathbf{X}$ with the stochastic representation $X_j = F_j^{-1}(F_1(X_1))$, $j = 2, \dots, m$.

If all univariate cdfs have some discontinuity points, then approximate F_1 by a sequence F_{1n} such that $F_{1n} \to_d F_1$ (this is always possible by, for example, convoluting X_1 with a $N(0, n^{-1})$ rv). From the preceding paragraph, $\min\{F_{1n}(x_1), F_2(x_2), \dots, F_m(x_m)\}$ is a cdf for all n, and the rectangle inequality (1.6) holds for it. Hence the rectangle inequality holds for $\min_j F_j(x_j)$ for all $(x_1, \dots, x_m, x_1 + h_1, \dots, x_m + h_m)$ such that $x_1, x_1 + h_1$ are continuity points of F_1. For the remaining case where x_1 (or $x_1 + h_1$)

is not a continuity point of F_1, a limit based on x_{1k} $(x_1 + h_{1k})$ decreasing to x_1 (x_1+h_1) can be used. The other necessary conditions for a multivariate cdf can easily be checked for $\min_j F_j(x_j)$, hence F_U is always a cdf. □

Theorem 3.3 *The Fréchet lower bound F_L is a cdf for $m = 2$.*

Proof. The proof is similar to that in the preceding theorem. Let $X_j \sim F_j$, $j = 1,2$. If at least one of the F_j is continuous, say F_1, then $F_L(x_1,x_2) = \max\{0, F_1(x_1) + F_2(x_2) - 1\}$ is the joint distribution of X_1, X_2 with the stochastic represention $X_2 = F_2^{-1}(1 - F_1(X_1))$. If neither F_1 nor F_2 is continuous, the idea in the second paragraph of the preceding proof works. Alternatively the rectangle inequality (1.5) can be checked directly for the few cases. □

With replacement by $U(0,1)$ margins and survival functions, the following two results are easily obtained.

Theorem 3.4 *The copula of the Fréchet upper bound is $C_U(\mathbf{u}) = \min\{u_1,\ldots,u_m\}$. For $m = 2$, the copula of the Fréchet lower bound is $C_L(\mathbf{u}) = \max\{0, u_1 + u_2 - 1\}$.*

Theorem 3.5 *The upper bound for $\mathcal{F}(\overline{F}_1,\ldots,\overline{F}_m)$ is $\overline{G}_U(\mathbf{x}) = \min_j\{\overline{F}_j(x_j)\}$ and the lower bound is $\overline{G}_L(\mathbf{x}) = \max\{0, \sum_j \overline{F}_j(x_j) - (m-1)\}$. $\overline{G}_U$ is the survival function of F_U, and when F_L is a proper cdf, $\overline{G}_L$ is the survival function for F_L.*

Proof. The bounds are proved in a similar way to before. The proof of the relationships between $\overline{G}_U, F_U$ and $\overline{G}_L, F_L$ is left as an exercise. An identity that can be used is

$$\max_{1\le j\le m} z_j = \sum_{S\in\mathcal{S}_m} (-1)^{|S|+1} \min_{i\in S} z_j.$$

□

For $m \geq 3$, F_L is in general not a proper cdf. An example for illustration is given next before further results are obtained.

Example 3.1 Consider the symmetric situation with F_1, F_2, F_3 each corresponding to a Bernouilli rv with parameter p. That is, $F_j(x) = 0$ if $x < 0$, $F_j(x) = q = 1-p$ if $0 \leq x < 1$ and $F_j(x) = 1$ if $x \geq 1$. Then F_L is a cdf if and only if $p \leq 1/3$ or $p \geq 2/3$.

Proof. For $p \geq 2/3$, the positive probability masses are

$$\Pr(\{(1,1,0)\}) = \Pr(\{(1,0,1)\}) = \Pr(\{(0,1,1)\}) = q$$

and $\Pr(\{(1,1,1)\}) = 1 - 3q$. For $p \leq 1/3$, the positive masses are

$$\Pr(\{(1,0,0)\}) = \Pr(\{(0,1,0)\}) = \Pr(\{(0,0,1)\}) = p$$

and $\Pr(\{(0,0,0)\}) = 1-3p$.

For $1/3 < p < 2/3$, $F = F_L$ satisfies $F(0,0,0) = 0$, $F(1,0,0) = F(0,1,0) = F(0,0,1) = \max\{0, 1-2p\}$, $F(1,1,0) = F(1,0,1) = F(0,1,1) = 1-p$, $F(1,1,1) = 1$. Hence $F(1,1,1) - F(1,1,0) - F(1,0,1) - F(0,1,1) + F(0,0,1) + F(0,1,0) + F(1,0,0)$ equals $1-3p$ if $p \le \frac{1}{2}$ and equals $3p-2$ if $p > \frac{1}{2}$. Since both of these quantities are negative, the rectangle inequality (1.6) does not hold. □

Continuing with this example, F_L can be shown to be sharp for $1/3 < p < 2/3$. For $p \ge \frac{1}{2}$, the distribution F with positive masses $\Pr(\{(0,1,0)\} = \Pr(\{(1,0,1)\}) = 1-p$ and $\Pr(\{(1,1,1)\}) = 2p-1$ takes care of the lower bound at six vertices of the cube: $F(1,1,0) = F(1,0,1) = F(0,1,1) = 1-p$, $F(1,0,0) = F(0,0,1) = 0$ and $F(0,0,0) = 0$. By permuting the indices, $F_L(0,1,0) = 0$ can also be achieved. Next suppose $p < \frac{1}{2}$. The distribution F with positive masses $\Pr(\{(0,1,0)\} = \Pr(\{(1,0,1)\}) = p$ and $\Pr(\{(0,0,0)\}) = 1-2p$ takes care of the lower bound at five vertices of the cube: $F(1,1,0) = F(1,0,1) = F(0,1,1) = 1-p$, $F(1,0,0) = F(0,0,1) = 1-2p$. By permuting the indices, $F_L(0,1,0) = 1-2p$ can also be achieved. Finally, the distribution F' with positive masses $\Pr(\{(0,1,0)\}) = \Pr(\{(1,0,0)\}) = \Pr(\{(0,0,1)\}) = (1-p)/2$ and $\Pr(\{(1,1,1)\}) = (3p-1)/2$ takes care of lower bound at (0,0,0): $F'(0,0,0) = 0$. □

Next we obtain conditions for F_L to be a cdf for $m = 3$ (and then we generalize the result to $m > 3$). Suppose all univariate margins F_j are not degenerate. Clearly, F_L is not a cdf if F_1, F_2, F_3 are continuous. (In this case, by applying the probability transform, F_j can be taken to be $U(0,1)$, and if $(U_1, U_2, U_3) \sim F_L$, then all three bivariate margins are two-dimensional Fréchet lower bounds. Hence $U_1 + U_2 = U_1 + U_3 = U_2 + U_3 = 1$ from Theorems 3.3 and 3.4, leading to a contradiction.) Similarly, if one of the three distributions is continuous, say F_1, then F_L cannot be a cdf. (If $(X_1, X_2, X_3) \sim F_L$, then $X_j = F_j^{-1}(1 - F_1(X_1))$, $j = 2, 3$; X_2, X_3 are then positively associated and this is a contradiction.) Hence a necessary condition is that each F_j has a discrete component.

Theorem 3.6 *A necessary and sufficient condition for the Fréchet lower bound F_L of $\mathcal{F}(F_1, F_2, F_3)$ to be a cdf is that either*

(a) $F_1(x_1) + F_2(x_2) + F_3(x_3) \le 1$ *whenever* $0 < F_j(x_j) < 1$, $j = 1, 2, 3$; *or*

(b) $F_1(x_1) + F_2(x_2) + F_3(x_3) \ge 2$ *whenever* $0 < F_j(x_j) < 1$, $j = 1, 2, 3$.

Note that (a) and (b) cannot both occur together.

Proof. Let $x_j < x'_j$, $p_j = F(x_j)$, $p'_j = F(x'_j)$, $j = 1, 2, 3$. Also let $(y)_+ = \max\{0, y\}$.

First we prove the sufficiency of (a). The rectangle condition (1.6) for F_L leads to

$$\begin{aligned}(p'_1 + p'_2 + p'_3 - 2)_+ - (p'_1 + p'_2 + p_3 - 2)_+ - (p'_1 + p_2 + p'_3 - 2)_+ \\ -(p_1 + p'_2 + p'_3 - 2)_+ + (p'_1 + p_2 + p_3 - 2)_+ + (p_1 + p'_2 + p_3 - 2)_+ \\ +(p_1 + p_2 + p'_3 - 2)_+ - (p_1 + p_2 + p_3 - 2)_+. \qquad (3.1)\end{aligned}$$

Assume that (x_1, x_2, x_3) satisfies condition (a). If $p'_1, p'_2, p'_3 < 1$, then (3.1) becomes 0 since each term is 0. If $p'_1 = 1$, $p'_2, p'_3 < 1$, then (3.1) is still 0 because $p'_2 + p'_3 \leq p_1 + p'_2 + p'_3 \leq 1$ from (a) and hence $p'_1 + p'_2 + p'_3 \leq 2$. If $p'_1 = p'_2 = 1$, $p'_3 < 1$, then (3.1) becomes $p'_3 - p_3 \geq 0$ since only the first two terms are non-zero. If $p'_1 = p'_2 = p'_3 = 1$, then (3.1) becomes $1 - p_3 - p_2 - p_1 \geq 0$. Each of the remaining cases is symmetric to one of these. Hence F_L is a cdf.

Next we prove the sufficiency of (b). Assume that (x'_1, x'_2, x'_3) satisfies condition (b). If $p_1, p_2, p_3 > 0$, then (3.1) becomes 0 since all of the terms are non-negative. If $p_1 = 0$, $p_2, p_3 > 0$, then the four terms with p_1 are 0 and (3.1) becomes 0. If $p_1 = p_2 = 0$, $p_3 > 0$, then only the first two terms may be non-zero and (3.1) becomes $p'_3 - p_3 \geq 0$. Each of the remaining cases is symmetric to one of these. Hence F_L is a cdf.

Finally we prove the necessity of (a) or (b). If neither (a) nor (b) holds then there is a vector (x_1, x_2, x_3) such that $1 < p_1 + p_2 + p_3 < 2$, $0 < p_j < 1$, $j = 1, 2, 3$. Let $p'_1 = p'_2 = p'_3 = 1$. Then (3.1) simplifies to

$$1 - p_1 - p_2 - p_3 + (p_1 + p_2 - 1)_+ + (p_1 + p_3 - 1)_+ + (p_2 + p_3 - 1)_+. \quad (3.2)$$

This will be shown to be negative in all cases and hence F_L is not a cdf. If $p_i + p_j \leq 1$ for all three pairs, then (3.2) is $1 - p_1 - p_2 - p_3 < 0$. If $p_1 + p_2 \geq 1$, $p_1 + p_3 \leq 1$, $p_2 + p_3 \leq 1$, then (3.2) becomes $-p_3 < 0$. If $p_1 + p_2 \geq 1$, $p_1 + p_3 \geq 1$, $p_2 + p_3 \leq 1$, then (3.2) becomes $p_1 - 1 < 0$. If $p_i + p_j \geq 1$ for all three pairs, then (3.2) becomes $p_1 + p_2 + p_3 - 2 < 0$. □

The generalization to higher dimensions is as follows. The ideas are clearer in the proof of the special case of $m = 3$.

Theorem 3.7 *A necessary and sufficient condition for the Fréchet lower bound F_L of $\mathcal{F}(F_1, \ldots, F_m)$ to be a cdf is that either* *(a)* $\sum_j F_j(x_j) \le 1$ *whenever* $0 < F_j(x_j) < 1$, $j = 1, \ldots, m$; *or*

(b) $\sum_j F_j(x_j) \ge m - 1$ *whenever* $0 < F_j(x_j) < 1$, $j = 1, \ldots, m$.

Proof. Let $x_j < x_j'$, $p_{j0} = F(x_j)$ and $p_{j1} = F(x_j')$, $j = 1, \ldots, m$. Let $(y)_+ = \max\{0, y\}$ as before.

First we prove the sufficiency of (a). The rectangle condition (1.6) for F_L leads to

$$\sum_{(\epsilon_1,\ldots,\epsilon_m):\epsilon_j=0 \text{ or } 1} (-1)^{m-\Sigma_j \epsilon_j} \left[\sum_j p_{j\epsilon_j} - (m-1)\right]_+. \tag{3.3}$$

Assume that $(x_1, \ldots, x_m)$ satisfies condition (a). Then by eliminating the zero terms in (3.3), we get:

$$\begin{aligned} &(p_{11} + \cdots + p_{m1} - (m-1))_+ \\ &- \quad (p_{10} + p_{21} + \cdots + p_{m1} - (m-1))_+ \\ &- \quad (p_{11} + p_{20} + p_{31} + \cdots + p_{m1} - (m-1))_+ - \cdots \\ &- \quad (p_{11} + \cdots + p_{m-1,1} + p_{m0} - (m-1))_+. \end{aligned} \tag{3.4}$$

(Note that from (a), a term is zero if two probabilities in it are less than 1.) If $p_{11} = \cdots = p_{m1} = 1$, then (3.4) becomes $1 - p_{10} - \cdots - p_{m0} \ge 0$. If $p_{11} = \cdots = p_{m-1,1} = 1$, $p_{m1} < 1$, then (3.4) becomes $p_{m1} - p_{m0} \ge 0$. If at least two of the p_{mj} are less than 1, then (3.4) is zero. Hence F_L is a cdf.

Next we prove the sufficiency of (b). Assume that $(x_1', \ldots, x_m')$ satisfies condition (b). If $p_{j0} > 0$ for all j, then (3.3) is zero. If at most $m-2$ of the p_{j0} are zero, then (3.3) is zero because the signs $(-1)^{m-\Sigma_j \epsilon_j}$ of p_{i0}, p_{i1} for the non-zero terms balance out for all i. If $p_{10} = \cdots = p_{m-1,0} = 0$ and $p_{m0} > 0$, then (3.3) becomes $p_{m1} - p_{m0} \ge 0$. If $p_{10} = \cdots = p_{m0} = 0$, then (3.3) becomes $p_{11} + \cdots + p_{m1} - (m-1) \ge 0$. Hence F_L is a cdf.

Finally, for the necessity of (a) or (b), we will use induction with the $m = 3$ case from the previous theorem as the starting point. Suppose $m \ge 4$ and the result is true for all dimensions less than m. Suppose there is a point $(y_1, \ldots, y_m)$ such that $1 < \sum_j F_j(y_j) < m - 1$ and $0 < F_j(y_j) < 1$ for all j.

If F_L is a cdf then its lower-dimensional margins are also Fréchet lower bounds. Hence, by the induction hypothesis, it is not possible that $1 < \sum_j F_j(y_j) - F_i(y_i) < m - 2$ for any i. Therefore for each i, either $\sum_j F_j(y_j) - F_i(y_i) \le 1$ or $\sum_j F_j(y_j) - F_i(y_i) \ge m - 2$. The same direction for the inequality must hold for all i because if, for example, $F_2(y_2) + \cdots + F_m(y_m) \ge m - 2$ with $i = 1$ then $\sum_{j=2}^m F_j(y_j) - F_{i'}(y_{i'}) > m - 3$ for $i' = 2, \ldots, m$ and hence it is impossible for $F_1(y_1)+\sum_{j=2}^m F_j(y_j)-F_{i'}(y_{i'})\leq 1$ to hold.

If $\sum_j F_j(y_j)-F_i(y_i)\leq 1$ for all i, then take $(x_1,\ldots,x_m)=(y_1,\ldots,y_m)$ and $(x_1',\ldots,x_m')=(\infty,\ldots,\infty)$. The rectangle condition (3.3) becomes $1-p_{10}-\cdots-p_{m0}=1-\sum_j F_j(y_j)<0$ since $p_{i0}+p_{i'0}\leq 1$ for all $i\neq i'$.

If $\sum_j F_j(y_j)-F_i(y_i)\geq m-2$ for all i and m is odd, then take $(x_1,\ldots,x_m)=(y_1,\ldots,y_m)$ and $(x_1',\ldots,x_m')=(\infty,\ldots,\infty)$. The only zero term in (3.3) is when $\epsilon_j=0$ for all j and hence (3.3) becomes $(-1)^m[(m-1)-p_{10}-\cdots-p_{m0}]<0$. If $\sum_j F_j(y_j)-F_i(y_i)\geq m-2$ for all i and m is even, then take $(x_1,\ldots,x_m)=(-\infty,y_2,\ldots,y_m)$ and $(x_1',\ldots,x_m')=(y_1,\infty,\ldots,\infty)$. All of the terms in (3.3) with $\epsilon_1=0$ are zero and the term with $\epsilon_1=1$, $\epsilon_j=0$, $j>2$, is also zero. Hence (3.3) becomes

$$\begin{aligned}\sum_{\substack{\epsilon:\epsilon_j=0 \text{ or } 1,\\ j>1,\epsilon_1=1}} & (-1)^{m-1-\Sigma_{j=2}^m\epsilon_j}\Big[\sum_j p_{j\epsilon_j}-(m-1)\Big]_+\\ &= (-1)^{m-1}[m-1-p_{11}-p_{20}-\cdots-p_{m0}]\\ &= (-1)^{m-1}[m-1-F_1(y_1)-\cdots-F_m(y_m)]<0.\end{aligned}$$

Hence the necessity of (a) or (b) has been proved for dimension m. □

An interpretation of the preceding theorem is the following. Condition (a) means that there is a finite upper support point ξ_j for F_j, and ξ_j is a point of mass of F_j, $j=1,\ldots,m$. The masses $1-F(\xi_j-)$ are such that $\sum_j[1-F(\xi_j-)]\geq m-1$. Similarly, condition (b) means that there is a finite lower support point ξ_j for F_j, $j=1,\ldots,m$, such that $\sum_j F(\xi_j)\geq m-1$.

The next theorem concerns the sharpness of the Fréchet lower bound for $m\geq 3$. The following lemma is used.

Lemma 3.8 *Let $A_1,\ldots,A_m$ be events such that $\Pr(A_i)=a_i$, $i=1,\ldots,m$. Then*

$$\max\Big\{0,\sum_j a_j-(m-1)\Big\}\leq \Pr(A_1\cap\cdots\cap A_m)\leq \min_j a_j$$

and the bounds are sharp.

Let A_i^0 be the complement of A_i and let $A_i^1=A_i$, $i=1,\ldots,m$. Then it is possible to assign probabilities to

$$A_1^{\epsilon_1}\cap\cdots\cap A_m^{\epsilon_m},\quad \epsilon_i=0 \text{ or } 1,\quad i=1,\ldots,m,$$

(in a continuous way over a_i) such that $\Pr(A_1\cap\cdots\cap A_m)=\max\{0,\sum_j a_j-(m-1)\}$ and $a_i=\sum_{\epsilon:\epsilon_i=1}\Pr(A_1^{\epsilon_1}\cap\cdots\cap A_m^{\epsilon_m})$.

Proof. The second inequality follows from the law of inclusion, and the first inequality follows from the law of addition applied to the complement: $\Pr(A_1 \cap \cdots \cap A_m) = 1 - \Pr(A_1^0 \cup \cdots \cup A_m^0) \geq 1 - \sum_j \Pr(A_j^0)$. The upper bound is sharp from the case where one event is a subset of the remaining events. There are two cases to consider for the sharpness of the lower bound. Let $b_i = 1 - a_i$, $i = 1, \ldots, m$.

The first case is $\sum_{i=1}^m a_i \geq m - 1$ or $\sum_{i=1}^m b_i \leq 1$. The events A_i^0 can then be made incompatible. Hence $\Pr(A_1^0 \cup \cdots \cup A_m^0) = \sum_{i=1}^m b_i$, $\Pr(A_1 \cap \cdots \cap A_m) = \sum_{i=1}^m a_i - (m-1)$, $\Pr(\cap_{i=1,\ldots,m,\, i\neq j} A_i) = \sum_{i=1}^m a_i - a_j - (m-2)$, $\Pr((\cap_{i=1,\ldots,m,\, i\neq j} A_i) \cap A_j^0) = 1 - a_j$, $j = 1, \ldots, m$, and $\Pr(A_1^{\epsilon_1} \cap \cdots \cap A_m^{\epsilon_m}) = 0$ if at least two of the ϵ_i are zero.

Next suppose there is an integer $1 \leq k < m$ such $\sum_{i=1}^k b_i \leq 1$, $\sum_{i=1}^{k+1} b_i > 1$. The events $A_1^0, \ldots, A_k^0$ can be made into incompatible events, and an event E, incompatible with $A_1^0, \ldots, A_k^0$, with probability $1 - \sum_{i=1}^k b_i$ can be added. Let D be an event, incompatible with E, with probability $\sum_{i=1}^{k+1} b_i - 1$. Then define $A_{k+1}^0 = E \cup D$; this has probability b_{k+1}. Define the remaining events $A_{k+2}, \ldots, A_n$ to be independent of each other and of $A_1, \ldots, A_{k+1}$. Then $\Pr(A_1 \cap \cdots \cap A_{k+1}) = \Pr(A_1 \cap \cdots \cap A_n) = 0$, $\Pr(A_1 \cap \cdots \cap A_k \cap A_{k+1}^0) = \Pr(E) = \sum_{i=1}^k a_i - (k-1)$, $\Pr((\cap_{i=1,\ldots,k+1,\, i\neq j} A_i) \cap A_j^0) = \Pr(A_j^0 \cap D^0)$, $j = 1, \ldots, k$, where D^0 is the complement of D, $\Pr((\cap_{i=1,\ldots,k,\, i\neq j} A_i) \cap A_j^0 \cap A_{k+1}^0) = \Pr(A_j^0 \cap D)$, $j = 1, \ldots, k$, and $\Pr(A_1^{\epsilon_1} \cap \cdots \cap A_{k+1}^{\epsilon_{k+1}}) = 0$ if $\sum_{i=1}^k (1 - \epsilon_i) \geq 2$. Finally, $\Pr(A_1^{\epsilon_1} \cap \cdots \cap A_m^{\epsilon_m}) = \Pr(A_1^{\epsilon_1} \cap \cdots \cap A_{k+1}^{\epsilon_{k+1}}) \cdot \prod_{i=k+2}^m a_i^{\epsilon_i} (1 - a_i)^{1-\epsilon_i}$. □

Theorem 3.9 *If the cdfs $F_1, \ldots, F_m$ are continuous, then the Fréchet lower bound $F_L(\mathbf{y})$ is sharp at each $\mathbf{y} \in \Re^m$.*

Proof. Without loss of generality, take F_i to be $U(0,1)$ for each i. (This is possible since one can always transform from the $U(0,1)$ cdf to a general continuous F_j.)

Let $U_1, \ldots, U_m$ be $U(0,1)$ rvs with a joint distribution to be defined constructively. Fix $(y_1, \ldots, y_m)$ with $0 < y_i < 1$. Let $A_i = \{U_i \leq y_i\}$, $i = 1, \ldots, m$. Other notation is the same as in Lemma 3.8. By Lemma 3.8, it is possible to assign probabilities to the regions $A_1^{\epsilon_1} \cap \cdots \cap A_m^{\epsilon_m}$ such that $\Pr(A_1 \cap \cdots \cap A_m) = \max\{0, \sum_j y_j - (m-1)\}$. Next assign probabilities uniformly within each region $A_1^{\epsilon_1} \cap \cdots \cap A_m^{\epsilon_m}$. Then the resulting univariate margins of the U_i are in fact uniform on (0,1) and hence F_L is sharp at $(y_1, \ldots, y_m)$. The uniform distribution can be established as follows for U_1, and then the other margins are also uniform by symmetry:

- $\Pr(U_1 \le x_1) = \sum_{\epsilon_2,\ldots,\epsilon_m}(x_1/y_1)\Pr(A_1 \cap A_2^{\epsilon_2} \cap \cdots \cap A_m^{\epsilon_m}) = (x_1/y_1)\Pr(A_1) = x_1$, $0 \le x_1 \le y_1$,
- $\Pr(U_1 \le x_1) = y_1 + \sum_{\epsilon_2,\ldots,\epsilon_m}[(x_1-y_1)/(1-y_1)]\Pr(A_1^0 \cap A_2^{\epsilon_2} \cap \cdots \cap A_m^{\epsilon_m}) = y_1 + [(x_1-y_1)/(1-y_1)]\Pr(A_1^0) = x_1$, $y_1 < x_1 \le 1$.

□

By approximating discrete distributions by a sequence of continuous distributions, the Fréchet lower bound should in general be sharp.

3.2 $\mathcal{F}(F_{12}, F_{13})$

In this section, we obtain some results on the class of trivariate distributions $\mathcal{F} = \mathcal{F}(F_{12}, F_{13})$ for which the (1,2) and (1,3) bivariate margins are given or fixed. We assume that the conditional distributions $F_{2|1}$ and $F_{3|1}$ are well defined. $\mathcal{F}$ is always non-empty since it contains the trivariate distribution which is such that the second and third variables are conditionally independent given the first, i.e., the cdf defined by $F(\mathbf{x}) = \int_{-\infty}^{x_1} F_{2|1}(x_2|y)F_{3|1}(x_3|y)\, dF_1(y)$. Similarly, perfect conditional positive and negative dependence lead to the Fréchet bounds in $\mathcal{F}$. The ideas in this section are extended to the method of mixtures of conditional distributions in Section 4.5.

Theorem 3.10 *The Fréchet upper bound of $\mathcal{F} = \mathcal{F}(F_{12}, F_{13})$ is given by $F_U(\mathbf{x}) = \int_{-\infty}^{x_1} \min\{F_{2|1}(x_2|y), F_{3|1}(x_3|y)\}\, dF_1(y)$ and the Fréchet lower bound is given by $F_L(\mathbf{x}) = \int_{-\infty}^{x_1} \max\{0, F_{2|1}(x_2|y) + F_{3|1}(x_3|y) - 1\}\, dF_1(y)$, and both of these bounds are proper cdfs.*

Proof. Let $F \in \mathcal{F}$. Then write $F(\mathbf{x}) = \int_{-\infty}^{x_1} F_{23|1}(x_2, x_3|y)\, dF_1(y)$, where $F_{23|1}$ is the bivariate conditional distribution of the second and third variables given the first. The results follow from Theorems 3.1 to 3.3. □

The above theorem extends to the Fréchet class $\mathcal{F}$ consisting of m-variate distributions given two different $(m-1)$-dimensional margins (these two marginal distributions have $m-2$ variables in common). For example, $F \in \mathcal{F}(F_{1\cdots m-1}, F_{1\cdots m-2,m})$, can be written in the form:

$$F(\mathbf{x}) = \int_{-\infty}^{x_1} \cdots \int_{-\infty}^{x_{m-2}} F_{m-1,m|1\cdots m-2}(x_{m-1}, x_m|y_1, \ldots, y_{m-2}) \cdot F_{1\cdots m-2}(dy_1, \ldots, dy_{m-2}).$$

The bivariate conditional distribution can be bounded by the bivariate Fréchet upper and lower bounds.

For the Fréchet class of survival functions $\mathcal{F}(\overline{F}_{12}, \overline{F}_{13})$ and its extensions, the Fréchet bounds $\overline{G}_L, \overline{G}_U$ satisfy $\overline{G}_L = \overline{F}_L$ and $\overline{G}_U = \overline{F}_U$.

3.3 $\mathcal{F}(F_{12}, F_3)$

In this section, we obtain some results on the class of trivariate distributions $\mathcal{F} = \mathcal{F}(F_{12}, F_3)$ for which the (1,2) bivariate margin and the univariate margin for third variable are given or fixed.

Similar to Theorem 3.1, an upper bound of the class $\mathcal{F}$ is $F_U(\mathbf{x}) = \min\{F_{12}(x_1, x_2), F_3(x_3)\}$. In general, F_U is not a proper distribution. The simple intuitive proof is that the (1,3) and (2,3) margins are both bivariate Fréchet upper bounds so that F_U cannot be a proper distribution unless F_{12} is a bivariate Fréchet upper bound distribution. As an illustration, we also use the rectangle condition of Section 1.4.2 to show that F_U is in general not a cdf, when F_{12} and F_3 are continuous. Let the arguments be $x_1 < x_1'$, $x_2 < x_2'$, $x_3 < x_3'$, and let $F_{12} = F_{12}(x_1, x_2)$, $F_{12'} = F_{12}(x_1, x_2')$, $F_{1'2} = F_{12}(x_1', x_2)$, $F_{1'2'} = F_{12}(x_1', x_2')$, $F_3 = F_3(x_3)$, $F_{3'} = F_3(x_3')$. We assume without loss of generality that $F_{3'} > F_3$ and $F_{1'2'} > F_{1'2} > F_{12'} > F_{12}$. Because $F_{12}(\cdot, \cdot)$ is a proper distribution, $F_{1'2'} - F_{1'2} - F_{12'} + F_{12} \geq 0$. There are 15 ways of ordering $F_{3'}, F_3, F_{1'2'}, F_{1'2}, F_{12'}, F_{12}$ that could be considered. One of them is $F_{1'2'} > F_{1'2} > F_{3'} > F_{12'} > F_3 > F_{12}$, in which case the rectangle condition leads to $F_{3'} - F_{3'} - F_{12'} - F_3 + F_3 + F_3 + F_{12} - F_{12} < 0$.

A lower bound for $\mathcal{F}$ is $F_L(\mathbf{x}) = \max\{0, F_{12}(x_1, x_2) + F_3(x_3) - 1\}$. It is left as an exercise to show that this is not a cdf in general. The intuitive argument is that the (1,3) and (2,3) margins of F_L are bivariate Fréchet lower bounds, so that F_L is a proper cdf only if F_{12} is the Fréchet upper bound.

These results clearly extend to $\mathcal{F}(F_{S_1}, \ldots, F_{S_k})$, where $S_1, \ldots, S_k$ is a partition of $\{1, \ldots, m\}$ with $|S_i| \geq 2$ for at least one i. The Fréchet bounds are generally not proper cdfs.

Next we go on to other results for $\mathcal{F} = \mathcal{F}(F_{12}, F_3)$ when $F_1 = F_3$ and F_{12} is continuous, such as the extremal elements of the set, and the concordance and more SI orderings on $\mathcal{F}$.

If F_{32} is chosen so that $F_{32} \prec_c F_{12}$ and F_{123} is the Fréchet upper bound given F_{12}, F_{32}, then $F_{123} \in \mathcal{F}(F_{12}, F_3)$ and F_{123} is smaller than $F_{123}'(x_1, x_2, x_3) = F_{12}(x_1 \wedge x_3, x_2)$ in the concordance ordering. Note that F_{123}' is the Fréchet upper bound given F_{12}, F_{32}

when $F_{32} = F_{12}$ and it is in $\mathcal{F}(F_{12}, F_3)$ when $F_1 = F_3$. The proof follows from $F_{123}(x_1, x_2, x_3) \le \min\{F_{12}(x_1, x_2), F_{32}(x_3, x_2)\}$ and $F_{32}(x_3, x_2) \le F_{12}(x_3, x_2)$, and similar inequalities for the survival functions. If F_{32} is chosen so that $F_{12} \prec_{SI} F_{32}$ and F_{123} is the Fréchet upper bound given F_{12}, F_{32}, then for $\mathbf{X} \sim F_{123}$, there is the stochastic relation $X_3 = g(X_1, X_2)$ with g (strictly) increasing in its two arguments. Note that $\Pr(X_1 \le x_1, X_2 \le x_2, g(X_1, X_2) \le x_3) = \Pr(X_1 \le x_1, X_2 \le x_2) = F_{12}(x_1, x_2)$ if $g(x_1, x_2) \le x_3$, and this is larger than $F_{12}(x_1 \wedge x_3, x_2)$ when $x_3 < x_1$. The inequality $g(x_1, x_2) < x_3 < x_1$ holds if x_2 is sufficiently small and x_1 is sufficiently large. Hence, the distribution of $(X_1, X_2, g(X_1, X_2))$ is not dominated by F'_{123}.

The above results suggest that when $F_1 = F_3$, maximal elements in $\mathcal{F}$ include those for which F_{32} is strictly larger than F_{12} in the $\prec_c$ or $\prec_{SI}$ ordering and F_{123} is the Fréchet upper bound in $\mathcal{F}(F_{12}, F_{23})$.

We show next that if $F_{12} \prec_{SI} F_{32} \prec_{SI} F'_{32}$ (with strict inequalities) and $F_{1|2}(\cdot|y)$, $F_{3|2}(\cdot|y)$, $F'_{3|2}(\cdot|y)$ are continuous for all y, then the Fréchet upper bounds

$$F_{123}(x_1, x_2, x_3) = \int_{-\infty}^{x_2} \min\{F_{1|2}(x_1|y), F_{3|2}(x_3|y)\} dF_2(y)$$

and

$$F'_{123}(x_1, x_2, x_3) = \int_{-\infty}^{x_2} \min\{F_{1|2}(x_1|y), F'_{3|2}(x_3|y)\} dF_2(y)$$

are not comparable in the $\prec_c$ ordering, i.e., neither function dominates the other uniformly over $\Re^3$. (If F_{32} and F'_{32} are both strictly more concordant than F_{12} but are themselves not ordered by concordance, then the trivariate distributions F_{123}, F'_{123} are clearly not ordered by concordance.) The assumptions imply that $F_{32}(x_3,x_2) \le F'_{32}(x_3, x_2)$ for all x_3, x_2, so it remains to show that there exists (x_1, x_2, x_3) such that $F'_{123}(x_1, x_2, x_3) < F_{123}(x_1, x_2, x_3)$. From Theorem 2.10, if x_1, x_3 are fixed, there exist y_0, y'_0, y''_0 such that

$$[F_{1|2}(x_1|y) - F_{3|2}(x_3|y)](y - y_0) \ge 0,$$

$$[F_{1|2}(x_1|y) - F'_{3|2}(x_3|y)](y - y'_0) \ge 0,$$

$$[F_{3|2}(x_3|y) - F'_{3|2}(x_3|y)](y - y''_0) \ge 0.$$

Therefore if y_0, y'_0, y''_0 are bounded away from the upper and lower end points of support, then $F_{1|2}(x_1|y) < F_{3|2}(x_3|y) < F'_{3|2}(x_3|y)$ for y small enough (less than $\min\{y_0, y'_0, y''_0\}$) and $F_{1|2}(x_1|y) >$

$F_{3|2}(x_3|y) > F'_{3|2}(x_3|y)$ for y large enough. If $y'_0 < y_0$, then

$$F_{123}(x_1, y_0, x_3) = \int_{-\infty}^{y'_0} F_{1|2}(x_1|y)\, dF_2(y) + \int_{y'_0}^{y_0} F_{1|2}(x_1|y)\, dF_2(y)$$

$$> \int_{-\infty}^{y'_0} F_{1|2}(x_1|y)\, dF_2(y) + \int_{y'_0}^{y_0} F'_{3|2}(x_3|y)\, dF_2(y) = F'_{123}(x_1, y_0, x_3).$$

The conditions where $y'_0 < y_0$ can be investigated.

For illustration, consider the special case of BVN distributions for F_{12}, F_{32}, F'_{32}. Let F_{12}, F_{32}, F'_{32} be BVSN with correlations $0 \leq \rho_{12} < \rho_{32} < \rho'_{32}$. Then for x_1, x_3 fixed, $y_0 = [x_3\sqrt{1-\rho_{12}^2} - x_1\sqrt{1-\rho_{32}^2}]/[\rho_{32}\sqrt{1-\rho_{12}^2} - \rho_{12}\sqrt{1-\rho_{32}^2}]$ and $y'_0 = [x_3\sqrt{1-\rho_{12}^2} - x_1\sqrt{1-\rho_{32}'^2}]/[\rho'_{32}\sqrt{1-\rho_{12}^2} - \rho_{12}\sqrt{1-\rho_{32}'^2}]$. Hence $y'_0 < y_0$ for some choices of x_1, x_3, e.g., $x_1 = 0 < x_3$.

The general approach for finding x_1, x_3 such that $y'_0 < y_0$ is as follows. $F_{1|2}(x_1|y)$ is increasing in x_1 for all y, so that if x_1 is chosen to be small enough relative to x_3, then $y_0, y'_0 > y''_0$. If $F_{3|2}(x_3|y) > F'_{3|2}(x_3|y)$ for $y > y''_0$ and $F_{1|2}(x_1|y) < F_{3|2}(x_3|y), F'_{3|2}(x_3|y)$ for $y < y''_0$, then $F_{1|2}(x_1|\cdot)$ must intersect $F'_{3|2}(x_3|\cdot)$ before $F_{3|2}(x_3|\cdot)$ or $y'_0 < y_0$.

If one weakens the assumption of $F_{12} \prec_{SI} F_{32} \prec_{SI} F'_{32}$ to $F_{12} \prec_c F_{32} \prec_c F'_{32}$, then the above argument is essentially still valid; details are a bit messier because the conditional distributions can have more crossings.

Similarly, minimal distributions in $\mathcal{F}$ can be studied.

3.4 $\mathcal{F}(F_{12}, F_{13}, F_{23})$

In this section, we study classes of trivariate distributions with three given bivariate margins, i.e., $\mathcal{F} = \mathcal{F}(F_{12}, F_{13}, F_{23})$. Here there is clearly a need to check for compatibility of the three bivariate margins, if they are arbitrary. For example, if the F_{12}, F_{13}, F_{23} are BVSN cdfs with respective correlations $\rho_{12}, \rho_{13}, \rho_{23}$, then the compatibility condition is that the correlation matrix with these ρ_{ij} is non-negative definite. Assuming that the three bivariate margins are compatible, we can obtain upper and lower bounds in a simple form and study them in the continuous case (all bivariate margins continuous). We show that these bounds are in general not proper cdfs, but we do obtain some conditions for which they are proper cdfs. We also obtain some conditions for which there is a unique $F \in \mathcal{F}$. Note also a difference compared to the class $\mathcal{F}(F_1, \ldots, F_m)$ in that an upper (lower) bound cdf for $\mathcal{F} = \mathcal{F}(F_{12}, F_{13}, F_{23})$ becomes a lower (upper) bound survival distribution for $\mathcal{F} = \mathcal{F}(\overline{F}_{12}, \overline{F}_{13}, \overline{F}_{23})$. Compatibility conditions, based on the bounds and other criteria, are also given. Some extensions are given in Section 3.6 on $\mathcal{F}(F_{ij}, 1 \le i < j \le m)$.

The results of this section and their extensions in Section 3.5 are crucial to the understanding of the construction method in Section 4.8.

3.4.1 Bounds

In this subsection, we obtain and analyse upper and lower bounds for $\mathcal{F} = \mathcal{F}(F_{12}, F_{13}, F_{23})$.

Theorem 3.11 *Let $a_1 = F_{12}$, $a_2 = F_{13}$, $a_3 = F_{23}$, $a_4 = 1 - F_1 - F_2 - F_3 + F_{12} + F_{13} + F_{23}$, $b_1 = F_{12} + F_{13} - F_1$, $b_2 = F_{12} + F_{23} - F_2$, $b_3 = F_{13} + F_{23} - F_3$. A lower bound is $F_L = \max\{0, b_1, b_2, b_3\}$ and an upper bound is $F_U = \min\{a_1, a_2, a_3, a_4\}$. For F_{12}, F_{13}, F_{23} to be compatible, $F_L \le F_U$ must hold everywhere.*

Proof. The last statement is obvious. Suppose F_{12}, F_{13}, F_{23} are compatible and let $F \in \mathcal{F}$. Clearly, $F \le \min\{F_{12}, F_{13}, F_{23}\}$. The fourth term a_4 in F_U comes from $\overline{F} = 1 - F_1 - F_2 - F_3 + F_{12} + F_{13} + F_{23} - F$; since $\overline{F} \ge 0$, $F \le a_4$. For the lower bound, suppose $\mathbf{X} \sim F$. Then for a permutation (i, j, k) of (1,2,3), $0 \le \Pr(X_i \le x_i, X_j > x_j, X_k > x_k) = F_i - F_{ij} - F_{ik} + F$ or $F \ge F_{ij} + F_{ik} - F_i$. $\square$

The bounds in Theorem 3.11 will be called the Fréchet bounds because they are simple and the analysis in Section 3.4.2 shows that for some choices of F_{12}, F_{13}, F_{23} they can both be equal to the unique distribution in $\mathcal{F}(F_{12}, F_{13}, F_{23})$.

Both of the bounds F_L, F_U have the right bivariate margins as $F_1 \to 1$, $F_2 \to 1$, $F_3 \to 1$. From analysis of the derivatives of first, second and third order, a requirement (in the continuous differentiable case) for the upper bound to be a proper distribution is that $F_{2|1} + F_{3|1} - 1 \ge 0$, $F_{1|2} + F_{3|2} - 1 \ge 0$, $F_{1|3} + F_{2|3} - 1 \ge 0$ when $a_4 \le \min\{F_{12}, F_{13}, F_{23}\}$. Similarly, a requirement for the lower bound to be a proper distribution is that $F_{2|1} + F_{3|1} - 1 \ge 0$ when $F_{12} + F_{13} - F_1 = F_L$, $F_{1|2} + F_{3|2} - 1 \ge 0$ when $F_{12} + F_{23} - F_2 = F_L$, and $F_{1|3} + F_{2|3} - 1 \ge 0$ when $F_{13} + F_{23} - F_3 = F_L$. These are necessary conditions, and other necessary conditions which come from the rectangle inequality (1.6) are studied below.

As an example, consider the case of pairwise independence ($F_{ij} = F_iF_j$ for $i < j$). For the upper bound, note that if $a_4 \le F_{12}$, then $(1-F_3)(1-F_1-F_2) \le 0$, which implies $F_1+F_2-1 = F_{1|3}+F_{2|3}-1 \ge 0$. Similarly the other two requirements are met when a_4 is the minimum. For the lower bound, supposing $F_L = F_1(F_2+F_3-1) \ge 0$, then $F_{2|1} + F_{3|1} - 1 = F_2 + F_3 - 1 \ge 0$; the other cases are covered by symmetry. However some other conditions are not met, with details given below.

A general analysis is given for the upper bound. Let $\mathbf{X} \sim F$ with $F \in \mathcal{F}$. The condition

$$1 - F_1 - F_2 - F_3 + F_{12} + F_{13} + F_{23} \le \min\{F_{12}, F_{13}, F_{23}\}, \quad (3.5)$$

when evaluated at (x_1, x_2, x_3), implies that

$$1 - F_i - F_j - F_k + F_{ij} + F_{ik} \le 0, \quad (3.6)$$

for all permutations (i, j, k) of (1,2,3). Inequality (3.6) is equivalent to

$$\frac{F_j - F_{ij}}{1 - F_i} + \frac{F_k - F_{ik}}{1 - F_i} - 1 \ge 0,$$

which in terms of probabilities for the rvs is

$$\Pr(X_j \le x_j \mid X_i > x_i) + \Pr(X_k \le x_k \mid X_i > x_i) - 1 \ge 0. \quad (3.7)$$

Note that (3.7) holds for all permutations (i, j, k) of (1,2,3) if x_i, x_j, x_k are sufficiently large. The conditions that (3.5) must imply in order for the Fréchet upper bound to be a distribution are

$$F_{j|i} + F_{k|i} - 1 \ge 0, \quad (3.8)$$

for all permutations (i, j, k) of (1,2,3). In terms of probabilities for the rvs, (3.8) is

$$\Pr(X_j \le x_j \mid X_i = x_i) + \Pr(X_k \le x_k \mid X_i = x_i) - 1 \ge 0. \quad (3.9)$$

If the bivariate margins are such that each conditional distribution $F_{i|j}$, $i \ne j$, is SI, then (3.7) implies (3.9). If they are such that each conditional distribution $F_{i|j}$, $i \ne j$, is strictly SD, then one can come up with values of (x_1, x_2, x_3) for which (3.7) does not imply (3.9). The reasoning is that if X_j is SI (SD) in X_i, then

$$\Pr(X_j \le x_j \mid X_i = x_i) \ \ge (\le) \ \Pr(X_j \le x_j \mid X_i > x_i).$$

The proof is that

$$\begin{aligned} \Pr(X_j \le x_j \mid X_i > x_i) &= [\overline{F}_i(x_i)]^{-1} \int_{x_i}^{\infty} F_{j|i}(x_j|y)\, dF_i(y) \\ &\le (\ge) \ F_{j|i}(x_j|x_i) \end{aligned}$$

if $F_{j|i}(x_j|y)$ is decreasing (increasing) in y.

A similar analysis can be given for the lower bound. F_L cannot be identically equal to 0. If $F_L = F_{ij} + F_{ik} - F_i \geq 0$, evaluated at (x_1, x_2, x_3), for some permutation (i, j, k) of $(1, 2, 3)$, then in terms of probabilities for the rvs,

$$\Pr(X_j \leq x_j \mid X_i \leq x_i) + \Pr(X_k \leq x_k \mid X_i \leq x_i) - 1 \geq 0. \quad (3.10)$$

The condition that (3.10) must imply in order for the Fréchet lower bound to be a distribution is (3.8) or (3.9). Using a similar argument to above, if the bivariate margins are such that each conditional distribution $F_{i|j}, i \neq j$, is SD, then (3.10) implies (3.9), and if each conditional distribution is SI, then one can come up with values of (x_1, x_2, x_3) for which (3.10) does not imply (3.9). This follows because $\Pr(X_j \leq x_j|X_i \leq x_i) = [F_i(x_i)]^{-1}\int_{-\infty}^{x_i} F_{j|i}(x_j|y)\, dF_i(y) \leq (\geq) F_{j|i}(x_j|x_i)$ if $F_{j|i}(x_j|y)$ is increasing (decreasing) in y.

A summary here is that if the bivariate margins are all SI (SD), then the Fréchet lower (upper) bound is not a proper distribution.

Next we study another condition which essentially shows that F_L and F_U are not proper distributions when all bivariate margins are PQD.

For the upper bound, consider first the case where the three bivariate margins are the same (and symmetric), and consider the rectangle condition based on the cube with corners (x, x, x) and (x', x', x') and $x' = x + \epsilon$ for a small $\epsilon > 0$. Let $F = F_1(x)$, $F' = F_1(x')$, $C_{12} = F_{12}(x, x)$, $C'_{12} = F_{12}(x, x') = F_{12}(x', x)$, $C''_{12} = F_{12}(x', x')$. Then $a_4 = 1 - 3F + 3C_{12} > C_{12}$ if $1 - 3F + 2C_{12} > 0$. If F_{12} is PQD, then $C_{12} \geq F^2$ and $1 - 3F + 2C_{12} \geq 1 - 3F + 2F^2 = (1-F)(1-2F) > 0$ if $F < \frac{1}{2}$. Suppose $F, F' < \frac{1}{2}$ and ϵ is sufficiently small; then the rectangle 'probability' is $C''_{12} - 3C'_{12} + 3C_{12} - C_{12} = (C''_{12} - 2C'_{12} + C_{12}) + (C_{12} - C'_{12}) < 0$, since the second negative term dominates when the expression is divided by ϵ^2 and $\epsilon \to 0$. For the more general case of arbitrary bivariate margins, take x_1, x_2, x_3 so that F_{12}, F_{13}, F_{23} are equal and small.

Now for the lower bound, consider again the symmetric case and use the same notation. The lower bound is greater than zero if $2C_{12} - F > 0$. If F_{12} is PQD, $2C_{12} - F \geq F(2F - 1) > 0$ if $F > \frac{1}{2}$. Suppose $F > \frac{1}{2}$ and ϵ is sufficiently small; then the rectangle 'probability' is $(2C''_{12} - F') - 3[C'_{12} + \max\{C''_{12} - F', C'_{12} - F\}] + 3[C'_{12} + \max\{C'_{12} - F', C_{12} - F\}] - (2C_{12} - F) = 2C''_{12} - 3C'_{12} + C_{12} - F' + F$ since $F - C_{12} \leq F' - C'_{12}$, $F - C'_{12} \leq F' - C''_{12}$ always hold (the probabilities correspond to $\Pr(X_1 \leq x, X_2 > x')$, $\Pr(X_1 \leq x', X_2 > x')$ in the latter case). Next, $2C''_{12} - 3C'_{12} + C_{12} -$

$F'+F = 2(C''_{12}-2C'_{12}+C_{12})+[(F-C_{12})-(F'-C'_{12})] < 0$, since the second negative term dominates when the expression is divided by ϵ^2 and $\epsilon \to 0$. For the more general case of arbitrary bivariate margins, take x_1, x_2, x_3 so that $F_{12}+F_{13}-F_1$, $F_{12}+F_{23}-F_2$, $F_{13}+F_{23}-F_3$ are equal or approximately equal and greater than 0.

3.4.2 Uniqueness

In this subsection, we study the cases where the three bivariate margins F_{12}, F_{13}, F_{23} uniquely determine a trivariate distribution F, and analyse whether the Fréchet bounds F_L and F_U are equal to F.

Examples are the following.

1. Suppose all three bivariate margins are Fréchet upper bounds. The unique trivariate distribution in $\mathcal{F}(F_{12}, F_{13}, F_{23})$ is $F = \min\{F_1, F_2, F_3\}$. For $F_1 \le F_2 \le F_3$, the upper bound is $F_U = \min\{F_1, F_1, F_2, 1-F_3+F_1\} = F_1$ and the lower bound is $F_L = \max\{0, F_1, F_1, F_1+F_2-F_3\} = F_1$. After considering all cases, by symmetry, $F = F_L = F_U$.
2. Suppose two bivariate margins are Fréchet lower bounds and one is a Fréchet upper bound. Assume $F_{12} = \min\{F_1, F_2\}$, $F_{13} = \max\{0, F_1+F_3-1\}$, $F_{23} = \max\{0, F_2+F_3-1\}$. The unique trivariate distribution in $\mathcal{F}$ is $F = \max\{0, \min\{F_1, F_2\}+F_3-1\}$. Cases to consider are:

   (a) $F_1 \le F_2 \le 1-F_3$: $F_U = \min\{F_1, 0, 0, 1-F_2-F_3\} = 0$ and $F_L = \max\{0, 0, F_1-F_2, -F_3\} = 0$;

   (b) $F_1 \le 1-F_3 \le F_2$: $F_U = \min\{F_1, 0, F_2+F_3-1, 0\} = 0$ and $F_L = \max\{0, 0, F_1+F_3-1, F_2-1\} = 0$;

   (c) $1-F_3 \le F_1 \le F_2$: $F_U = \min\{F_1, F_1+F_3-1, F_2+F_3-1, F_1+F_3-1\} = F_1+F_3-1$ and $F_L = \max\{0, F_1+F_3-1, F_1+F_3-1, F_1+F_2+F_3-2\} = F_1+F_3-1$.

   Other cases are covered by symmetry. Hence $F = F_L = F_U$.

General results on when there is a unique $F \in \mathcal{F}$ are given next. We assume that the univariate margins are all continuous, so that they could be taken to be $U(0,1)$ without loss of generality. Let $x_1 < x'_1$, $x_2 < x'_2$, $x_3 < x'_3$ be support points of the univariate margins F_1, F_2, F_3, respectively. Let $F = F_{123} \in \mathcal{F}(F_{12}, F_{13}, F_{23})$. Let $p_{123}, p_{1'23}, \ldots, p_{1'2'3'}$ be the masses or accumulated density values in neighbourhoods of $(x_1, x_2, x_3), (x'_1, x_2, x_3), \ldots, (x'_1, x'_2, x'_3)$, respectively. The eight triplets can be viewed as being on the corners of a cube. If all of the eight probabilities are positive, then one can make small shifts of masses or density to get a new trivariate disribution with the same bivariate margins. The shifts are:

$$p_{123} \to p_{123} + \epsilon, \quad p_{1'2'3} \to p_{1'2'3} + \epsilon, \quad p_{1'23'} \to p_{1'23'} + \epsilon,$$

$$p_{12'3'} \to p_{12'3'} + \epsilon, \quad p_{1'23} \to p_{1'23} - \epsilon, \quad p_{12'3} \to p_{12'3} - \epsilon,$$

$$p_{123'} \to p_{123'} - \epsilon, \quad p_{1'2'3'} \to p_{1'2'3'} - \epsilon,$$

where ϵ is small enough in absolute value that all of the new probabilities exceed zero.

If at least one of $p_{123}, p_{1'2'3}, p_{1'23'}, p_{12'3'}$ is zero and at least one of $p_{1'23}, p_{12'3}, p_{123'}, p_{1'2'3'}$ is zero, no matter what are the choices of x_j, x_j', $j = 1, 2, 3$, then F is the unique distribution with the given bivariate margins. For $U(0,1)$ margins, examples where this occur are:

(a) the mass is totally on one of the four diagonals of the unit cube (corresponding to the case where each bivariate margin is a Fréchet upper or lower bound);

(b) the mass is on one of the planes $x_i = x_j$ or $x_i + x_j = 1$, $i \neq j$ (corresponding to the case where one of the bivariate margins is the Fréchet upper or lower bound and one of the bivariate margins could be any copula);

(c) the mass is on the surface $y_3 = g(y_1, y_2)$, where g is strictly increasing in y_1, y_2 (this happens in the case where F_{12}, F_{32} are such that $F_{12} \prec_{\mathrm{SI}} F_{32}$ and F is the Fréchet upper bound given F_{12}, F_{32}; the function $g(y_1, y_2)$ is $F_{3|2}^{-1}(F_{1|2}(y_1|y_2)|y_2)$), or g is strictly decreasing in y_1, y_2;

(d) the mass is on the surface $y_3 = g(y_1, y_2)$, where g is strictly decreasing in y_1 and strictly increasing in y_2 (this happens in the case where F_{12}, F_{32} are such that F_{12}, F_{32} are SI and F is the Fréchet lower bound given F_{12}, F_{32}; the function $g(y_1, y_2)$ is $F_{3|2}^{-1}(1 - F_{1|2}(y_1|y_2)|y_2)$), or g is strictly increasing in y_1 and strictly decreasing in y_2.

Consider the cube with the eight points from $\{x_1, x_1'\} \times \{x_2, x_2'\} \times \{x_3, x_3'\}$. The proof for example (c) with strict increase is as follows: if (x_1, x_2, x_3) and (x_1', x_2', x_3') are on the surface, then there can be no mass or density on the other six vertices of the cube; if (x_1, x_2, x_3), (x_1', x_2, x_3') and possibly (x_1, x_2', x_3'), by appropriate choice of x_2', are on the surface, then there can be no mass or density on the other five vertices of the cube. The proof for example (d) with strict increase in y_2 and strict decrease in y_1 is: if (x_1, x_2, x_3) and (x_1, x_2', x_3') are on the surface, then there can be no mass or density on (x_1, x_2', x_3), (x_1', x_2, x_3), (x_1, x_2, x_3'), (x_1', x_2, x_3') and (x_1', x_2', x_3'); if (x_1', x_2, x_3) and (x_1', x_2', x_3') are on the surface, then there can be no mass or density on (x_1, x_2, x_3), (x_1, x_2', x_3), (x_1', x_2', x_3), (x_1', x_2, x_3') and (x_1, x_2', x_3').

In the case in which the mass is on the surface $y_3 = g(y_1, y_2)$ and no monotonicity in y_2 exists, no general conclusion can be drawn. The specific g would have to be studied to check what happens on cubes. This is because if there are masses at (x_1, x_2, x_3) and (x_1', x_2', x_3), then there is the possibility of masses at both (x_1', x_2, x_3') and (x_1, x_2', x_3').

Now we look at the Fréchet bounds where there is a unique $F \in \mathcal{F}$. In the case $F_{13} = \min\{F_1, F_3\}$ with $U(0,1)$ univariate margins, $F_{12} = F_{32}$. If $F_1 \le F_3$, then $F_U = \min\{F_{12}, F_{23}, F_1 \wedge F_3, 1 - F_2 - F_3 + F_{23} + F_{12}\} = F_{12}$ and $F_L = \max\{0, F_{12}, F_{12} + (F_{23} - F_2), F_1 + F_{23} - F_3\} = F_{12}$ since $F_{12} - F_1 - F_{23} + F_3 = \Pr(X_1 \le x_3, X_2 > x_2) - \Pr(X_1 \le x_1, X_2 > x_2) \ge 0$. The case $F_1 \ge F_3$ can be handled symmetrically. Hence $F_L = F_U$ and these must equal the unique trivariate distribution in $\mathcal{F}$.

Next consider the case $F_{13} = \max\{0, F_1 + F_3 - 1\}$ with $U(0,1)$ margins, so that $F_{23}(x_2, x_3) = F_2(x_2) - F_{12}(1 - x_3, x_2)$. If $F_1 + F_3 - 1 \le 0$, then $F_U = F_{13} = 0$ and $F_L = 0$ since $F_{12} + F_{23} - F_2 = F_{12}(x_1, x_2) - F_{12}(1 - x_3, x_2) \le 0$. If $F_1 + F_3 - 1 > 0$, then $F_U = \min\{F_1 + F_3 - 1 = x_1 + x_3 - 1, F_{12}, F_2(x_2) - F_{12}(1 - x_3, x_2), F_{12} + F_{23} - F_2 = F_{12}(x_1, x_2) - F_{12}(1 - x_3, x_2)\} = F_{12}(x_1, x_2) - F_{12}(1 - x_3, x_2)$ and $F_L = \max\{0, F_{12} + F_3 - 1, F_{23} + F_1 - 1, F_{12}(x_1, x_2) - F_{12}(1 - x_3, x_2)\} = F_{12}(x_1, x_2) - F_{12}(1 - x_3, x_2)$ since $F_{12} + F_{23} - F_2 \ge F_{23} + F_1 - 1$ if $F_1 + F_2 - 1 \le F_{12}$. Again $F_L = F_U$ and these must equal the unique trivariate distribution in $\mathcal{F}$.

Next consider the case $F_{12} \ne F_{32}$, with F_{13} coming from the Fréchet upper bound given F_{12}, F_{32}. From case (c) above, there is a unique $F \in \mathcal{F}$. If $F_{1|2} \prec_{\mathrm{SI}} F_{3|2}$, then from Theorem 2.10, for fixed x_1, x_3, there exists $y_0(x_1, x_3)$ such that

$$F_{1|2}(x_1|y) \le F_{3|2}(x_3|y), \quad \text{for } y \le y_0(x_1, x_3), \tag{3.11}$$

$$F_{1|2}(x_1|y) \ge F_{3|2}(x_3|y), \quad \text{for } y \ge y_0(x_1, x_3). \tag{3.12}$$

Let

$$F(\mathbf{x}) = \int_{-\infty}^{x_2} \min\{F_{1|2}(x_1|y),\ F_{3|2}(x_3|y)\}\, F_2(dy).$$

Then from (3.11) and (3.12), $F(\mathbf{x}) = F_{12}(x_1, x_2)$ if $x_2 \le y_0(x_1, x_3)$ and

$$\begin{aligned} F(\mathbf{x}) &= \int_{-\infty}^{y_0} F_{1|2}\, dF_2 + \int_{y_0}^{x_2} F_{3|2}\, dF_2 \\ &= \int_{-\infty}^{y_0} F_{1|2}\, dF_2 + \int_{y_0}^{\infty} F_{3|2}\, dF_2 - \int_{x_2}^{\infty} F_{3|2}\, dF_2 \\ &= F_{13}(x_1, x_3) - [F_3(x_3) - F_{23}(x_2, x_3)] \end{aligned}$$

if $x_2 \ge y_0(x_1, x_3)$. Hence $F_U = F_{12}(x_1, x_2)$ if $x_2 \le y_0(x_1, x_3)$ and $F_L = F_{13}(x_1, x_3) + F_{23}(x_2, x_3) - F_3(x_3)$ if $x_2 \ge y_0(x_1, x_3)$, and one of the bounds is always reached. If $x_2 > y_0(x_1, x_3)$, then $F = F_{13} + F_{23} - F_3$ is less than F_{13}, F_{23} and $F_{12} = \int_{-\infty}^{x_2} F_{2|1} dF_2$. Also $F_{13} + F_{23} - F_3 < 1 - F_1 - F_2 - F_3 + F_{12} + F_{13} + F_{23}$ is equivalent to $0 < 1 - F_1 - F_2 + F_{12}$ so that $F = F_L < F_U$ in this case. If $x_2 < y_0(x_1, x_3)$, then $F = F_{12}$ is greater than $F_{12} + F_{13} - F_1$ and $F_{12} + F_{23} - F_2$. Also $F > F_{13} + F_{23} - F_3$ is equivalent to $\Pr(X_1 > x_1, X_2 > x_2, X_3 \le x_3) > 0$ so that $F_L < F_U = F$ in this case. This illustrates examples where there is a unique $F \in \mathcal{F}$ but neither F_L nor F_U is equal to F.

3.4.3 Compatibility conditions

For $\mathcal{F} = \mathcal{F}(F_{12}, F_{13}, F_{23})$, compatibility conditions for F_{12}, F_{13}, F_{23} are obtained by considering two of the three bivariate margins to be arbitrary, and the third bivariate margin to have constraints given the other two. Throughout this subsection, (i, j, k) will denote a permutation of $(1, 2, 3)$. Methods for obtaining compatibility conditions are:

(i) comparison of F_{jk} with the Fréchet bounds in $\mathcal{F}(F_{ij}, F_{ik})$;

(ii) sets of bivariate Kendall's tau $(\tau_{12}, \tau_{13}, \tau_{23})$ in the continuous case;

(iii) sets of bivariate tail dependence parameters $(\lambda_{12}, \lambda_{13}, \lambda_{23})$.

For method (i), we make use of results from Section 3.2 to get (three) inequalities of the form:

$$\int_{-\infty}^{\infty} \max\{0, F_{i|j} + F_{k|j} - 1\}\, dF_j \le F_{ik} \le \int_{-\infty}^{\infty} \min\{F_{i|j}, F_{k|j}\}\, dF_j.$$

In practice, for given F_{12}, F_{13}, F_{23}, these inequalities can only be checked numerically over a grid of points.

Next consider the range of possible vectors $(\zeta_{12}, \zeta_{23}, \zeta_{13})$, where ζ_{jk} is a measure of dependence for the (j, k) bivariate margin. We derive results only for $\zeta = \tau$, Kendall's tau, and $\zeta = \lambda$, the upper tail dependence parameter, but the ideas can be applied to other bivariate measures of association.

Theorem 3.12 *Let $F \in \mathcal{F}(F_{12}, F_{23}, F_{13})$ and suppose F_{jk}, $j < k$, are continuous. Let $\tau_{jk} = \tau_{kj}$ be the value of Kendall's tau for F_{jk}, $j \neq k$. Then the inequality*

$$-1 + |\tau_{ij} + \tau_{jk}| \leq \tau_{ik} \leq 1 - |\tau_{ij} - \tau_{jk}|. \tag{3.13}$$

holds for all permutations (i,j,k) of $(1,2,3)$ and the bounds are sharp. Therefore if F_{ij} are bivariate margins such that τ_{ij} is the Kendall tau value for F_{ij}, $1 \leq i < j \leq 3$, and (3.13) does not hold for some (i,j,k), then the three bivariate margins are not compatible.

Proof. The proof of (3.13) will be given in the case of $(i,j,k) = (1,2,3)$. The other inequalities follow by permuting indices. Let (X_{s1}, X_{s2}, X_{s3}), $s = 1,2$, be independent random vectors from F. Then $\tau_{jk} = 2\eta_{jk} - 1$, $1 \leq j < k \leq 3$, where

$$\eta_{jk} = \Pr((X_{1j} - X_{2j})(X_{1k} - X_{2k}) > 0).$$

Then

$$\begin{aligned}
\eta_{13} &= \Pr((X_{11} - X_{21})(X_{12} - X_{22})^2(X_{13} - X_{23}) > 0) \\
&= \Pr((X_{11} - X_{21})(X_{12} - X_{22}) > 0, (X_{12} - X_{22})(X_{13} - X_{23}) > 0) \\
&\quad + \Pr((X_{11} - X_{21})(X_{12} - X_{22}) < 0, (X_{12} - X_{22})(X_{13} - X_{23}) < 0).
\end{aligned}$$

Hence, from Lemma 3.8, an upper bound for η_{13} is $\min\{\eta_{12}, \eta_{23}\} + \min\{1 - \eta_{12}, 1 - \eta_{23}\}$ and a lower bound is $\max\{0, \eta_{12} + \eta_{23} - 1\} + \max\{0, (1 - \eta_{12}) + (1 - \eta_{23}) - 1\}$. After substituting for τ_{jk} and simplifying, inequality (3.13) results. The sharpness follows from the special trivariate normal case given next.

For the BVN distribution, $\tau = (2/\pi)\arcsin(\rho)$ (Exercise 2.14). Hence, for the trivariate normal distributions, the constraint $-1 \leq \rho_{13\cdot 2} \leq 1$ (for the partial correlation) is the same as

$$\rho_{12}\rho_{23} - [(1-\rho_{12}^2)(1-\rho_{23}^2)]^{1/2} \leq \rho_{13} \leq \rho_{12}\rho_{23} + [(1-\rho_{12}^2)(1-\rho_{23}^2)]^{1/2}$$

or

$$-\cos(\tfrac{1}{2}\pi(\tau_{12} + \tau_{23})) \leq \sin(\tfrac{1}{2}\pi\tau_{13}) \leq \cos(\tfrac{1}{2}\pi(\tau_{12} - \tau_{23}))$$

or equivalently (3.13) with $(i,j,k) = (1,2,3)$. □

Theorem 3.13 *Let (X_{s1}, X_{s2}, X_{s3}), $s = 1,2$, be independent random vectors from the continuous distribution F. For a permutation (i,j,k) of (1,2,3), let the events E_1, E_2 be defined as $\{(X_{1i} -$*

$X_{2i})(X_{1j} - X_{2j}) > 0\}$ and $\{(X_{1k} - X_{2k})(X_{1j} - X_{2j}) > 0\}$, respectively. The upper bound in (3.13) is attained if $E_1 \subset E_2$ or $E_2 \subset E_1$, and the lower bound in (3.13) is attained if $E_1 \subset E_2^c$ or $E_2 \subset E_1^c$ or $E_1^c \subset E_2$ or $E_2^c \subset E_1$ (equivalently, $E_1 \cap E_2 = \emptyset$ or $E_1^c \cap E_2^c = \emptyset$).

Proof. The proof in the preceding theorem for (3.13) is based on

$$\begin{aligned} &\max\{0, \Pr(E_1) + \Pr(E_2) - 1\} + \max\{0, \Pr(E_1^c) + \Pr(E_2^c) - 1\} \\ &\quad \le \Pr(E_1 \cap E_2) + \Pr(E_1^c \cap E_2^c) \\ &\quad \le \min\{\Pr(E_1), \Pr(E_2)\} + \min\{\Pr(E_1^c), \Pr(E_2^c)\}. \end{aligned}$$

The conclusion follows. □

Unlike Kendall's tau, it does not appear possible to obtain closed-form sharp bounds for the range of λ_{ik} given $\lambda_{ij}, \lambda_{jk}$. However, simple bounds can be obtained.

Theorem 3.14 *Let $F \in \mathcal{F}(F_{12}, F_{23}, F_{13})$. Let $\lambda_{jk} = \lambda_{kj}$ be the upper tail dependence parameter value for F_{jk}, $j \ne k$. Then the inequality*

$$\max\{0, \lambda_{ij} + \lambda_{jk} - 1\} \le \lambda_{ik} \le 1 - |\lambda_{ij} - \lambda_{jk}| \tag{3.14}$$

holds for all permutations (i, j, k) of $(1, 2, 3)$. Hence if F_{ij} are bivariate margins such that λ_{ij} is the upper tail dependence value for F_{ij}, $1 \le i < j \le 3$, and (3.14) does not hold for some (i, j, k), then the three bivariate margins are not compatible.

Proof. The proof of (3.14) will be given in the case of $(i, j, k) = (1, 2, 3)$. The other inequalities follow by permuting indices. Let C be the copula associated with F and, for $i < j$, let C_{ij} be the (i, j) bivariate margin of C. Let $\mathbf{U} = (U_1, U_2, U_3) \sim C$. Then

$$\begin{aligned} &\Pr(U_3 > u \mid U_1 > u) \\ &\quad = \Pr(U_3 > u, U_2 > u \mid U_1 > u) + \Pr(U_3 > u, U_2 \le u \mid U_1 > u) \\ &\quad = \Pr(U_3 > u, U_1 > u \mid U_2 > u) + \Pr(U_3 > u, U_2 \le u \mid U_1 > u) \\ &\quad \le \min\{\Pr(U_1 > u \mid U_2 > u), \Pr(U_3 > u \mid U_2 > u)\} \\ &\qquad +1 - \Pr(U_2 > u \mid U_1 > u). \end{aligned}$$

By taking limits as $u \to 1$, $\lambda_{13} \le \min\{\lambda_{12}, \lambda_{23}\} + 1 - \lambda_{12}$. Similarly, by interchanging the subscripts 1 and 3, $\lambda_{13} \le \min\{\lambda_{12}, \lambda_{23}\} + 1 - \lambda_{23}$. From combining these two upper bounds, $\lambda_{13} \le 1 - |\lambda_{12} - \lambda_{23}|$. For the lower bound on λ_{13},

$$\begin{aligned}\Pr(U_3 > u \mid U_1 > u) &= \Pr(U_3 > u, U_1 > u)/\Pr(U_2 > u) \\ &\geq \Pr(U_3 > u, U_1 > u \mid U_2 > u) \\ &\geq \max\{0, \Pr(U_3 > u \mid U_2 > u) + \Pr(U_1 > u \mid U_2 > u) - 1\}.\end{aligned}$$

By taking the limit as $u \to 1$, the lower bound in (3.14) obtains. □

To end this subsection, we give two examples of applications of the preceding two theorems.

Example 3.2 Consider the bivariate family B4 of copulas in Section 5.1: $C(u, v; \delta) = (u^{-\delta} + v^{-\delta} - 1)^{-1/\delta}$, $0 \leq \delta < \infty$. From Theorem 4.3, the Kendall tau associated with this copula is $\tau = \delta/(\delta+2)$. Now consider $C_{ij} = C(\cdot; \delta_{ij})$, $1 \leq i < j \leq 3$, so that $\tau_{ij} = \delta_{ij}/(\delta_{ij} + 2)$ or $\delta_{ij} = 2\tau_{ij}/(1 - \tau_{ij})$. From Theorem 3.12, if $\delta_{12} = \delta_{23} = \delta$, then bounds on δ_{13} for compatibility are $\max\{0, \delta/2-1\} \leq \delta_{13} < \infty$. In Section 5.3, there is a construction of a trivariate family of copulas with bivariate margins in B4 with parameters of the form $(\delta_{12}, \delta_{13}, \delta_{23}) = (\delta, \theta, \delta)$ with $\theta \geq \delta$; so far, no trivariate distribution with $\max\{0, \delta/2 - 1\} \leq \theta \leq \delta$ has been constructed. □

Example 3.3 Consider the bivariate family B5 of copulas in Section 5.1: $C(u, v; \delta) = 1 - (\overline{u}^\delta + \overline{v}^\delta - [\overline{uv}]^\delta)^{1/\delta}$, $1 \leq \delta < \infty$, where $\overline{u} = 1 - u$, $\overline{v} = 1 - v$. From Example 2.3, the upper tail dependence parameter is $\lambda = 2-2^{1/\delta}$. Now consider $C_{ij} = C(\cdot; \delta_{ij})$, $1 \leq i < j \leq 3$, so that $\lambda_{ij} = 2-2^{1/\delta_{ij}}$ or $\delta_{ij} = [\log 2]/[\log(2-\lambda_{ij})]$. From Theorem 3.14, if $\delta_{12} = \delta_{23} = \delta$, then bounds on δ_{13} for compatibility are $\max\{1, [\log 2]/[\log(2^{1+1/\delta} - 1)]\} \leq \delta_{13} < \infty$. The construction in Section 5.3 covers the multivariate extension of B5 as well, so there is a trivariate distribution with bivariate margins in B5 with parameters of the form $(\delta_{12}, \delta_{13}, \delta_{23}) = (\delta, \theta, \delta)$ with $\theta \geq \delta$. □

3.5 $\mathcal{F}(F_{123}, F_{124}, F_{134}, F_{234})$ *

The results of Section 3.4 can be extended to Fréchet class of 4-variate distributions with given trivariate margins. Bounds of $\mathcal{F} = \mathcal{F}(F_{123}, F_{124}, F_{134}, F_{234})$ are a simple extension of the bounds in Section 3.4, and they extend to the Fréchet class of m-variate distributions given the set of m margins of dimension $(m-1)$. The 2^{m-1} terms involved in each of the bounds F_L, F_U appear with the method in Section 4.8 for constructing a multivariate distribution based on bivariate margins. As in Section 3.4, the bounds F_L, F_U

are referred to as the Fréchet bounds.

The bounds can be obtained based on the non-negativity of the 16 orthant probabilities in four dimensions. Let $a_1 = F_1 - F_{12} - F_{13} - F_{14} + F_{123} + F_{124} + F_{134}$, $a_2 = F_2 - F_{12} - F_{23} - F_{24} + F_{123} + F_{124} + F_{234}$, $a_3 = F_3 - F_{13} - F_{23} - F_{34} + F_{123} + F_{134} + F_{234}$, $a_4 = F_4 - F_{14} - F_{24} - F_{34} + F_{124} + F_{134} + F_{234}$. The Fréchet upper bound F_U is

$$F_U = \min\{F_{123}, F_{124}, F_{134}, F_{234}, a_1, a_2, a_3, a_4\}. \tag{3.15}$$

Let $a_{ij} = F_{ijk} + F_{ijl} - F_{ij}$, for $i < j$ and $k \ne l \ne i, j$, and let $a_0 = F_{123} + F_{124} + F_{134} + F_{234} - F_{12} - F_{13} - F_{14} - F_{23} - F_{24} - F_{34} + F_1 + F_2 + F_3 + F_4 - 1$. The Fréchet lower bound F_L is

$$F_L = \max\{0, a_{12}, a_{13}, a_{14}, a_{23}, a_{24}, a_{34}, a_0\}. \tag{3.16}$$

For $F_{123}, F_{124}, F_{134}, F_{234}$ to be compatible, $F_L \le F_U$ must hold everywhere. Let (i, j, k, l) be a permutation of $(1, 2, 3, 4)$. Conditions for the upper bound to be a proper distribution (in the continuous differentiable case) include:

- if $a_i = F_U$, then $1 - F_{j|i} - F_{k|i} - F_{l|i} + F_{jk|i} + F_{jl|i} + F_{kl|i} \ge 0$ and $F_{k|ij} + F_{l|ij} - 1 \ge 0$, $F_{j|ik} + F_{l|ik} - 1 \ge 0$, $F_{j|il} + F_{k|il} - 1 \ge 0$.

Conditions for the lower bound to be a proper distribution include:

- if $a_{ij} = F_L > 0$, then $F_{jk|i} + F_{jl|i} - F_{j|i} \ge 0$, $F_{ik|j} + F_{il|j} - F_{i|j} \ge 0$ and $F_{k|ij} + F_{l|ij} - 1 \ge 0$;
- if $a_0 = F_L > 0$, then for all permutations (i, j, k, l), $F_{jk|i} + F_{jl|i} + F_{kl|i} - F_{j|i} - F_{k|i} - F_{l|i} + 1 \ge 0$, $F_{k|ij} + F_{l|ij} - 1 \ge 0$.

3.6 $\mathcal{F}(F_{ij}, 1 \le i < j \le m)$ *

In this section, we consider extensions from Section 3.4 to m-variate distributions given the set of $m(m-1)/2$ bivariate margins. The Fréchet class is $\mathcal{F} = \mathcal{F}(F_{ij}, 1 \le i < j \le m)$. There are many conditions needed for the compatibility of F_{ij}, $1 \le i < j \le m$. Clearly for each triple (i_1, i_2, i_3) from $\{1, \ldots, m\}$, the compatibility conditions from Section 3.4 must hold for $\mathcal{F}(F_{i_1 i_2}, F_{i_1 i_3}, F_{i_2 i_3})$. For example, if $\tau_{jk} = \tau_{kj}$ is the Kendall tau value for F_{jk}, $j < k$, then

$$-1 + |\tau_{i_1 i_2} + \tau_{i_2 i_3}| \le \tau_{i_1 i_3} \le 1 - |\tau_{i_1 i_2} - \tau_{i_2 i_3}| \tag{3.17}$$

for all triples (i_1, i_2, i_3).

However we have not successfully obtained extensions of (3.17) or (3.14) that depend on $r(r-1)/2$ bivariate margins with $4 \le r \le m$.

3.7 General $\mathcal{F}(F_S : S \in \mathcal{S}^*)$, $\mathcal{S}^* \subset \mathcal{S}_m$ *

In this section, let $\mathcal{S}^* \subset \mathcal{S}_m$. We state a condition on $\mathcal{S}^*$ which ensures that $\mathcal{F}(F_S : S \in \mathcal{S}^*)$ is non-empty. Let $k = |\mathcal{S}^*| \geq 2$. The condition is that there is an enumeration $S_1, \ldots, S_k$ of $\mathcal{S}^*$ such that

$$S_j \cap (\cup_{i<j} S_i) \in \cup_{i<j} 2^{S_i}, \quad j = 2, \ldots, k,$$

where 2^{S_i} denotes a power set (all subsets of the set in the exponent).

Example 3.4 Cases to illustrate this result are the following.

1. Let $m = 5$, $k = 3$, $S_1 = \{1,2,3\}$, $S_2 = \{2,3,4\}$, $S_3 = \{1,4,5\}$. Then $S_2 \cap S_1 = \{2,3\} \subset S_1$ but $S_3 \cap (S_1 \cup S_2) = \{1,4\}$ is not a subset of either S_1 or S_2. A similar situation occurs for the other two relevant indexings of the three sets: $S_2 \cap (S_1 \cup S_3) = \{2,3,4\}$ is not a subset of S_1 or S_3, and $S_1 \cap (S_2 \cup S_3) = \{1,2,3\}$ is not a subset of S_2 or S_3. Hence the compatibility of $F_{S_1}, F_{S_2}, F_{S_3}$ need not hold in general, and further checks would have to be done for specific given margins.

2. Let $m = 7$, $k = 4$, $S_1 = \{1,2,3\}$, $S_2 = \{2,3,4,5\}$, $S_3 = \{3,5,6\}$, $S_4 = \{3,6,7\}$. Then $S_3 \cap (S_1 \cup S_2) = \{3,5\} \subset S_2$ and $S_4 \cap (S_1 \cup S_2 \cup S_3) = \{3,6\} \subset S_3$. A simple construction of $F_{1\cdots7} \in \mathcal{F}(F_{S_i}, i = 1,2,3,4)$ is as follows. First take the conditional distributions $F_{1|23}$ and $F_{45|23}$ and use conditional independence to construct $F_{12345} = \int F_{1|23} F_{45|23}\, dF_{23}$. From this F_{12345} and F_{356}, next take $F_{124|35}$ and $F_{6|35}$ and use conditional independence to construct $F_{123456} = \int F_{124|35} F_{6|35}\, dF_{35}$. Finally, take the conditional distributions $F_{1245|36}$ and $F_{7|36}$ to construct $F_{1234567} = \int F_{1245|36} F_{7|36}\, dF_{36}$.

In the first case, one cannot simply construct F_{1234} from the conditional independence of $F_{1|23}$ and $F_{4|23}$ because there is no guarantee that the (1,4) margin of this distribution is the same as the (1,4) margin of F_{145}. □

3.8 Bibliographic notes

Early papers on Fréchet classes and bounds include Fréchet (1935; 1951; 1957). For an alternative direct proof of Theorem 3.2 based on the rectangle inequality, see Dall'Aglio (1972). The main idea there is based on $F_j = F_j(x_j)$, $F_j' = F_j(x_j')$, $x_j < x_j'$, $j = 1, \ldots, m$, with $F_1 \leq F_2 \leq \cdots \leq F_m$. The notation is a bit cumbersome. Theorems 3.4 and 3.5 and their proofs are based on Dall'Aglio

(1972). The results in Section 3.4.3 on sets of bivariate Kendall tau or tail dependence parameters are from Joe (1996a). The result in Section 3.7 is from Kellerer (1964). Results such as Theorem 3.9 and results in Sections 3.3, 3.4 and 3.5 are probably new (at least in the method of presentation).

3.9 Exercises

3.1 If $F_1 = F_2$ are continuous univariate cdfs of rvs that are symmetric about zero, what is the stochastic representation for the Fréchet lower bound in $\mathcal{F}(F_1, F_2)$?

3.2 For $j = 1, 2$, let F_j be the cdf for the Binomial$(2, p_j)$ distribution. For some special cases of p_1, p_2, deduce the pmf for the Fréchet upper and lower bounds. Of the nine probability masses, what is the minimum number of zeros in the Fréchet bounds?

3.3 Let $U \sim U(0,1)$. Compute the correlation of U and $a(U)$ for the following real-valued monotone functions a: $a(u) = u^\alpha$, $\alpha > 0$; $a(u) = (1-u)^\beta$, $\beta > 0$; $a(u) = e^{\gamma u}$, $-\infty < \gamma < \infty$, $\gamma \neq 0$. Note that $(U, a(U))$ has the Fréchet upper or lower bound copula.

3.4 For the bivariate Fréchet upper bound (respectively, the lower bound), show by taking appropriate univariate margins F_1, F_2 that the correlation can take any value in the interval (0,1] (respectively, $[-1, 0)$).

3.5 Establish the identity

$$\max_{1 \le j \le m} z_j = \sum_{S \in \mathcal{S}_m} (-1)^{|S|+1} \min_{i \in S} z_j ,$$

and prove Theorem 3.5.

3.6 In the bivariate case, show that the Fréchet upper bound has TP_2 cdf and survival function, and the Fréchet lower bound has RR_2 cdf and survival function.

3.7 Do some checks (possibly numerical) on compatibility conditions for parameters for trivariate distributions with bivariate margins in one of the families B2–B7 in Section 5.1.

3.8 Show that the lower Fréchet bound for $\mathcal{F} = \mathcal{F}(F_{12}, F_3)$, $F_L(\mathbf{x}) = \max\{0, F_{12}(x_1, x_2) + F_3(x_3) - 1\}$, is not a cdf in general.

3.9 For a bivariate copula C, with $(U_1, U_2) \sim C$, Blomqvist's q can be defined as $2\Pr((U_1 - \frac{1}{2})(U_2 - \frac{1}{2}) > 0) - 1$. Now let $F \in \mathcal{F}(C_{12}, C_{13}, C_{23})$, where C_{ij}, $i < j$, are compatible bivariate copulas. Let q_{ij} be Blomqvist's q for C_{ij}, $i < j$. Show that

$$-1 + |q_{ij} + q_{jk}| \le q_{ik} \le 1 - |q_{ij} - q_{jk}|$$

for all permutations (i, j, k) of (1,2,3).

3.10 For the bounds of $\mathcal{F}(F_{123}, F_{124}, F_{134}, F_{234})$, do an analysis similar to that on the bounds of $\mathcal{F}(F_{12}, F_{13}, F_{23})$.

3.11 Generalize the bounds for the classes $\mathcal{F}(F_{12}, F_{13}, F_{23})$ and $\mathcal{F}(F_{123}, F_{124}, F_{134}, F_{234})$ to the Fréchet class of m-variate distributions given the set of m margins of dimension $(m-1)$.

3.10 Unsolved problems

3.1 Improve on the Fréchet lower bound for exchangeable rvs. See Scarsini (1985) for partial results.

3.2 Improve on the Fréchet lower bound for (strongly) stationary sequences of rvs. That is, given a stationary sequence $X_1, X_2, \ldots$ with univariate margin F_0, what is an improved lower bound for the cdf of X_i, X_{i+k}, $k \ge 1$?

3.3 Consider $\mathcal{F}(F_{ij} : 1 \le i < j \le m)$. For $i < j$, let τ_{ij} be the Kendall tau value for F_{ij} and let λ_{ij} be the upper tail dependence value for F_{ij}. Obtain better results for bounds or compatibility for sets of τ_{ij} and λ_{ij}, $1 \le i < j \le m$, $m > 3$.

3.4 Obtain better compatibility conditions for $\{F_{ij} : 1 \le i < j \le m\}$ to be the set of bivariate margins of an m-variate distribution.

3.5 Suppose $\mathcal{S}^*$ does not satisfy the condition in Section 3.7. Obtain some compatibility conditions (preferably checkable) for $\{F_S : S \in \mathcal{S}^*\}$ to correspond to marginal distributions of a multivariate distribution.

CHAPTER 4

Construction of multivariate distributions

In this chapter, some methods for construction of multivariate distributions are given, with different approaches separated into different sections. As mentioned in Chapter 1, one cannot just write down a parametric family of functions with the right boundary properties and expect them to satisfy the rectangle condition of a multivariate cdf. Generally a family of multivariate distributions must be constructed through methods such as mixtures, stochastic representations and limits. The methods in Sections 4.2 to 4.6 and Section 4.9 are based on mixtures and stochastic representations. Families of multivariate extreme value distributions (see Chapter 6) are often obtained through the *extreme value* limit. The methods in Sections 4.7 and 4.8 consist of constructions of m-variate objects given the set of $m(m-1)/2$ bivariate margins.

Families of bivariate distributions are easier to obtain and the bivariate rectangle condition can usually be checked analytically. There has been much more in the statistics and probability literature on families of bivariate distributions (e.g., the book on bivariate continuous distributions by Hutchinson and Lai 1990, and the book on bivariate discrete distributions by Kocherlakota and Kocherlakota 1992), but many of the families do not have obvious or nice multivariate generalizations. Parametric families of bivariate copulas that have nice properties are summarized in Section 5.1. Nice properties include a range of dependence covering independence and the Fréchet upper bound, and extendibility to higher dimensions. Desirable properties are given in Section 4.1 below.

A summary of the highlights of this chapter is the following.

- Laplace transforms play an important role in construction of copulas; they are linked to the mixing distributions.

- Section 4.3 has mixtures of max-id distributions that lead to parametric families of copulas with wide dependence structure and closed-form cdfs.

- Copulas with closed-form cdfs and having the property of all bivariate margins in the same one-parameter family have limited dependence structures (Section 4.2).
- Extensions of LT families are used in Section 4.4 to extend families of copulas to include negative dependence.
- Mixtures of conditional distributions in Section 4.5 generalize the MVN family.
- Objects with given bivariate margins are proposed in Sections 4.7 and 4.8, although it is unknown under what conditions they are proper distributions.

4.1 Desirable properties of a multivariate model °

Some desirable properties for a parametric family of multivariate distributions are:

A. interpretability, which could mean something like a mixture, stochastic or latent variable representation;

B. the closure property under the taking of margins, in particular the bivariate margins belonging to the same parametric family (this is especially important if, in statistical modelling, one thinks first about appropriate univariate margins, then bivariate and sequentially to higher-order margins);

C. a flexible and wide range of dependence (with type of dependence structure depending on applications);

D. a closed-form representation of the cdf and density (a closed-form cdf is useful if the data are discrete and a continuous latent random vector is used), and if not closed-form, then a cdf and density that are computationally feasible to work with.

A stronger version of property B is that not only do margins belong to the same parametric (dimension-independent) family but also all parameters are associated with or are expressed in some marginal distribution. We refer to this as property B′.

Generally, it is not possible to satisfy all of these desirable properties, in which case one must decide on the relative importance of the properties and give up one or more of them. Actually another property that could be given up is the additive property of the probability measure associated with a multivariate distribution;

see Section 4.7 for multivariate objects that have given bivariate margins.

The properties are discussed below for some examples.

1. MVN distributions satisfy properties A, B and C but do not have closed-form cdfs. The latter is an inconvenience when a MVN latent vector is used such as in the multivariate probit model (see Sections 7.1.7 and 7.3). For high dimensions, good approximations can be used in place of time-consuming numerical integrations for evaluation of the MVN cdf.

2. The families of partially exchangeable copulas given in Section 4.2 do not satisfy property C, but satisfy the other properties.

3. The copulas in Section 4.5 satisfy property C and can lead to extreme value copulas, but have only partial closure under the taking of margins, and can be computationally harder to work with as the dimension increases.

4. The multivariate Poisson and other distributions in Section 4.6 satisfy property B but not property B$'$.

There is no known multivariate family that has all of the properties but the family of MVN distributions may be the closest. Because of its wide range of dependence, it is used as a latent distribution in multivariate models.

Using theory from this chapter, some parametric families of multivariate copulas with a wide range of dependence are given in Chapter 5. The starting point for construction of multivariate copulas are the MVN copula and bivariate families of copulas, and things like mixtures, latent variables and stochastic representations. The models in Sections 4.3, 4.5, 4.7 and 4.8 build on bivariate families of copulas. Within each section or construction method, it is stated which of the properties A to D are satisfied.

4.2 Laplace transforms and mixtures of powers °

To illustrate the main ideas in the use of Laplace transforms (LTs) and mixtures of powers of univariate cdfs or survival functions to construct multivariate distributions, we start with the univariate and bivariate cases. Let M be a univariate cdf of a positive rv (so that $M(0) = 0$) and let ϕ be the **Laplace transform** of M, i.e.,

$$\phi(s) = \int_0^\infty e^{-sw}\, dM(w), \quad s \geq 0.$$

Some properties of LTs are given in the Appendix. Note that throughout this book, unless otherwise stated, LTs correspond to positive rvs (no mass at 0). The reasoning is given in the next paragraph.

For an arbitrary univariate cdf F, there exists a unique cdf G such that

$$F(x) = \int_0^\infty G^\alpha(x)\, dM(\alpha) = \phi(-\log G(x)). \tag{4.1}$$

Rewriting (4.1) leads to $G = \exp\{-\phi^{-1}(F)\}$. (If M has positive mass π_0 at 0, then $\lim_{s\to\infty}\phi(s) = \pi_0$, and (4.1) cannot be solved for $G(x)$ when $F(x) < \pi_0$.) There is a similar relationship for survival functions:

$$\overline{F}(x) = \int_0^\infty \overline{H}^\alpha(x)\, dM(\alpha) = \phi(-\log \overline{H}(x)),$$

if $\overline{H} = \exp\{-\phi^{-1}(\overline{F})\}$.

Next consider the bivariate class $\mathcal{F}(F_1, F_2)$. For $j = 1, 2$, let $G_j = \exp\{-\phi^{-1}(F_j)\}$. Then the following is a cdf in $\mathcal{F}(F_1, F_2)$:

$$\int_0^\infty G_1^\alpha G_2^\alpha\, dM(\alpha) = \phi(-\log G_1 - \log G_2) = \phi(\phi^{-1}(F_1) + \phi^{-1}(F_2)). \tag{4.2}$$

The copula (which obtains from taking $U(0,1)$ cdfs for F_1, F_2) is

$$C(u_1, u_2) = \phi(\phi^{-1}(u_1) + \phi^{-1}(u_2)). \tag{4.3}$$

This rather simple form has been called an **Archimedean** copula. For (4.3) to have closed form, both ϕ and ϕ^{-1} need to have closed forms; there are plenty of examples satisfying this and other nice properties (see the Appendix). One could construct other bivariate distributions from general mixtures of the form

$$\int G_1(x_1, \alpha)\, G_2(x_2; \alpha)\, dM(\alpha),$$

but other than $G_j(\cdot; \alpha)$ in the form of powers, closed-form copulas are generally not obtained. These mixtures usually result in distributions with positive dependence; see Exercise 4.1 for some conditions.

Similarly, one could work with survival functions to get

$$\int_0^\infty \overline{H}_1^\alpha \overline{H}_2^\alpha dM(\alpha) = \phi(-\log \overline{H}_1 - \log \overline{H}_2) = \phi(\phi^{-1}(\overline{F}_1) + \phi^{-1}(\overline{F}_2)),$$

where $\overline{H}_j = \exp\{-\phi^{-1}(\overline{F}_j)\}$, $j = 1, 2$. This is an associated copula of (4.3); see Section 1.6. For multivariate extensions, we use the cdfs rather than survival functions, but keep in mind that there are the associated copulas.

With m univariate cdfs $F_1, \ldots, F_m$, a simple extension is to the multivariate cdf $F = \phi(\sum_{j=1}^m \phi^{-1}(F_j))$, with **Archimedean** copula

$$C(\mathbf{u}) = \phi\Big(\sum_{j=1}^{m} \phi^{-1}(u_j)\Big). \tag{4.4}$$

This multivariate copula is permutation-symmetric in the m arguments so that it is a distribution for exchangeable $U(0,1)$ rvs.

To get more general types of dependence, extensions which are written in the form of copulas are:

$$C(\mathbf{u}) = \int_0^\infty K(\exp\{-\alpha\phi^{-1}(u_1)\}, \ldots, \exp\{-\alpha\phi^{-1}(u_m)\})\, dM(\alpha), \tag{4.5}$$

$$C(\mathbf{u}) = \int_0^\infty \cdots \int_0^\infty K(G_1^{\alpha_1}, \ldots, G_m^{\alpha_m})\, M_m(d\alpha_1, \ldots, d\alpha_m), \tag{4.6}$$

where K is a multivariate copula, $G_j(u_j) = \exp\{-\phi_j^{-1}(u_j)\}$ and M_m is a multivariate distribution such that jth univariate margin has LT ϕ_j, $j = 1, \ldots, m$.

Because we do not have multivariate families of copulas with a wide range of dependence structure, (4.5) is not immediately useful. So with K in (4.6) being the independence copula, for (4.6) to be useful in getting copulas that are not permutation-symmetric, convenient choices of M_m must be made in order to simplify the multivariate integral and get a simple form for the copula.

Examples of these choices are given next. The result is that if LTs are chosen so that certain conditions are satisfied, then multivariate copulas can be obtained such that each bivariate margin has the form of (4.3) for some LT. However, the number of distinct LTs among the $m(m-1)/2$ bivariate margins is only $m-1$, so that the resulting dependence structure is one of partial exchangeability.

The construction makes use of Theorem A.1 in the Appendix. Let $\mathcal{L}_\infty^*$ be the class of infinitely differentiable increasing functions from $[0, \infty)$ onto $[0, \infty)$, with alternating signs for the derivatives; see (1.2) in Section 1.3.

The general multivariate result is notationally complex, so we indicate the pattern and conditions from the trivariate and 4-variate extensions of (4.3). The trivariate generalization of (4.3) is:

$$C(\mathbf{u}) = \psi(\psi^{-1} \circ \phi[\phi^{-1}(u_1) + \phi^{-1}(u_2)] + \psi^{-1}(u_3)), \tag{4.7}$$

where ψ, ϕ are LTs and $\nu = \psi^{-1} \circ \phi \in \mathcal{L}_\infty^*$. Note that (4.7) has (1,2) bivariate margin of the form (4.3) with LT ϕ, and (1,3) and (2,3) bivariate margins of the form (4.3) with LT ψ. Also (4.4) is a special case of (4.7) when $\psi = \phi$. The mixture representation for (4.7) that generalizes (4.2) is

$$C(\mathbf{u}) = \int_0^\infty \int_0^\infty G_1^\beta(u_1)\, G_2^\beta(u_2)\, dM_2(\beta;\alpha)\; G_3^\alpha(u_3)\, dM_1(\alpha), \quad (4.8)$$

where $G_1 = G_2 = \exp\{-\phi^{-1}\}$ and $G_3 = \exp\{-\psi^{-1}\}$, M_1 is the distribution corresponding to ψ, $M_2(\cdot;\alpha)$ being the distribution with LT χ_α, and χ_α is defined by $\chi_\alpha^{-1}(z) = \nu^{-1}(-\alpha^{-1}\log z)$.

The derivation is as follows. Equation (4.7) has the formal representation

$$\int_0^\infty G_{12}^\alpha(u_1, u_2)\, G_3^\alpha(u_3)\, dM_1(\alpha),$$

where M_1 and G_3 are as defined above, and

$$G_{12}(u_1, u_2) = \exp\{-\nu[\phi^{-1}(u_1) + \phi^{-1}(u_2)]\}.$$

In the bivariate case, the power of a cdf need not be a cdf so that it must be proved that G_{12}^α is a cdf. We show this by giving $F_\alpha = G_{12}^\alpha$ the mixture representation in (4.8). The univariate margins of F_α are $F_{j\alpha}(u_j) = \exp\{-\alpha\psi^{-1}(u_j)\}$, $j = 1, 2$. Hence $u_j = \psi(-\alpha^{-1}\log F_{j\alpha})$, $j = 1, 2$,

$$F_\alpha = \exp\{-\alpha\,\nu[\nu^{-1}(-\alpha^{-1}\log F_{1\alpha}) + \nu^{-1}(-\alpha^{-1}\log F_{2\alpha})]\}$$

and

$$\nu^{-1}(-\alpha^{-1}\log F_\alpha) = \nu^{-1}(-\alpha^{-1}\log F_{1\alpha}) + \nu^{-1}(-\alpha^{-1}\log F_{2\alpha}).$$

With χ_α defined as above,

$$F_\alpha = \chi_\alpha[\chi_\alpha^{-1}(F_{1\alpha}) + \chi_\alpha^{-1}(F_{2\alpha})] \quad (4.9)$$

Therefore $\chi_\alpha = e^{-\alpha\nu} = \chi^\alpha$, where $\chi = \chi_1$. From (4.2) and (4.3), (4.9) is a cdf for all $\alpha > 0$ if χ_α is a LT for all $\alpha > 0$ or if $\nu \in \mathcal{L}_\infty^*$ (using Theorem A.1). The representation (4.8) holds with $M_2(\cdot;\alpha)$ being the distribution with LT χ^α and, for all $\alpha > 0$ and $j = 1, 2$, $G_j = \exp\{-\chi_\alpha^{-1}(F_{j\alpha})\} = \exp\{-\nu^{-1}(-\alpha^{-1}\log F_{j\alpha})\} = \exp\{-\nu^{-1} \circ \psi^{-1}\} = \exp\{-\phi^{-1}\}$.

There are two generalizations or nestings of LTs for four dimensions. In higher dimensions, there are many possible nestings. At each level of nesting of LTs, ϕ_r within ϕ_s say, the condition $\phi_s^{-1} \circ \phi_r \in \mathcal{L}_\infty^*$ must be satisfied in order for the result to be a multivariate distribution. Let ψ, ϕ, ζ be LTs. The first LT representation is

$$C(\mathbf{u}) = \psi[\psi^{-1}\circ\phi(\phi^{-1}\circ\zeta(\zeta^{-1}(u_1)+\zeta^{-1}(u_2))+\phi^{-1}(u_3))+\psi^{-1}(u_4)], \tag{4.10}$$

where $\psi^{-1}\circ\phi$ and $\phi^{-1}\circ\zeta$ are in $\mathcal{L}_\infty^*$. A second distinct LT representation is

$$C(\mathbf{u}) = \psi(\psi^{-1}\circ\phi[\phi^{-1}(u_1)+\phi^{-1}(u_2)]+\psi^{-1}\circ\zeta[\zeta^{-1}(u_3)+\zeta^{-1}(u_4)]), \tag{4.11}$$

where $\psi^{-1}\circ\phi$ and $\psi^{-1}\circ\zeta$ are in $\mathcal{L}_\infty^*$. Note that all trivariate margins of (4.10) and (4.11) have form (4.7) and all bivariate margins of (4.10) and (4.11) have form (4.3). Clearly the idea of (4.10) and (4.11) generalizes to higher dimensions.

The mixture represention for (4.10) and (4.11) have the respective forms:

$$\int_0^\infty \int_0^\infty \int_0^\infty G_1^\gamma G_2^\gamma \, dM_3(\gamma;\beta)\, G_3^\beta \, dM_2(\beta;\alpha)\, G_4^\alpha \, dM_1(\alpha)$$

and

$$\int_0^\infty \int_0^\infty G_1^\beta G_2^\beta \, dM_2(\beta;\alpha) \int_0^\infty G_3^\gamma G_4^\gamma \, dM_3(\gamma;\alpha)\, dM_1(\alpha).$$

With reference to the properties in Section 4.1, (4.4), (4.7), (4.10) and (4.11) are interpretable from their mixture form, they are closed under the taking of margins, but have a limited type of positive dependence. They have closed-form cdfs if families for the LTs are chosen appropriately; see Sections 5.1 to 5.3.

*4.2.1 Dependence properties **

In this subsection, we analyse how the various bivariate margins of (4.7), (4.10), (4.11), etc., compare with each other in the concordance ordering. One key result is that all of these constructions produce positively dependent random vectors only. (Generalizations to achieve negative dependence are given in Section 4.4.) Other positive dependence results for the various constructions are obtained, and some results on the multivariate concordance and PFD orderings for (4.7), (4.10), (4.11) and their generalizations are given.

Theorem 4.1 *Let* $C_i(u_1, u_2) = \phi_i(\phi_i^{-1}(u_1) + \phi_i^{-1}(u_2))$, *where* ϕ_i *is a LT,* $i = 1, 2$. *Then* $C_1 \prec_c C_2$ *if and only if* $\omega = \phi_2^{-1} \circ \phi_1$ *is superadditive* ($\omega(x + y) \geq \omega(x) + \omega(y)$ *for all* $x, y \geq 0$). *Similarly,* *for the multivariate extensions,* $C_{im}(\mathbf{u}) = \phi_i(\sum_{j=1}^{m} \phi_i^{-1}(u_j))$, $i = 1, 2$, $C_{1m} \prec_{cL} C_{2m}$ *if and only if* ω *is superadditive. Since* $\omega(0) = 0$, *sufficient conditions for* ω *to be superadditive are* ω *convex and* ω *star-shaped with respect to the origin* ($\omega(x)/x$ *increasing in* x).

Proof. Let $u_j = \phi_1(x_j)$, $j = 1, 2$. Then

$$
\begin{aligned}
&C_1 \prec_c C_2 \\
&\quad \Leftrightarrow\ \omega(\phi_1^{-1}(u_1) + \phi_1^{-1}(u_2)) \ge \phi_2^{-1}(u_1) + \phi_2^{-1}(u_2)\ \forall 0 \le u_1, u_2 \le 1 \\
&\quad \Leftrightarrow\ \omega(x_1 + x_2) \ge \omega(x_1) + \omega(x_2)\ \forall x_1, x_2 \ge 0.
\end{aligned}
$$

The sufficient conditions for superadditivity are left as an exercise. □

Corollary 4.2 *Let* $C_i(u_1, u_2) = \phi_i(\phi_i^{-1}(u_1) + \phi_i^{-1}(u_2))$, *where* ϕ_i *is a LT,* $i = 1, 2$. *Suppose* $\nu = \phi_1^{-1} \circ \phi_2 \in \mathcal{L}_\infty^*$, *then* $C_1 \prec_c C_2$.

Proof. ν has non-negative first derivative and non-positive second derivative, and satisfies $\nu(0) = 0$. Therefore ν is concave and $\nu^{-1} = \phi_2^{-1} \circ \phi_1$ is convex, and the preceding theorem applies. □

As a result of this corollary, the trivariate copula in (4.7) has a (1,2) bivariate margin copula which is more concordant than the (1,3) and (2,3) bivariate margin copulas (which are identical). However, there are applications, such as the construction of a second-order Markov chain time series (see Section 8.1), in which one would like the two bivariate margins that are identical to be more concordant than the third margin. Trivariate copulas with this property are constructed in Section 4.3. Similarly, for (4.10), (4.11) and their multivariate extensions, bivariate copulas, associated with LTs that are more nested, are larger in concordance than those that are less nested. For example, for (4.11), the (1,2) and (3,4) bivariate margins are more concordant than the remaining four bivariate margins (but there need not be any concordance ordering between these two margins).

The types of dependence that are possible from (4.7), (4.10), (4.11) and their generalizations are similar to those from hierarchical or random effects normal models. Analogies to (4.7), (4.10), (4.11) are respectively:

(i) $Y_1 = \mu + \xi_1 + \epsilon_1$, $Y_2 = \mu + \xi_1 + \epsilon_2$, $Y_3 = \mu + \xi_2 + \epsilon_3$, with $\mu, \xi_1, \xi_2, \epsilon_1, \epsilon_2, \epsilon_3$ independent zero mean normal rvs with respective variances $\alpha, \beta, \beta, 1-\alpha-\beta, 1-\alpha-\beta, 1-\alpha-\beta$. The correlation matrix for (Y_1, Y_2, Y_3) is $\begin{bmatrix} 1 & \alpha+\beta & \alpha \\ \alpha+\beta & 1 & \alpha \\ \alpha & \alpha & 1 \end{bmatrix}$.

(ii) $Y_1 = \mu + \xi_1 + \lambda_1 + \epsilon_1$, $Y_2 = \mu + \xi_1 + \lambda_1 + \epsilon_2$, $Y_3 = \mu + \xi_1 + \lambda_2 + \epsilon_3$, $Y_4 = \mu + \xi_2 + \lambda_3 + \epsilon_4$, with μ, ξ_i, λ_i, ϵ_i being independent zero mean normal rvs, μ with variance α, ξ_i with variance β, λ_i with variance γ and ϵ_i with variance $1-\alpha-\beta-\gamma$. The correlation matrix for (Y_1, Y_2, Y_3, Y_4) is

$$\begin{bmatrix} 1 & \alpha+\beta+\gamma & \alpha+\beta & \alpha \\ \alpha+\beta+\gamma & 1 & \alpha+\beta & \alpha \\ \alpha+\beta & \alpha+\beta & 1 & \alpha \\ \alpha & \alpha & \alpha & 1 \end{bmatrix}.$$

(iii) $Y_1 = \mu + \xi_1 + \epsilon_1$, $Y_2 = \mu + \xi_1 + \epsilon_2$, $Y_3 = \mu + \xi_2 + \epsilon_3$, $Y_4 = \mu + \xi_2 + \epsilon_4$, with μ, ξ_i, ϵ_i being independent zero mean normal rvs, μ with variance α, ξ_1 with variance β_1, ξ_2 with variance β_2 and ϵ_i with variance $1-\alpha-\beta_j$ ($j=1$ for $i=1,2$ and $j=2$ for $i=3,4$). The correlation matrix for (Y_1, Y_2, Y_3, Y_4) is

$$\begin{bmatrix} 1 & \alpha+\beta_1 & \alpha & \alpha \\ \alpha+\beta_1 & 1 & \alpha & \alpha \\ \alpha & \alpha & 1 & \alpha+\beta_2 \\ \alpha & \alpha & \alpha+\beta_2 & 1 \end{bmatrix}.$$

Next we give some results on association, Kendall's tau, positive dependence and dependence orderings.

Theorem 4.3 *For the copula (4.3), Kendall's tau can be written as the one-dimensional integral:*

$$\tau = 4\int_0^1 \frac{\phi^{-1}(t)}{(\phi^{-1})'(t)}dt + 1 = 1 - 4\int_0^\infty s[\phi'(s)]^2 ds.$$

Proof. This proof is a modification of that in Genest and MacKay (1986), making use of the present notation. From Section 2.1.9, Kendall's tau is

$$\tau = 4\int_0^1\int_0^1 C(u_1, u_2)\, dC(u_1, u_2) - 1.$$

The double integral (without the factor of 4) becomes

$$\int_0^1\int_0^1 \phi(\phi^{-1}(u_1) + \phi^{-1}(u_2)) \cdot \phi''(\phi^{-1}(u_1) + \phi^{-1}(u_2))$$

$$\cdot[\phi' \circ \phi^{-1}(u_1)]^{-1}[\phi' \circ \phi^{-1}(u_2)]^{-1} du_1 du_2 \qquad (4.12)$$

Make the transformation $z = \phi(\phi^{-1}(u_1) + \phi^{-1}(u_2))$, $t = u_1$ with Jacobian $\frac{\partial(z,t)}{\partial(u_1,u_2)}$ equal to $\phi'(\phi^{-1}(u_1) + \phi^{-1}(u_2)) \cdot [\phi' \circ \phi^{-1}(u_2)]^{-1}$. Since $z \le t$ from the Fréchet upper bound, (4.12) becomes

$$\int_0^1 \int_z^1 z\,\phi'' \circ \phi^{-1}(z)[\phi' \circ \phi^{-1}(z)]^{-1}[\phi' \circ \phi^{-1}(t)]^{-1}\,dt\,dz$$
$$= \int_0^1 z\,\phi'' \circ \phi^{-1}(z)[\phi' \circ \phi^{-1}(z)]^{-1}[-\phi^{-1}(z)]\,dz$$
$$= \int_0^\infty s\,\phi(s) \cdot \phi''(s)\,ds.$$

With integration by parts, this becomes

$$s\,\phi(s) \cdot \phi'(s)\Big|_0^\infty - \int_0^\infty [\phi(s) + s\,\phi'(s)]\phi'(s)\,ds = \tfrac{1}{2} - \int_0^\infty s\,[\phi'(s)]^2 ds.$$

The conclusion of the theorem follows. □

Theorem 4.4 *The copula in (4.3) has TP_2 density. Hence it is also PQD.*

Proof. This follows easily from the mixture representation. With $U(0,1)$ margins in (4.2), G_1, G_2 are differentiable, say with respective densities g_1, g_2. Hence $C(u_1, u_2) = \int_0^\infty G_1^\alpha(u_1)G_2^\alpha(u_2)\,dM(\alpha)$ has density

$$c(u_1, u_2) = \alpha^2 g_1(u_1)g_2(u_2) \int_0^\infty G_1^{\alpha-1}(u_1)G_2^{\alpha-1}(u_2)\,dM(\alpha).$$

In the integrand, $G_1^{\alpha-1}(u_1)$ and $G_2^{\alpha-1}(u_2)$ are TP_2, so the integral is TP_2 from the total positivity results in Karlin (1968). ($D(x,z) = \int A(x,y)B(y,z)\,d\sigma(y)$ is TP_2 in x, z if A is TP_2 in x, y and B is TP_2 in y, z and σ is a sigma-finite measure.) The property of PQD follows from Theorem 2.3. □

Theorem 4.5 *The copula in (4.6) is associated if K and M_m are cdfs of associated rvs.*

Proof. $K(\mathbf{x})$ associated implies $K(G_1^{\alpha_1}, \ldots, G_m^{\alpha_m})$ is associated for all $(\alpha_1, \ldots, \alpha_m) \in (0,\infty)^m$, because association is invariant to (strictly) increasing transforms on rvs. Let $\mathbf{U} \sim K$ and $\mathbf{A} \sim M_m$, with $\mathbf{U}, \mathbf{A}$ independent. Then a stochastic representation of a random vector with the distribution in (4.6) is

$$(X_1, \ldots, X_m) = (G_1^{-1}(U_1^{1/A_1}), \ldots, G_m^{-1}(U_m^{1/A_m})).$$

We show that $\mathbf{X}$ is associated using the usual conditioning argument. Let $h_1(\mathbf{X}), h_2(\mathbf{X})$ be increasing functions in $\mathbf{X}$. Assuming the covariance exists, then

$$\mathrm{Cov}\,(h_1(\mathbf{X}), h_2(\mathbf{X})) = \mathrm{E}\left[\mathrm{Cov}\,(h_1(\mathbf{X}), h_2(\mathbf{X}) \mid \mathbf{A})\right]$$
$$+ \mathrm{Cov}\,(\mathrm{E}\,[h_1(\mathbf{X}) \mid \mathbf{A}], \mathrm{E}\,[h_1(\mathbf{X}) \mid \mathbf{A}]).$$

The first term is non-negative given each $\mathbf{A}$, because of the first statement in this proof. Since $G_j^{-1}(U_j^{1/A_j})$ is increasing in A_j for each j and h_1, h_2 are increasing, $\mathrm{E}[h_i(\mathbf{X})|\mathbf{A}]$, $i = 1, 2$, are increasing in $\mathbf{A}$. Therefore the second term is non-negative because $\mathbf{A}$ is associated. □

Theorem 4.6 *Suppose ψ, ϕ are LTs such that $\psi^{-1} \circ \phi \in \mathcal{L}_\infty^*$. Let $C_1(\mathbf{u}) = \psi(\sum_{j=1}^m \psi^{-1}(u_j))$ and $C_2(\mathbf{u}) = \phi(\sum_{j=1}^m \phi^{-1}(u_j))$ be as defined in (4.4). Then $C_1 \prec_{\text{pfd}} C_2$. If $\psi(s) = e^{-s}$, so that $C_1(\mathbf{u}) = C_I(\mathbf{u}) = \prod_{j=1}^m u_j$, then $C_I \prec_{\text{pfd}} C_2$ (without needing further conditions).*

Proof. Let $G_1(u) = \exp\{-\psi^{-1}(u)\}$, $G_2(u) = \exp\{-\phi^{-1}(u)\}$, $0 \le u \le 1$, with respective densities g_1, g_2. Let M_ψ, M_ϕ be the distributions with respective LTs ψ, ϕ, and let $M_\nu(\cdot;\alpha)$ be the distribution with LT $\exp\{-\alpha\nu\}$, where $\nu = \psi^{-1} \circ \phi$. Then it is straightforward to verify that

$$C_2(\mathbf{u}) = \int_0^\infty \int_0^\infty \prod_j G_2^\beta(u_j)\, dM_\nu(\beta;\alpha)\, dM_\psi(\alpha)$$

and

$$C_1(\mathbf{u}) = \int_0^\infty \prod_j G_1^\alpha(u_j)\, dM_\psi(\alpha),$$

with $G_1^\alpha(u) = \int_0^\infty G_2^\beta(u)\, dM_\nu(\beta;\alpha)$ (see Theorem A.3). The densities c_1, c_2 of C_1, C_2 are obtained by replacing G_2^β by $\beta G_2^{\beta-1} g_2$ in the integrals defining C_2 and G_1^α. Let h be a bounded function on [0,1], with an extra non-negativity constraint if m is odd; let $\mathbf{U}$ have density c_1 and $\mathbf{U}'$ have density c_2. Then

$$\mathrm{E}\left[\prod_{i=1}^m h(U_i)\right] = \int_0^\infty \left[\int_0^\infty h^*(\beta)\, dM_\nu(\beta;\alpha)\right]^m dM_\psi(\alpha) \qquad (4.13)$$

where $h^*(\beta) = \beta \int_0^1 G_2^{\beta-1}(u)\, g_2(u)\, h(u)\, du$. From Theorem 2.18 and Example 2.11, the right-hand side of (4.13) is dominated by

$$\int_0^\infty \int_0^\infty [h^*(\beta)]^m dM_\nu(\beta;\alpha)\, dM_\psi(\alpha) = \mathrm{E}\left[\prod_{i=1}^m h(U_i')\right].$$

The case of h unbounded and integrable can be handled by taking a limit. Hence $\mathbf{U} \prec_{\text{pfd}} \mathbf{U}'$ or $C_1 \prec_{\text{pfd}} C_2$.

Finally, let $\psi(s) = e^{-s}$. Then the above condition becomes $-\log\phi \in \mathcal{L}_\infty^*$. However, the condition is actually not needed. Let $\mathbf{U} \sim C_2$. It suffices to prove that $\mathrm{E}[\prod_{i=1}^m h(U_i)] \ge \{\mathrm{E}[h(U_1)]\}^m$ for all bounded h (with an extra non-negativity constraint if m is odd), or

$$\mathrm{E}\left[\prod_{i=1}^{m} h(U_i)\right] = \int_0^\infty k^m(\alpha)\, dM_\phi(\alpha) \geq \left[\int_0^\infty k(\alpha)\, dM_\phi(\alpha)\right]^m,$$

where $k(\alpha) = \alpha \int_0^1 G_2^{\alpha-1}(u) g_2(u) h(u)\, du$. This follows from Example 2.11 and inequality (2.33) (with independent rvs on the left-hand side of (2.33)).

□

Theorem 4.7 *Suppose ψ, ϕ are LTs such that $\psi^{-1} \circ \phi \in \mathcal{L}_\infty^*$. Let C_1, C_2 be as defined in the preceding theorem. Then $C_1 \prec_c C_2$.*

Proof. Let $\mathbf{U} \sim C_1$, $\mathbf{U}' \sim C_2$. To establish the concordance ordering, it is necessary to show that

$$\Pr(U_j \leq a_j, j = 1, \ldots, m) \leq \Pr(U_j' \leq a_j, j = 1, \ldots, m) \quad (4.14)$$

and

$$\Pr(U_j > a_j, j = 1, \ldots, m) \leq \Pr(U_j' > a_j, j = 1, \ldots, m) \quad (4.15)$$

for all $\mathbf{a} \in (0,1)^m$. From the representations in the preceding proof, the left-hand sides of (4.14) and (4.15) have the form

$$\int_0^\infty \left[\prod_{j=1}^{m} \int_0^\infty k_j(\beta) dM_\nu(\beta; \alpha)\right] dM_\psi(\alpha) \quad (4.16)$$

and the right-hand sides have the form

$$\int_0^\infty \int_0^\infty \left[\prod_{j=1}^{m} k_j(\beta)\right] dM_\nu(\beta; \alpha)\, dM_\psi(\alpha), \quad (4.17)$$

where $k_j(\beta) = G_2^\beta(a_j)$ for (4.14) and $k_j(\beta) = 1 - G_2^\beta(a_j)$ for (4.15). The $k_j(\beta)$ are decreasing in $\beta \geq 0$ in the former case and increasing in the latter case. A (bounded) decreasing function in $[0, \infty)$ can be taken as a limit of finite sums of the form $\sum_i c_i I_{[0,z_i]}$ for positive constants c_i, z_i; similarly, an increasing function can be taken as a limit of finite sums of the form $\sum_i c_i I_{(z_i,\infty)}$ for positive constants c_i, z_i. Since the quantities in (4.16) and (4.17) are linear in each $k_j(\beta)$, to show that the quantity in (4.16) is less than or equal to that in (4.17), it suffices to prove the inequality with $k_j(\beta)$ replaced by $I_{[0,y_j]}(\beta)$, $j = 1, \ldots, m$, or by $I_{(y_j,\infty)}(\beta)$, $j = 1, \ldots, m$, where y_j are positive constants. Fix α and let $B_1, \ldots, B_m$ be iid with distribution $M_\nu(\cdot; \alpha)$. Then (4.14) and (4.15) follow since $\Pr(B_j \leq y_j, j = 1, \ldots, m) \leq \Pr(B_1 \leq \min_j y_j) = \min_j \Pr(B_j \leq y_j)$ and

$\Pr(B_j > y_j, j = 1, \ldots, m) \le \Pr(B_1 > \max_j y_j) = \min_j \Pr(B_j > y_j)$ by the Fréchet upper bound inequality. □

Theorem 4.8 *Suppose ψ, ϕ_1, ϕ_2 are LTs such that (a) $\nu_i = \psi^{-1} \circ \phi_i \in \mathcal{L}_\infty^*$, $i = 1, 2$, and (b) $\phi_1^{-1} \circ \phi_2 \in \mathcal{L}_\infty^*$. Let C_i be as defined in (4.7) with $\phi = \phi_i$, $i = 1, 2$. Then $C_1 \prec_c C_2$.*

Proof. For $\xi = \psi, \phi_1, \phi_2$, let $G_\xi = \exp\{-\xi^{-1}\}$ and let M_ξ denote the cdf with LT ξ. Let $\nu = \phi_1^{-1} \circ \phi_2$ and let $M_\nu(\cdot; \beta)$ denote the distribution with LT $\exp\{-\beta\nu\}$. For $\mu = \nu_1, \nu_2$, let $M_\mu(\cdot; \alpha)$ denote the distribution with LT $\exp\{-\alpha\mu\}$. Then using Theorem A.3, representations for C_1, C_2 are:

$$\int_0^\infty \int_0^\infty G_{\phi_1}^\beta(u_1)\, G_{\phi_1}^\beta(u_2)\, dM_{\nu_1}(\beta; \alpha)\; G_\psi^\alpha(u_3)\, dM_\psi(\alpha),$$

$$\int_0^\infty \int_0^\infty G_{\phi_2}^\gamma(u_1)\, G_{\phi_2}^\gamma(u_2)\, dM_{\nu_2}(\gamma; \alpha)\; G_\psi^\alpha(u_3)\, dM_\psi(\alpha),$$

with $G_\psi^\alpha(u) = \int_0^\infty G_{\phi_2}^\gamma dM_{\nu_2}(\gamma; \alpha)$, $G_{\phi_1}^\beta(u) = \int_0^\infty G_{\phi_2}^\gamma dM_\nu(\gamma; \beta)$ and

$$\int_0^\infty \int_0^\infty h^\gamma\, dM_\nu(\gamma; \beta)\, dM_{\nu_1}(\beta; \alpha) = \int_0^\infty h^\gamma\, dM_{\nu_2}(\gamma; \alpha)$$

for all positive constants h.

Let $\mathbf{U} \sim C_1$, $\mathbf{U}' \sim C_2$. The concordance ordering is proved in a similar way to the preceding theorem; it is necessary to show

$$\Pr(U_j \le a_j, j = 1, 2, 3) \le \Pr(U_j' \le a_j, j = 1, 2, 3) \tag{4.18}$$

and

$$\Pr(U_j > a_j, j = 1, 2, 3) \le \Pr(U_j' > a_j, j = 1, 2, 3) \tag{4.19}$$

for all $\mathbf{a} \in (0, 1)^3$. From the representations given above, the left-hand sides of (4.18) and (4.19) have the form

$$\int_0^\infty \left[\int_0^\infty \left(\int k_1(\gamma) dM_\nu(\gamma; \beta)\right)\left(\int k_2(\gamma) dM_\nu(\gamma; \beta)\right) dM_{\nu_1}(\beta; \alpha)\right] \cdot \left(\int_0^\infty k_3(\gamma)\, dM_{\nu_2}(\gamma; \alpha)\right) dM_\psi(\alpha),$$

and the right-hand sides have the form

$$\int_0^\infty \left[\int_0^\infty \left(\int_0^\infty k_1(\gamma)\, k_2(\gamma)\, dM_\nu(\gamma; \beta)\right) dM_{\nu_1}(\beta; \alpha)\right] \cdot \left(\int_0^\infty k_3(\gamma)\, dM_{\nu_2}(\gamma; \alpha)\right) dM_\psi(\alpha),$$

where $k_j(\gamma) = G^{\gamma}_{\phi_2}(a_j)$ for (4.18) and $k_j(\gamma) = 1 - G^{\gamma}_{\phi_2}(a_j)$ for (4.19). The $k_j(\gamma)$ are decreasing in $\gamma \geq 0$ in the former case and increasing in the latter case. Inequalities (4.18) and (4.19) now follow from

$$\left(\int_0^\infty k_1(\gamma)\, dM_\nu(\gamma;\beta)\right)\left(\int_0^\infty k_2(\gamma)\, dM_\nu(\gamma;\beta)\right)$$
$$\leq \int_0^\infty k_1(\gamma)\, k_2(\gamma)\, dM_\nu(\gamma;\beta)$$

for all β; this last inequality comes from $(Z_1, Z_2) \prec_c (Z_1, Z_1)$, where Z_1, Z_2 are iid with distribution $M_\nu(\cdot;\beta)$. □

Theorem 4.9 *Suppose ψ_1, ψ_2, ϕ are LTs such that (a) $\omega_i = \psi_i^{-1} \circ \phi \in \mathcal{L}^*_\infty$, $i = 1, 2$, and (b) $\psi_1^{-1} \circ \psi_2 \in \mathcal{L}^*_\infty$. Let C_i be as defined in (4.7) with $\psi = \psi_i$, $i = 1, 2$. Then $C_1 \prec_c C_2$.*

Proof. The proof follows the methods of the preceding theorem. Using the same notation, a result that is needed for the upper orthant probabilities is that

$$h_1(\gamma) = 1 - G^{\gamma}_{\psi_2}(a_1) - G^{\gamma}_{\psi_2}(a_2) + \exp\{-\gamma\omega_2(\phi^{-1}(a_1) + \phi^{-1}(a_2))\} \tag{4.20}$$

is increasing in γ. A proof of this that generalizes to higher dimensions is the following.

One can write $h_1(\gamma) = \int_0^\infty (1 - G^{\beta}_{\phi}(a_1))(1 - G^{\beta}_{\phi}(a_2))\, dM_{\omega_2}(\beta;\gamma)$. Since $(1-G^{\beta}_{\phi}(a_1))(1-G^{\beta}_{\phi}(a_2))$ is increasing in $\beta > 0$ and $M_{\omega_2}(\beta;\gamma)$ is stochastically increasing in $\gamma > 0$ (Theorem A.3), then $h_1(\gamma)$ is increasing in $\gamma > 0$. □

Theorem 4.10 *Suppose $\psi_1, \psi_2, \phi, \zeta$ are LTs such that $\omega_i = \psi_i^{-1} \circ \phi \in \mathcal{L}^*_\infty$ $i = 1, 2$, $\nu = \psi_1^{-1} \circ \psi_2 \in \mathcal{L}^*_\infty$ and $\phi^{-1} \circ \zeta \in \mathcal{L}^*_\infty$. Let C_i be as defined in (4.10) with $\psi = \psi_i$, $i = 1, 2$. Then $C_1 \prec_c C_2$.*

Proof. We use similar notation to the preceding two proofs. Representations for C_j are

$$C_1(\mathbf{u}) = \int_0^\infty \left(\int_0^\infty \int_0^\infty \int_0^\infty G^{\gamma}_{\zeta}(u_1) G^{\gamma}_{\zeta}(u_2)\, dM_{\phi^{-1}\circ\zeta}(\gamma;\beta)\, G^{\beta}_{\phi}(u_3)\right.$$
$$\left. dM_{\omega_2}(\beta;\eta)\, dM_\nu(\eta;\alpha)\right) \cdot \left(\int_0^\infty G^{\eta}_{\psi_2}(u_4)\, dM_\nu(\eta;\alpha)\right) dM_{\psi_1}(\alpha),$$
$$C_2(\mathbf{u}) = \int_0^\infty \int_0^\infty \left(\int_0^\infty \int_0^\infty G^{\gamma}_{\zeta}(u_1)\, G^{\gamma}_{\zeta}(u_2)\, dM_{\phi^{-1}\circ\zeta}(\gamma;\beta)\, G^{\beta}_{\phi}(u_3)\right.$$
$$\left. dM_{\omega_2}(\beta;\eta)\right) \cdot G^{\eta}_{\psi_2}(u_4)\, dM_\nu(\eta;\alpha)\, dM_{\psi_1}(\alpha).$$

Let $\mathbf{U} \sim C_1$, $\mathbf{U}' \sim C_2$. The harder step for establishing the concordance ordering is the inequality for the upper orthant probabilities. These are:

$$\Pr(U_j > a_j, j = 1,2,3,4) = \qquad (4.21)$$
$$\int_0^\infty \left(\int_0^\infty h_1(\eta) dM_\nu(\eta;\alpha)\right)\left(\int_0^\infty h_2(\eta) dM_\nu(\eta;\alpha)\right) dM_{\psi_1}(\alpha),$$

$$\Pr(U_j' > a_j, j = 1,2,3,4) = \qquad (4.22)$$
$$\int_0^\infty \left(\int_0^\infty h_1(\eta) h_2(\eta) dM_\nu(\eta;\alpha)\right) dM_{\psi_1}(\alpha),$$

where $h_2(\eta) = 1 - G_{\psi_2}^\eta(a_4)$ and

$$h_1(\eta) = \int_0^\infty \int_0^\infty \left(1 - G_\zeta^\gamma(a_1)\right)\left(1 - G_\zeta^\gamma(a_2)\right) dM_{\phi^{-1}\circ\zeta}(\gamma;\beta)$$
$$\cdot\left(1 - G_\phi^\beta(a_3)\right) dM_{\omega_2}(\beta;\eta).$$

Similar to the proof that (4.20) is increasing, $\int_0^\infty (1 - G_\zeta^\gamma(a_1))(1 - G_\zeta^\gamma(a_2))\, dM_{\phi^{-1}\circ\zeta}(\gamma;\beta)$ is increasing in β. Also $1 - G_\phi^\beta(a_3)$ is increasing in β, so that $h_1(\eta)$ is increasing in $\eta > 0$. Trivially, $h_2(\eta)$ is increasing in $\eta > 0$. Therefore, by Lemma 2.1,

$$\int_0^\infty h_1(\eta)\, dM_\nu(\eta;\alpha) \int_0^\infty h_2(\eta)\, dM_\nu(\eta;\alpha)$$
$$\leq \int_0^\infty h_1(\eta)\, h_2(\eta)\, dM_\nu(\eta;\alpha),$$

and (4.21) is less than or equal to (4.22) for all $\mathbf{a}$. □

The ideas in the above theorems generalize for multivariate extensions of (4.7), (4.10) and (4.11) (even though the notation is messy for the general case).

4.2.2 Frailty and proportional hazards

If survival functions $\overline{F}_j$ are substituted into the arguments in (4.4), then one gets

$$C(\overline{F}_1(x_1), \ldots, \overline{F}_m(x_m)) = \int_0^\infty \prod_{j=1}^m H^\alpha(\overline{F}_j(x_j))\, dM(\alpha).$$

Similarly, survival functions can be substituted into (4.7), (4.10), (4.11) and their extensions. These models have been used for multivariate survival data in a familial or cluster setting; the parameter α is interpreted as a frailty parameter. These models also have a proportional hazards interpretation since $H^\alpha(\overline{F}_i(x_i)) = \exp\{-\alpha\psi^{-1}(\overline{F}_i(x_i))\}$ is a family with cumulative hazard proportional to $\Lambda(x_i) = \psi^{-1}(\overline{F}_i(x_i))$, where ψ is the LT of M; the proportionality constant α of the proportional hazards is random with distribution M. If there are $m = 2$ subjects in a cluster with survival times T_1, T_2, then $\Pr(T_1 > t_1, T_2 > t_2)$ becomes

$$\int_0^\infty e^{-\alpha[\Lambda(t_1)+\Lambda(t_2)]}\, dM(\alpha) = \psi(\Lambda(t_1) + \Lambda(t_2)).$$

4.3 Mixtures of max-id distributions

Let M be the cdf of a positive rv and let its LT be ψ. Another extension of (4.2) and (4.4) is:

$$F = \int_0^\infty H^\alpha dM(\alpha) = \psi(-\log H) \tag{4.23}$$

where H is a max-id m-variate distribution. (See Section 2.1.8 for max-id.) Section 4.3.1 consists of conditions for max-id so that we know about possible choices of H in (4.23). Section 4.3.2 is devoted to dependence properties of (4.23). Some special cases are given next before these subsections.

We look at cases of (4.23) that can lead to parametric families of multivariate distributions or copulas with closed-form cdfs, flexible dependence structure and partial closure under the taking of margins. Specific parametric families are given in Section 5.5.

Let K_{ij}, $1 \le i < j \le m$, be bivariate copulas that are max-id. Let $H_1, \ldots, H_m$ be univariate cdfs. Consider the mixture:

$$\int_0^\infty \prod_{1\le i<j\le m} K_{ij}^\alpha(H_i, H_j) \prod_{i=1}^m H_i^{\nu_i\alpha} dM(\alpha)$$

$$= \psi\Big(-\sum_{1\le i<j\le m} \log K_{ij}(H_i, H_j) - \sum_{i=1}^m \nu_i \log H_i\Big), \tag{4.24}$$

where usually the ν_i are non-negative, though they can be negative if some of the copulas K_{ij} correspond to independence. This special construction builds on bivariate copulas, of which there are several parametric families with nice properties in Section 5.1. The univariate margins of (4.24) are

$$F_i = \psi\big(-(\nu_i + m - 1)\log H_i\big).$$

Hence (4.24) is a copula, if $H_i(u_i)$ is chosen to be $\exp\{-p_i\psi^{-1}(u_i)\}$ with $p_i = (\nu_i + m - 1)^{-1}$, $i = 1, \ldots, m$. With these substitutions, the copula is

$$C(\mathbf{u}) = \psi\Big(-\sum_{i<j} \log K_{ij}(e^{-p_i\psi^{-1}(u_i)}, e^{-p_j\psi^{-1}(u_j)}) + \sum_{i=1}^{m} \nu_i p_i \psi^{-1}(u_i)\Big). \tag{4.25}$$

An interpretation is that the LT ψ leads to a minimal level of (pairwise) dependence, the copulas K_{ij} add some individual pairwise dependence beyond the global dependence, and the parameters ν_i lead to bivariate and multivariate asymmetry (the asymmetries are represented through $\nu_i/(\nu_i+\nu_j)$, $i \neq j$). Also the parameters ν_i are included in order that the family (4.25) is closed under margins. For example, if $H_m \to 1$ in (4.24), then (4.24) becomes

$$\int_0^\infty \prod_{1\le i<j\le m-1} K_{ij}^\alpha(H_i, H_j) \prod_{i=1}^{m-1} H_i^{(\nu_i+1)\alpha} dM(\alpha)$$

$$= \psi\Big(-\sum_{1\le i<j\le m-1} \log K_{ij}(H_i, H_j) - \sum_{i=1}^{m-1} (\nu_i + 1)\log H_i\Big),$$

and the resulting marginal copula is

$$C_{1\ldots m-1}(\mathbf{u}) = \psi\Big(-\sum_{1\le i<j\le m-1} \log K_{ij}(e^{-p_i\psi^{-1}(u_i)}, e^{-p_j\psi^{-1}(u_j)}) + \sum_{i=1}^{m-1} (\nu_i + 1) p_i \psi^{-1}(u_i)\Big).$$

(Hence the 'parameters' K_{ij} remain the same but the parameters ν_j change with taking margins; this can be shown notationally by $\nu_i^{(m-1)} = \nu_i^{(m)} + 1$ and $\nu_i^{(2)} = \nu_i^{(m)} + m - 2$.) The (i,j) bivariate marginal copula of (4.25) is

$$C_{ij}(u_i, u_j) = \psi\big(-\log K_{ij}(e^{-p_i\psi^{-1}(u_i)}, e^{-p_j\psi^{-1}(u_j)}) + (\nu_i + m - 2)p_i\psi^{-1}(u_i) + (\nu_j + m - 2)p_j\psi^{-1}(u_j)\big). \tag{4.26}$$

The copula (4.26) is more concordant than

$$C_\psi(u_i, u_j) = \psi(\psi^{-1}(u_i) + \psi^{-1}(u_j)), \tag{4.27}$$

and it increases in concordance as K_{ij} increases in concordance; this explains the above interpretation for ψ and K_{ij}. These results are proved in Section 4.3.2.

Interesting and useful special cases of (4.23) and (4.24) are the following.

1. Let $m = 3$, $K_{13}(x,y) = xy$, $\nu_1 = \nu_2 = \nu_3 = 0$, $K_{12}(u,v) = K_{23}(u,v) = K(u,v)$, where K is symmetric in u, v. Then (4.24) becomes
$$\int_0^\infty K^\alpha(H_1,H_2)K^\alpha(H_3,H_2)\, H_1^\alpha H_3^\alpha \, dM(\alpha), \tag{4.28}$$
with copula
$$C(\mathbf{u}) = \psi(-\log K(e^{-0.5\psi^{-1}(u_1)}, e^{-0.5\psi^{-1}(u_2)})$$
$$-\log K(e^{-0.5\psi^{-1}(u_3)}, e^{-0.5\psi^{-1}(u_2)}) + \tfrac{1}{2}\psi^{-1}(u_1) + \tfrac{1}{2}\psi^{-1}(u_3)); \tag{4.29}$$
if H_1, H_2, H_3 are chosen appropriately so that the univariate margins of (4.28) are all the same, then the (1,2) and (2,3) bivariate margins of (4.28) are the same and are more concordant than the (1,3) margin. Hence this model would be appropriate for generating a second-order stationary Markov chain.

2. Let $m = 3$, $K_{13}(x,y) = xy$, $\nu_1 = \nu_3 = -1$, $\nu_2 = 0$, $K_{12}(u,v) = K_{23}(v,u) = K(u,v)$. Then (4.24) becomes
$$\int_0^\infty K^\alpha(H_1,H_2)K^\alpha(H_3,H_2)\, dM(\alpha),$$
with copula
$$C(\mathbf{u}) = \psi(-\log K(e^{-\psi^{-1}(u_1)}, e^{-0.5\psi^{-1}(u_2)})$$
$$-\log K(e^{-\psi^{-1}(u_3)}, e^{-0.5\psi^{-1}(u_2)})). \tag{4.30}$$
The (1,2) and (3,2) bivariate margins of (4.30) are the same and are more concordant than the (1,3) margin; with $j = 1, 3$, they are
$$\psi(-\log K(e^{-\psi^{-1}(u_j)}, e^{-0.5\psi^{-1}(u_2)}) + \tfrac{1}{2}\psi^{-1}(u_2)).$$

3. Let $\nu_1 = \nu_2 = -1$, $\nu_3 = \cdots = \nu_m = 0$, $K_{12}(x,y) = xy$, $p_1 = p_2 = (m-2)^{-1}$, $p_3 = \cdots = p_m = (m-1)^{-1}$. Then (4.24) becomes
$$\int_0^\infty \prod_{\substack{1\le i<j\le m,\\ (i,j)\ne(1,2)}} K_{ij}^\alpha(H_i,H_j)\, dM(\alpha)$$

with copula

$$C(\mathbf{u}) = \psi\Big(- \sum_{\substack{1\le i<j\le m,\\ (i,j)\ne(1,2)}} \log K_{ij}(e^{-p_i\psi^{-1}(u_i)}, e^{-p_j\psi^{-1}(u_j)})\Big). \tag{4.31}$$

If ψ is a one-parameter family of LTs and each K_{ij} is a one-parameter family of copulas, then this is a family with $m(m-1)/2$ parameters. The labelling is such that the indices 1,2 are assigned to the pair of variables with the least amount of pairwise dependence. The (1,2) bivariate margin has the copula in (4.27).

With reference to the properties in Section 4.1, (4.25) is interpretable from its mixture form, is in part closed under the taking of margins, has a wide range of positive dependence, and has closed-form cdf if parametric families for ψ and K_{ij} are chosen appropriately (see Section 5.5).

4.3.1 Max-infinite divisibility conditions

In this subsection, we obtain conditions for the distributions in Section 4.2 to be max-id, so that we have candidates for (4.23) and (4.24).

First, with LT ψ, we consider the Archimedean copula $C(\mathbf{u}) = \psi(\sum_{j=1}^m \psi^{-1}(u_j))$. Then $C^\gamma(\mathbf{u}) = \exp\{\gamma\sigma(\sum_{j=1}^m \chi(u_j))\}$, where $\chi = \psi^{-1}$ and $\sigma = \log\psi$. Note that $\sigma' = \psi'/\psi$, $\sigma'' = (\psi''\psi - \psi'^2)/\psi^2$, $\sigma''' = [2(\psi')^3 - 3\psi\psi'\psi'' + \psi^2\psi''']/\psi^3$, $\sigma^{(4)} = [-6(\psi')^4 + 12\psi(\psi')^2\psi'' - 4\psi^2\psi'\psi''' - 3\psi^2(\psi'')^2 + \psi^3\psi^{(4)}]/\psi^4$. With $\chi_i' = \chi'(u_i)$, the mth-order mixed derivatives $c_{1\cdots m}$ of C^γ for $m = 2, 3, \ldots$ are:

- $c_{12} = e^{\gamma\sigma}\chi_1'\chi_2'[\gamma^2\sigma'^2 + \gamma\sigma'']$,
- $c_{123} = e^{\gamma\sigma}\chi_1'\chi_2'\chi_3'[\gamma^3\sigma'^3 + 3\gamma^2\sigma'\sigma'' + \gamma\sigma''']$, etc.

From the pattern of the derivatives, C^γ is max-id for up to dimension m if $-\sigma \in \mathcal{L}_m^*$, and C^γ is max-id for all m if $-\sigma \in \mathcal{L}_\infty^*$, where $\mathcal{L}_m^*$, $m \ge 1$, are defined in (1.2) of Section 1.3.

Next we turn to max-id for the partially symmetric copulas in Section 4.2. For the trivariate case, let

$$\begin{aligned} C(\mathbf{u}) &= \psi(\psi^{-1}\circ\phi(\phi^{-1}(u_1) + \phi^{-1}(u_2)) + \psi^{-1}(u_3)) \\ &\stackrel{\text{def}}{=} \psi\big(\omega(\chi(u_1) + \chi(u_2)) + \chi(u_3)\big). \end{aligned}$$

Let $H = C^\gamma$ and let $\sigma = \log\psi$, so that

$$H(\mathbf{u}) = \exp\{\gamma\,\sigma\big(\omega(\chi(u_1) + \chi(u_2)) + \chi(u_3)\big)\}.$$

Suppose $\omega \in \mathcal{L}_2^*$ and $-\log\psi \in \mathcal{L}_3^*$. Then the mixed derivatives up to third order are non-negative since each term of the derivatives is non-negative. The derivatives are:

- $\partial H/\partial u_3 = \gamma H \chi_3' \sigma'$,
- $\partial^2 H/\partial u_1 \partial u_3 = \gamma H \chi_1' \chi_3' [\gamma \sigma'^2 \omega' + \sigma'' \omega']$,
- $\partial^3 H/\partial u_1 \partial u_2 \partial u_3 = \gamma H \chi_1' \chi_2' \chi_3' [\gamma^2 \sigma'^3 \omega'^2 + 3\gamma\sigma'\sigma''\omega'^2 + \gamma\sigma'^2\omega'' + \sigma'''\omega'^2 + \sigma''\omega'']$.

For higher-dimensional copulas in this class, write

$$C(\mathbf{u}) = \exp\{\gamma\sigma(\omega_1 \circ \omega_2 \circ \cdots \circ \omega_k(\cdots) + \cdots)\}$$

and let $\chi_i' = \chi_i'(u_i)$. Suppose $-\log\psi = -\sigma \in \mathcal{L}_m^*$ and the ω_i are in $\mathcal{L}_{n_i}^*$ for sufficiently large n_i (greater than or equal to the number of terms in the argument of ω_i). Then the copula is max-id. As above, differentiation of a term will lead to terms that are each non-negative. For example, differentiation of $H = C^\gamma$ in a term with respect to u_i leads to a factor like $\gamma H \sigma' \omega_1' \cdots \omega_k' \chi_i' \geq 0$, differentiation of $[\sigma^{(j)}]^\ell$ in a term leads to a factor $\ell[\sigma^{(j)}]^{\ell-1}\sigma^{(j+1)}\omega_1' \cdots \omega_k' \chi_i'$ which has the same sign as $[\sigma^{(j)}]^\ell$, and differentiation of $\omega_\ell^{(j)}$ in a term leads to a factor $\omega_\ell^{(j+1)}\omega_{\ell+1}' \cdots \omega_k' \chi_i'$ which has the same sign as $\omega_\ell^{(j)}$.

More generally, consider

$$F(\mathbf{u}) = \psi(-\log K(\mathbf{u}))$$

where K is max-id and $-\log\psi \in \mathcal{L}_m^*$. We use Theorem 2.7 to prove that F is also max-id. Let $\sigma = \log\psi$ and $\kappa = \log K$, so that $F = \psi(-\log K) = \exp\{\sigma(-\kappa)\}$. Let $H = \exp\{\gamma\sigma(-\kappa)\}$, and let κ_S denote the partial derivative of κ with respect to u_i, $i \in S$. Then

- $\partial H/\partial u_1 = -\gamma H \sigma' \kappa_1 \geq 0$,
- $\partial^2 H/\partial u_1 \partial u_2 = \gamma^2 H \sigma'^2 \kappa_1 \kappa_2 + \gamma H \sigma'' \kappa_1 \kappa_2 - \gamma H \sigma' \kappa_{12} \geq 0$, etc.,

since each term is non-negative. The pattern of derivatives of each term being non-negative continues for higher-order partial derivatives of H. For example, differentiation of H in a term with respect to u_i leads to a factor like $-\gamma H \sigma' \kappa_i \geq 0$, differentiation of $[\sigma^{(j)}]^\ell$ in a term leads to a factor $-\ell[\sigma^{(j)}]^{\ell-1}\sigma^{(j+1)}\kappa_i$ which has the same sign as $[\sigma^{(j)}]^\ell$, and differentiation of κ_S in a term leads to a non-negative factor.

4.3.2 Dependence properties *

This section consists mainly of results on concordance and tail dependence.

Theorem 4.11 *The bivariate copula (4.26) is increasing in* $\prec_c$ *as* K_{ij} *increases in* $\prec_c$. *Hence (a) (4.25) is increasing in* $\prec_c^{pw}$, *and (b) the bivariate copula given in (4.26) is more concordant than that given in (4.27).*

Proof. The proof of the first statement is easy. Then (a) follows from results in Sections 2.2.1 and 2.2.3, and (b) follows because K_{ij} max-id implies it is TP_2 and hence PQD (see Section 2.1.7). □

Let K be a bivariate copula and ψ be a LT. With $(i,j) = (1,2)$ and $m = 2$, (4.26) becomes

$$C(u_1,u_2) = \psi(-\log K(e^{-p_1\psi^{-1}(u_1)}, e^{-p_2\psi^{-1}(u_2)}) + \nu_1 p_1 \psi^{-1}(u_1) + \nu_2 p_2 \psi^{-1}(u_2)), \qquad (4.32)$$

where $\nu_1, \nu_2 \geq 0$ are arbitrary and $p_i = (\nu_i + 1)^{-1}$, $i = 1, 2$.

Theorem 4.12 *If* $\psi'(0)$ *is finite, the copula*

$$C_\psi(u,v) = \psi(\psi^{-1}(u) + \psi^{-1}(v))$$

in (4.27) does not have upper tail dependence. If C_ψ *has upper tail dependence, then* $\psi'(0) = -\infty$ *and the tail dependence parameter is*

$$\lambda_U = 2 - 2 \lim_{s\to 0}[\psi'(2s)/\psi'(s)]. \qquad (4.33)$$

Proof. We begin by writing:

$$\lim_{u\to 1} \overline{C}_\psi(u,u)/(1-u) = \lim_{u\to 1}[1 - 2u + \psi(2\psi^{-1}(u))]/(1-u)$$

$$= 2 - 2 \lim_{u\to 1} \psi'(2\psi^{-1}(u))/\psi'(\psi^{-1}(u)) = 2 - 2 \lim_{s\to 0}[\psi'(2s)/\psi'(s)].$$

If $\psi'(0) \in (-\infty, 0)$, then the limit is zero and C_ψ does not have upper tail dependence. $\psi'(0)$ cannot equal 0 because it is the negative of the expectation of a positive rv. The rest of the result follows. □

Theorem 4.13 *The copula (4.32) has upper tail dependence only if either* $\psi'(0) = -\infty$ *or* K *has upper tail dependence or both. (The details of the tail dependence parameter are in the proof.)*

Proof. Suppose that the copula K in (4.32) has upper tail dependence parameter $\beta \in [0,1]$ ($\beta = 0$ implies no tail dependence). We consider first the case $p_1 = p_2$ or $\nu_1 = \nu_2$. Subsequently, for the case of $p_1 \neq p_2$, bounds will be obtained.

For x less than and close to 1, $\overline{K}(x,x) \sim \beta(1-x)$ so that $K(x,x) \sim 2x - 1 + \beta(1-x) = 1 - (2-\beta)(1-x)$. Let $p_1 = p_2 = p = (\nu+1)^{-1}$. Then for u near 1,

$$\begin{aligned} &-\log\bigl[K(e^{-p\psi^{-1}(u)}, e^{-p\psi^{-1}(u)})\bigr] + 2\nu p\,\psi^{-1}(u) \\ &\sim \quad -\log[1-(2-\beta)(1-e^{-p\psi^{-1}(u)})] + 2\nu p\,\psi^{-1}(u) \\ &\sim \quad -\log[1-(2-\beta)p\,\psi^{-1}(u)] + 2\nu p\,\psi^{-1}(u) \\ &\sim \quad (2-\beta)p\,\psi^{-1}(u) + 2\nu p\,\psi^{-1}(u) = \gamma\,\psi^{-1}(u), \end{aligned}$$

where $\gamma = [2(\nu+1)-\beta]p = 2-\beta/(\nu+1) \in [1,2]$. Hence, for u near 1,

$$\begin{aligned} [1-2u+C(u,u)]/(1-u) &\sim [1-2u+\psi(\gamma\psi^{-1}(u))]/(1-u) \\ &\sim 2-\gamma\psi'(\gamma\psi^{-1}(u))/\psi'(\psi^{-1}(u)) \end{aligned}$$

and the upper tail dependence parameter of C in (4.32) is $\lambda_U = 2-\gamma\lim_{s\to 0}\psi'(\gamma s)/\psi'(s)$.

If C_ψ does not have upper tail dependence, then $\lambda_U = 2-\gamma = \beta/(\nu+1)$ and C has upper tail dependence if and only if K has upper tail dependence (and the tail dependence parameter of K is larger since $\nu \ge 0$).

If C_ψ has upper tail dependence, then $\gamma\lim_{s\to 0}\psi'(\gamma s)/\psi'(s)$ should be increasing and λ_U decreasing as γ increases or as ν increases (this follows from Theorem 4.14 below and Theorem 2.3(d)). If $\beta = 0$ so that $\gamma = 2$, then $\delta_U = 2-2\lim_{s\to 0}\psi'(2s)/\psi'(s)$ is the tail dependence parameter of C_ψ. If $\beta = 1$ and $\nu = 0$ so that $\gamma = 1$, then $\lambda_U = 1$. Hence the tail dependence parameter of (4.32) is greater than or equal to that of C_ψ.

For the asymmetric case with $p_1 \le p_2$ ($\nu_1 \ge \nu_2$),

$$\begin{aligned} K\bigl(e^{-p_2\psi^{-1}(u)}, e^{-p_2\psi^{-1}(u)}\bigr) &\le K\bigl(e^{-p_1\psi^{-1}(u)}, e^{-p_2\psi^{-1}(u)}\bigr) \\ &\le K\bigl(e^{-p_1\psi^{-1}(u)}, e^{-p_1\psi^{-1}(u)}\bigr) \end{aligned}$$

so that from above, the tail dependence parameter λ_U is bounded as follows:

$$2-\gamma_2\lim_{s\to 0}\psi'(\gamma_2 s)/\psi'(s) \le \lambda_U \le 2-\gamma_1\lim_{s\to 0}\psi'(\gamma_1 s)/\psi'(s),$$

where $\gamma_i = (2-\beta)/(\nu_i+1) + \nu_1/(\nu_1+1) + \nu_2/(\nu_2+1)$, $i = 1,2$. Note that $\gamma_1 \le \gamma_2$. □

For the next theorem, LTD1 refers to the second variable LTD in the first variable and LTD2 refers to the first variable LTD in the second variable.

Theorem 4.14 *Let C be as given in (4.32). (a) Then C increases in concordance as p_1 increases (from 0 to 1) and ν_1 decreases if K satisfies the LTD1 property. (b) Also C increases in concordance as p_2 increases and ν_2 decreases if K satisfies the LTD2 property. (c) If $p_1 = p_2 = p$ and $\nu_1 = \nu_2 = \nu$, then C increases in concordance as p increases if K satisfies both LTD1 and LTD2.*

Proof. Details will mainly be given for case (a). Let $0 < p_1' = (\nu_1' + 1)^{-1} < p_1 \le 1$ and $\nu_1' > \nu_1 \ge 0$. Then with $y = e^{-p_2\psi^{-1}(u_2)}$,

$$\psi\left(-\log K(e^{-p_1'\psi^{-1}(u_1)}, y) + \nu_1' p_1' \psi^{-1}(u_1) + \nu_2 p_2 \psi^{-1}(u_2)\right)$$

$$\le \psi\left(-\log K(e^{-p_1\psi^{-1}(u_1)}, y) + \nu_1 p_1 \psi^{-1}(u_1) + \nu_2 p_2 \psi^{-1}(u_2)\right),$$

for all u_1, u_2, if

$$K(e^{-p_1'\psi^{-1}(u_1)}, y)\, e^{-\nu_1' p_1' \psi^{-1}(u_1)} \le K(e^{-p_1\psi^{-1}(u_1)}, y)\, e^{-\nu_1 p_1 \psi^{-1}(u_1)},$$

for all u_1, y, or if (with $x = e^{-\psi^{-1}(u_1)}$ and $\nu = \nu_1$, $\nu' = \nu_1'$)

$$K(x^{1/(\nu'+1)}, y)\, x^{\nu'/(\nu'+1)} \le K(x^{1/(\nu+1)}, y)\, x^{\nu/(\nu+1)}, \ \forall x, y \in (0, 1).$$

This is the same as $K(x^{1/(\nu+1)}, y)\, x^{\nu/(\nu+1)}$ decreasing in $\nu \ge 0$ for all x, y or $K(x^{1-\xi}, y)\, x^{\xi} = [K(x^{1-\xi}, y)/x^{1-\xi}]\, x$ decreasing in $\xi \in [0, 1]$. Finally, this is the same as the LTD1 condition of $K(z, y)/z$ decreasing in z for all y.

For (c), the concordance ordering is equivalent to

$$\{K(x^{1-\xi}, y^{1-\xi})/[x^{1-\xi}y^{1-\xi}]\}\, xy \quad \downarrow \xi \in [0, 1]$$

for all x, y. This follows from the LTD1 and LTD2 conditions because if $0 \le \xi < \xi' \le 1$, then the conditions imply

$$\begin{aligned} K(x^{1-\xi'}, y^{1-\xi'})\, x^{\xi'} y^{\xi'} &\le K(x^{1-\xi}, y^{1-\xi'})\, x^{\xi} y^{\xi'} \\ &\le K(x^{1-\xi}, y^{1-\xi})\, x^{\xi} y^{\xi}. \end{aligned}$$

□

Note that from Theorems 2.6 and 2.3(d), if K is max-id, then it satisfies the LTD1 and LTD2 conditions.

Analagous results for lower tail dependence are given next.

Theorem 4.15 *The copula $C_\psi(u, v) = \psi(\psi^{-1}(u) + \psi^{-1}(v))$ has lower tail dependence parameter equal to*

$$\lambda_L = 2 \lim_{s\to\infty} [\psi'(2s)/\psi'(s)]. \tag{4.34}$$

Proof. The proof is similar to that of Theorem 4.12 and is left as an exercise. □

Theorem 4.16 *If $p_1 = p_2 = p$ (and $\nu_1 = \nu_2 = \nu$) in (4.32) and the lower tail dependence parameter of K is $\beta \in (0,1]$, then the lower tail dependence parameter of C in (4.32) is*

$$\lambda_L = \gamma \lim_{s\to\infty} \psi'(-\log\beta + \gamma s)/\psi'(s), \tag{4.35}$$

where $\gamma = p(1+2\nu) \geq 1$. If the lower tail dependence parameter of K is 0, then the lower tail dependence parameter of C is less than the right-hand side of (4.35) for all $\beta > 0$ (with $\gamma = p(1+2\nu)$). If the behaviour at the lower tail is $K(x,x) \sim \beta x^r$ as $x \to 0$ with $r > 1$, then the lower tail dependence parameter of C is given by (4.35) with $\gamma = p(r+2\nu) \geq 1$.

Proof. This is left as an exercise. □

Illustrations of tail dependence for the LT families LTA to LTD (in the Appendix) are given in the following examples.

Example 4.1 Upper tail dependence for (4.3) with different families of LTs.

LTA. $\psi'(s) = -\theta^{-1}s^{1/\theta-1}\exp\{-s^{1/\theta}\}$ and $\psi'(0) = -\infty$. The limit in (4.33) is $\lambda_U = 2 - 2\lim_{s\to 0}[\psi'(2s)/\psi'(s)] = 2 - 2\cdot 2^{1/\theta-1} = 2 - 2^{1/\theta}$.

LTB. $\psi'(s) = -\theta^{-1}(1+s)^{-1/\theta-1}$ and $\psi'(0) = -\theta^{-1}$. So $\lambda_U = 0$.

LTC. $\psi'(s) = -\theta^{-1}(1-e^{-s})^{1/\theta-1}e^{-s}$ and $\psi'(0) = -\infty$. The limit in (4.33) is $\lambda_U = 2 - 2^{1/\theta}$.

LTD. $\psi'(s) = -\theta^{-1}(1-e^{-\theta})e^{-s}/[1-(1-e^{-\theta})e^{-s}]$ and $\psi'(0) = -\theta^{-1}e^{\theta}(1-e^{-\theta})$. So $\lambda_U = 0$.

Example 4.2 Lower tail dependence for (4.3) with different families of LTs.

LTA. From (4.34), $\lambda_L = 2\lim_{s\to\infty}[\psi'(2s)/\psi'(s)] = \lim_{s\to\infty} 2^{1/\theta}\exp\{-(2^{1/\theta}-1)s^{1/\theta}\} = 0$.

LTB. $\lambda_L = \lim_{s\to\infty} 2(1+s[1+s]^{-1})^{-1/\theta-1} = 2^{-1/\theta}$.

LTC. $\lambda_L = \lim_{s\to\infty} 2(1+e^{-s})^{1/\theta-1}e^{-s} = 0$.

LTD. $\lambda_L = \lim_{s\to\infty} 2e^{-s}[1-(1-e^{-\theta})e^{-s}]/[1-(1-e^{-\theta})e^{-2s}] = 0$.

□

Example 4.3 Lower tail dependence for (4.32) with $\nu_1 = \nu_2 = \nu$ for different families of LTs.

LTA. The limit in (4.35) is $\lim_s \gamma[-s^{-1}\log\beta + \gamma]^{1/\theta-1}\exp\{s^{1/\theta}\} \cdot\exp\{-[-\log\beta+\gamma s]^{1/\theta}\} = \lambda_L$. If $\nu > 0$ so that $\gamma > 1$ then $\lambda_L = 0$, and if $\gamma = 1$ (and $\nu = 0$, $r = 1$) and $\beta > 0$, then $\lambda_L = 1$ for $\theta > 1$ and $\lambda_L = \beta$ for $\theta = 1$.

LTB. The limit in (4.35) becomes $\lim_s \gamma[\gamma + (1-\gamma-\log\beta)(1+s)^{-1}]^{-1/\theta-1} = \gamma^{-1/\theta} = \lambda_L$. If $\nu = 0$ and $r = 1$, then $\gamma = 1$ and $\lambda_L = 1$. If $r = 2$, as for the case of the independence copula, then $\gamma = 2$ and $\lambda_L = 2^{-1/\theta}$, the same as the copula B4 with parameter θ. If $1 \le r < 2$, then $1 \le \gamma < 2$ and there is more lower tail dependence than the copula B4 with parameter θ. For example, let K be the copula B6 with parameter $\delta \ge 1$; then $K(x,x) \sim x^r$ with $r = 2^{1/\delta}$ so that $\lambda_L = 2^{-1/(\theta\delta)}$ if $\nu = 0$ and $p = 1$ (and $\gamma = r$). If K is the copula B7 with parameter $\delta > 0$, then $K(x,x) \sim x^r$ with $r = 2 - 2^{-1/\delta}$. If K is the copula B3 with parameter $-\infty < \delta < \infty$, then $K(x,x) \sim -\delta^{-1}\log[1 - \delta^2 x^2/(1 - e^{-\delta})] \sim \delta x^2/(1-e^{-\delta})$ and $\gamma = 2$. Note that the copula families B3, B4, B6, B7 are in Section 5.1.

LTC. The limit in (4.35) is $\lambda_L = \lim_s \gamma[(1-\beta e^{-\gamma s})/(1-e^{-s})]^{1/\theta-1} \cdot \beta e^{-(\gamma-1)s}$. This is 0 if $\gamma > 1$ and β if $\gamma = 1$.

LTD. The limit in (4.35) is $\lambda_L = \lim_s \beta e^{-(\gamma-1)s}[1-(1-e^{-\theta})e^{-s}]/[1-(1-e^{-\theta})\beta e^{-\gamma s}]$. This is 0 if $\gamma > 1$ and β if $\gamma = 1$.

□

From Theorem 4.11, the (1,2) and (3,2) bivariate margins of (4.29) and (4.30) are more concordant than the (1,3) margin. This is different from (4.7) in which the bivariate margin that is different is more concordant than the other two. In (4.30), the dependence of the (1,2) and (3,2) margins increases in concordance as K increases in concordance. Letting K be the Fréchet upper bound leads to the greatest possible dependence for these two bivariate margins. In fact, in this limiting case, the upper bound of the inequality (3.13) on Kendall's tau is attained.

Theorem 4.17 *Let K be the bivariate Fréchet upper bound in (4.30), to obtain*

$$C(\mathbf{u}) = \psi\big(\max\{\psi^{-1}(u_1), \tfrac{1}{2}\psi^{-1}(u_2)\} + \max\{\psi^{-1}(u_3), \tfrac{1}{2}\psi^{-1}(u_2)\}\big).$$

Let τ_{ij} be the Kendall tau value for the (i,j) bivariate margin. Then $\tau_{12} = \tau_{23} > \tau_{13}$ and the upper bound $\tau_{12} = 1 - (\tau_{23} - \tau_{13})$ of (3.13) is attained.

Proof. In Theorem 3.13, let $(i,j,k) = (1,3,2)$ and let $\mathbf{X}, \mathbf{X}'$ be independent triples with cdf C. Let $E_1 = \{(X_1 - X_1')(X_3 - X_3') > 0\}$ and $E_2 = \{(X_2 - X_2')(X_3 - X_3') > 0\}$. Let $\tau = \tau_{12} = \tau_{23}$. By Theorem 3.13, the upper bound in (3.13), $\tau = 1 - \tau + \tau_{13}$, is attained if $E_1 \subset E_2$. A representation for C is

$$\int_0^\infty \min\{H_1^\alpha(u_1), H_2^\alpha(u_2)\} \min\{H_3^\alpha(u_3), H_2^\alpha(u_2)\}\, dM(\alpha),$$

with $G_1(\cdot;\alpha) \stackrel{\text{def}}{=} H_1^\alpha = H_3^\alpha = \exp\{-\alpha\psi^{-1}\}$ and $G_2(\cdot;\alpha) \stackrel{\text{def}}{=} H_2^\alpha = \exp\{-\frac{1}{2}\alpha\psi^{-1}\}$. Note that

$$G_1^{-1}(G_2(u;\alpha);\alpha) = \psi(\tfrac{1}{2}\psi^{-1}(u)) \stackrel{\text{def}}{=} b(u),$$

independently of α. Hence for $\mathbf{X}, \mathbf{X}'$, there is the representation

$$X_1 = b(X_{21}), \quad X_3 = b(X_{22}), \quad X_2 = \max\{X_{21}, X_{22}\},$$

$$X_1' = b(X_{21}'), \quad X_3' = b(X_{22}'), \quad X_2' = \max\{X_{21}', X_{22}'\},$$

where X_{21}, X_{22} are independent with distribution H_2^α given α, X_{21}', X_{22}' are independent with distribution $H_2^{\alpha'}$ given α', and α, α' are independent rvs with distribution M. Since b is strictly increasing, $X_1 > X_1'$, $X_3 > X_3'$ implies $X_{21} > X_{21}'$, $X_{22} > X_{22}'$, which implies $X_2 > X_2'$ and $(X_2 - X_2')(X_3 - X_3') > 0$. The same conclusion holds starting from $X_1 < X_1'$, $X_3 < X_3'$ and hence $E_1 \subset E_2$. □

4.4 Generalizations of functional forms

Extensions to multivariate distributions with negative dependence usually come from extending functional forms, especially when mixture and stochastic representations do not extend. An example to illustrate this is the MVN distribution with exchangeable dependence structure. If the equicorrelation parameter is ρ, then, for $\rho \geq 0$, a stochastic representation is

$$Y_j = \sqrt{\rho}\, Z_0 + \sqrt{1-\rho}\, Z_j, \quad j = 1, \ldots, m,$$

where $Z_0, Z_1, \ldots, Z_m$ are iid $N(0,1)$ rvs. This does not extend to the range of negative dependence.

Larger classes of functions, generalizing LTs, are $\mathcal{L}_m$, $m \geq 1$, which are defined in (1.1) in Section 1.3. Related classes of functions are $\mathcal{L}_n^*$, $n \geq 1$, which were also used in Section 4.3.1. With the classes $\mathcal{L}_m$ and $\mathcal{L}_n^*$, multivariate distributions with some negative dependence can be obtained. The multivariate distributions in the two preceding sections have positive dependence only.

We extend the permutation-symmetric copulas, then the partially symmetric copulas of Section 4.2, and finally the copulas with general dependence structures in Section 4.3. We get copulas with negatively dependent bivariate margins. Hence we get families of multivariate copulas with a wider range of dependence; however, the extensions do not necessarily have a mixture representation.

Consider a copula of the general form (4.4):

$$C(\mathbf{u}) = \phi\Big(\sum_{i=1}^{m} \phi^{-1}(u_i)\Big), \tag{4.36}$$

where $\phi : [0,\infty) \to [0,1]$ is strictly decreasing and continuously differentiable (of all orders), and $\phi(0) = 1$, $\phi(\infty) = 0$. Then it is easily verified that a necessary and sufficient condition for (4.36) to be a proper distribution is that $(-1)^j \phi^{(j)} \geq 0$, $j = 1,\ldots,m$, i.e., the derivatives are alternating in sign up to order m, or $\phi \in \mathcal{L}_m$. If (4.36) is a copula for all m, then ϕ must be completely monotone and hence be a LT.

One can extend (4.36) further by allowing strictly decreasing functions ϕ which are defined from $[0,B]$ onto $[0,1]$ for some $0 < B < \infty$ and satisfy $\phi(B) = 0$. For (4.36), ϕ is defined to be 0 on (B,∞). If ϕ is continuously differentiable in $(0,B)$ and the derivatives alternate in sign up to order m, then (4.36) is a proper distribution. In Section 5.4, there is one example of a family with ϕ in this extended class. However this extension of $\mathcal{L}_m$ is not useful for applications because the support of (4.36) is not all of $(0,1)^m$. (Note that $\phi(\sum_{i=1}^m \phi^{-1}(u_i)) = 0$ if $\sum_{i=1}^m \phi^{-1}(u_i) > B$ or if $u_1,\ldots,u_m$ are all sufficiently close to 0.)

In (4.36), C has negative lower orthant dependence (NLOD) if

$$\sum_{j=1}^{m} \phi^{-1}(u_j) \geq \phi^{-1}(u_1 \cdots u_m)$$

or if

$$\sum_{j=1}^{m} \phi^{-1}(e^{-z_j}) \geq \phi^{-1}(\exp\{-z_1 - \cdots - z_m\})$$

for $z_j \geq 0$, $j = 1,\ldots,m$. Let $\eta(z) = \phi^{-1}(e^{-z})$, $z > 0$; then the condition on ϕ is equivalent to η being subadditive. Since $\eta(0) = 0$ and η is increasing, the subadditivity condition will be satisfied if η is concave. The concavity of η is equivalent to the convexity of $\eta^{-1} = -\log\phi$ and the subadditivity of η is equivalent to the superadditivity of $-\log\phi$. If C is NLOD, then note that all of its bivariate margins are NQD. Similarly, C is PLOD if η is superadditive or if $-\log\phi$ is subadditive.

Now let C_1, C_2 be two copulas of the form (4.36) based on different functions ϕ_1, ϕ_2. In terms of the functions ϕ_i, C_2 is more

PLOD than C_1 if and only if $\omega = \phi_2^{-1} \circ \phi_1$ is superadditive. (The result in Theorem 4.1 still holds.)

Next consider copulas with the functional forms of (4.7), (4.10), (4.11), where the functions ϕ, ψ, ζ are in $\mathcal{L}_n$ for some n (to be determined). What other conditions are needed for these to be proper copulas? These can be seen from taking derivatives up to the dimension of the copula.

From the derivatives, sufficient conditions are that the composition of the form $\psi^{-1} \circ \phi$ is in $\mathcal{L}_n^*$ with n being the number of summands in the argument of this function. Specifically, for (4.7) in the notation $C(\mathbf{u}) = \psi(\omega(\chi_1(u_1) + \chi_1(u_2)) + \chi_2(u_3))$,

- $\partial C/\partial u_3 = \psi' \chi_2'$,
- $\partial^2 C/\partial u_2 \partial u_3 = \psi'' \chi_2' \omega' \chi_1'(u_2)$, and
- $\partial^3 C/\partial u_1 \partial u_2 \partial u_3 = [\psi''' \chi_2' \omega'^2 + \psi'' \chi_2' \omega''] \chi_1'(u_2) \chi_1'(u_1)$.

These derivatives are non-negative if $\psi \in \mathcal{L}_3$, $\phi \in \mathcal{L}_1$, and $\omega = \psi^{-1} \circ \phi \in \mathcal{L}_2^*$. Similarly, (4.11) is a copula if $\psi \in \mathcal{L}_4$, $\phi, \zeta \in \mathcal{L}_1$, and $\psi^{-1} \circ \phi, \psi^{-1} \circ \zeta \in \mathcal{L}_2^*$, and (4.10) is a copula if $\psi \in \mathcal{L}_4$, $\phi \in \mathcal{L}_3$, $\zeta \in \mathcal{L}_1$, $\psi^{-1} \circ \phi \in \mathcal{L}_3^*$ and $\phi^{-1} \circ \zeta \in \mathcal{L}_2^*$.

Let $\psi, \phi \in \mathcal{L}_1$. If $\psi^{-1} \circ \phi \in \mathcal{L}_2^*$, $\psi \in \mathcal{L}_3$ and $-\log \psi$ is convex or superadditive, then the (1,3) and (2,3) bivariate margins of (4.7) are NQD. If also $-\log \phi$ is convex or superadditive, then the (1,2) bivariate margin is NQD but it is more concordant than the (1,3) margin. That is, a non-permutation-symmetric trivariate copula with all NQD bivariate margins results. Similar analyses apply in higher dimensions. Note that if ψ, ϕ are not in $\mathcal{L}_\infty$ or if $\psi^{-1} \circ \phi$, etc., are not in $\mathcal{L}_\infty^*$, then (4.7), (4.10), (4.11) and their extensions do not have representations as mixtures.

More generally, from (4.23) in Section 4.3, $F(\mathbf{u}) = \psi(-\log H(\mathbf{u}))$ is a multivariate cdf if H is max-id and $-\log \psi \in \mathcal{L}_m^*$. This form includes (4.7), (4.10), (4.11) and their multivariate extensions as special cases. Copulas with all bivariate margins distinct can be obtained and copulas with some NQD bivariate margins can result if $-\log \psi$ is convex. Hence the general form allows a fairly wide range of dependence structures.

Specific parametric families illustrating the ideas in this section are given in Section 5.4. These families better satisfy property C in Section 4.1, because of attaining a wider range of dependence including negative dependence; this is done at the expense of property A, as the mixture representation is lost.

Next we consider multivariate copulas that are mixtures of integer powers of multivariate distributions. Suppose ψ is the LT of a distribution with support on the positive integers. Then the multivariate family, $F = \psi(-\log H)$, in (4.23) extends to arbitrary H (not necessary max-id) since a representation is $\sum_{n=1}^{\infty} \pi_n H^n$, where π_n is the probability mass at the integer n for the distribution M with LT ψ. Some of the LTs in the Appendix have support on the positive integers.

Now take the form of (4.24). If K_{ij} is chosen to be NQD, then the (i,j) bivariate margin of (4.24) is less concordant than $C_\psi(u,v) = \psi(\psi^{-1}(u) + \psi^{-1}(v))$. That is, bivariate margins of (4.24) can be either more dependent or less dependent than C_ψ instead of being just more dependent for the general LT. This possibility should add extra flexibility for modelling, but from the results in Section 4.3.2 on tail dependence no new multivariate extreme value copulas come from allowing negative dependence in the K_{ij}.

When $m = 2$, $p_1 = p_2 = 1$, $\nu_1 = \nu_2 = 0$ and $K = K_{12}$ is the Fréchet lower bound, (4.24) becomes

$$C(u_1, u_2) = \psi(-\log(e^{-\psi^{-1}(u_1)} + e^{-\psi^{-1}(u_2)} - 1)_+),$$

where $(y)_+ = \max\{0, y\}$. This is NQD if and only if

$$(e^{-\psi^{-1}(u_1)} + e^{-\psi^{-1}(u_2)} - 1)_+ \leq e^{-\psi^{-1}(u_1 u_2)}$$

for all $0 \leq u_1, u_2 \leq 1$. (Note that PQD is not possible because $C(u,u) = 0$ if u is such that $0 < e^{-\psi^{-1}(u)} < \frac{1}{2}$.) Let $g(z) = \exp\{-\psi^{-1}(e^{-z})\}$; then the NQD condition becomes $g(z_1)+g(z_2) \leq g(z_1+z_2)+1$ for all $z_1, z_2 > 0$. Finally, let $h(z) = 1 - g(z)$ so that $h(0) = 0$ and $h(\infty) = 1$ and h is increasing. The condition becomes $h(z_1+z_2) \leq h(z_1)+h(z_2)$ for all $z_1, z_2 > 0$ or h is subadditive. The condition of subadditivity is satisfied here if h is concave or anti-star-shaped (the region below the curve $y = h(x)$ is star-shaped with respect to the origin). The anti-star-shaped condition can be written as $z^{-1}h(z)$ decreasing in z.

An example is the family LTD in which $\psi(s) = -\theta^{-1}\log[1 - ce^{-s}]$, where $c = 1 - e^{-\theta}$, $\theta > 0$. Then $\exp\{-\psi^{-1}(t)\} = c^{-1}(1 - e^{-\theta t})$ and $h(z) = c^{-1}[\exp\{-\theta e^{-z}\} - e^{-\theta}]$. The second derivative is $h''(z) = c^{-1}\theta e^{-z}\exp\{-\theta e^{-z}\}(\theta e^{-z} - 1)$ and this is uniformly non-positive if $0 < \theta \leq 1$.

4.5 Mixtures of conditional distributions

One main object in this section is to construct families of k-variate distributions based on two given $(k-1)$-dimensional margins (which must have $k-2$ variables in common), e.g., $\mathcal{F}(F_{12}, F_{23})$,

$\mathcal{F}(F_{1,\ldots,k-1}, F_{2,\ldots,k})$. The families can be made to interpolate between perfect conditional negative dependence and perfect conditional positive dependence with conditional independence in the middle. That is, this is an extension of Theorem 3.10 in Section 3.2. If one is given $F_{12}, F_{23}, \ldots, F_{m-1,m}$, $m \geq 3$, one can build an m-variate distribution starting with trivariate distributions $F_{i,i+1,i+2} \in \mathcal{F}(F_{i,i+1}, F_{i+1,i+2})$, then 4-variate distributions from $F_{i,\ldots,i+3} \in \mathcal{F}(F_{i,i+1,i+2}, F_{i+1,i+2,i+3})$, etc. There is a bivariate copula C_{ij} associated with the (i,j) bivariate margin of the m-variate distributon. For (i,j) with $|j-i| > 1$, C_{ij} measures the amount of conditional dependence in the ith and jth variables, given those variables with indices in between. Hence this is another construction method that builds on bivariate copulas.

For $m = 3$, the trivariate family is

$$F_{123}(\mathbf{y}) = \int_{-\infty}^{y_2} C_{13}(F_{1|2}(y_1|z_2), F_{3|2}(y_3|z_2))\, F_2(dz_2), \qquad (4.37)$$

where $F_{1|2}, F_{3|2}$ are conditional cdfs obtained from F_{12}, F_{23}. By construction, (4.37) is a proper trivariate distribution with univariate margins F_1, F_2, F_3, (1,2) bivariate margin F_{12}, and (2,3) bivariate margin F_{23}. C_{13} can be interpreted as a copula representing the amount of conditional dependence (in the first and third univariate margins given the second). $C_{13}(u_1, u_3) = u_1 u_3$ corresponds to conditional independence and $C_{13}(u_1, u_3) = \min\{u_1, u_3\}$ corresponds to perfect conditional positive dependence.

For $m = 4$, define F_{234} in a similar way to F_{123} (by adding 1 to all subscripts in (4.37)). Note that F_{123}, F_{234} have a common bivariate margin F_{23}. The 4-variate family is

$$F_{1234}(\mathbf{y}) = \int_{-\infty}^{y_2}\int_{-\infty}^{y_3} C_{14}(F_{1|23}(y_1|\mathbf{z}), F_{4|23}(y_4|\mathbf{z}))\, F_{23}(dz_2, dz_3), \qquad (4.38)$$

where $F_{1|23}, F_{4|23}$ are conditional cdfs obtained from F_{123}, F_{234}, and $\mathbf{z} = (z_2, z_3)$.

This can be extended recursively. Assuming $F_{1\cdots m-1}, F_{2\cdots m}$ have been defined with a common $(m-2)$-dimensional margin $F_{2\cdots m-1}$, the m-variate family is

$$F_{1\cdots m}(\mathbf{y}) = \int_{-\infty}^{y_2}\cdots\int_{-\infty}^{y_{m-1}} C_{1m}\big(F_{1|2\cdots m-1}(y_1|z_2,\ldots,z_{m-1}),$$

$$F_{m|2\cdots m-1}(y_m|z_2,\ldots,z_{m-1})\big)\cdot F_{2\cdots m-1}(dz_2,\ldots,dz_{m-1}), \qquad (4.39)$$

where $F_{1|2\cdots m-1}$, $F_{m|2\cdots m-1}$ are conditional cdfs obtained from

$F_{1\cdots m-1}, F_{2\cdots m}$.

Similar to (4.37)–(4.39), one can define a family of m-variate distributions through survival functions, $\overline{F}_S$. Let $\overline{F}_j = 1 - F_j$ be the univariate survival functions. The bivariate margins with consecutive indices are $\overline{F}_{j,j+1}(y_j, y_{j+1}) = C^*_{j,j+1}(\overline{F}_j(y_j), \overline{F}_{j+1}(y_{j+1}))$, where $C^*_{j,j+1}$ is the copula linking the univariate survival functions to the bivariate survival function. The m-variate case is like (4.39) with all F replaced by $\overline{F}$ and the integrals having lower limits y_j, $j = 2, \ldots, m-1$, and upper limits ∞. This leads to

$$\overline{F}_{1\cdots m}(\mathbf{y}) = \int_{y_2}^{\infty} \cdots \int_{y_{m-1}}^{\infty} C^*_{1m}\big(\overline{F}_{1|2\cdots m-1}(y_1|z_2, \ldots, z_m-1),$$

$$\overline{F}_{m|2\cdots m-1}(y_m|z_2, \ldots, z_{m-1})\big) \cdot F_{2\cdots m-1}(dz_2, \ldots, dz_{m-1}). \quad (4.40)$$

It is straightforward to show that this family is the same as that from (4.37)–(4.39) with $C^*_{jk}(u,v) = u + v - 1 + C_{jk}(1-u, 1-v)$ or $C_{jk}(u,v) = u + v - 1 + C^*_{jk}(1-u, 1-v)$.

Models (4.39) and (4.40) are a unifying method for constructing multivariate distributions with a given copula for each bivariate margin. The MVN family is a special case of (4.39). Other special cases are given in Sections 5.5 and 6.3.

Example 4.4 (MVN.) Let $F_j = \Phi$, $j = 1, \ldots, m$, where Φ is the standard normal cdf. For $j < k$, let $C_{jk} = F_{\theta_{jk}}(\Phi^{-1}(u_j), \Phi^{-1}(u_k))$, where $F_{\theta_{jk}}$ is the BVSN cdf with correlation θ_{jk}. Then for (4.39), with $k - j > 1$, $\theta_{jk} = \rho_{jk\cdot(j+1,\ldots,k-1)}$ is the partial correlation of variables j and k given variables $j+1, \ldots, k-1$. This type of parametrization, which is not unique because of the indexing, of the MVN distribution may be useful for some applications because each θ_{jk} can be in the range $(-1, 1)$ and there is no constraint similar to that of a positive definite matrix.

Proof. Starting with $F_{j,j+1}$ being BVN, we show that if $F_{j,\ldots,j+n-2}$ $(2 < n \le m,\ j \le m-n+2)$ are $(n-1)$-dimensional MVN, then $F_{j,\ldots,j+n-1}$ $(j \le m-n+1)$ are n-dimensional MVN. It suffices to show that $F_{1,\ldots,m}$ in (4.39) is MVN assuming that $F_{1,\ldots,m-1}$ and $F_{2,\ldots,m}$ are MVN, for $m \ge 3$.

Let Φ_Ω, ϕ_Ω respectively denote the MVN cdf and pdf with zero mean vector and covariance matrix Ω. Let

$$R \stackrel{\text{def}}{=} \begin{bmatrix} 1 & \rho_{1m} \\ \rho_{1m} & 1 \end{bmatrix}, \quad \begin{bmatrix} 1 & \Sigma_{12} \\ \Sigma_{21} & \Sigma_{22} \end{bmatrix} \quad \text{and} \quad \begin{bmatrix} \Sigma_{22} & \Sigma_{2m} \\ \Sigma_{m2} & 1 \end{bmatrix}$$

be the covariance matrices associated with C_{1m}, $F_{1,\ldots,m-1}$ and $F_{2,\ldots,m}$, respectively. Also let $a_{11} = [1 - \Sigma_{12}\Sigma_{22}^{-1}\Sigma_{21}]^{1/2}$, $a_{mm} =$

$[1-\Sigma_{m2}\Sigma_{22}^{-1}\Sigma_{2m}]^{1/2}$ and $\mathbf{z}_2 = (z_2,\ldots,z_{m-1})$, $\mathbf{z} = (z_1,\ldots,z_m)$. With BVN copulas and univariate standard normal margins, (4.39) simplifies to

$$\int_{-\infty}^{x_2}\cdots\int_{-\infty}^{x_{m-1}} \Phi_R\left(\frac{x_1-\mathbf{z}_2\Sigma_{21}}{a_{11}}, \frac{x_m-\mathbf{z}_2\Sigma_{2m}}{a_{mm}}\right)\phi_{\Sigma_{22}}(\mathbf{z}_2)\,d\mathbf{z}_2. \tag{4.41}$$

Writing Φ_R as an integral, (4.41) becomes

$$(a_{11}a_{mm})^{-1}\int_{-\infty}^{x_1}\cdots\int_{-\infty}^{x_m} \phi_R\left(\frac{z_1-\mathbf{z}_2\Sigma_{21}}{a_{11}}, \frac{z_m-\mathbf{z}_2\Sigma_{2m}}{a_{mm}}\right)\phi_{\Sigma_{22}}(\mathbf{z}_2)d\mathbf{z}. \tag{4.42}$$

Clearly, the integrand of (4.42) is a constant multipied by the exponential of a quadratic form in $z_1,\ldots,z_m$, so that (4.42) corresponds to an m-dimensional MVN cdf. Let the covariance matrix of the resulting MVN distribution be denoted by

$$\begin{bmatrix} 1 & \Sigma_{12} & \sigma_{1m} \\ \Sigma_{21} & \Sigma_{22} & \Sigma_{2m} \\ \sigma_{1m} & \Sigma_{m2} & 1 \end{bmatrix}.$$

The squared reciprocal in (4.42) of $(2\pi)^{m/2}$ times the constant is $|\Sigma_{22}|(1-\rho_{1m}^2)a_{11}^2a_{mm}^2$; it is also equal to

$$|\Sigma_{22}|\cdot\left|\begin{bmatrix} 1 & \sigma_{1m} \\ \sigma_{m1} & 1 \end{bmatrix} - \begin{bmatrix} \Sigma_{12} \\ \Sigma_{m2} \end{bmatrix}\Sigma_{22}^{-1}[\Sigma_{21} \quad \Sigma_{2m}]\right|.$$

Hence $(1-\rho_{1m}^2)a_{11}^2a_{mm}^2 = a_{11}^2a_{mm}^2-(\sigma_{1m}-\Sigma_{12}\Sigma_{22}^{-1}\Sigma_{2m})^2$ or $\rho_{1m}^2 = \{(\sigma_{1m}-\Sigma_{12}\Sigma_{22}^{-1}\Sigma_{2m})/[a_{11}a_{mm}]\}^2$. Since σ_{1m} must be increasing as ρ_{1m} increases, $\rho_{1m} = (\sigma_{1m}-\Sigma_{12}\Sigma_{22}^{-1}\Sigma_{2m})/[a_{11}a_{mm}]$, which is the partial correlation of the variables 1 and m given variables $2,\ldots,m-1$. □

Example 4.5 A special case consists of the multivariate distributions arising from a first-order Markov chain based on a copula C and a marginal distribution F. That is, $C_{j,j+1} = C$ for all j and C_{jk} corresponds to the independence copula if $k-j>1$. In this case, for $m \ge 4$, (4.39) can be more simply written as

$$F_{1,\ldots,m}(\mathbf{y}) = \int_{-\infty}^{y_2}\cdots\int_{-\infty}^{y_{m-1}} F_{1|2}(y_1|z_2)$$
$$\cdot F_{m|m-1}(y_m|z_{m-1})\,F_{2,\ldots,m-1}(dz_2,\ldots,dz_{m-1}),$$

with transition distribution $F_{i|i-1}(x_i|x_{i-1}) = B(F(x_{i-1}),F(x_i))$, where $B(u,v) = \partial C(u,v)/\partial u$. These Markov chains are studied further in Section 8.1. □

Special parametric families from this construction method are given in Section 5.5. With reference to the properties in Section 4.1, these families have interpretability and a wide range of dependence, closure only for some margins, densities without integrals through the recursions but no simple forms for cdfs (the behaviour is similar to MVN distributions).

*4.5.1 Dependence properties **

It should be clear from the construction method of (4.37)–(4.39) (and from the MVN example) that a wide range of dependence can result, by allowing the copulas C_{jk}, $j < k$, to range from the Fréchet lower bound to the Fréchet upper bound. A result for the trivariate case shows that the bounds for the Kendall tau values $\tau_{12}, \tau_{13}, \tau_{23}$ in (3.13) of Section 3.4.3 can be achieved when the Fréchet bounds are used in (4.37).

Theorem 4.18 *Let F_{123} be defined as in (4.37). Let τ_{jk} be the Kendall tau value for the (j,k) bivariate margin, $j < k$. If C_{13} in (4.37) is the Fréchet upper bound copula and $F_{12} \prec_{\mathrm{SI}} F_{32}$ (that is, $F_{3|2}^{-1}(F_{1|2}(y_1|y_2)|y_2)$ is (strictly) increasing in y_2), then $\tau_{13} = 1 - |\tau_{12} - \tau_{23}|$. Similarly, if C_{13} is the Fréchet lower bound copula and $F_{3|2}^{-1}(1 - F_{1|2}(y_1|y_2)|y_2)$ is (strictly) increasing in y_2, then $\tau_{13} = -1 + |\tau_{12} + \tau_{23}|$.*

Proof. Let $(X_{t1}, X_{t2}, X_{t3}), t = 1, 2$, be independent random vectors from the distribution F_{123}. With C_{13} being the Fréchet upper and lower bound, (4.37) becomes respectively

$$F_U(\mathbf{y}) = \int_{-\infty}^{y_2} \min\{F_{1|2}(y_1|z), F_{3|2}(y_3|z)\}\, F_2(dz)$$

and

$$F_L(\mathbf{y}) = \int_{-\infty}^{y_2} \max\{F_{1|2}(y_1|z) + F_{3|2}(y_3|z) - 1, 0\}\, F_2(dz).$$

For F_U, representations for the two vectors are $X_{13} = r(X_{11}, X_{12})$ and $X_{23} = r(X_{21}, X_{22})$ where $r(x_1, x_2) = F_{3|2}^{-1}(F_{1|2}(x_1|x_2)|x_2)$. Let E_1, E_2 be as defined in Theorem 3.13 with $(i,j,k) = (1,2,3)$. The function r is increasing in x_1, and if r is also increasing in x_2, then $(X_{11} - X_{21})(X_{12} - X_{22}) > 0$ implies $(X_{13} - X_{23})(X_{12} - X_{22}) > 0$ or $E_1 \subset E_2$, and the upper bound in (3.13) is attained. For F_L, representations are $X_{13} = s(X_{11}, X_{12})$ and $X_{23} = s(X_{21}, X_{22})$, where $s(x_1, x_2) = F_{3|2}^{-1}(1 - F_{1|2}(x_1|x_2)|x_2)$. If s is increasing in x_2, then

$(X_{11}-X_{21})(X_{12}-X_{22})<0$ implies $(X_{13}-X_{23})(X_{12}-X_{22})>0$. That is, $E_1^c \subset E_2$, and hence the lower bound in (3.13) is attained. A sufficient condition for s to satisfy the given condition is that both $F_{1|2}(\cdot|y)$ and $F_{3|2}(\cdot|y)$ are SI. More generally, the condition on s is equivalent to the $\prec_{\mathrm{SI}}$ ordering on F_{12}^* and F_{32}, where $F_{12}^*(x,y)=F_2(y)-F_{12}(F_1^{-1}(1-F_1(x)),y)$. □

The next results concern concordance and tail dependence.

Theorem 4.19 *As C_{jk} increases in concordance, with other bivariate margins held fixed, then $F_{j\cdots k}$ increases in the $\prec_{\mathrm{cL}}$ ordering and hence F_{jk} increases in concordance.*

Proof. This is obvious. □

It can be checked (for example, with the MVN family) that a stronger concordance property such as 'F_{13} increases in concordance as C_{12} increases in concordance' does not hold.

Theorem 4.20 *For the trivariate distribution given in (4.37), if C_{12} and C_{23} have upper tail dependence and some regularity conditions hold, then F_{13} has upper tail dependence. For the general m-dimensional distribution in (4.39), if $C_{j,j+1}$, $j=1,\ldots,m-1$, have upper tail dependence and some regularity conditions hold, then F_{jk}, $k-j>1$, all have upper tail dependence. (The tail dependence conditions appear in the proof).*

Proof. To stress ideas and concepts, we assume the existence of derivatives and other regularity conditions as needed. Some equivalent conditions for bivariate tail dependence are given first. Sometimes it is more convenient to be working with exponential margins than uniform margins. For a bivariate copula C, let

$$G(x,y)=C(1-e^{-x},1-e^{-y}). \tag{4.43}$$

The definition of upper tail dependence in Section 2.1.10 becomes

$$e^x\overline{G}(x,x)\to\lambda\in(0,1], \quad x\to\infty.$$

Now assuming that G has derivatives up to second order, let $G_{1|2}(x|y)=e^y\frac{\partial G(x,y)}{\partial y}$ and $\overline{G}_{1|2}=1-G_{1|2}$. Then

$$e^x\overline{G}(x,x)=e^x\int_x^\infty \overline{G}_{1|2}(x|y)e^{-y}\,dy=\int_0^\infty \overline{G}_{1|2}(x|x+v)e^{-v}\,dv.$$

Assuming that $e^x\overline{G}(x,x)$ converges as $x\to\infty$ and that

$$\overline{G}_{1|2}(x|x+v)\to a(v) \tag{4.44}$$

for all v ($v < 0$ is needed below), where a is continuous and $a \leq 1$, then by the bounded convergence theorem,

$$e^x \overline{G}(x, x) \to \int_0^\infty a(v) e^{-v} dv.$$

Tail dependence holds if and only if a is not identically 0 almost surely (a.s.) on $(0, \infty)$.

Now let g be the density of G. Then

$$\begin{aligned} e^x \overline{G}(x, x) &= e^x \int_x^\infty \int_x^\infty g(y_1, y_2)\, dy_1\, dy_2 \\ &= \int_0^\infty \int_0^\infty e^x g(x + v_1, x + v_2)\, dv_1\, dv_2. \quad (4.45) \end{aligned}$$

Assuming that

$$e^x g(x + v_1, x + v_2) \to b(v_1, v_2)$$

and that the Lebesgue dominated convergence theorem can be used in (4.45),

$$e^x \overline{G}(x, x) \to \int_0^\infty \int_0^\infty b(v_1, v_2)\, dv_1\, dv_2$$

and tail dependence holds if and only if b is not identically 0 (a.s.) on $(0, \infty)^2$.

Next suppose that C_{12}, C_{23} have upper tail dependence and that the Lebesgue dominated convergence theorem can be applied to (4.46) below. Then F_{13} in (4.37) has upper tail dependence and the tail dependence parameter is given in (4.47).

Let F_{12}, F_{23} be defined as in (4.43) with C_{12}, C_{23}, respectively. Let a be defined as in (4.44) with subscripts 12 or 32 for the (1,2) or (3,2) bivariate margin, respectively. Putting exponential margins in (4.37) leads to

$$F_{123}(\mathbf{y}) = \int_0^{y_2} C_{13}(F_{1|2}(y_1|z_2), F_{3|2}(y_3|z_2))\, e^{-z_2}\, dz_2$$

and

$$\begin{aligned} \overline{F}_{13}(x, x) &= 1 - F_1(x) - F_3(x) + F_{13}(x, x) \\ &= 1 - \int_0^\infty F_{1|2}(x|z)\, e^{-z} dz - \int_0^\infty F_{3|2}(x|z)\, e^{-z}\, dz \\ &\quad + \int_0^\infty C_{13}(F_{1|2}(x|z), F_{3|2}(x|z))\, e^{-z}\, dz \\ &= \int_0^\infty \overline{C}_{13}(F_{1|2}(x|z), F_{3|2}(x|z))\, e^{-z}\, dz. \end{aligned}$$

Hence

$$e^x \overline{F}_{13}(x,x) = \int_{-x}^{\infty} \overline{C}_{13}(F_{1|2}(x|x+v), F_{3|2}(x|x+v))\, e^{-v}\, dv \quad (4.46)$$

$$\to \int_{-\infty}^{\infty} \overline{C}_{13}(1-a_{12}(v), 1-a_{32}(v))\, e^{-v}\, dv \quad (4.47)$$

assuming the Lebesgue dominated convergence theorem can be used and a_{12}, a_{32} are the limits for $\overline{F}_{1|2}, \overline{F}_{3|2}$ as in (4.44).

Next for the multivariate extension. Suppose that $C_{j,j+1}$, $j = 1,\ldots,m-1$, have upper tail dependence and that all copulas C_{jk} have densities. For (4.39) with exponential univariate margins, suppose for j,k, with $k > j$, that the following pointwise convergences hold as $x \to \infty$:

(a) $\overline{F}_{j|j+1,\ldots,k}(x|x+v_{j+1},\ldots,x+v_k) \to a_{j,j+1,\ldots,k}(v_{j+1},\ldots,v_k)$,

(b) $\overline{F}_{k|j,\ldots,k-1}(x|x+v_j,\ldots,x+v_{k-1}) \to a_{k,j,\ldots,k-1}(v_j,\ldots,v_{k-1})$,

(c) $e^x f_{j,\ldots,k}(x+v_j,\ldots,x+v_k) \to b_{j,\ldots,k}(v_j,\ldots,v_k)$,

and that the functions on the right-hand sides of (a), (b), (c) are not identically 0 (a.s.). The remainder of the proof, which is omitted, makes use of these limits in an inductive manner. The Lebesgue dominated convergence theorem is assume to hold for the limits of integrals with terms from (a), (b) and (c). □

For the result in the above theorem, positive dependence for the copula C_{jk}, $k-j>1$, is not necessary. For example, even if C_{13} corresponds to the Fréchet lower bound, F_{13} can have upper tail dependence — under regularity conditions, the tail dependence parameter is $\lambda_{13} = \int_{-\infty}^{\infty} e^{-v} \max\{a_{12}(v) + a_{32}(v) - 1, 0\}\, dv$. The assumptions given in above theorem are not really too strong since they do hold in special cases such as those in Section 6.3.1.

4.6 Convolution-closed infinitely divisible class °

For parametric families of univariate distributions that are convolution-closed and infinitely divisible, there is a multivariate extension that makes use of these properties. It leads to positively dependent multivariate distributions only. These distributions are applied to time series models for count data, etc., in Section 8.4.

A family F_θ is **convolution-closed** if $Y_i \sim F_{\theta_i}$, $i = 1,2$, and Y_1, Y_2 independent implies the convolution $Y_1 + Y_2 \sim F_\eta$, where η is a function of θ_1, θ_2, usually the sum. A univariate distribution F is **infinitely divisible** if $Y \sim F$ and for all positive integers

n, there exists a distribution $F^{(n)}$ and iid rvs $Y_{n1}, \ldots, Y_{nn}$ with distribution $F^{(n)}$ such that $Y \stackrel{d}{=} Y_{n1} + \cdots + Y_{nn}$. If F_θ, $\theta > 0$, is a convolution-closed infinitely divisible parametric family such that $F_{\theta_1} * F_{\theta_2} = F_{\theta_1+\theta_2}$, where $*$ is the convolution operator, then $F_\theta^{(n)}$ can be taken to be $F_{\theta/n}$. It is assumed that F_0 corresponds to the degenerate distribution at 0.

Examples of convolution-closed infinitely divisible parametric families are Poisson(θ), Gamma(θ, σ) with σ fixed, and $N(0, \theta)$; others are given in Section 8.4. A convolution-closed parametric family that is not infinitely divisible is Binomial(θ, p) with p fixed. A parametric family that is infinitely divisible but not convolution-closed is the lognormal family.

Definition. Let Z_S, $S \in \mathcal{S}_m$, be $2^m - 1$ independent rvs in the family F_θ such that Z_S has parameter $\theta_S \geq 0$ (if the parameter is zero, the random variable is also zero). The stochastic representation for a family of **multivariate distributions with univariate margins in a given convolution-closed infinitely divisible class**, parametrized by $\{\theta_S : S \in \mathcal{S}_m\}$, is

$$X_j = \sum_{S: j \in S} Z_S, \quad j = 1, \ldots, m; \tag{4.48}$$

X_j has distribution F_{η_j}, where $\eta_j = \sum_{S \in \mathcal{S}_m : j \in S} \theta_S$.

In the bivariate case, the stochastic representation becomes

$$X_1 = Z_1 + Z_{12}, \quad X_2 = Z_2 + Z_{12}, \tag{4.49}$$

where Z_1, Z_2, Z_{12} are independent rvs in the family F_θ with respective parameters $\theta_1, \theta_2, \theta_{12}$. For $j = 1, 2$, the distribution of X_j is $F_{\theta_j + \theta_{12}}$.

The parameters of the above family can be interpreted as multiples of multivariate cumulants, which are defined next.

Definition. Let $(X_1, \ldots, X_m) \sim H$, with moment generating function $M(t_1, \ldots, t_m) = \mathrm{E}(\exp[t_1 X_1 + \cdots + t_m X_m])$ and cumulant generating function $K(\mathbf{t}) = \log M(\mathbf{t})$. The **multivariate mixed cumulant** of mth order is

$$\kappa_{12\cdots m} = \frac{\partial^m K}{\partial t_1 \cdots \partial t_m}\Big(0, \ldots, 0\Big).$$

When $m = 2$, the bivariate mixed cumulant κ_{12} is a covariance. Similarly, if S is a non-empty subset of $\{1, \ldots, m\}$, one can obtain the mixed cumulant κ_S of order $|S|$ from the marginal distribution H_S. The set $\{\kappa_S : S \in \mathcal{S}_m\}$ contains information about the dependence in H.

For a convolution-closed infinitely divisible family of cdfs F_θ, let K_θ be the corresponding family of cumulant functions. Assuming that enough moments exist, the rth cumulant of F_θ is $\kappa_\theta^{(r)} \stackrel{\text{def}}{=} K_\theta^{(r)}(0)$, the rth derivative evaluated at 0. Since $K_\theta(t) = NK_{\theta/N}(t)$ for all positive integers N, there are constants γ_r such that $\kappa_\theta^{(r)} = \gamma_r\theta$, $r = 1, 2, \ldots$. The joint cumulant generating function of (X_1, X_2) in (4.49) is $K(t_1, t_2) = K_{\theta_1,\theta_2,\theta_{12}}(t_1, t_2) = K_{\theta_1}(t_1) + K_{\theta_2}(t_2) + K_{\theta_{12}}(t_1 + t_2)$. Hence the bivariate cumulant is $\kappa_{12} = \kappa_{\theta_{12}}^{(2)}$, the second cumulant of Z_{12}. For (4.48), the joint cumulant generating function of $(X_1, \ldots, X_m)$ is $K(\mathbf{t}) = K(\mathbf{t}; \theta_S, S \in \mathcal{S}_m) = \sum_S K_{\theta_S}(\sum_{i \in S} t_i)$. Hence the mth-order mixed cumulant is $\kappa_{1\cdots m} = \kappa_{\theta_{1\cdots m}}^{(m)}$, the mth cumulant of $Z_{1\cdots m}$.

With reference to the properties in Section 4.1, these families of distributions are interpretable, closed under the taking of margins, have a wide range of positive dependence, but the densities and cdfs involve multi-dimensional sums or integrals that grow quickly in dimension as m increases. See Exercise 4.16 on an algorithm to find rvs of the form (4.48) that yield a given (non-negative) covariance matrix, when all univariate margins are in a given family in the convolution-closed infinitely divisible class.

Replacing the summation symbol $+$ by the minimum symbol $\wedge$, one can define multivariate families from F_θ that are closed under independent minima. This is the main idea in Marshall and Olkin (1967), for exponential distributions, and Arnold (1967). See also Chapter 6 on min-stable multivariate exponential distributions.

4.7 Multivariate distributions given bivariate margins

In preceding sections, families of multivariate distributions, some with a wide range of dependence structure, have been constructed. However, no method has been given that constructs a multivariate cdf from the set of bivariate margins. In this section we mention some approximate methods. One method, based on regression with binary variables, is a formula for constructing a multivariate object that has the right bivariate margins, but it cannot be shown analytically that the object is a proper cdf (it has been shown numerically in some cases). A second method is based on maximum entropy given the bivariate densities; this method generally does not have closed form and must be computed numerically as an approximation. These two methods are outlined in the following two subsections. A third method is given in Section 4.8.

4.7.1 Regression with binary variables

Let $m \geq 3$ be the dimension and let $\mathbf{X} \sim F$ with only the bivariate margins F_{ij}, $1 \leq i < j \leq m$, known. What is constructed in this subsection can be thought of as an approximation to F based on the set of bivariate margins. The approximation should be good when F is approximately maximum entropy given the bivariate margins or the information about F is contained almost entirely in the bivariate margins. For multivariate probabilities concerning $\mathbf{X}$, we need rectangle probabilities of the form

$$\Pr(w_1 < X_1 \leq x_1, \ldots, w_m < X_m \leq x_m). \tag{4.50}$$

This can be decomposed as the product of conditional probabilities:

$$\begin{aligned} &\Pr(w_1 < X_1 \leq x_1, w_2 < X_2 \leq x_2) \\ &\quad \cdot \textstyle\prod_{k=3}^{m} \Pr(w_k < X_k \leq x_k \mid w_j < X_j \leq x_j, j = 1, \ldots, k-1). \end{aligned} \tag{4.51}$$

Let $I_i = I(w_i < X_i \leq x_i)$, $i = 1, \ldots, m$, where $I(A)$ denotes the indicator of the event A. Note that $\mathrm{E}(I_i) = F_i(x_i) - F_i(w_i)$.

The first step is an approximation of

$$\begin{aligned} \Pr(w_k < X_k \leq x_k \mid w_1 < X_1 \leq x_1, \ldots, w_{k-1} < X_{k-1} \leq x_{k-1}) \\ = \mathrm{E}(I_k \mid I_1 = 1, \ldots, I_{k-1} = 1) \end{aligned} \tag{4.52}$$

by

$$\mathrm{E}(I_k) + \Omega_{21}\Omega_{11}^{-1}(1 - \mathrm{E}(I_1), \ldots, 1 - E(I_{k-1}))^T, \tag{4.53}$$

where Ω_{21} is a row vector consisting of the entries $\mathrm{Cov}(I_k, I_i) = \mathrm{E}(I_k I_i) - \mathrm{E}(I_k)\mathrm{E}(I_i)$, $i = 1, \ldots, k-1$, and Ω_{11} is a $(k-1) \times (k-1)$ matrix with (i,j) element $\mathrm{Cov}(I_i, I_j) = \mathrm{E}(I_i I_j) - \mathrm{E}(I_i)\mathrm{E}(I_j)$, $1 \leq i, j \leq k-1$. Note that $\mathrm{E}(I_i I_j) = \Pr(w_i < X_i \leq x_i, w_j < X_j \leq x_j) = F_{ij}(x_i, x_j) - F_{ij}(x_i, w_j) - F_{ij}(w_i, x_j) + F_{ij}(w_i, w_j)$. It is easily verified that (4.52) and (4.53) are identical if $k = 2$. The use of (4.53) as an approximation to (4.52) is analogous to the formula

$$\mathrm{E}(\mathbf{Y}_2 | \mathbf{Y}_1 = \mathbf{y}_1) = \boldsymbol{\mu}_2 + \Sigma_{21}\Sigma_{11}^{-1}(\mathbf{y}_1 - \boldsymbol{\mu}_1)$$

for a MVN random vector $(\mathbf{Y}_1^T, \mathbf{Y}_2^T)^T$ with mean vector $(\boldsymbol{\mu}_1^T, \boldsymbol{\mu}_2^T)^T$ and covariance matrix $\begin{bmatrix} \Sigma_{11} & \Sigma_{12} \\ \Sigma_{21} & \Sigma_{22} \end{bmatrix}$.

Expression (4.53) can be substituted into (4.51) to get one approximation to (4.50). However, the decomposition into conditional probabilities is not unique and different decompositions lead to different approximations in general. That is, (4.50) is also equivalent to

$$\Pr(w_{i_1} < X_{i_1} \le x_{i_1}, w_{i_2} < X_{i_2} \le x_{i_2})$$
$$\cdot \prod_{k=3}^{m} \Pr(w_{i_k} < X_{i_k} \le x_{i_k} \mid w_{i_j} < X_{i_j} \le x_{i_j}, j = 1, \ldots, k-1),$$

where $(i_1, \ldots, i_m)$ is a permutation of $(1, \ldots, m)$ with $i_1 < i_2$. There are $m!/2$ permutations that could be considered. For each permutation, an approximation of the form (4.53) can be used for each conditional probability. An overall approximation, denoted by $P = P(\mathbf{w}, \mathbf{x})$, for (4.50) is the average of the $m!/2$ approximations. For a permutation, if (4.53) happens to exceed 1 or be less than 0, it is replaced by 1 or 0, respectively.

A conjecture is that the approximation should be reasonable if the dependence is not too large and if the multivariate distribution is close to maximum entropy given the bivariate margins.

Another use of the approximation formula given a set of bivariate margins is for computing multivariate probabilities. The bivariate margins of P are as given (take w_k, x_k to be $-\infty, \infty$ except for $k = i, j$), but the additivity property of a probability measure is not satisfied. Hence, to get a formula here for a multivariate object given the set of bivariate margins, we are giving up the additivity property. For some applications, this may be acceptable.

Next let $w_j \to -\infty$, $j = 1, \ldots, m$, so that $P = P(\mathbf{x})$ is an approximate cdf; it has the bivariate margins F_{ij}, $i < j$, but need not correspond to a proper probability measure. The formula P in the trivariate case for lower orthant probabilities can be written explicitly by expanding the matrix inverse.

With $m = k = 3$ in (4.51)–(4.53), one gets

$$\begin{aligned} D(F_{12}, F_{13}, F_{23}) &\stackrel{\text{def}}{=} F_{12}F_3 \\ &+ F_{12}\big[(F_{13} - F_1F_3)(1 - F_2)(F_2 - F_{12}) \\ &\quad +(F_{23} - F_2F_3)(1 - F_1)(F_1 - F_{12})\big] \Big/ \\ &\quad \big[F_1F_2(1 - F_1 - F_2 + F_{12}) + F_{12}(F_1F_2 - F_{12})\big], \end{aligned}$$

where the arguments are x_1, x_2, x_3. This looks like a perturbation of $F_{12}F_3$; it can be shown that convergence as $F_i \to 1, i = 1, 2, 3$, leads to the right bivariate margins (use l'Hospital's rule for $i = 1, 2$). This can now be averaged over the three permutations to get

$$P = \big[D(F_{12}, F_{13}, F_{23}) + D(F_{13}, F_{12}, F_{23}) + D(F_{23}, F_{12}, F_{13})\big]/3. \tag{4.54}$$

This is a reasonably simple formula for a trivariate object with bivariate margins F_{12}, F_{13}, F_{23}.

It can be shown analytically that (4.54) is a trivariate distribution in some cases. For example, it is correct for independence and Fréchet upper bound margins, and combinations of Fréchet upper/lower bounds for the bivariate margins. Some numerical computations in the trivariate case seem to suggest that P is a proper cdf if F_{ij} is not too far away from F_iF_j for all $i < j$. Expression (4.54) can have negative rectangle 'probabilities' for small rectangles, but the use of $P(\mathbf{w}, \mathbf{x})$ leads always to non-negative 'probabilities' (that are not additive).

4.7.2 Maximum entropy given bivariate margins *

Generally maximum entropy problems refer to maximizing

$$-\int f \log f \, d\nu,$$

subject to some constraints on f, where f is a density with respect to the measure ν. The maximum entropy density can be interpreted as the density that is 'smoothest' given the constraints. In this subsection, we apply ideas of maximum entropy to the case where constraints are the given bivariate margins. The Appendix has some background results on maximum entropy.

For simplicity of notation, consider first the trivariate discrete case. Let p_{ijk}, $1 \le i \le I$, $1 \le j \le J$, $1 \le k \le K$, be a trivariate discrete pmf. The solution (if one exists) to the maximum entropy problem of maximizing $-\sum_{i,j,k} p_{ijk} \log p_{ijk}$ subject to $\sum_i p_{ijk} = m_{+jk} \forall j, k$, $\sum_j p_{ijk} = m_{i+k} \forall i, k$, $\sum_k p_{ijk} = m_{ij+} \forall i, j$, has the form

$$p_{ijk} = \exp\{\lambda_{+jk}\lambda_{i+k}\lambda_{ij+}\}$$

for some constants $\lambda_{+jk}, \lambda_{i+k}, \lambda_{ij+}$. This can easily be shown using the Lagrange multiplier method. The extension to the multivariate discrete case is straightforward.

To generalize to the continuous multivariate pdfs with given compatible bivariate margins, the method of calculus of variations can be used. The maximum entropy problem becomes that of maximizing $-\int f(\mathbf{x}) \log f(\mathbf{x}) \, d\mathbf{x}$ subject to $\int f(\mathbf{x}) \, d\mathbf{x}_{-i,-j} = f_{ij}(x_i, x_j)$ for all $1 \le i < j \le m$, where f_{ij} is the (i, j) bivariate marginal pdf and $\mathbf{x}_{-i,-j}$ is $\mathbf{x}$ without the ith and jth components.

To determine the solution, let $J(\epsilon) = \int (f(\mathbf{x}) + \epsilon g(\mathbf{x})) \log(f(\mathbf{x}) + \epsilon g(\mathbf{x})) \, d\mathbf{x}$. Then for the maximum entropy solution $f = f^*$, we must have $J'(0) = 0$ for all functions $g(\mathbf{x})$ satisfying $\int g(x) \, d\mathbf{x}_{-i,-j} = 0$

for all i, j. This reduces to $\int g(\mathbf{x}) \log f^*(\mathbf{x})\, d\mathbf{x} = 0$ for all such g. (Note that $J''(0) = \int [g^2(\mathbf{x})/f^*(\mathbf{x})]\, d\mathbf{x}$.) It is then easily verified that the solution f^* has the form $f^*(\mathbf{x}) = \prod_{i,j} h_{ij}(x_i, x_j)$ for a set of functions h_{ij}, where h_{ij} is positive whenever f_{ij} is positive.

An example (possibly the only one) where the conditions can be verified is the MVN density. The MVSN density with correlations ρ_{ij}, $1 \le i < j \le m$, is a maximum entropy density given BVSN margins f_{ij} with respective correlations ρ_{ij} (that result in a positive definite correlation matrix), since it can be decomposed into the form of the density in the preceding paragraph. For example, with $\Sigma^{-1} = (\rho^{ij})$ and $f(\mathbf{x}) = B \exp\{-\frac{1}{2} \sum \rho^{ij} x_i x_j\}$, take $h_{ij}(x_i, x_j) = B^{2/[m(m-1)]} \exp\{-\frac{1}{2}(m-1)^{-1}[\rho^{ii} x_i^2 + \rho^{jj} x_j^2] - \rho^{ij} x_i x_j\}$ for $1 \le i < j \le m$.

In the discrete trivariate case, the maximum entropy solution can be put in the form $p_{ijk} = \alpha_{ij} \beta_{ik} \gamma_{jk}$ for non-negative constants $\alpha_{ij}, \beta_{ik}, \gamma_{jk}$. These can be solved numerically with the proportional iterative rescaling method. The iteration has the form:

$$\alpha_{ij}^{(r+1)} = p_{ij+} \Big/ \sum_k \beta_{ik}^{(r)} \gamma_{jk}^{(r)}, \qquad \beta_{ik}^{(r+1)} = p_{i+k} \Big/ \sum_j \alpha_{ij}^{(r+1)} \gamma_{jk}^{(r)},$$

$$\gamma_{jk}^{(r+1)} = p_{+jk} \Big/ \sum_i \alpha_{ij}^{(r+1)} \beta_{ik}^{(r+1)},$$

starting with $\alpha_{ij}^{(0)} = \beta_{ij}^{(0)} = \gamma_{ij}^{(0)} = N^{-1}$, for example. For a set of three bivariate margins C_{12}, C_{13}, C_{23} in the form of copulas, discretization can be applied to bivariate copulas to get an approximation to the continuous (trivariate) maximum entropy density. Numerical experience is that the convergence is usually fast, with more iterations required as the dependence in C_{12}, C_{13}, C_{23} increases or as these copulas become more different. This numerical approximation generalizes to higher dimensions, but computational complexity increases exponentially with the dimension m.

4.8 Molenberghs and Lesaffre construction

In this section, we extend the ideas in Molenberghs and Lessafre (1994). Let $F_{ij}, 1 \le i < j \le m$, be given (compatible) bivariate margins, not necessary Plackett distributions as in the cited reference. We build up trivariate objects $F_{i_1 i_2 i_3}$, $1 \le i_1 < i_2 < i_3 \le m$, first and then extend to multivariate objects in higher dimensions, one dimension at a time. There are applications for multivariate binary and ordinal data (see Chapters 7 and 11).

Let $a_1 = F_{12}$, $a_2 = F_{13}$, $a_3 = F_{23}$, $a_4 = 1 - F_1 - F_2 - F_3 + F_{12} + F_{13} + F_{23}$, $b_1 = F_{12} + F_{13} - F_1$, $b_2 = F_{12} + F_{23} - F_2$, $b_3 = F_{13} + F_{23} - F_3$. A trivariate object F_{123} can be constructed from the F_{12}, F_{13}, F_{23} as a solution to the 'product ratio'

$$\psi = \psi_{123} = \frac{F_{123}(F_{123} - b_1)(F_{123} - b_2)(F_{123} - b_3)}{(a_1 - F_{123})(a_2 - F_{123})(a_3 - F_{123})(a_4 - F_{123})}. \quad (4.55)$$

The eight terms in (4.55) must be non-negative, so that a constraint on the bivariate margins is that $b \le a$, where $b = \max\{0, b_1, b_2, b_3\}$ and $a = \min\{a_1, a_2, a_3, a_4\}$. Note that these are the Fréchet bounds for $\mathcal{F}(F_{12}, F_{13}, F_{23})$ (see Section 3.4). Interpretations for the ratio (4.55) are given below.

Note that (4.55) must be solved pointwise for each $\mathbf{x} \in \Re^3$ in order to get $F_{123}(\mathbf{x})$. If the solution yields a proper cdf F_{123} and $\mathbf{X} \sim F_{123}$, then for a given $\mathbf{x}$, the ratio can be expressed in terms of orthant probabilities. With $I_j = I(X_j > x_j)$, $j = 1, 2, 3$, let

$$p_{0++} = F_1 = F_1(x_1) = \Pr(I_1 = 0),$$
$$p_{+0+} = F_2 = F_2(x_2),$$
$$p_{++0} = F_3 = F_3(x_3),$$
$$p_{00+} = F_{12} = F_{12}(x_1, x_2) = \Pr(I_1 = I_2 = 0),$$
$$p_{0+0} = F_{13},\ p_{+00} = F_{23}, \text{ and}$$
$$p_{000} = z = F_{123} = \Pr(I_1 = I_2 = I_3 = 0).$$

Then

$$p_{100} = \Pr(I_1 = 1, I_2 = I_3 = 0) = F_{23} - z,$$
$$p_{010} = \Pr(I_2 = 1, I_1 = I_3 = 0) = F_{13} - z,$$
$$p_{001} = \Pr(I_3 = 1, I_1 = I_2 = 0) = F_{12} - z,$$
$$p_{110} = \Pr(I_1 = I_2 = 1, I_3 = 0) = F_3 - F_{13} - F_{23} + z,$$
$$p_{101} = \Pr(I_1 = I_3 = 1, I_2 = 0) = F_2 - F_{12} - F_{23} + z,$$
$$p_{011} = \Pr(I_2 = I_3 = 1, I_1 = 0) = F_1 - F_{12} - F_{13} + z, \text{ and}$$
$$p_{111} = \Pr(I_1 = I_2 = I_3 = 1) = 1 - F_1 - F_2 - F_3 + F_{12} + F_{13} + F_{23} - z.$$

Hence (4.55) is the same as

$$\psi_{123} = [p_{000}p_{011}p_{101}p_{110}]/[p_{001}p_{010}p_{100}p_{111}].$$

Equation (4.55) can be written so that $z = F_{123}$ is the root of

$$\psi(a_1 - z)(a_2 - z)(a_3 - z)(a_4 - z) - z(z - b_1)(z - b_2)(z - b_3) \stackrel{\text{def}}{=} h(z).$$

$h(z)$ has exactly one root in the interval $[b, a]$, by checking the end-point values and the monotonicity. Clearly, $h(b) > 0$, $h(a) < 0$ and $h'(z) < 0$, if $b < a$. However, what has not been shown is that the solution results in a proper cdf, i.e., the non-negativity of the rectangle evaluations in (1.6). Even the monotonicity of F_{123} has not been shown.

In (4.55), the indices $(1, 2, 3)$ can be replaced by (i_1, i_2, i_3), so that, for example, the parameters $\psi_{123}, \psi_{124}, \psi_{134}, \psi_{234}$ in (4.55) correspond to the four trivariate margins of a 4-variate distribution. For $m \geq 4$, the extension can be made for the multivariate object $F_{1\cdots m}$ given m compatible $(m-1)$-dimensional marginals, defined through the 'product ratio', which has 2^{m-1} orthant probabilities in the numerator and 2^{m-1} orthant probabilities in the denominator. Assuming that the result is a proper cdf and $\mathbf{X} \sim F_{1\cdots m}$, define $I_j = I(X_j > x_j)$, $j = 1, \ldots, m$. The terms in the numerator have $r \equiv m \bmod 2$ I_j equal to 0 and the terms in the denominator have $r \equiv (m-1) \bmod 2$ I_j equal to 0.

For example, for $m = 4$, with $z = F_{1234}$, the 'product ratio' is:

$$\psi_{1234} = \frac{z(z - a_0)\prod_{1 \leq i < j \leq 4}(z - a_{ij})}{(F_{123} - z)(F_{124} - z)(F_{134} - z)(F_{234} - z)\prod_{i=1}^{4}(a_i - z)},$$

where the a_i and a_{ij} are defined near the equations (3.15) and (3.16).

Next we turn to some interpretations of (4.55); these extend to higher dimensions but the notation is cumbersome.

A first interpretation of (4.55) is in terms of cross-product ratios for bivariate Bernoulli distributions: with probabilities $\pi_{rs} = \Pr(Y_1 = r, Y_2 = s)$, $r, s = 0, 1$, for a random binary pair (Y_1, Y_2), the cross-product (odds) ratio is $\pi_{00}\pi_{11}/[\pi_{01}\pi_{10}]$. A continuous bivariate distribution can be discretized into bivariate binary distributions for pairs of cutoff points so that there is a cross-product ratio associated with each pair. ψ_{123} is the ratio of cross-product ratios of conditional bivariate distributions, and F_{123} is defined so that this ratio is constant over $\mathbf{x}$, i.e.,

$$\psi_{123} \equiv \frac{F_{12|3}[1 - F_{1|3} - F_{2|3} + F_{12|3}]}{[F_{1|3} - F_{12|3}][F_{2|3} - F_{12|3}]} \Bigg/ \frac{F_{12|3'}[1 - F_{1|3'} - F_{2|3'} + F_{12|3'}]}{[F_{1|3'} - F_{12|3'}][F_{2|3'} - F_{12|3'}]}, \tag{4.56}$$

where the first (numerator) cross-product ratio is conditional on $X_3 \leq x_3$ and the second (denominator) cross-product ratio is conditional on $X_3 > x_3$. The second ratio can be simplified to

$$\frac{\frac{F_{12}-F_{123}}{1-F_3}\left[1-\frac{F_1-F_{13}}{1-F_3}-\frac{F_2-F_{23}}{1-F_3}+\frac{F_{12}-F_{123}}{1-F_3}\right]}{\left[\frac{F_1-F_{13}}{1-F_3}-\frac{F_{12}-F_{123}}{1-F_3}\right]\left[\frac{F_2-F_{23}}{1-F_3}-\frac{F_{12}-F_{123}}{1-F_3}\right]}$$

$$= \frac{(F_{12}-F_{123})(1-F_1-F_2-F_3+F_{12}+F_{13}+F_{23}-F_{123})}{(F_1-F_{12}-F_{13}+F_{123})(F_2-F_{12}-F_{23}+F_{123})}$$

The first cross-product ratio simplifies to

$$\frac{F_{123}(F_3-F_{13}-F_{23}+F_{123})}{(F_{13}-F_{123})(F_{23}-F_{123})}$$

so that (4.56) simplifies to (4.55). Note that that (4.55) is symmetric in the three bivariate margins and (4.56) is not.

A second, maximum entropy, interpretation of ψ in (4.55) can be given, based on the binary variables I_1, I_2, I_3. Let $p_{r_1r_2r_3}$, $r_j = 0, 1$, $j = 1, 2, 3$, be defined as earlier. Consider the problem of maximizing the entropy $H(z) = -\sum_{r_1,r_2,r_3=0,1} p_{r_1r_2r_3} \log p_{r_1r_2r_3}$ subject to the constraints of the three bivariate margins. (See the Appendix for some background on maximum entropy.) z is constrained so that each of the eight probabilities $p_{r_1r_2r_3}$ is non-negative. Then $H'(z) = 0$ if and only if $\log z - \log p_{001} - \log p_{010} - \log p_{100} + \log p_{011} + \log p_{101} + \log p_{110} - \log p_{111} = \log\psi = 0$ or $\psi = 1$.

Hence the interpretation of the parameter $\psi = \psi_{123}$ is that $\psi = 1$ for the maximum entropy trivariate Bernoulli distribution given the three bivariate binary margins, and $\psi > 1$ [$\psi < 1$] for a larger [smaller] p_{000} (and a smaller [larger] p_{111}) compared with the maximum entropy distribution. Note that a different maximum entropy trivariate Bernoulli distribution results for each $\mathbf{x}$. This intepretation extends to higher dimensions. For the ratio ψ_{1234} involving 4-variate probabilities, $\psi_{1234} = 1$ for the maximum entropy 4-variate Bernoulli distribution given the four trivariate margins.

In (4.55), the Fréchet lower and upper bounds of $\mathcal{F}(F_{12}, F_{13}, F_{23})$ are attained as $\psi \to 0$ and $\psi \to \infty$, respectively. It is analytically shown in Section 3.4 that these Fréchet bounds are generally not proper distributions, and this is also true for the multivariate extension. This suggests that (4.55) and its extensions do not yield proper distributions if ψ is too small or too large (and this has been verified numerically). In any case, the useful thing about these formulas is that they are multivariate objects with margins equal to those given even if rectangle inequalities are not always satisfied. For example, (4.55) could be considered as a formula of an object that has bivariate margins F_{12}, F_{13}, F_{23}.

With reference to the properties in Section 4.1, these distributions are partially interpretable, can have closure under the taking of margins, have a wide range of dependence, and have cdfs that are not in closed form but not time-consuming to compute from recursions. It may be difficult to code the recursion for higher dimensions. The 'objects' from the construction can be used for multivariate binary and ordinal data but not continuous data as the densities do not have a simple form.

4.9 Spherically symmetric families: univariate margins

In this section, we work in an opposite direction from the other sections in this chapter. Given the easily constructed class of spherically symmetric distributions (which extend to elliptically contoured distributions), we study the possible univariate margins in this class (note that LTs reoccur). This then provides some information on when the class might be useful. With reference to the properties in Section 4.1, they are all satisfied except for a closed-form cdf, but the class of possible univariate margins is limited.

Spherically symmetric distributions are mixtures of distributions that are uniform on surfaces of hyperspheres (with varying radii). Their densities have contours that are spheres. If a density exists with respect to Lebesgue measure, then the density has the form $h(\mathbf{x}^T\mathbf{x})$ for a non-negative function h.

The first result below is on the univariate margins of uniform distributions on unit hyperspheres.

Theorem 4.21 *Suppose that* $\mathbf{Z}$ *is uniform on the surface of the unit hypersphere* $\{\mathbf{z} : z_1^2 + \ldots + z_m^2 = 1\}$*. Then the marginal distribution of* Z_1 *has density*

$$g_m(u) = \left[B(\tfrac{1}{2}, \tfrac{m-1}{2})\right]^{-1}(1-u^2)^{(m-3)/2}, \quad |u| \le 1, \qquad (4.57)$$

where B *is the beta function. More generally, for* $1 \le k < m$*,* $(Z_1, \ldots, Z_k)$ *has density*

$$g_{m,k}(u_1, \ldots, u_k) = \frac{\Gamma(m/2)}{\Gamma^k(\frac{1}{2})\Gamma((m-k)/2)}(1-u_1^2-\cdots-u_k^2)^{(m-k-2)/2}. \qquad (4.58)$$

Proof. Consider first the marginal distribution of Z_1^2. This is the same as the conditional distribution of Z_1^2 given $Z_1^2+\cdots+Z_m^2 = 1$, when $Z_1, \ldots, Z_m$ are iid $N(0,1)$ rvs. Since Z_i^2 has the chi-square distribution with one degree of freedom or the Gamma$(\frac{1}{2}, 2)$ distribution, the conditional distribution is Beta $(\frac{1}{2}, \frac{m-1}{2})$. Hence the density of Z_1 is

$$\left[B(\tfrac{1}{2},\tfrac{m-1}{2})\right]^{-1}[u^2]^{-1/2}(1-u^2)^{(m-3)/2}\,u.$$

The generalization is left as an exercise. □

The density (4.57) is increasing in $u>0$ if $m=2$, constant in u if $m=3$, and decreasing in $u>0$ if $m\geq 4$. Therefore, univariate marginal densities of spherically symmetric distributions are mixtures of densities of the form $r^{-1}g_m(u/r)I_{(-r,r)}(u)$, with g_m given in (4.57). Let $\mathcal{M}_m$ be the **set of possible univariate margins of spherically symmetric distributions** of dimension m; then a density $f_{m,1}$ in $\mathcal{M}_m$ has the form

$$f_{m,1}(x)=\int_0^\infty r^{-1}g_m(\tfrac{x}{r})\,I_{(-r,r)}(x)\,dG(r)=\int_{|x|}^\infty r^{-1}g_m(\tfrac{x}{r})\,dG(r), \tag{4.59}$$

where $G(r)$ is the probability that the radius is less than or equal to r (we are assuming that G has no mass at zero).

Let G be the distribution of the radial direction of the spherically symmetric distribution. If the spherically symmetric distribution has density $\phi_m(\mathbf{x}^T\mathbf{x})$, where $\mathbf{x}=(x_1,\ldots,x_m)^T$, then $G(r)=\int_0^r\phi_m(t^2)S_m t^{m-1}dt$, where $S_m=2\pi^{m/2}/\Gamma(m/2)$ is the surface area of the unit hypersphere in $\Re^m$. Hence the necessary condition (on ϕ_m) of

$$\int_0^\infty \phi_m(x^2)\,x^{m-1}dx=\int_0^\infty \phi_m(y)\,y^{m/2-1}\,dy<\infty$$

arises. Let R have the distribution G and let $\mathbf{U}$ be uniform on the surface of the unit hypersphere in $\Re^m$; then a stochastic representation for $\mathbf{X}$ with density $\phi_m(\mathbf{x}^T\mathbf{x})$ is $\mathbf{X}=R\mathbf{U}$. From this representation, X_1 has moments of order k if R has moments of order k ($k>0$ can be a non-integer). The necessary condition is $\int_0^\infty r^k\,dG(r)<\infty$ or $\int_0^\infty r^{k+m-1}\phi_m(r^2)\,dr<\infty$.

Next we return to the study of $\mathcal{M}_m$ for all $m\geq 2$. Lower-dimensional marginals of a spherically symmetric distribution are spherically symmetric so that $\mathcal{M}_m\subset\mathcal{M}_{m-1}$ for $m\geq 3$. Also lower-dimensional marginals always have densities with respect to Lebesgue measure, even if the spherically symmetric distribution has mass on some surfaces of hyperspheres. If there is a density in $\Re^m$ and if $\phi_j(x_1^2+\cdots+x_j^2)$ is the marginal density of $(X_1,\ldots,X_j)$

for $2 \le j < m$, then

$$\phi_j(y^2) = \int_{-\infty}^{\infty} \cdots \int_{-\infty}^{\infty} \phi_m(y^2 + x_{j+1}^2 + \cdots + x_m^2)\, dx_{j+1} \cdots dx_m. \tag{4.60}$$

If the spherically symmetric distribution has mass on some surfaces of m-dimensional hyperspheres, say mass p_i for the radius of r_i, $i = 1, 2, \ldots$, and if ϕ_m is the absolutely continuous part of the density, then

$$\phi_j(y^2) = \int_{-\infty}^{\infty} \cdots \int_{-\infty}^{\infty} \phi_m(y^2 + x_{j+1}^2 + \cdots + x_m^2)\, dx_{j+1} \cdots dx_m$$

$$+ \sum_i p_i\, r_i^{-j} g_{m,j}^*(y^2/r_i^2)\, I_{[0,r_i^2]}(y^2),$$

where $g_{m,k}^*(z)$ obtains from the right-hand side of (4.58), with argument z in place of $u_1^2 + \cdots + u_k^2$.

From (4.59) and (4.57), densities in $\mathcal{M}_m$ for $m \ge 3$ are decreasing on $[0, \infty)$ (and symmetric about zero). For $m = 3$, all symmetric densities that are decreasing on $[0, \infty)$ are in $\mathcal{M}_3$. From (4.57) and its derivation, $g_m(u) = [B(\frac{1}{2}, \frac{m-1}{2})]^{-1}(1-u^2)^{(m-3)/2} I_{[-1,1]}(u)$ is in $\mathcal{M}_m$ but not in $\mathcal{M}_{m+1}$. More generally, if a spherically symmetric distribution has mass on some surfaces of m-dimensional hyperspheres, then its univariate marginal density is not in $\mathcal{M}_{m+1}$.

An interesting problem is the characterization of

$$\mathcal{M}_\infty = \cap_{n=1}^{\infty} \mathcal{M}_n.$$

This can be studied using a recursion formula for ϕ_{m-2} from ϕ_m. Let A be the upper bound of support for the radial variable; A could be finite or infinite. From the above, we can suppose that there is no mass at the point A. For $j = m - 2$ in (4.60), making a polar coordinate transform from (x_{m-1}, x_m) to (s, θ) leads to

$$\phi_{m-2}(y^2) = 2\pi \int_0^{\sqrt{A^2-y^2}} \phi_m(y^2 + s^2)\, s\, ds = 2\pi \int_y^A \phi_m(u^2)\, u\, du.$$

Hence $\phi'_{m-2}(y^2) = -\pi\, \phi_m(y^2)$ or $\phi_m(y^2) = -\pi^{-1} \phi'_{m-2}(y^2)$. If $m = 2j + 1$ is an odd integer greater than 2, then by recursion,

$$\phi_{2j+1} = (-1)^j \pi^{-j} \phi_1^{(j)}, \tag{4.61}$$

where $\phi_1^{(j)}$ is the jth derivative of ϕ_1 and $\phi_1(y^2) = f(y)$ is the univariate marginal density. (Note that the recursion in (4.61) still holds for lower-dimensional margins if a spherically symmetric distribution has mass on some surfaces of hyperspheres.) For example, if $\phi_1(w) = (2\pi)^{-1/2}e^{-w/2}$ for the standard normal density, then $\phi_{2j+1}(w) = (2\pi)^{-j-1/2}e^{-w/2}$, and if $\phi_1(w) = c_\nu(1 + w/\nu)^{-(\nu+1)/2}$ for the t distribution with ν degrees of freedom, then $\phi_{2j+1}(w) = c_\nu\pi^{-j}[\prod_{i=1}^{j}(\frac{1}{2} + \frac{2i-1}{2\nu})](1 + w/\nu)^{-(\nu+2j+1)/2}$.

Since the left-hand side ϕ_{2j+1} of (4.61) is non-negative for a proper density, if $f(y) = \phi_1(y^2)$ is in $\mathcal{M}_\infty$, then ϕ_1 is completely monotone (see the Appendix for the definition). If $A = \infty$ and $\phi_1 > 0$ on $[0, \infty)$, then it is a multiple of a LT of a non-negative rv and has the form $\phi_1(w) = \phi_1(0)\int_0^\infty e^{-xw}\,dP(x)$, where P is the cdf of the non-negative rv. Hence

$$f(y) = \phi_1(0)\int_0^\infty e^{-xy^2}\,dP(x) \tag{4.62}$$

is a scale mixture of normal densities with mean 0. There are some conditions for the mixing distribution P in order that $f \in \mathcal{M}_\infty$. From (4.62) and the necessary condition for the radial density, $\int_0^\infty w^{j-1/2}\phi_{2j+1}(w)\,dw < \infty$ implies

$$\begin{aligned}&\int_0^\infty x^j \int_0^\infty w^{j-1/2}\,e^{-xw}\,dw\,dP(x)\\ &\quad = \textstyle\int_0^\infty x^j\,\Gamma(j+\frac{1}{2})x^{-j-1/2}\,dP(x) < \infty\end{aligned}$$

or $\int_0^\infty x^{-1/2}dP(x) < \infty$. Also $\phi_{2j+1}(w) = \pi^{-j}\int_0^\infty x^j e^{-xw}\,dP(x) < \infty$ for all $j \geq 1$, which implies that the jth integer moment of the mixing distribution P must exist in order for $\phi_{2j+1}(0)$ to be finite. Equation (4.62) can be written more clearly as a scale mixture of normal densities, i.e.,

$$\phi_1(w) = (2\pi)^{-1/2}\int_0^\infty \exp\{-\tfrac{1}{2}wa^2\}\,a\,dQ(a), \tag{4.63}$$

for a distribution Q. If X has density $f(y) = \phi_1(y^2)$, (4.63) corresponds to the stochastic representation $X = Z/S$, where Z is standard normal and S is a positive rv with distribution Q. Now

$$\phi_{2j+1}(w) = \pi^{-j}(2\pi)^{-1/2}\int_0^\infty \exp\{-\tfrac{1}{2}wa^2\}\,(a^2/2)^j\,a\,dQ(a)$$

and $\phi_{2j+1}(0)$ is finite only if the moment of order $2j+1$ of Q is finite. The condition $\int_0^\infty w^{j-1/2}\phi_{2j+1}(w)\,dw < \infty$ becomes

$$\int_0^\infty (a^2/2)^j\,a\,\Gamma(j+\tfrac{1}{2})\,(a^2/2)^{-j-1/2}dQ(a) = \sqrt{2}\,\Gamma(j+\tfrac{1}{2}) < \infty$$

so that it is always satisfied.

Hence a density in $\mathcal{M}_\infty$ which is positive everywhere must be a scale mixture of normal densities. It is easily verified directly that if $\mathbf{X} = \mathbf{Z}/S$ where $Z_1, \ldots, Z_m$ are iid $N(0,1)$ and S is a positive rv, then $\mathbf{X}$ has a spherically symmetric distribution and the univariate marginal density is that of Z_1/S for all m. The density of $\mathbf{X}$ is

$$(2\pi)^{-m/2} \int_0^\infty a \exp\Big\{-\tfrac{1}{2}a^2 \sum_{i=1}^m x_i^2\Big\}\, dF_S(a),$$

where F_S is the cdf of S and the univariate marginal is

$$(2\pi)^{-1/2} \int_0^\infty a \exp\{-\tfrac{1}{2}a^2 x^2\}\, dF_S(a).$$

For example, for the t distribution with ν degrees of freedom, $S = \sqrt{U/\nu}$, where $U \sim \chi^2_\nu$. Since $U/\nu \sim \text{Gamma}(\nu/2, 2/\nu)$, the density of S is $f_S(a) = 2\,[\Gamma(\nu/2)]^{-1}(\nu/2)^{\nu/2}a^{\nu-1}e^{-\nu a^2/2}$.

Now suppose $0 < A < \infty$. What are the completely monotone functions ϕ_1 in this case? It turns out there are none if the requirements of $(-1)^j\phi^{(j)} > 0$ on $[0, A)$ and $\phi^{(j)}(A) = 0$, $j \geq 1$, are to be met. We next show that $\phi^{(j)}(A) = 0$ is needed to extend to densities of higher dimensions by making use of the recursion formula.

If the m-variate density ϕ_m is given, the $(m-1)$-variate density ϕ_{m-1} satisfies

$$\phi_{m-1}(v) = 2\int_0^{\sqrt{A^2-v}} \phi_m(v + x^2)\, dx.$$

Taking the derivative,

$$\phi'_{m-1}(v) = 2\int_0^{\sqrt{A^2-v}} \phi'_m(v + x^2)\, dx - \phi_m(A^2)\,(A^2 - v)^{-1/2}.$$

Then $\phi_{m+2} = -\phi'_m/\pi \geq 0$ and $\phi_{m+1} = -\phi'_{m-1}/\pi \geq 0$ are together possible only if $\phi_m(A^2) = 0$. (This condition is automatically satisfied if $A = \infty$.) Therefore if one tries to extend for a function ϕ_1 on $[0, A]$ by defining $\phi_{2j+1} = (-1)^j\pi^{-j}\phi_1^{(j)}$ as in (4.61), then a necessary additional condition is $\phi_1^{(j)}(A) = 0$, $j \geq 1$.

Example 4.6 A few cases are listed to illustrate the ideas in the two preceding paragraphs.

1. (Uniform on surface of hypersphere of dimension $m > 3$.) Let $\phi_1(w) = [B(\frac{1}{2}, \frac{m-1}{2})]^{-1}(1-w)^{(m-3)/2}$. If $m = 2n + 1$ is odd,

then the derivatives of ϕ_1 alternate in sign up to the $(n-1)$th derivative and the nth derivative is a constant, and hence (4.61) cannot be extended beyond dimension m. If $m = 2n$ is even, then the derivatives of ϕ_1 alternate in sign up to the $(n-1)$th derivative and then do not change sign, $\phi_1^{(j)}(1) = 0$ for $j < n-1$, and $\phi_1^{(n-1)}(1) = \infty$, and hence (4.61) cannot be extended beyond dimension m.

2. Let $\phi_1(w) = c_\beta(1-w)^\beta$, $0 < w < 1$, where $\beta > 0$. The derivatives of ϕ_1 alternate in sign up to the $[\beta]$th derivative, where $[\beta]$ is the ceiling integer function. Also $\phi_1^{(j)}(1) = 0$ for $j < [\beta]$. Hence $f(y) = \phi_1(y^2)$ is in $\mathcal{M}_{2[\beta]+1}$ and at most $\mathcal{M}_{2[\beta]+3}$.
3. Let $\phi_1(w) = c_\alpha(1-w^\alpha)$, $0 < w < 1$, where $0 < \alpha < 1$. The derivatives of ϕ_1 alternate in sign, but $\phi_1'(w) = -\alpha c_\alpha w^{\alpha-1} = -\alpha c_\alpha \neq 0$ when $w = 1$.

□

Next we return to some conditions for a density to be in $\mathcal{M}_m$. Because of the boundary condition for density with support on a bounded interval, consider only densities that are continuously differentiable up to some order (for bounded support, this means the density and its derivatives at the end point of support are 0). From (4.61), $f(y) = \phi_1(y^2)$ is in $\mathcal{M}_{2n+1}$ but not $\mathcal{M}_{2n+3}$ if $(-1)^j\phi_1^{(j)} \geq 0$, $j = 1, \ldots, n$, and $(-1)^{n+1}\phi_1^{(n+1)}(w) < 0$ for some w. Since one also has the recursion $\phi_{2j+2} = (-1)^j\pi^{-j}\phi_2^{(j)}$ for a spherically symmetric distribution in $2j+2$ dimensions, $f(y) = \phi_1(y^2) \in \mathcal{M}_2$ is in $\mathcal{M}_{2n}$ but not $\mathcal{M}_{2n+2}$ if $(-1)^j\phi_2^{(j)} \geq 0$, $j = 1, \ldots, n-1$, and $(-1)^n\phi_2^{(n)}(w) < 0$ for some w.

Example 4.7 For some symmetric densities f which are decreasing on $[0, \infty)$, we check for the largest m such that $f \in \mathcal{M}_m$.

1. $f_\alpha(x) = c_\alpha \exp\{-|x|^\alpha\}$, $-\infty < x < \infty$, where $\alpha > 0$, and $\phi_1(w) = c_\alpha \exp\{-w^{\alpha/2}\}$, $w \geq 0$. For $0 \leq \alpha \leq 2$, ϕ_1 is a multiple of a LT so that $f_\alpha \in \mathcal{M}_\infty$. In particular, for $\alpha = 1$, the double exponential density is in $\mathcal{M}_\infty$. The second derivative of ϕ_1 is $c_\alpha(\alpha/2)w^{\alpha/2-2}\exp\{-w^{\alpha/2}\}[\alpha w^{\alpha/2}/2 + (1-\alpha/2)]$; it can be negative if $\alpha > 2$ and w is near 0. Therefore for $\alpha > 2$, f_α is not in $\mathcal{M}_5$. ϕ_2 can be obtained from ϕ_3 using (4.60) and then $\phi_4 = -\pi^{-1}\phi_2'$. It has been checked numerically that $\phi_4(w)$ is not non-negative for all $w \geq 0$ when $\alpha > 2$, so that f_α is also not in $\mathcal{M}_4$ for $\alpha > 2$.
2. (Logistic.) $f(x) = e^{-x}/(1+e^{-x})^2$ is in $\mathcal{M}_\infty$, and $\phi_1(w) = (e^{\sqrt{w}/2} + e^{-\sqrt{w}/2})^{-2} = (2 + e^{\sqrt{w}} + e^{-\sqrt{w}})^{-1}$ is a LT. If X has a

standard logistic distribution and $Z \sim N(0,1)$, then $X \stackrel{d}{=} Z/V$, where V has density

$$g(v) = 2\sum_{k=1}^{\infty}(-1)^{k-1}k^2v^{-3}\exp\{-k^2/(2v^2)\}.$$

$(2V)^{-1}$ has the asymptotic distribution of the Kolmogorov distance statistic.

□

4.10 Other approaches

Other approaches that have been used for constructing multivariate families but which are not discussed or used in this book are:

(i) multivariate generalizations of univariate moment or probability generating functions, e.g., several families of multivariate gamma distributions in Krishnaiah (1985) and compound bivariate Poisson distributions in Kocherlakota (1988);

(ii) multivariate characteristic functions, e.g., multivariate stable distributions in Press (1972);

(iii) multivariate functional equations generalizing those satisfied by univariate distributions, e.g., families of multivariate exponential survival functions in Ghurye and Marshall (1984) and Marshall and Olkin (1991);

(iv) infinite series expansions — there are some for the bivariate case without multivariate extensions, e.g., the bivariate gamma distribution of Kibble (1941) and other distributions in Lancaster (1969).

4.11 Bibliographic notes

Variations of the property of closure of multivariate models under the taking of margins are presented in Xu (1996). This includes the concept of model parameters being marginally expressible, which is given in Section 4.1, as well as the concept of parameters being expressible from or appearing in univariate and bivariate margins.

The families in Section 4.2 are from Marshall and Olkin (1988) and Joe (1993); the dependence results in Theorems 4.6 to 4.10 are new (thanks are due to T. Hu for help in the completion of these proofs). See Genest and MacKay (1986) for some background on bivariate Archimedean copulas and results on orderings of these copulas. In the literature for Archimedean copulas, sometimes ϕ (or some other symbol) denotes a LT (or a function in $\mathcal{L}_1$) and sometimes it denotes the inverse of a function in this class, so the reader should be careful in using these results. Frailty models for a special type of multivariate survival data in the familial or cluster setting are studied in Oakes (1989), Hougaard (1986) and Hougaard, Harvald and Holm (1992).

References are Joe and Hu (1996) for Section 4.3, Joe (1996a) for Section 4.5, Joe (1996b) for Section 4.6 and Joe (1995) for Section 4.7.1. The special case of the multivariate Poisson distribution is given in Teicher (1954). The application in Joe (1995) is for the MVN distribution; an improved approximation based on having all of the trivariate and 4-variate margins is also given. A reference for Section 4.8 is Molenberghs and Lesaffre (1994); this does not prove that the multivariate extension is a proper distribution, and neither does Plackett (1965) for the bivariate case. A reference for Section 4.9 is Kelker (1970); see also Chapter 2 of Fang, Kotz and Ng (1990) for a different treatment. A reference for the infinite divisibility of the lognormal distribution is Thorin (1977). See Andrews and Mallows (1974) and Stefanski (1991) for the logistic distribution as a scale mixture of normals.

For results and construction methods for multivariate distributions with given non-overlapping multivariate margins, see Marco and Ruiz-Rivas (1992), Genest, Molina and Lallena (1995) and Li, Scarsini and Shaked (1996).

4.12 Exercises

4.1 Let M be a univariate cdf. Let $G_1(\cdot;\alpha)$ and $G_2(\cdot;\alpha)$ be families of distributions indexed by a real-valued parameter α. Define $F(x_1,x_2) = \int G_1(x_1;\alpha)G_2(x_2;\alpha)\,dM(\alpha)$.

(a) Show that if G_1 and G_2 are both stochastically increasing or both stochastically decreasing as α increases, then F is positively dependent in several senses (e.g., association, PQD).

(b) Show that if G_1 increases stochastically and G_2 decreases stochastically as α increases, then F is negatively dependent in the sense of NQD.

What results generalize to m dimensions with

$$F(\mathbf{x}) = \int G_1(x_1;\alpha)\cdots G_m(x_m;\alpha)\,dM(\alpha)?$$

4.2 Let H be a univariate cdf. Show that H^α is stochastically increasing in $\alpha > 0$.

4.3 Show that (4.4) is invariant to scale changes in the LT. That is, if in (4.4) $\phi(s)$ is replaced by the function $\phi^*(s) = \phi(s/\sigma)$ for $\sigma > 0$, then the same copula results.

4.4 In Section 4.6, take F_θ to be the family of Poisson distributions. Obtain the pmfs for the bivariate and trivariate Poisson distributions with the representations given by (4.48) and (4.49).

4.5 Show that the multivariate Poisson distribution (Section 4.6) satisfies property B of Section 4.1 but not property B′.

4.6 In (4.37), let F_{12}, F_{23} be copulas in the family B10 with parameters θ_{12}, θ_{23} respectively and let C_{13} be the independence copula. Obtain F_{123} and its (1,3) bivariate margin. Extend this to a result for (4.39). (Joe 1996a)

4.7 For the partially symmetric copulas in Section 4.2, show that there are three distinct forms for $m = 5$ that generalize (4.7), (4.10) and (4.11). How many distinct forms are there for dimension $m = 6$?

4.8 In (4.3), substitute in the Poisson LT $\psi(s) = e^{-\theta}\exp\{\theta e^{-s}\}$, $\theta > 0$. Even though ϕ^{-1} is defined only on $[e^{-\theta}, 1]$, show that (4.3) leads to the function

$$u_1 u_2 \exp\{\theta^{-1} \log u_1 \log u_2\}.$$

Show that this is not a proper cdf, even though it has the $U(0,1)$ margins (compare (5.18) in Section 5.4).

4.9 Show that the distribution (4.40) is equivalent to that from (4.37)–(4.39) with $C^*_{jk}(u,v) = u + v - 1 + C_{jk}(1-u, 1-v)$.

4.10 For the distribution in (4.37), show that a stronger concordance property such as 'F_{13} increases in concordance as C_{12} increases in concordance' does not hold.

4.11 Derive (4.58).

4.12 Let g be an increasing function on $[0,\infty)$ satisfying $g(0) = 0$. Show that g convex implies that g is star-shaped ($x^{-1}g(x)$ increasing in x) which in turns implies that g is superadditive ($g(x_1 + x_2) \ge g(x_1) + g(x_2)$ for all $x_1, x_2 \ge 0$).

4.13 Obtain the tail dependence parameters for (4.3) for other families of LTs (in the Appendix).

4.14 Analyse the concordance properties of (4.4) with $m = 2$ and the LT family LTM in (4.4).

4.15 Consider the function

$$F(u,v,w) = \psi_\alpha(\psi_\alpha^{-1} \circ \psi_\delta(\psi_\delta^{-1}(u) + \psi_\delta^{-1}(v))$$
$$+\psi_\alpha^{-1} \circ \psi_\beta(\psi_\beta^{-1}(u) + \psi_\beta^{-1}(w)) - \psi_\alpha^{-1}(u)),$$

where ψ_θ is a family of LTs. Let $C(u,v;\theta) = \psi_\theta(\psi_\theta^{-1}(u) + \psi_\theta^{-1}(v))$. Show that $F(u,v,w)$ is a formula with bivariate margins $C(v,w;\alpha)$, $C(u,w;\beta)$, $C(u,v;\delta)$ as u,v,w respectively tend to 1, but that it is not a proper copula in general.

4.16 Suppose we want to generate rvs with a multivariate distribution of the form (4.48) with a given (feasible) covariance matrix Σ when all univariate margins are in a family F_θ that is in the convolution-closed infinitely divisible class. Assume that F_θ has been rescaled so that the variance is 1 when $\theta = 1$, or that F_θ is parametrized by the variance. Let $X_i \sim F_{\alpha_i}$, $i = 1,\ldots,m$, and let $Z_S \sim F_{\theta_S}$, $S \in \mathcal{S}_m$. Then $\alpha_i = \theta_i + \sum_{S:i\in S,|S|\geq 2} \theta_S$ and $\sigma_{ii'} = \sum_{S:i,i'\in S} \theta_S$ for $i \neq i'$. The algorithm which follows yields the desired constants as well as determining whether a given covariance matrix is possible with form (4.48). Verify the details.

(a) Let $\Omega = \Sigma = (\omega_{ii'})$. Set $\gamma \leftarrow \min_{i<i'} \omega_{ii'}$. If $\gamma > 0$, set $\theta_{\{1,\ldots,m\}} = \gamma$, $S = \{1,\ldots,m\}$, and go to step (c).

(b) If $\min_{i<i'} \omega_{ii'} = 0$, set $\gamma \leftarrow \min\{\omega_{ii'} : \omega_{ii'} > 0, i < i'\}$. If $\gamma = \omega_{kk'} > 0$, let S be a maximal set that contains k, k', with maximal meaning that if $j, j' \in S$, then $\omega_{jj'} > 0$. Set $\theta_S = \gamma$. If $\gamma = 0$, go to step (d).

(c) Set $\Omega \leftarrow \Omega - \gamma C$, where $C = (c_{ii'})$ with $c_{ii} = I(i \in S)$, and $c_{ii'} = I(i,i' \in S)$ for $i \neq i'$. If Ω is such that $\omega_{ii'} > [\omega_{ii}\omega_{i'i'}]^{1/2}$, for some $i \neq i'$, then the initial matrix Σ is not feasible. Otherwise, go to step (b).

(d) Set $\alpha_i = \omega_{ii}$, $i = 1,\ldots,m$.

4.17 If $f \in \mathcal{M}_m$ has cdf F and finite second moments, and $h(\mathbf{x};\Sigma) = |\Sigma|^{-1}\phi_m(\mathbf{x}^T\Sigma^{-1}\mathbf{x})$ is the corresponding family of elliptically contoured distributions, then the full range of correlation matrices (of order m) is possible. If the distribution F cannot result from the location–scale transform of a density in $\mathcal{M}_m$, then the full range of correlation matrices, in the class $\mathcal{F}(F,\ldots,F)$ of m-variate distributions with all univariate margins equal to F, cannot be achieved. The full range of correlation matrices for all dimensions is achievable only if $f \in \mathcal{M}_\infty$. As an example, the full range for $U(-1,1)$ and $U(0,1)$ margins is possible for dimension 3 but not $m > 3$.

4.18 Modify Theorem 4.17 to obtain a family of trivariate copulas such that (i) the (1,2) and (3,2) bivariate margins are the same and are negatively dependent; (ii) the (1,3) bivariate margin is positively dependent; and (iii) the Kendall tau value $\tau_{12} = \tau_{23}$ is the most negative possible given τ_{13}.

4.19 Prove the results in Theorems 4.15 and 4.16 on lower tail dependence. (Joe and Hu 1996)

4.20 Let $C(\mathbf{u}) = \phi(\sum_{j=1}^m \phi^{-1}(u_j))$ as in (4.4).

(a) Show that C is MTP_2 if and only if $\log \phi$ is L-superadditive (see Unsolved Problem 2.4 for the definition of L-superadditive). Note that because ϕ is differentiable, the condition is equivalent to $\partial^2 \log(\phi(x+y))/\partial x \partial y \geq 0$.

(b) Show that the density of C is MTP_2 if and only if $(-1)^m \phi^{(m)}(x+y)$ is TP_2 in x, y.

(c) Check whether the LT families in the Appendix lead to families of copulas with the MTP_2 property for either the cdf or the density.

4.21 Do some analysis on (4.54) with various bivariate margins including the Fréchet bounds.

4.13 Unsolved problems

4.1 Find parametric families of copulas that satisfy all of the desirable properties in Section 4.1.

4.2 Obtain conditions for (4.54) and extensions to be proper cdfs.

4.3 Obtain conditions for (4.55) and extensions to be proper cdfs.

4.4 Prove or disprove the $\prec_{\mathrm{pfd}}$ ordering for C_1 and C_2 in Theorems 4.8, 4.9 and 4.10.

CHAPTER 5

Parametric families of copulas

This chapter is intended as a *reference* of useful parametric families of copulas together with their properties. The inclusion of properties is important because, in a given situation or application, the choice of appropriate models can depend on the properties.

Many of the parametric families of multivariate copulas make use of the theory in Chapter 4, and are useful for multivariate models in subsequent chapters. Some of the families are also referred to earlier in this book. A summary of the sections, including the highlights, is the following. Section 5.1 consists of bivariate one-parameter families of copulas with nice dependence properties and Section 5.2 consists of two-parameter families of copulas; these families can be used to build multivariate copulas. Section 5.3 has multivariate extensions to symmetric and partially symmetric copulas; these are the only known class of parametric families of copulas that have closed-form cdfs, are closed under the taking of margins (in the stronger sense of property B′ in Section 4.1), and extrapolate between the independence and Fréchet upper bound copulas. Section 5.4 has extensions of families in Sections 5.1 to 5.3 to include negative dependence. Section 5.5 consists of parametric families of copulas that cover general dependence structures, including some that have closed-form cdfs.

5.1 Bivariate one-parameter families °

Listed in the first part of this section are known simple one-parameter families of copulas that: (i) interpolate between independence and Fréchet upper bound; (ii) are absolutely continuous; and (iii) have support on all of $(0,1)^2$. Also these families are symmetric in the two arguments. If these conditions are relaxed, there are (infinitely) many other one-parameter families, and a few are listed in the last part of this section. One-parameter families of copulas are parsimonious models that are good starting points for modelling. They are useful for bivariate data, as well as a component of the multivariate copulas in Sections 4.3, 4.5, 4.7 and 4.8, and of the first-order Markov time series in Sections 8.1 and 11.6.

The notation $C(u, v; \delta)$ is used for a family of copulas, with the dependence parameter δ increasing as the dependence increases. The original source of each family of copulas is given, as well as the density and some properties. Besides dependence properties, other properties include reflection symmetry, extreme value copula, existence of a LT so that the family has form (4.3), and multivariate extendibility. **Reflection symmetry** for a copula C means that if $(U, V) \sim C$, then $(1-U, 1-V) \sim C$; this property is convenient for latent variable models for multivariate binary data. A bivariate copula C is an **extreme value copula** if $C(u^t, v^t) = C^t(u, v)$ for all $t > 0$. After transferring to unit exponential survival margins, with $\overline{G}(x, y) = C(e^{-x}, e^{-y})$, the extreme value copulas are easily recognized from $A(x, y) = -\log G(x, y)$ being homogeneous of order 1, i.e., $A(tx, ty) = tA(x, y)$ for all $t > 0$. (With details in Chapter 6, a G of this form is a **min-stable bivariate exponential** distribution.) Some families of extreme value copulas are obtained as the extreme value limits of other families (see Chapter 6). The **extreme value limits** from the lower and upper orthant tails are the copulas associated with the limits of

$$[C^*(1-n^{-1}e^{-x}, 1-n^{-1}e^{-y})]^n \quad \text{and} \quad [C(1-n^{-1}e^{-x}, 1-n^{-1}e^{-y})]^n$$

respectively, where $C^*(u, v) = u + v - 1 + C(1-u, 1-v)$.

The verification of the dependence properties and the limits at the end points of the parameter space are left as exercises. Dependence properties that are conjectured but not proved are listed in the section on unsolved problems. A visual representation of what tail dependence means for the contours of the density with $N(0, 1)$ margins is given in Figures 5.1 and 5.2 for the copulas B3 and B6 with parameter values corresponding to a Kendall tau value of 0.5.

The following notation is used in several families: $\overline{u} = 1 - u$, $\overline{v} = 1 - v$, $\tilde{u} = -\log u$, $\tilde{v} = -\log v$. Also C_U, C_I, C_L are used for the Fréchet upper bound, independence and Fréchet lower bound copulas, respectively.

Family B1. Bivariate normal. For $0 \leq \delta \leq 1$, $C(u, v; \delta) = \Phi_\delta(\Phi^{-1}(u), \Phi^{-1}(v))$, where Φ is the $N(0, 1)$ cdf, Φ^{-1} is the functional inverse of Φ and Φ_δ is the BVSN cdf with correlation δ. With

$x = \Phi^{-1}(u)$, $y = \Phi^{-1}(v)$, the density is

$$c(u,v;\delta) = (1-\delta^2)^{-1/2}\exp\{-\tfrac{1}{2}(1-\delta^2)^{-1}[x^2+y^2-2\delta xy]\} \cdot \exp\{\tfrac{1}{2}[x^2+y^2]\}.$$

Properties. Increasing in $\prec_c$, increasing in $\prec_{SI}$, TP_2 density, reflection symmetry, multivariate extension, extension to negative dependence. C_U for $\delta = 1$, C_I for $\delta = 0$, C_L for $\delta = -1$. A non-standard upper extreme value limit leads to family B8.

Family B2. Plackett (1965). For $0 \le \delta < \infty$,

$$C(u,v;\delta) = \tfrac{1}{2}\eta^{-1}\{1+\eta(u+v) - [(1+\eta(u+v))^2 - 4\delta\eta uv]^{1/2}\},$$

where $\eta = \delta - 1$. The density is

$$c(u,v;\delta) = [(1+\eta(u+v))^2 - 4\delta\eta uv]^{-3/2}\delta[1+\eta(u+v-2uv)].$$

Properties. Increasing in $\prec_c$, increasing in $\prec_{SI}$, SI, reflection symmetry, extension to negative dependence. C_U for $\delta \to \infty$, C_I for $\delta \to 1$, C_L for $\delta \to 0$.

Family B3. Frank (1979). For $0 \le \delta < \infty$,

$$C(u,v;\delta) = -\delta^{-1}\log\big([\eta - (1-e^{-\delta u})(1-e^{-\delta v})]/\eta\big),$$

where $\eta = 1 - e^{-\delta}$. The density is

$$c(u,v;\delta) = \delta\eta\, e^{-\delta(u+v)}/[\eta - (1-e^{-\delta u})(1-e^{-\delta v})]^2.$$

Properties. Increasing in $\prec_c$, increasing in $\prec_{SI}$, TP_2 density, reflection symmetry, partial multivariate extension, extension to negative dependence, mixture of powers with LT $\psi(s;\delta) = -\delta^{-1}\log[1-(1-e^{-\delta})e^{-s}]$ (family LTD in the Appendix). C_U for $\delta \to \infty$, C_I for $\delta \to 0$, C_L for $\delta \to -\infty$.

Family B4. Kimeldorf and Sampson (1975). For $0 \le \delta < \infty$,

$$C(u,v;\delta) = (u^{-\delta} + v^{-\delta} - 1)^{-1/\delta}.$$

The density is

$$c(u,v;\delta) = (1+\delta)[uv]^{-\delta-1}(u^{-\delta}+v^{-\delta}-1)^{-2-1/\delta}.$$

Properties. Increasing in $\prec_c$, increasing in $\prec_{SI}$, TP_2 density, lower tail dependence, partial multivariate extension, extension to negative dependence, mixture of powers with LT $\psi(s;\delta) = (1+s)^{-1/\delta}$ (family LTB). C_U for $\delta \to \infty$, C_I for $\delta \to 0$. The lower extreme value limit leads to family B7.

Family B5. Joe (1993). For $1 \le \delta < \infty$,

$$C(u,v;\delta) = 1 - (\overline{u}^{\delta} + \overline{v}^{\delta} - \overline{u}^{\delta}\overline{v}^{\delta})^{1/\delta}.$$

The density is

$$c(u,v;\delta) = (\overline{u}^\delta + \overline{v}^\delta - \overline{u}^\delta \overline{v}^\delta)^{-2+1/\delta} \overline{u}^{\delta-1} \overline{v}^{\delta-1} [\delta - 1 + \overline{u}^\delta + \overline{v}^\delta - \overline{u}^\delta \overline{v}^\delta].$$

Properties. Increasing in $\prec_c$, increasing in $\prec_{SI}$, TP_2 density, upper tail dependence, partial multivariate extension, mixture of powers with LT $\psi(s;\delta) = 1 - (1 - e^{-s})^{1/\delta}$ (family LTC). C_U for $\delta \to \infty$, C_I for $\delta = 1$. The upper extreme value limit leads to family B6.

Family B6. Gumbel (1960a). For $1 \le \delta < \infty$,

$$C(u,v;\delta) = \exp\{-(\tilde{u}^\delta + \tilde{v}^\delta)^{1/\delta}\}.$$

The density is

$$c(u,v;\delta) = C(u,v;\delta)(uv)^{-1} \frac{(\tilde{u}\tilde{v})^{\delta-1}}{(\tilde{u}^\delta + \tilde{v}^\delta)^{2-1/\delta}} [(\tilde{u}^\delta + \tilde{v}^\delta)^{1/\delta} + \delta - 1].$$

Properties. Increasing in $\prec_c$, increasing in $\prec_{SI}$, TP_2 density, upper tail dependence, extreme value copula, partial multivariate extension, mixture of powers with LT $\psi(s;\delta) = \exp\{-s^{1/\delta}\}$ (family LTA). C_U for $\delta \to \infty$, C_I for $\delta = 1$.

Family B7. Galambos (1975). For $0 \le \delta < \infty$,

$$C(u,v;\delta) = uv \exp\{(\tilde{u}^{-\delta} + \tilde{v}^{-\delta})^{-1/\delta}\}.$$

The density is

$$c(u,v;\delta) = [C(u,v;\delta)/uv] \cdot [1 - (\tilde{u}^{-\delta} + \tilde{v}^{-\delta})^{-1-1/\delta}(\tilde{u}^{-\delta-1} + \tilde{v}^{-\delta-1})$$
$$+ (\tilde{u}^{-\delta} + \tilde{v}^{-\delta})^{-2-1/\delta}(\tilde{u}\tilde{v})^{-\delta-1}\{1 + \delta + (\tilde{u}^{-\delta} + \tilde{v}^{-\delta})^{-1/\delta}\}].$$

Properties. Increasing in $\prec_c$, SI, upper tail dependence, extreme value copula, partial multivariate extension. C_U for $\delta \to \infty$, C_I for $\delta \to 0$.

Family B8. Hüsler and Reiss (1989). Let Φ be defined as in family B1. For $\delta \ge 0$,

$$C(u,v;\delta) = \exp\{-\tilde{u}\Phi(\delta^{-1} + \tfrac{1}{2}\delta\log[\tilde{u}/\tilde{v}]) - \tilde{v}\Phi(\delta^{-1} + \tfrac{1}{2}\delta\log[\tilde{v}/\tilde{u}])\}.$$

With $z = \tilde{u}/\tilde{v}$, the density is

$$c(u,v;\delta) = (uv)^{-1}C(u,v;\delta) \cdot [\Phi(\delta^{-1} + \tfrac{1}{2}\delta\log z^{-1})\Phi(\delta^{-1} + \tfrac{1}{2}\delta\log z)$$
$$+ \tfrac{1}{2}\delta\tilde{v}^{-1}\phi(\delta^{-1} + \tfrac{1}{2}\delta\log z)],$$

where ϕ is the standard normal univariate density.

Properties. Increasing in $\prec_c$, SI, upper tail dependence, extreme value copula, multivariate extension. C_U for $\delta \to \infty$, C_I for $\delta \to 0$.

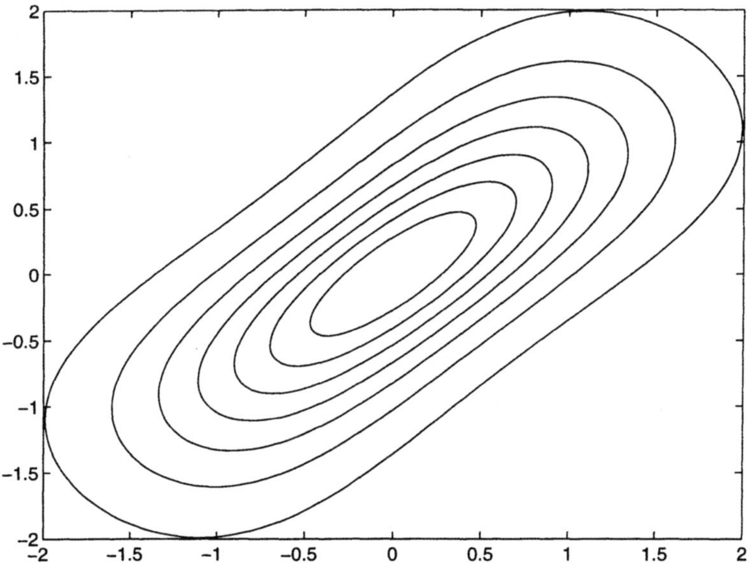

Figure 5.1. *Contours of density for family B3 with normal margins, $\delta = 5.7$.*

A few remarks on the families are the following.

1. A proof of reflection symmetry for the family B3 is as follows. The property is equivalent to $c(u, v; \delta) = c(1 - u, 1 - v; \delta)$, $0 < u, v < 1$. Let $x = e^{-\delta u}$, $y = e^{-\delta v}$, $\gamma = e^{-\delta}$. The non-constant part of the density is $xy/[x + y - xy - \gamma]^2$. The changes $u \to 1 - u$ and $v \to 1 - v$ become $x \to \gamma x^{-1}$ and $y \to \gamma y^{-1}$ and it is straightforward to check that $\gamma^2 x^{-1} y^{-1}/[\gamma x^{-1} + \gamma y^{-1} - \gamma^2 x^{-1} y^{-1} - \gamma]^2 = xy/[x + y - xy - \gamma]^2$. Frank (1979) showed that the family B3 of copulas are the only ones of the form $\phi(\phi^{-1}(u) + \phi^{-1}(v))$ that have the reflection symmetry property. The reflection symmetry property does not hold for the permutation-symmetric multivariate extension of the family B3 (see Section 7.1.7).

2. Suppose (U, V) is a bivariate $U(0, 1)$ random pair; the Plackett

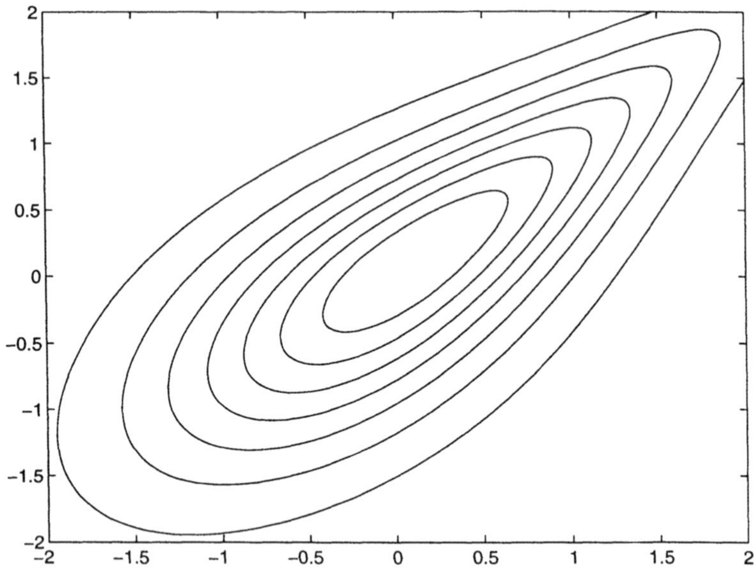

Figure 5.2. *Contours of density for family B6 with normal margins, $\delta = 2$.*

family B2 is constructed from the cross-product ratio:

$$\frac{\Pr(U \le u, V \le v)\Pr(U > u, V > v)}{\Pr(U \le u, V > v)\Pr(U > u, V \le v)}$$
$$= \frac{C(u,v)[1-u-v+C(u,v)]}{[u-C(u,v)][v-C(u,v)]} \equiv \delta,$$

for all $u, v \in (0,1)$, where $\delta > 0$. Perhaps amazingly, this leads to a distribution for all $\delta > 0$, with special cases as given earlier. The above equation is quadratic in C; the appropriate root of the quadratic is given in the preceding list. The proof that the second-order mixed derivative is non-negative is left as an exercise. The property of reflection symmetry for this copula is clear from the above derivation from the cross-product ratio.

3. We illustrate extreme value limits in some cases. To check that

a limit is a proper distribution, one needs to check the boundary conditions but not the rectangle condition in Section 1.4.1.

B1. Let a_n, b_n be sequences of reals such that $\Phi^n(a_n + b_n x) \to \exp\{-e^{-x}\}$. Then

$$\Phi_\rho^n(a_n + b_n x_1, a_n + b_n x_2) \to \prod_{j=1}^{2} \exp\{-e^{-x_j}\}$$

if the correlation ρ is less than 1. An interpretation is that for BVN distributions, the tails are asymptotically independent.

B5. With exponential margins, $F(x, y) = C(1-e^{-x}, 1-e^{-y}; \delta) = 1 - [e^{-\delta x} + e^{-\delta y} - e^{-\delta x}e^{-\delta y}]^{1/\delta}$ and

$$F^n(x + \log n, y + \log n) \sim [1 - n^{-1}(e^{-\delta x} + e^{-\delta y})^{1/\delta}]^n$$

$$\to \exp\{-(e^{-\delta x} + e^{-\delta y})^{1/\delta}\}$$

The copula of the limiting BEV distribution is in the family B6; in min-stable bivariate exponential form $\overline{G}(x, y; \delta) = \exp\{-A(x, y; \delta)\}$, with $A(x, y; \delta) = (x^\delta + y^\delta)^{1/\delta}$.

B4. With exponential margins, $F(x, y) = C(1-e^{-x}, 1-e^{-y}; \delta) = [(1 - e^{-x})^{-\delta} + (1 - e^{-y})^{-\delta} - 1]^{-1/\delta}$ and

$$F^n(x + \log n, y + \log n) \sim [1 + \delta n^{-1}(e^{-x} + e^{-y})]^{-n/\delta}$$

$$\to \exp\{-e^{-x} - e^{-y}\}.$$

Hence the upper extreme value limit is the independence copula. For the lower extreme value limit, we apply the copula to exponential survival margins, i.e., $F(x, y) = 1 - e^{-x} - e^{-y} + (e^{\delta x} + e^{\delta y} - 1)^{-1/\delta}$. Then $F^n(x + \log n, y + \log n) \sim [1 - n^{-1}e^{-x} - n^{-1}e^{-y} + n^{-1}(e^{\delta x} + e^{\delta y})^{-1/\delta}]^n \to \exp\{-e^{-x} - e^{-y} + (e^{\delta x} + e^{\delta y})^{-1/\delta}\}$, and the copula of the limiting BEV distribution is in the family B7.

B8. This is obtained from a non-standard extreme value limit for the BVN distribution. To get a limit which is not the independence copula, the correlation in B1 is allowed to increase to 1; i.e., $\lim \Phi_{\rho_n}^n(a_n + b_n x_1, a_n + b_n x_2)$, with ρ_n increasing to 1 at an appropriate rate, leads to the family B8. There is a multivariate extension from this extreme value limit for the MVN distribution and it has a parameter for each bivariate margin.

Table 5.1. *Parameter values corresponding to given Kendall tau values*

τ	B1	B2	B3	B4	B5	B6	B7	B8
0	0	1	0	0	1	1	0	0
0.1	0.156	1.57	0.91	0.22	1.19	1.11	0.34	0.66
0.2	0.309	2.48	1.86	0.50	1.44	1.25	0.51	0.87
0.3	0.454	4.00	2.92	0.86	1.77	1.43	0.70	1.11
0.4	0.588	6.60	4.16	1.33	2.21	1.67	0.95	1.41
0.5	0.707	11.4	5.74	2.00	2.86	2.00	1.28	1.80
0.6	0.809	21.1	7.93	3.00	3.83	2.50	1.79	2.39
0.7	0.891	44.1	11.4	4.67	5.46	3.33	2.62	3.34
0.8	0.951	115.	18.2	8.00	8.77	5.00	4.29	5.24
0.9	0.988	530.	20.9	18.0	14.4	10.0	9.30	10.9
1	1	∞	∞	∞	∞	∞	∞	∞

To give an indication of the amount of dependence that exists as the δ increases for the families B1 to B8, the values of δ that correspond to Kendall tau and Spearman rho values from 0 to 1 in steps of 0.1 are given in Tables 5.1 and 5.2, respectively. Theorem 4.3 applies to yield a simple form for τ for the families B4 and B6, with $\tau = \delta/(\delta+2)$, $\delta \geq 0$, and $\tau = (\delta-1)/\delta$, $\delta \geq 1$, respectively. Also there is a simple form for τ and ρ_S for the BVN distribution (see Exercise 2.14). All other values of τ and ρ_S were obtained by one- or two-dimensional numerical integration, or by Monte Carlo simulation. These tables suggest that Spearman's rho is greater than Kendall's tau for these families (see Exercise 2.18).

We list below the conditional distributions $C_{2|1}(v|u) = \frac{\partial C}{\partial u}(u,v)$ corresponding to the families B2 to B8. These are useful for simulating random pairs from the copula families, among other things. If U, Q are independent random $U(0,1)$ variates, then $(U,V) = (U, C_{2|1}^{-1}(Q|U))$ has distribution C. If $C_{2|1}^{-1}$ does not exist in closed form, then $v = C_{2|1}^{-1}(q|u)$ can be obtained from the equation $q = C_{2|1}(v|u)$ using a numerical root-finding routine. Of the families B2 to B8, $C_{2|1}^{-1}$ has closed form only for the families B3 and B4. With the notation $\tilde{u} = -\log u$, $\tilde{v} = -\log v$, the list of conditional distributions $C_{2|1}$ is:

B2. $C_{2|1}(v|u;\delta) = \frac{1}{2} - \frac{1}{2}[\eta u + 1 - (\eta+2)v]/[(1+\eta(u+v))^2 - 4\delta\eta uv]^{1/2}$;

Table 5.2. *Parameter values corresponding to given Spearman rho values*

ρ_S	B1	B2	B3	B4	B5	B6	B7	B8
0	0	1	0	0	1	1	0	0
0.1	0.105	1.35	0.60	0.14	1.12	1.07	0.28	0.58
0.2	0.209	1.84	1.22	0.31	1.27	1.16	0.40	0.73
0.3	0.313	2.52	1.88	0.51	1.46	1.26	0.51	0.88
0.4	0.416	3.54	2.61	0.76	1.69	1.38	0.65	1.05
0.5	0.518	5.12	3.45	1.06	1.99	1.54	0.81	1.24
0.6	0.618	7.76	4.47	1.51	2.39	1.75	1.03	1.50
0.7	0.717	12.7	5.82	2.14	3.00	2.07	1.34	1.86
0.8	0.813	24.2	7.90	3.19	4.03	2.58	1.86	2.45
0.9	0.908	66.1	12.2	5.56	6.37	3.73	3.01	3.73
1	1	∞	∞	∞	∞	∞	∞	∞

B3. $C_{2|1}(v|u;\delta) = e^{-\delta u}[(1-e^{-\delta})(1-e^{-\delta v})^{-1} - (1-e^{-\delta u})]^{-1}$,
$C_{2|1}^{-1}(q|u;\delta) = -\delta^{-1}\log\{1-(1-e^{-\delta})/[(q^{-1}-1)e^{-\delta u}+1]\}$;

B4. $C_{2|1}(v|u;\delta) = [1+u^{\delta}(v^{-\delta}-1)]^{-1-1/\delta}$,
$C_{2|1}^{-1}(q|u;\delta) = [(q^{-\delta/(1+\delta)}-1)u^{-\delta}+1]^{-1/\delta}$;

B5. $C_{2|1}(v|u;\delta) = [1+(1-v)^{\delta}(1-u)^{-\delta}-(1-v)^{\delta}]^{-1+1/\delta}[1-(1-v)^{\delta}]$;

B6. $C_{2|1}(v|u;\delta) = u^{-1}\exp\{-(\tilde{u}^{\delta}+\tilde{v}^{\delta})^{1/\delta}\}\cdot[1+(\tilde{v}/\tilde{u})^{\delta}]^{-1+1/\delta}$;

B7. $C_{2|1}(v|u;\delta) = v\exp\{(\tilde{u}^{-\delta}+\tilde{v}^{-\delta})^{-1/\delta}\}\{1-[1+(\tilde{u}/\tilde{v})^{\delta}]^{-1-1/\delta}\}$;

B8. $C_{2|1}(v|u;\delta) = C(u,v;\delta)\cdot u^{-1}\Phi(\delta^{-1}+\frac{1}{2}\delta\log(\tilde{u}/\tilde{v}))$.

A few one-parameter families of bivariate copulas that are not as simple or do not have properties that are as nice as those already listed are given in the remainder of this section.

Family B9. Raftery (1984; 1985). For $0 \le \delta \le 1$,

$$C(u,v;\delta) = B(u \wedge v, u \vee v; \delta),$$

where

$$B(x,y;\delta) = x - \left[\frac{1-\delta}{1+\delta}\right] x^{1/(1-\delta)}[y^{-\delta/(1-\delta)} - y^{1/(1-\delta)}].$$

The density is $c(u,v;\delta) = b(u \wedge v, u \vee v; \delta)$ where

$$b(x,y;\delta) = (1-\delta^2)^{-1}x^{\delta/(1-\delta)}(\delta y^{-1/(1-\delta)} + y^{\delta/(1-\delta)}).$$

Properties. Increasing in $\prec_c$, lower tail dependence, C_I for $\delta \to 0$, C_U for $\delta = 1$.

Family B10. Morgenstern (1956). For $-1 \le \delta \le 1$,

$$C(u, v; \delta) = uv[1 + \delta(1-u)(1-v)].$$

The density is

$$c(u, v; \delta) = 1 + \delta(1-2u)(1-2v).$$

Properties. Increasing in $\prec_c$, increasing in $\prec_{SI}$, TP_2 density for $\delta \ge 0$, reflection symmetry, positive dependence for $\delta > 0$ and negative dependence for $\delta < 0$, C_I for $\delta = 0$, multivariate extension.

Family B11. For $0 \le \delta \le 1$,

$$C(u, v; \delta) = \delta \min\{u, v\} + (1-\delta)uv.$$

Properties. Increasing in $\prec_c$, reflection symmetry, C_I for $\delta = 0$, C_U for $\delta = 1$, multivariate extension, singular component with mass δ.

Family B12. For $0 \le \delta \le 1$,

$$C(u, v; \delta) = [\min\{u, v\}]^{\delta}[uv]^{1-\delta}.$$

Properties. Increasing in $\prec_c$, reflection symmetry, C_I for $\delta = 0$, C_U for $\delta = 1$, multivariate extension, singular component with mass $\delta/(2-\delta)$.

Remarks on these additional families are the following.

1. The family B9 has not been included with the preceding eight one-parameter families of bivariate copulas because its form is not quite as simple (it is a function of $u \vee v$ and $u \wedge v$).

   The derivation of the family B9 is from a bivariate exponential distribution based on the stochastic representation:

$$X_1 = (1-\delta)Z_1 + IZ_{12}, \quad X_2 = (1-\delta)Z_2 + IZ_{12}, \tag{5.1}$$

   where I, Z_1, Z_2, Z_{12} are independent rvs, the Zs have unit exponential distributions, and $I \sim$ Bernoulli (δ). This leads to the bivariate exponential survival function:

$$G(x, y; \delta) = \begin{cases} e^{-x} - \frac{1-\delta}{1+\delta} e^{-x/(1-\delta)}[e^{y\delta/(1-\delta)} - e^{-y/(1-\delta)}], & x \ge y, \\ e^{-y} - \frac{1-\delta}{1+\delta} e^{-y/(1-\delta)}[e^{x\delta/(1-\delta)} - e^{-x/(1-\delta)}], & x \le y. \end{cases}$$

   and it becomes the given copula after the transform to $U(0,1)$ margins. The derivation of G from the representation comes from the sum of integrals: $(1-\delta)e^{-(x+y)/(1-\delta)} + \delta \int_0^x \exp\{-(x -$

$z)/(1-\delta)\}\exp\{-(y-z)/(1-\delta)\}e^{-z}\,dz+\delta\int_x^y \exp\{-(y-z)/(1-\delta)\}e^{-z}\,dz+\delta\int_y^\infty e^{-z}\,dz$, if $x \le y$.

By extending the stochastic representation, the $\prec_c$ ordering can be proved (see Exercise 5.8 for some details).

The multivariate extension in Raftery (1984) covers a wide range of dependence but the parameters are not easily interpretable.

2. The family B10 is convenient for illustrating dependence concepts because of its simple form. However, because of its limited range of dependence (see Example 2.4), it is not a useful model for data. A convenient multivariate extension is:

$$C(\mathbf{u};\delta_{jk},1\le j<k\le m)=u_1\cdots u_m\Big[1+\sum_{j<k}\delta_{jk}(1-u_j)(1-u_k)\Big]. \tag{5.2}$$

Constraints include $|\delta_{jk}|\le 1$ for all j,k and there are also other joint inequality constraints in the parameters to achieve a nonnegative density (see Exercise 5.7).

3. The families B11 and B12 have limited applications because of the singular component. Possibly they have more uses for discrete univariate margins. The family B12 is the copula associated with the Marshall–Olkin bivariate exponential distribution when both univariate margins have the same mean. The mass of the singularity can be computed using Theorem 1.1.

5.2 Bivariate two-parameter families

Two-parameter families might be used to capture more than one type of dependence. Examples are one parameter for upper tail dependence and one for concordance, or one parameter for upper tail dependence and one for lower tail dependence. One general approach for two-parameter families is the use of (4.32) with $\nu_1=\nu_2=0$ (see (5.3) below). From the tail dependence results in Section 4.3.2, the use of the LT families LTA, LTB, LTC leads to copulas with tail dependence; the properties of the copulas for LTC are similar to those for LTA. The examples given in this section show various possible types of behaviour for upper and lower tail dependence.

Below are some two-parameter bivariate families of the form

$$C(u,v)=\psi(-\log K(e^{-\psi^{-1}(u)},e^{-\psi^{-1}(v)})), \tag{5.3}$$

where K is max-id and ψ is a LT. Two-parameter families result if K is parametrized by a parameter δ and ψ is parametrized by a parameter θ (denoted by ψ_θ). If K is increasing in concordance as δ increases, then clearly C increases in concordance as δ increases with θ fixed. The concordance ordering for δ fixed and θ varying is harder to check. If K has the form of an Archimedean copula (4.3), then, from (5.3), C also has the form of an Archimedean copula. That is, if $K(x,y;\delta) = \phi_\delta(\phi_\delta^{-1}(x) + \phi_\delta^{-1}(y))$ for a family ϕ_δ, then

$$\begin{aligned} C(u,v;\theta,\delta) &= \psi_\theta\left(-\log\phi_\delta\left[\phi_\delta^{-1}(e^{-\psi_\theta^{-1}(u)}) + \phi_\delta^{-1}(e^{-\psi_\theta^{-1}(v)})\right]\right) \\ &= \eta_{\theta,\delta}\left(\eta_{\theta,\delta}^{-1}(u) + \eta_{\theta,\delta}^{-1}(v)\right), \end{aligned} \tag{5.4}$$

where $\eta_{\theta,\delta}(s) = \psi_\theta(-\log\phi_\delta(s))$. For δ fixed and $\theta_2 > \theta_1$ with $\eta_i = \eta_{\theta_i,\delta}$, $i = 1,2$, the concordance ordering of $C(\cdot;\theta_1,\delta)$ and $C(\cdot;\theta_2,\delta)$ could be established by showing that $\omega = \eta_2^{-1} \circ \eta_1$ is superadditive (Theorem 4.1).

Family BB1. In (5.3), let K be the family B6 and let ψ be the family LTB. Then the resulting two-parameter family of the form (5.4) is

$$\begin{aligned} C(u,v;\theta,\delta) &= \left\{1 + \left[(u^{-\theta}-1)^\delta + (v^{-\theta}-1)^\delta\right]^{1/\delta}\right\}^{-1/\theta} \\ &= \eta(\eta^{-1}(u) + \eta^{-1}(v)), \quad \theta > 0,\ \delta \geq 1, \end{aligned} \tag{5.5}$$

where $\eta(s) = \eta_{\theta,\delta}(s) = (1 + s^{1/\delta})^{-1/\theta}$ (family LTE in the Appendix).

Some properties of the family of copulas (5.5) are:

(a) The family B4 is a subfamily when $\delta = 1$, and the family B6 is obtained as $\theta \to 0$. C_I obtains as $\theta \to 0$ and $\delta \to 1$ and C_U obtains as $\theta \to \infty$ or $\delta \to \infty$.

(b) The lower tail dependence parameter is $2^{-1/(\delta\theta)}$, while the upper tail dependence parameter is $2 - 2^{1/\delta}$, independent of θ. The extreme value limits from the lower and upper tails are the families B7 and B6, respectively.

(c) Concordance increases as θ increases because $\omega(s)/s$ is increasing, where $\omega(s) = \eta_{\theta_2,\delta}^{-1}(\eta_{\theta_1,\delta}(s)) = [(1+s^{1/\delta})^\rho - 1]^\delta$, $\theta_1 < \theta_2$, and $\rho = \theta_2/\theta_1 > 1$.

Family BB2. In (5.3), let K be the family B4 and let ψ be the family LTB. Then the two-parameter family of the form (5.4) is

$$\begin{aligned} C(u,v;\theta,\delta &\quad \left[1 + \delta^{-1}\log\left(e^{\delta(u^{-\theta}-1)} + e^{\delta(v^{-\theta}-1)} - 1\right)\right]^{-1/\theta} \\ &= \eta(\eta^{-1}(u) + \eta^{-1}(v)), \quad \theta,\delta > 0, \end{aligned} \tag{5.6}$$

where $\eta(s) = \eta_{\theta,\delta}(s) = [1 + \delta^{-1}\log(1 + s)]^{-1/\theta}$ (family LTF in the Appendix).

Some properties of the family of copulas (5.6) are:

(a) C_I obtains as $\theta \to 0$, C_U obtains as $\theta \to \infty$ or $\delta \to \infty$. The limit as $\delta \to 0$ leads to the family B4.

(b) The lower tail dependence parameter is 1, while there is no upper tail dependence.

(c) Concordance increases as θ increases because ω is convex, where $\omega(s) = \eta^{-1}_{\theta_2,\delta}(\eta_{\theta_1,\delta}(s)) = \exp\{\delta([1 + \delta^{-1}\log(1 + s)]^{\rho} - 1)\} - 1$, $\theta_1 < \theta_2$, and $\rho = \theta_2/\theta_1 > 1$.

Family BB3. In (5.3), let K be the family B4 and let ψ be the family LTA. Then the two-parameter family of the form (5.4) is

$$\begin{aligned} C(u, v; \theta, \delta) &= \exp\{-[\delta^{-1}\log(e^{\delta\tilde{u}^{\theta}} + e^{\delta\tilde{v}^{\theta}} - 1)]^{1/\theta}\} \\ &= \eta(\eta^{-1}(u) + \eta^{-1}(v)), \quad \theta \geq 1, \delta > 0, \quad (5.7) \end{aligned}$$

where $\eta(s) = \eta_{\theta,\delta}(s) = \exp\{-[\delta^{-1}\log\,(1 + s)]^{1/\theta}\}$ (family LTG in the Appendix), $\tilde{u} = -\log u$ and $\tilde{v} = -\log v$.

Some properties of the family of copulas (5.7) are:

(a) The family B4 is a subfamily when $\theta = 1$, and the family B6 is obtained as $\delta \to 0$. C_U obtains as $\theta \to \infty$ or $\delta \to \infty$.

(b) The lower tail dependence parameter is $2^{-1/\delta}$ when $\theta = 1$ and 1 when $\theta > 1$, while the upper tail dependence parameter is $2 - 2^{1/\theta}$, independent of δ. The upper extreme value limit is the family B6.

(c) Concordance increases as θ increases if and only if

$$-D\delta^{-1}\log(D/\delta) + [e^{\delta x}x\log x + e^{\delta y}y\log y]/(e^{\delta x} + e^{\delta y} - 1) \leq 0$$

for all $x, y > 0$ and $\delta > 0$, where $D = \log(e^{\delta x} + e^{\delta y} - 1)$. This condition holds from numerical checks but has not been confirmed analytically. With a change of parametrization to (θ, α) with $\alpha = \delta^{1/\theta}$, the family of copulas has been shown to be increasing in concordance with both parameters θ and α.

Family BB4. In (5.3), let K be the family B7 and let ψ be the family LTB. Then the two-parameter family is

$$C(u, v; \theta, \delta) = \left(u^{-\theta} + v^{-\theta} - 1 - [(u^{-\theta} - 1)^{-\delta} + (v^{-\theta} - 1)^{-\delta}]^{-\frac{1}{\delta}}\right)^{-\frac{1}{\theta}},$$

$$\theta \geq 0, \delta > 0. \quad (5.8)$$

Some properties of the family of copulas (5.8) are:

(a) The family B4 is obtained when $\delta \to 0$, and the family B7 obtains as $\theta \to 0$. C_U obtains as $\theta \to \infty$ or $\delta \to \infty$.

(b) The lower tail dependence parameter is $(2-2^{-1/\delta})^{-1/\theta}$, while the upper tail dependence parameter is $2^{-1/\delta}$, independent of θ. The lower extreme value limit leads to the min-stable bivariate exponential family $\exp\{-A(x,y)\}$, with $A(x,y) = x + y - [x^{-\theta} + y^{-\theta} - (x^{\theta\delta} + y^{\theta\delta})^{-1/\delta}]^{-1/\theta}$ and this is a two-parameter extension of the family B7. The upper extreme value limit is the family B7.

(c) Concordance increases as θ increase if and only if $[x+y-1-((x-1)^{-\delta}+(y-1)^{-\delta})^{-1/\delta}]\log[x+y-1-((x-1)^{-\delta}+(y-1)^{-\delta})^{-1/\delta}]-x\log x-y\log y+[(x-1)^{-\delta}+(y-1)^{-\delta}]^{-1/\delta-1}[(x-1)^{-\delta}x\log x+(y-1)^{-\delta}y\log y] \geq 0$ for all $x, y > 1$ and $\delta > 0$. This condition holds from numerical checks but has not been confirmed analytically.

Family BB5. In (5.3), let K be the family B7 and let ψ be the family LTA. Then the two-parameter family is

$$C(u,v;\theta,\delta) = \exp\{-[\tilde{u}^{\theta} + \tilde{v}^{\theta} - (\tilde{u}^{-\theta\delta} + \tilde{v}^{-\theta\delta})^{-1/\delta}]^{1/\theta}\}, \quad (5.9)$$

$\theta \geq 1$, $\delta > 0$, where $\tilde{u} = -\log u$, $\tilde{v} = -\log v$.

Some properties of the family of copulas (5.9) are:

(a) The family B6 is obtained when $\delta \to 0$ and the family B7 is obtained when $\theta = 1$. C_U obtains as $\theta \to \infty$ or $\delta \to \infty$.

(b) The lower tail dependence parameter is 0 and the upper tail dependence parameter is $2 - (2 - 2^{-1/\delta})^{1/\theta}$. The upper extreme value limit leads to the min-stable bivariate exponential family $\exp\{-A(x,y)\}$, with $A(x,y) = [x^{\theta} + y^{\theta} - (x^{-\theta\delta} + y^{-\theta\delta})^{-1/\delta}]^{1/\theta}$, and this is a two-parameter extension of the family B6.

(c) Concordance increases as θ increases if and only if $[x+y-(x^{-\delta}+y^{-\delta})^{-1/\delta}]\log[x+y-(x^{-\delta}+y^{-\delta})^{-1/\delta}]-x\log x-y\log y+(x^{-\delta}+y^{-\delta})^{-1/\delta-1}(x^{-\delta}\log x+y^{-\delta}\log y) \geq 0$ for all $x, y > 0$ and $\delta > 0$. This condition holds from numerical checks but has not been confirmed analytically.

Family BB6. In (5.3), let K be the family B6 and let ψ be the family LTC. Then the two-parameter family of form (5.4) is

$$C(u,v;\theta,\delta) = 1 - \left(1 - \exp\left\{-[(-\log(1-\overline{u}^{\theta}))^{\delta} + (-\log(1-\overline{v}^{\theta}))^{\delta}]^{\frac{1}{\delta}}\right\}\right)^{\frac{1}{\theta}}$$
$$= \eta(\eta^{-1}(u) + \eta^{-1}(v)), \quad \theta \geq 1, \delta \geq 1, \quad (5.10)$$

where $\overline{u} = 1 - u$, $\overline{v} = 1 - v$ and $\eta(s) = \eta_{\theta,\delta}(s) = 1 - [1 - \exp\{-s^{1/\delta}\}]^{1/\theta}$ (family LTH in the Appendix).

Some properties of the family of copulas (5.10) are:

(a) The family B6 is obtained when $\theta = 1$, and the family B5 is obtained when $\delta = 1$. C_U obtains as $\theta \to \infty$ or $\delta \to \infty$.

(b) The lower tail dependence parameter is 0, and the upper tail dependence parameter is $2 - 2^{1/(\theta\delta)}$. The upper extreme value limit is the family B6.

(c) Concordance increases as θ increases because $\omega(s)$ is convex in $s > 0$, where $\omega(s) = \eta^{-1}_{\theta_2,\delta}(\eta_{\theta_1,\delta}(s)) = [\sigma(s^{1/\delta})]^{\delta}$, $\theta_1 < \theta_2$, and $\sigma(t) = -\log(1 - [1 - e^{-t}]^{\rho})$, with $\rho = \theta_2/\theta_1 > 1$.

Family BB7. In (5.3), let K be the family B4 and let ψ be the family LTC. Then the two-parameter family of the form (5.4) is

$$\begin{aligned} C(u,v;\theta,\delta) &= 1 - \left(1 - \left[(1-\overline{u}^{\theta})^{-\delta} + (1-\overline{v}^{\theta})^{-\delta} - 1\right]^{-1/\delta}\right)^{1/\theta} \\ &= \eta(\eta^{-1}(u) + \eta^{-1}(v)), \quad \theta \geq 1,\, \delta > 0, \qquad (5.11) \end{aligned}$$

where $\eta(s) = \eta_{\theta,\delta}(s) = 1 - [1 - (1+s)^{-1/\delta}]^{1/\theta}$ (family LTI in the Appendix).

Some properties of the family of copulas (5.11) are:

(a) The family B4 is obtained when $\theta = 1$, and the family B5 is obtained as $\delta \to 0$. C_U obtains as $\theta \to \infty$ or $\delta \to \infty$.

(b) The lower tail dependence parameter is $2^{-1/\delta}$, independent of θ, and the upper tail dependence parameter is $2 - 2^{1/\theta}$, independent of δ. The extreme value limits from the lower and upper tails are, respectively, the families B7 and B6.

(c) Concordance increases as θ increases when $\delta \leq 1$; the proof is non-trivial. It is conjectured that concordance is also increasing in θ when $\delta > 1$.

Other two-parameter families of copulas of the form (4.3) are based on two-parameter families of LTs that do not come from a composition of the form (5.4).

Family BB8. A family of copulas based on a two-parameter family of LTs, $\phi(s) = \delta^{-1}[1 - \{1 - [1-(1-\delta)^{\theta}]e^{-s}\}^{1/\theta}]$, $\theta \geq 1$, $0 < \delta \leq 1$ (family LTJ in the Appendix), is:

$$C(u,v;\theta,\delta) = \delta^{-1}[1 - \{1 - [1-(1-\delta)^{\theta}]^{-1}[1-(1-\delta u)^{\theta}][1-(1-\delta v)^{\theta}]\}^{\frac{1}{\theta}}], \qquad (5.12)$$

$\theta \geq 1$, $0 \leq \delta \leq 1$. Some properties of the family of copulas (5.12) are:

(a) C_I obtains as $\delta \to 0$ or $\theta \to 1$. The family B5 is obtained when $\delta = 1$, and the family B3 is obtained as $\theta \to \infty$ with $\gamma = 1 - (1-\delta)^\theta$ held constant (or with $\delta = 1 - (1-\gamma)^{1/\theta}$).

(b) The family is derived as a power mixture family with the above LT. The family does not have tail dependence except when $\delta = 1$. It extends to the multivariate case for each fixed δ.

(c) Concordance increases as θ or δ increases. The proof of the concordance ordering is non-trivial.

Family BB9. From the two-parameter family of LTs, $\phi(s) = \exp\{-(\alpha^\theta + s)^{1/\theta} + \alpha\}$, $\alpha \geq 0$, $\theta \geq 1$ (family LTL in the Appendix), the two-parameter family of copulas is

$$C(u, v; \theta, \alpha) = \exp\{-[(\alpha - \log u)^\theta + (\alpha - \log v)^\theta - \alpha^\theta]^{1/\theta} + \alpha\}, \tag{5.13}$$

$\theta \geq 1$, $\alpha > 0$. Some properties of the family of copulas (5.13) are:

(a) C_I obtains as $\alpha \to \infty$ or for $\theta = 1$, and C_U obtains as $\theta \to \infty$. The family B6 is a subfamily when $\alpha = 0$.

(b) Concordance increases as either θ increases or α decreases.

The next example is one of a two-parameter family in which concordance is not monotone in both parameters. It is not clear if the concordance ordering in two parameters can be obtained after making a parameter change. For example, with a reparametrization of LTL to $\exp\{-(\delta + s)^{1/\theta} + \delta^{1/\theta}\}$, the reparametrization of (5.13) is not always increasing in concordance in θ for fixed δ.

Family BB10. The LT of the negative binomial distribution is $\phi(s) = [(1-\theta)e^{-s}/(1-\theta e^{-s})]^\alpha = [(1-\theta)/(e^s - \theta)]^\alpha$, where $0 \leq \theta < 1$ and $\alpha > 0$ (family LTM in the Appendix). The inverse is $\phi^{-1}(t) = \log[(1-\theta)t^{-1/\alpha} + \theta]$. The family of copulas is

$$C(u, v; \theta, \alpha) = uv[1 - \theta(1 - u^{1/\alpha})(1 - v^{1/\alpha})]^{-\alpha}, \tag{5.14}$$

$0 \leq \theta \leq 1$, $\alpha > 0$. Some properties of the family of copulas (5.14) are:

(a) C_I obtains as $\alpha \to \infty$; C_U obtains as $\alpha \to 0$ when $\theta = 1$, but C_I obtains as $\alpha \to 0$ for $0 < \theta < 1$.

(b) Concordance increases as θ increases for a fixed α. The concordance is decreasing in α for $\theta = 1$. For $0 < \theta < 1$ fixed, there is no concordance ordering as α increases.

5.3 Multivariate copulas with partial symmetry

A multivariate parametric family of copulas is an extension of a (one-parameter) bivariate family if: (i) all bivariate marginal copulas of the multivariate copula are in the given bivariate family; and (ii) all multivariate marginal copulas of order 3 to $m-1$ have the same multivariate form.

There is no proven general multivariate extension of a bivariate parametric family that has a dependence parameter for each bivariate margin (but see Section 4.8). This section has the multivariate extension for copulas having the forms in Section 4.2. After possibly permuting the indices, the form of the multivariate extension is that the (i,j) bivariate margin has parameter δ_{ij}, with $\delta_{ij} = \beta_{a_i,a_j}$ if $\{a_i,\ldots,a_j\}$ is the smallest cluster of consecutive indices that contain indices i and j. There are $m-1$ parameters with the form $\beta_{a,b}$, $a<b$, and the $\beta_{a,b}$ satisfy $\beta_{a_1,b_1} \geq \beta_{a_2,b_2}$ if $a_2 \leq a_1 < b_1 \leq b_2$. There are no constraints for β_{a_1,b_1} and β_{a_2,b_2} if $b_1 < a_2$ or $b_2 < a_1$. The labelling for the $m-1$ parameters $\beta_{a,b}$ is such that $a_1 < a_2 \leq b_1 < b_2$ or $a_2 < a_1 \leq b_2 < b_1$ is not allowed. That is, clusters (of indices) are hierarchical or nested, and cannot overlap. (Compare the data analysis technique of hierarchical clustering; see also the comparison with hierarchical normal models in Section 4.2.1.)

Examples are:

- $m=3$. Parameters $\beta_{1,2} \geq \beta_{1,3}$ with $\delta_{12} = \beta_{1,2}$, $\delta_{13} = \delta_{23} = \beta_{1,3}$.
- $m=4$. Parameters $\beta_{1,2} \geq \beta_{1,3} \geq \beta_{1,4}$ with $\delta_{12} = \beta_{1,2}$, $\delta_{13} = \delta_{23} = \beta_{1,3}$, and $\delta_{14} = \delta_{24} = \delta_{34} = \beta_{1,4}$.
- $m=4$. Parameters $\beta_{1,2}, \beta_{3,4} \geq \beta_{1,4}$ with $\delta_{12} = \beta_{1,2}$, $\delta_{34} = \beta_{3,4}$, and $\delta_{13} = \delta_{14} = \delta_{23} = \delta_{24} = \beta_{1,4}$.
- $m \geq 5$. There are three different structural forms for $m=5$. The number of different clustering forms increases rapidly with m.

The permutation-symmetric subcase obtains when all of the $\beta_{a,b}$ are the same.

Below we list some (trivariate) families with this partially symmetric dependence structure; this will suggest the form of the higher-dimensional copulas, without the tedious notation of the latter. In referring to properties of parametric families of copulas, an m-variate copula is a **multivariate extreme value** (MEV) copula if

$$C(u_1^t,\ldots,u_m^t) = C^t(u_1,\ldots,u_m) \quad \forall t>0.$$

After converting to unit exponential survival margins, with $\overline{G}(\mathbf{x}) = C(e^{-x_1}, \ldots, e^{-x_m})$, the MEV copulas are easily recognized from $A(\mathbf{x}) = -\log G(\mathbf{x})$ being homogeneous of order 1, i.e., $A(t\mathbf{x}) = tA(\mathbf{x})$ for all $t > 0$. **Extreme value limits** (upper and lower) refer to the limit of

$$[K(1 - n^{-1}e^{-x_1}, \ldots, 1 - n^{-1}e^{-x_m})]^n, \quad n \to \infty,$$

where K is either C or the associated copula C^* when C is applied to survival functions (see Section 1.6).

For the families LTA, LTB, LTC and LTD of LTs ϕ_θ, the property of $\phi_{\theta_1}^{-1} \circ \phi_{\theta_2} \in \mathcal{L}_\infty^*$, $\theta_1 < \theta_2$, is satisfied using results in the Appendix. Therefore the constructions in Section 4.2 apply to yield copulas of the form (4.7), (4.10) and (4.11) and their extensions. The increasing in concordance property for these parametric families follows from theorems in Section 4.2.1. Hence the $\prec_c$ ordering holds but its verification is not easy; however, the $\prec_c^{pw}$ ordering follows easily from the $\prec_c$ ordering of the bivariate margins.

For the trivariate case, with $\theta_1 = \beta_{1,3} \le \beta_{1,2} = \theta_2$, copulas have the form

$$\phi_{\theta_1}(\phi_{\theta_1}^{-1} \circ \phi_{\theta_2}(\phi_{\theta_2}^{-1}(u_1) + \phi_{\theta_2}^{-1}(u_2)) + \phi_{\theta_1}^{-1}(u_3)), \tag{5.15}$$

Theorems 4.8 and 4.9 then imply that (5.15) is increasing in concordance as θ_1 or θ_2 increase with $\theta_1 \le \theta_2$.

Family M3. A generalization of family B3. Let $\phi_\theta(s) = -\theta^{-1}\log[1 - (1 - e^{-\theta})e^{-s}]$, $\theta \ge 0$ (family LTD in the Appendix). For $\theta_1 < \theta_2$, the family (5.15) becomes

$$\begin{aligned} C(\mathbf{u}; \theta_1, \theta_2) &= -\theta_1^{-1}\log\{1 - c_1^{-1}(1 - [1 - c_2^{-1}(1 - e^{-\theta_2 u_1}) \\ &\qquad \cdot(1 - e^{-\theta_2 u_2})]^{\theta_1/\theta_2})\,(1 - e^{-\theta_1 u_3})\}, \end{aligned}$$

where $c_1 = 1 - e^{-\theta_1}$ and $c_2 = 1 - e^{-\theta_2}$.

Family M4. A generalization of family B4. Let $\phi_\theta(s) = (1 + s)^{-1/\theta}$, $\theta \ge 0$ (family LTB in the Appendix). For $\theta_1 \le \theta_2$, the family (5.15) becomes

$$C(u_1, u_2, u_3; \theta_1, \theta_2) = [(u_1^{-\theta_2} + u_2^{-\theta_2} - 1)^{\theta_1/\theta_2} + u_3^{-\theta_1} - 1]^{-1/\theta_1}$$

The lower extreme value limit of this family is the family M7 which generalizes B7 (see Section 6.3.1).

Family M5. Generalization of families B5 and BB8. With δ fixed in (0,1), the family BB8 has the same multivariate generalization. Let $\phi_\theta(s) = \delta^{-1}[1 - (1 - c(\theta)e^{-s})^{1/\theta}]$, $\theta \ge 1$, where

$c(\theta) = 1 - (1-\delta)^\theta$ (family LTJ in the Appendix). For $\theta_1 < \theta_2$, the family (5.15) becomes

$$C(\mathbf{u};\theta_1,\theta_2) = \delta^{-1}\Big(1 - \big[1 - \{1 - [1 - A(u_1,\theta_2) \cdot A(u_2,\theta_2)/c(\theta_2)]^{\theta_1/\theta_2}\}A(u_3,\theta_1)/c(\theta_1)\big]^{1/\theta_1}\Big),$$

where $A(u,\theta) = 1 - (1-\delta u)^\theta$. The limit as $\delta \to 1$ leads to

$$C(\mathbf{u};\theta_1,\theta_2) = 1 - \big\{[v_1^{\theta_2}(1-v_2^{\theta_2}) + v_2^{\theta_2}]^{\theta_1/\theta_2}(1-v_3^{\theta_1}) + v_3^{\theta_1}\big\}^{1/\theta_1}, \tag{5.16}$$

where $v_j = 1 - u_j$, $j = 1,2,3$. The copula resulting from the extreme value limit of (5.16) is the family M6.

Family M6. A generalization of family B6. Let $\phi_\theta(s) = \exp\{-s^{1/\theta}\}$, $\theta \geq 1$ (family LTA in the the Appendix). For $\theta_1 < \theta_2$, the family (5.15) becomes

$$C(\mathbf{u};\theta_1,\theta_2) = \exp\big\{-\big([(-\log u_1)^{\theta_2} + (-\log u_2)^{\theta_2}]^{\theta_1/\theta_2} + (-\log u_3)^{\theta_1}\big)^{1/\theta_1}\big\}.$$

The generalization of the family B7 to M7 is given in Section 6.3.1.

5.4 Extensions to negative dependence *

In this section, we use the theory of Section 4.4 to extend some families in the previous two sections to negative dependence. The extension comes from extending families of LTs ϕ_θ to functions in $\mathcal{L}_n$ for different n (see (1.1) in Section 1.3). From Section 4.4, a condition needed for negative dependence is the subadditivity of the function $\eta(z) = \phi^{-1}(e^{-z})$, with concavity of η or equivalently convexity of $\eta^{-1} = -\log\phi$ being a sufficient condition.

Details are shown for several parametric families to study the range of negative dependence that can be achieved with theory of Section 4.4. It is not known what is the most negative dependence that can be obtained with this approach. A number of the tedious calculations were done with the aid of symbolic manipulation software.

Family M4E. Extension of B4, M4 and LTB. Write the gamma family of LTs as $\phi(s) = \phi_\theta(s) = (1+\theta s)^{-1/\theta}$, $\theta > 0$. The LT $\phi_0(s) = e^{-s}$ is obtained as $\theta \to 0$, and the family ϕ_θ extends to $\theta <$

Table 5.3. *Most negative values of τ and ρ_S for family M4E and MVN.*

m	λ_m	τ	ρ_S	τ(MVN)	ρ_S(MVN)
2	1	−1	−1	−1	−1
3	1/2	−0.3333	−0.4667	−0.3333	−0.4826
4	1/3	−0.2000	−0.2930	−0.2163	−0.3198
5	1/4	−0.1429	−0.2119	−0.1609	−0.2394
6	1/5	−0.1111	−0.1656	−0.1282	−0.1913
7	1/6	−0.0909	−0.1359	−0.1066	−0.1593

0 by writing $\phi_\theta(s) = (1+\theta s)_+^{-1/\theta}$, $\theta < 0$, where $(z)_+ = \max\{0, z\}$. (The extended family ϕ_θ is the family of generalized Pareto survival functions.) Let $\lambda = -\theta$ so that $\phi_\theta(s) = (1-\lambda s)^{1/\lambda}$, $0 \le s \le \lambda^{-1}$, $\lambda > 0$. The function $\eta(z) = \phi_\theta^{-1}(e^{-z})$ equals $\lambda^{-1}(1 - e^{-\lambda z})$ and has second derivative $\eta''(z) = -\lambda e^{-\lambda z} \le 0$, so that it is concave. It is easily verified that

$$(-1)^j \phi_\theta^{(j)}(s) = \left[\prod_{k=0}^{j-1}(1-k\lambda)\right](1-\lambda s)^{-j+1/\lambda}, \quad j = 1, 2, \ldots,$$

and this is non-negative if and only if $\lambda \le (j-1)^{-1}$. Hence $\phi_\theta \in \mathcal{L}_m$ and (4.4) exists if $\theta \ge -(m-1)^{-1} = -\lambda_m$. The Fréchet lower bound obtains for $\lambda = 1$ ($\theta = -1$).

Table 5.3 lists the bivariate Kendall tau and Spearman rho values for the most negatively dependent permutation-symmetric copula of form (4.4) from this family of ϕ_θ in the cases $m = 2, \ldots, 7$. The Kendall tau value is $-\lambda/(2-\lambda)$. For comparison, the values of τ, ρ_S for the most negatively dependent exchangeable MVN distributions are given in the last two columns of Table 5.3. The formulas are $\tau = \frac{2}{\pi}\arcsin(\rho_m)$ and $\rho_S = \frac{6}{\pi}\arcsin(\rho_m/2)$, with $\rho_m = -(m-1)^{-1}$. The extended family has a good range of exchangeable negative dependence compared with the MVN distributions but the drawback is that zero density exists in a certain region (see Section 4.4).

Family M3E. Extension of B3, M3 and LTD. The logarithmic series family of LTs is $\phi_\theta(s) = -\theta^{-1}\log[1-(1-e^{-\theta})e^{-s}]$ for $\theta > 0$. The limit as $\theta \to 0$ is $\phi_0(s) = e^{-s}$. The family extends to functions in $\mathcal{L}_1$ for all negative parameter values; in this case, write $\phi_\theta(s) = \lambda^{-1}\log[1+(e^\lambda - 1)e^{-s}]$, with $\lambda = -\theta > 0$. Then $\phi_\theta^{-1}(t) =$

Table 5.4. *Most negative values of τ and ρ_S for family M3E.*

m	y_m	λ_m	τ	ρ_S
2	∞	∞	-1	-1
3	1	0.69315	-0.0766	-0.1148
4	0.26795	0.23740	-0.0264	-0.0395
5	0.10102	0.09624	-0.0107	-0.0160
6	0.04310	0.04219	-0.0047	-0.0070

$-\log[(e^{\lambda t}-1)/(e^{\lambda}-1)]$ and $\eta(z) = \phi_\theta^{-1}(e^{-z}) = -\log[(\exp\{\lambda e^{-z}\}-1)/(e^{\lambda}-1)]$. It is straightforward to obtain $\eta'(z) = (\exp\{\lambda e^{-z}\} - 1)^{-1}\lambda e^{-z}\exp\{\lambda e^{-z}\} \geq 0$ and $\eta''(z) = (\exp\{\lambda e^{-z}\} - 1)^{-2}\lambda e^{-z} \cdot \exp\{\lambda e^{-z}\}[1 + \lambda e^{-z} - \exp\{\lambda e^{-z}\}] \leq 0$. Therefore η is concave.

Next we study $(-1)^j\phi_\theta^{(j)}(s)$. Let $y = (e^{\lambda}-1)e^{-s}$ so that $dy/ds = -y$. Then

$$(-1)^j\phi_\theta^{(j)}(s) = \lambda^{-1}A_j(y)/(1+y)^j, \quad j = 1, 2, \ldots,$$

where $A_j(y)$ is a polynomial in y (of degree $j-1$ for $j \geq 2$). The recursion for the A_j is

$$A_{j+1}(y) = y[(1+y)A_j'(y) - jA_j(y)],$$

with $A_1(y) = A_2(y) = y$. If $A_j(y) = \sum_{k=1}^{j-1} a_{jk}y^k$, $j \geq 2$, it can be shown that $a_{j+1,1} = a_{j1} = 1$, $a_{j+1,j} = -a_{j,j-1} = (-1)^j$ by induction, and $a_{j+1,k} = ka_{jk} - (j+1-k)a_{j,k-1}$ for $2 \leq k \leq j-1$. The polynomials $A_j(y)$ are positive for y near zero, with the first positive root of A_j decreasing as j increases. Let y_j be this root and let $\lambda_j = \log(1+y_j)$. Then $\phi_\theta \in \mathcal{L}_m$ if $\lambda \leq \lambda_m$. For $m = 2$, $A_2(y) = y$ so that $y_2 = \lambda_2 = \infty$. That is, bivariate copulas of the form (4.3) exist for the entire extended parameter range. The Fréchet lower bound copula obtains as $\theta \to -\infty$ ($\lambda \to \infty$). In Table 5.4, for $m = 2, \ldots, 6$, are listed y_m, λ_m and the bivariate Kendall tau and Spearman rho values for the most negatively dependent permutation-symmetric copula of the form (4.4) from this family of ϕ_θ. From the table, this family does not allow much extension into the negatively dependent range for $m \geq 3$; the range is much smaller than that of the family M4E.

Next we consider partially symmetric copulas of the form (4.7), (4.10) and (4.11). Let $\omega = \phi_{\theta_1}^{-1} \circ \phi_{\theta_2}$ with $\theta_1 < \theta_2$. With $\rho = \theta_1/\theta_2$,

Table 5.5. *Lower bounds (LB) on parameters for functions in $\mathcal{L}_3^*$ (family M3E).*

θ_1:	-1	-2	-3	-4	-5	-6
θ_2(LB):	-1	-1.11	-1.00	-0.89	-0.81	-0.75

$c_i = 1 - e^{-\theta_i}$, $i = 1, 2$,

$$\omega(s) = -\log\{[1 - (1 - c_2 e^{-s})^\rho]/c_1\}.$$

For $\theta_1 \geq 0, \omega$ is in $\mathcal{L}_\infty^*$. For $\theta_1 = -\lambda < 0$, consider the three cases of: (i) $\theta_2 > 0$; (ii) $\theta_2 = 0$; and (iii) $\theta_2 < 0$. For (i), $\rho < 0, 0 < c_2 < 1$ and $\omega(s) = -\log[(1-y)^\rho - 1] + \log|c_1|$, with $y = c_2 e^{-s} \leq c_2$ $(dy/ds = -y)$. For (ii), $\omega(s) = -\log(e^y - 1) + \log(e^\lambda - 1)$, with $y = \lambda e^{-s} \leq \lambda$. For (iii), $\rho > 1$, $c_1, c_2 < 0$ and $\omega(s) = -\log[(1+y)^\rho - 1] + \log|c_1|$, with $y = -c_2 e^{-s} \leq |c_2|$.

The analysis of ω is not simple. For (ii), $\omega'(s) = ye^y/(e^y - 1) = y/(1 - e^{-y}) \geq 0$, $\omega''(s) = y(1 - e^{-y})^{-2}[(y + 1)e^{-y} - 1]$, $\omega'''(s) = y(1-e^{-y})^{-3}[e^{-2y}(y^2+3y+1)+e^{-y}(y^2-3y-2)+1]$. $\omega''(s) \leq 0$ since $y + 1 \leq e^y$. $w'''(s) \geq 0$ if and only if $e^{-2y}(y^2 + 3y + 1) + e^{-y}(y^2 - 3y - 2) + 1 \geq 0$, or $g(y) = y^2 + 3y + 1 + e^y(y^2 - 3y - 2) + e^{2y} \geq 0$. Since $g(0) = 0$ and $g'(y) \geq 2y + 3 + e^y(2y^2 + y - 3) \stackrel{\text{def}}{=} g_1(y)$, it suffices to show $g_1(y) \geq 0$. This is true since $g_1(0) = g_1'(0) = 0$ and $g_1'' > 0$.

For (i), let $\omega_1 = \phi_{\theta_1}^{-1} \circ \phi_0$ and $\omega_2 = \phi_0^{-1} \circ \phi_{\theta_2}$, where $\phi_0(s) = e^{-s}$. Then $\omega = \omega_1 \circ \omega_2$. Let $y = -\theta_1 e^{-s}$. Then $\omega_1' = y/(1 - e^{-y}) \geq 0$, $\omega_1'' = -ye^y(e^y - 1)^{-2}[e^y - y - 1] \leq 0$, $\omega_1''' = ye^y(e^y - 1)^{-3}[y^2(1 + e^y) + 3y(1-e^y) + (1-e^y)^2] \geq 0$ and $(-1)^{j-1}\omega_2^{(j)} \geq 0, j \geq 1$. Hence it follows that $\omega' = \omega_1'(\omega_2)\omega_2' \geq 0$, $\omega'' = \omega_1''(\omega_2)(\omega_2')^2 + \omega_1'(\omega_2)\omega_2'' \leq 0$ and $\omega''' \geq 0$.

For (iii), $\omega'(s) = \rho y(1+y)^{\rho-1}/[(1+y)^\rho - 1] \geq 0$, $\omega''(s) = -\rho y(1+y)^{\rho-2}[(1+y)^\rho - 1]^{-2}[(1+y)^\rho - 1 - \rho y]$ and $\omega'''(s) = \rho y(1+y)^{\rho-3}[(1+y)^\rho - 1]^{-3}[(1-y)z^2 + z(\rho^2y^2 + (2-3\rho)y - 2) + (\rho^2y^2 + (3\rho-1)y + 1)]$ where $z = (1+y)^\rho$. $\omega''(s) \leq 0$ since $(1+y)^\rho - 1 - \rho y = \rho(\rho-1)y_0^2/2 \geq 0$ for some $0 \leq y_0 \leq y$. $\omega'''(s) \to 0$ as $y \to 0$ (or $s \to \infty$) and $\omega'''(s) > 0$ for y near 0. A conjecture based on some numerical computations is that $\omega'''(s) \geq 0$ for all $s \geq 0$ if $\omega'''(0) \geq 0$.

For fixed θ_1, there should be a lower bound on the possible value of θ_2 so that $\omega \in \mathcal{L}_3^*$. These values are given in Table 5.5 for a few cases; they were obtained numerically.

Extension of B6, M6 and LTA. The positive stable LTs $\phi_\theta(s) = \exp\{-s^{1/\theta}\}$, $\theta \geq 1$, extend to decreasing functions in $\mathcal{L}_1$ for $0 < \theta < 1$. The second derivative of ϕ_θ is

$$\phi_\theta''(s) = \exp\{-s^{1/\theta}\}\lambda s^{\lambda-2}[\lambda s^\lambda - \lambda + 1],$$

where $\lambda = \theta^{-1}$. This derivative can be negative near $s = 0$ if $\theta < 1$ or $\lambda > 1$, so that $\phi_\theta \notin \mathcal{L}_2$ for $0 < \theta < 1$. Hence no copulas of the form (4.3) result from the extension.

Extension of B5, M5 and LTC. Let the LT family be $\phi_\theta(s) = 1-(1-e^{-s})^{1/\theta}$, $\theta \geq 1$. This can be extended to decreasing functions in $\mathcal{L}_1$ for $0 < \theta < 1$. Then $\eta(z) = \phi^{-1}(e^{-z}) = -\log[1-(1-e^{-z})^\theta]$, $\eta'(z) = \theta(1-e^{-z})^{\theta-1}e^{-z}/[1-(1-e^{-z})^\theta]$ and

$$\eta''(z) = [1-(1-e^{-z})^\theta]^{-2}\theta(1-e^{-z})^{\theta-2}e^{-z}\{(1-e^{-z})^\theta + \theta e^{-z} - 1\}.$$

The term $g(w) = (1-w)^\theta + \theta w - 1$ is non-positive, for $w = e^{-z} \in [0,1]$, since $g(0) = 0$ and $g'(w) = \theta[1-(1-w)^{\theta-1}] \leq 0$ for $0 < w, \theta < 1$. Therefore η is concave. With $y = e^{-s}$, $dy/ds = -y$ and $\lambda = \theta^{-1}$, the second derivative of ϕ_θ is

$$\phi_\theta''(s) = \lambda y(1-y)^{\lambda-2}(1-\lambda y),$$

and this can be less than 0 if $\lambda > 1$ $(\theta < 1)$. Therefore for this family of ϕ_θ, there are no negatively dependent copulas for (4.3).

We study one more family that has more negative dependence than the family M3E.

Family MB9E. Extension of family BB9 and LTL. A two-parameter LT family is $\phi(s) = \phi_{\theta,\alpha}(s) = \exp\{-(\alpha^\theta + s)^{1/\theta} + \alpha\}$, $\theta \geq 1$, $\alpha \geq 0$. This can be extended into a family in $\mathcal{L}_1$ for $\theta > 0$, $\alpha \geq 0$. Let $\lambda = \theta^{-1}$. $H(s) = -\log\phi(s) = (\alpha^\theta + s)^\lambda - \alpha$ is convex for $0 < \theta < 1$. With $y = \alpha^\theta + s \geq \alpha^\theta$, the first four derivatives of ϕ are:

$$\phi'(s) = -\phi(s)\lambda y^{\lambda-1} \leq 0; \qquad \phi''(s) = \phi(s)\lambda y^{\lambda-2}[\lambda y^\lambda - \lambda + 1];$$

$$\phi'''(s) = -\phi(s)\lambda y^{\lambda-3}[\lambda^2 y^{2\lambda} - 3\lambda(\lambda-1)y^\lambda + (\lambda-1)(\lambda-2)];$$

$$\phi^{(4)}(s) = \phi(s)\lambda y^{\lambda-4}[\lambda^3 y^{3\lambda} - 6\lambda^2(\lambda-1)y^{2\lambda} + \lambda(\lambda-1)(7\lambda-11)y^\lambda - (\lambda-1)(\lambda-2)(\lambda-3)].$$

$\phi''(s) \geq 0$ for all $s \geq 0$, if $\lambda(\alpha-1)+1 \geq 0$. If $0 < \lambda \leq 1$, then $\phi_{\theta,\alpha} \in \mathcal{L}_2$ for all $\alpha > 0$, and if $\lambda > 1$, then $\phi_{\theta,\alpha} \in \mathcal{L}_2$ for $\alpha \geq 1-\theta = 1-\lambda^{-1}$. The third and fourth derivatives are harder to analyse. Let $z = y^\lambda$. The roots of $\lambda^2 z^2 - 3\lambda(\lambda-1)z + (\lambda-1)(\lambda-2)$ in ϕ''' are $[3(\lambda-1) \pm \sqrt{(\lambda-1)(5\lambda-1)}]/(2\lambda)$. If the roots are real

Table 5.6. *Lower boundary values of α for given θ for family MB9E.*

θ^{-1}:	1	2	3	4	5	∞
$\mathcal{L}_2$:	0	0.5	0.6667	0.75	0.80	1
$\mathcal{L}_3$:	0	1.5	1.8819	2.0687	2.1798	2.6180
$\mathcal{L}_4$:	0	2.7247	3.3333	3.6287	3.8037	4.4909

($\lambda < 1/5$ or $\lambda \geq 1$), let z_0 be larger root with the + sign. Then $\phi''' \leq 0$ for all $s \geq 0$ if $\alpha \geq z_0$ or if both roots are complex. If $0 < \lambda \leq 1/5$, $z_0 < 0 \leq \alpha$. If $\lambda > 1$, z_0 is positive so that there is a lower bound on α in order that $\phi_{\theta,\alpha} \in \mathcal{L}_3$. Similarly, the lower bound on α for $\mathcal{L}_4$ can be computed numerically. For some selected values of $\lambda > 1$, a table of the lower bounds on α for $\phi_{\theta,\alpha}$ to be in $\mathcal{L}_2, \mathcal{L}_3, \mathcal{L}_4$ is given in Table 5.6.

For $0 < \theta < 1$ and $\alpha \geq 1 - \theta$, $C(u,v) = \phi(\phi^{-1}(u) + \phi^{-1}(v))$ is a copula with negative dependence. The formula is

$$C(u,v) = \exp\{-[(\alpha - \log u)^\theta + (\alpha - \log v)^\theta - \alpha^\theta]^{1/\theta} + \alpha\}, \quad (5.17)$$

and as $\theta \to 0$, the copula becomes

$$C(u,v) = uv\exp\{-\alpha^{-1}(\log u)(\log v)\}, \quad \alpha \geq 1. \quad (5.18)$$

The distribution in (5.17) is increasing in concordance as θ increases for all fixed α, and it is decreasing (increasing) in concordance as α increases for fixed $\theta \geq 1$ ($\theta \leq 1$). Expression (5.18) represents the family of copulas for the family of bivariate exponential distributions in Example 2.5. Note that (5.17) does not depend on α for $\theta = 1$ (the independence copula). The distribution in (5.18) is increasing in concordance as α increases and the independence copula obtains as $\alpha \to \infty$; (5.18) has the form of (4.3) with $\psi_\alpha(s) = \exp\{\alpha(1 - e^s)\}$, $\alpha \geq 1$. This family of functions in $\mathcal{L}_2$ comes from a limit of $\phi_{\theta,\alpha}$ as $\theta \to 0$ if scaling is done with the limit, i.e., it is the limit of $\phi_{\theta,\alpha}(\alpha^\theta \theta s) = \exp\{-\alpha(1 + \theta s)^{1/\theta} + \alpha\}$.

For (5.18), the Kendall tau and Spearman rho values corresponding to the α values in the last column of Table 5.6 are given in Table 5.7. These are the smallest possible values for the permutation-symmetric copulas (4.4) with ϕ in this family. For $m = 3, 4$, a greater range of negative dependence obtains compared with family M3E.

It is not known how to find LT families with extensions such that (4.4) has positive density on $(0,1)^m$ and has more negative

Table 5.7. *Most negative values of τ and ρ_S for family MB9E.*

m	α	τ	ρ_S
2	1	−0.3613	−0.5240
3	2.6180	−0.1637	−0.2435
4	4.4909	−0.1010	−0.1511

dependence than the examples given in this section. Also unknown are results that can quantify the range of negative dependence that can be achieved with the approach of Section 4.4.

5.5 Multivariate copulas with general dependence

In this section, we list five families of parametric multivariate copulas that have flexible general positive dependence structure. Some are extensions of families in Section 5.2. The ideas of Sections 4.4, 4.5 and 5.4 can be used to extend some of them into the range of negatively dependent bivariate margins. The first three families are based on the theory of Secton 4.3 and have closed-form cdfs; two are families of extreme value copulas, and in the third a limiting family of extreme value copulas is obtained in Section 6.3.1. Repeating from Section 5.3, an m-variate copula C is a (multivariate) extreme value copula if $C(u_1^t, \ldots, u_m^t) = C^t(u_1, \ldots, u_m)$ for all $t > 0$. The last two families are based on Section 4.5 and their extreme value limits are also given in Section 6.3.1.

Family MM1. In (4.25), let K_{ij} be the family B6 with parameter δ_{ij} and let ψ be the family LTA with parameter θ. The result is the family: for $\theta \geq 1$ and $\delta_{ij} \geq 1$, $i < j$,

$$C(\mathbf{u}) = \exp\Big\{-\Big[\sum_{i<j}\big((p_i z_i^\theta)^{\delta_{ij}} + (p_j z_j^\theta)^{\delta_{ij}}\big)^{1/\delta_{ij}} + \sum_{j=1}^{m} \nu_j p_j z_j^\theta\Big]^{1/\theta}\Big\}, \tag{5.19}$$

where $z_j = -\log u_j$, $\nu_j \geq 0$, $p_j = (\nu_j + m - 1)^{-1}$, $j = 1, \ldots, m$. This is a family of extreme value copulas since the exponent in (5.19) is homogeneous of order 1 as a function of $z_1, \ldots, z_m$. The bivariate margins are:

$$\begin{aligned} C_{ij}(u_i, u_j) &= \exp\{-[((p_i z_i^\theta)^{\delta_{ij}} + (p_j z_j^\theta)^{\delta_{ij}})^{1/\delta_{ij}} \\ &\quad + (\nu_i + m - 2)p_i z_i^\theta + (\nu_j + m - 2)p_j z_j^\theta]^{1/\theta}\}. \end{aligned}$$

When $p_i = p_j = 1$ or $\nu_i = \nu_j = 2 - m$, this bivariate copula corresponds to the family B6. The tail dependence parameter is $\lambda_{ij} = 2 - [(p_i^{\delta_{ij}} + p_j^{\delta_{ij}})^{1/\delta_{ij}} + (\nu_i + m - 2)p_i + (\nu_j + m - 2)p_j]^{1/\theta}$; it increases as δ_{ij} or θ increases. Further tail dependence analysis of this family is given in Section 6.3.1.

Family MM2. In (4.25), let K_{ij} be the family B7 with parameter δ_{ij} and let ψ be the family LTB with parameter θ. Let $\hat{u}_j = p_j(u_j^{-\theta} - 1)$, $p_j = (\nu_j + m - 1)^{-1}$, $\nu_j \geq 0$, $j = 1, \ldots, m$. The result is the family: for $\theta > 0$ and $\delta_{ij} > 0$, $i < j$,

$$C(\mathbf{u}) = \Big[\sum_{j=1}^{m} u_j^{-\theta} - (m-1) - \sum_{1 \leq i < j \leq m} (\hat{u}_i^{-\delta_{ij}} + \hat{u}_j^{-\delta_{ij}})^{-1/\delta_{ij}}\Big]^{-1/\theta}. \tag{5.20}$$

The special case $[\sum_{i=1}^{m} u_i^{-\theta} - (m-1)]^{-1/\theta}$ arises as $p_j \to 0$, $j = 1, \ldots, m$. The bivariate margins are:

$$C_{ij}(u_i, u_j) = [u_i^{-\theta} + u_j^{-\theta} - 1 - (\hat{u}_i^{-\delta_{ij}} + \hat{u}_j^{-\delta_{ij}})^{-1/\delta_{ij}}]^{-1/\theta}.$$

When $p_i = p_j = 1$, this bivariate copula corresponds to the family BB4. This copula has both lower and upper tail dependence. Using the approximations $u_j^{-\theta} - 1 \approx \theta(1 - u_j)$ and $\hat{u}_j \approx p_j\theta(1 - u_j)$ as $u_j \to 1$, $j = 1, \ldots, m$, the upper tail dependence parameters can be computed as $\lambda_{ij,U} = (p_i^{-\delta_{ij}} + p_j^{-\delta_{ij}})^{-1/\delta_{ij}}$. The lower tail dependence parameters are $\lambda_{ij,L} = [2 - (p_i^{-\delta_{ij}} + p_j^{-\delta_{ij}})^{-1/\delta_{ij}}]^{-1/\theta}$. The tail dependence parameters are increasing as δ_{ij} increases. The upper tail dependence parameters do not depend on θ; the lower tail dependence parameters increase as θ increases.

The upper tail extreme value limit is

$$\exp\Big\{-\sum_i e^{-x_i} + \sum_{i<j} \big(p_i^{-\delta_{ij}} e^{\delta_{ij}x_i} + p_j^{-\delta_{ij}} e^{\delta_{ij}x_j}\big)^{-1/\delta_{ij}}\Big\}.$$

It is not so interesting as it does not depend on θ. The lower tail extreme value limit is more interesting. It is similar to (5.19) and generalizes the family B7 (see the family MM8 in Section 6.3.1).

Family MM3. In (4.25), let K_{ij} be the family B7 with parameter δ_{ij} and let ψ be the family LTA with parameter θ. Let $z_j = -\log u_j$, $p_j = (\nu_j + m - 1)^{-1}$, $\nu_j \geq 0$, $j = 1, \ldots, m$. The result is the family of extreme value copulas:

$$C(\mathbf{u}) = \exp\Big\{-\Big[\sum_{j=1}^{m} z_j^{\theta} - \sum_{i<j} \big(p_i^{-\delta_{ij}} z_i^{-\theta\delta_{ij}} + p_j^{-\delta_{ij}} z_j^{-\theta\delta_{ij}}\big)^{-1/\delta_{ij}}\Big]^{1/\theta}\Big\}, \tag{5.21}$$

$\theta > 0$ and $\delta_{ij} > 0$, $i < j$.

The bivariate margins are:

$$C_{ij}(u_i, u_j) = \exp\{-[z_i^\theta + z_j^\theta - (p_i^{-\delta_{ij}} z_i^{-\theta\delta_{ij}} + p_j^{-\delta_{ij}} z_j^{-\theta\delta_{ij}})^{-1/\delta_{ij}}]^{1/\theta}\}. \tag{5.22}$$

When $p_i = p_j = 1$, this bivariate copula corresponds to the family BB5. The upper tail dependence parameter for (5.22) is $\lambda_{ij} = 2 - [2 - (p_i^{-\delta_{ij}} + p_j^{-\delta_{ij}})^{-1/\delta_{ij}}]^{1/\theta}$. It is increasing as δ_{ij} or θ increases.

For $m = 3$, with $\delta_{13} \to 0$, $\delta_{12} = \delta_{23} = \delta$, $\nu_1 = \nu_3 = -1$, $\nu_2 = 0$, $p_1 = p_3 = 1$, $p_2 = \frac{1}{2}$, (5.21) becomes

$$C(\mathbf{u}) = \exp\{-[z_1^\theta + z_2^\theta + z_3^\theta - (z_1^{-\theta\delta} + 2^\delta z_2^{-\theta\delta})^{-\frac{1}{\delta}} - (z_3^{-\theta\delta} + 2^\delta z_2^{-\theta\delta})^{-\frac{1}{\delta}}]^{\frac{1}{\theta}}\}. \tag{5.23}$$

The bivariate margins of (5.23) are $C_{j2}(u_j, u_2) = \exp\{-[z_j^\theta + z_2^\theta - (z_j^{-\theta\delta} + 2^\delta z_2^{-\theta\delta})^{-1/\delta}]^{1/\theta}\}$, $j = 1, 3$, and $C_{13}(u_1, u_3) = \exp\{-(z_1^\theta + z_3^\theta)^{1/\theta}\}$. The tail dependence parameters are $\lambda_{j2} = 2 - [2 - (1 + 2^\delta)^{-1/\delta}]^{1/\theta}$, $j = 1, 3$, and $\lambda_{13} = 2 - 2^{1/\theta}$. As $\delta \to \infty$, $\lambda_{j2} \to 2 - (1.5)^{1/\theta}$.

Family MM4. Consider the construction in (4.37)–(4.39) for mixtures of conditional distributions with the associated copulas in the family B4, i.e., $C_{ij}(u_i, u_j) = C(u_i, u_j; \delta_{ij}) = u_i + u_j - 1 + [(1 - u_i)^{-\delta_{ij}} + (1 - u_j)^{-\delta_{ij}} - 1]^{-1/\delta_{ij}}$, $\delta_{ij} > 0$, $1 \le i < j \le m$. From Theorem 4.20, the resulting copula family has upper tail dependence for each bivariate margin and has a wide range of dependence. This family leads to a family of MEV distributions (see Section 6.3.1).

Family MM5. Consider the construction in (4.37)–(4.39) for mixtures of conditional distributions with the family B5 for the bivariate copulas, i.e., $C_{ij} = C(\cdot; \delta_{ij})$, $\delta_{ij} \ge 1$, is in the family B5 for all $1 \le i < j \le m$. The resulting copula family has similar dependence properties to the family MM4, and it also leads to a family of MEV distributions.

From Theorem 4.11, 4.14 and results in Section 5.1, the families MM1–MM3 are increasing in the $\prec_c^{\text{pw}}$ ordering as the δ_{ij} increase or ν_j decrease; it is conjectured (with support from numerical checks) that they also increase in $\prec_c^{\text{pw}}$ as θ increases. For the families MM4–MM5, for $i < j$ with $j - i \ge 2$, the conditional dependence of the ith and jth variables, given variables $i + 1, \ldots, j - 1$, increases as δ_{ij} increases; the $\prec_c^{\text{pw}}$ ordering need not hold as the δ_{ij} increase, but the (i, j) bivariate margin increases in concordance for δ_{ij} increasing (Theorem 4.19).

As can be seen from the parametric families in this section, one motivation for the development of the theory in Chapter 4 has been the construction of parametric families of MEV distributions with general dependence structure. There is more on the topic of MEV distributions in the next chapter.

5.6 Bibliographic notes

For the families B1 to B8, refer to Joe (1993) for some background and multivariate extensions. Without all of the properties of Section 5.1, there are many parametric families of bivariate copulas; see Hutchinson and Lai (1990) for a fairly exhaustive compilation. B3 has been studied in Nelsen (1986) and Genest (1987), and B4 has been studied in Clayton (1978), Cook and Johnson (1981), and Oakes (1982). B12 is studied in Cuadras and Augé (1981) and Nelsen (1991). Of the properties in Section 5.1, one of the hardest to check is the SI ordering. This property is shown in Fang and Joe (1992) for some families of copulas, and is included in the examples in Section 2.2.7 for others; for B2, Theorem 2.14 can be applied with symbolic manipulation software.

The families in Sections 5.2 to 5.5 are mainly from Joe and Hu (1996) and Joe (1994; 1996a). The extension to negative dependence for the bivariate case has been known for B4; see Ruiz-Rivas (1981) and Genest and MacKay (1986). Regarding M3E, this model has been used in Meester and MacKay (1994), but without mentioning that there is a limit to the range of the parameter space for negative dependence. For a more general multivariate version of B10 than (5.2), see Johnson and Kotz (1975) and Shaked (1975).

5.7 Exercises

5.1 Verify directly the SI property for the families B2 to B7 (in the range of positive dependence).

5.2 For the family B2, show that $u-C(u,v;\theta) = C(u,1-v;\theta^{-1})$. For the family B3, show that $u-C(u,v;\theta) = C(u,1-v;-\theta)$. (See Sections 7.1.7 and 1.6 for the interpretation.)

5.3 Verify the density for the Plackett copula in the family B2 and show that it is non-negative.

5.4 For the family B11 with copula $C(u,v;\delta) = \delta(u \wedge v) + (1-\delta)uv$, $0 \leq \delta \leq 1$, show that if $(U,V) \sim C(\cdot;\delta)$ and $Y_1 = I(U \leq x)$, $Y_2 = I(V \leq x)$, then the correlation of Y_1, Y_2

is δ for any $0 < x < 1$. Does this property hold for other families?

5.5 Verify the properties listed for the families B1–B8. Also verify the copulas (independence or Fréchet bounds) that are obtained at the end points of the parameter ranges.

5.6 Verify the properties listed for the families BB1–BB10.

5.7 For the multivariate copula in (5.2), show for $m = 3$ that the constraints on the parameters are:

$$-1 \le \delta_{12}, \delta_{13}, \delta_{23} \le 1,$$
$$-1 + |\delta_{12} + \delta_{23}| \le \delta_{13} \le 1 - |\delta_{12} - \delta_{23}|.$$

Determine the constraints on the parameters for $m = 4$.

5.8 Prove that the family B9 is increasing in $\prec_c$, by filling in details in the following (proof due to T. Hu). Let Z_1, Z_2, Z_3, Z_4, $Z_{12}, Z_{12}^*, W_{12}, W_{12}^*$ be rvs with unit exponential distributions. Let $I_\delta, I_{\delta'}, I_\eta, I_\eta^*$ be Bernoulli rvs with respective parameters $\delta, \delta', \eta, \eta$. Let $F(\cdot;\delta)$ be the bivariate exponential cdf for the stochastic representation in (5.1):

$$X_1 = (1-\delta)Z_1 + I_\delta Z_{12}, \quad X_2 = (1-\delta)Z_2 + I_\delta Z_{12}, \quad (5.24)$$

where $I_\delta, Z_1, Z_2, Z_{12}$ are independent. Suppose $\delta' > \delta > 0$ and set $\eta = (\delta' - \delta)/(1-\delta)$. Let $(Y_1, Y_2) \sim F(\cdot;\delta')$ with stochastic representation:

$$Y_1 = (1-\delta')Z_3 + I_{\delta'} Z_{12}^*, \quad Y_2 = (1-\delta')Z_4 + I_{\delta'} Z_{12}^*,$$

where $I_{\delta'}, Z_3, Z_4, Z_{12}^*$ are independent. Substitute $Z_1 = (1-\eta)Z_3 + I_\eta W_{12}$, $Z_2 = (1-\eta)Z_4 + I_\eta^* W_{12}^*$ into (5.24) to get

$$\begin{aligned} X_1 &= (1-\delta')Z_3 + [(1-\delta)I_\eta W_{12} + I_\delta Z_{12}], \\ X_2 &= (1-\delta')Z_4 + [(1-\delta)I_\eta^* W_{12}^* + I_\delta Z_{12}], \quad (5.25) \end{aligned}$$

with $I_\delta, I_\eta, I_\eta^*, Z_3, Z_4, W_{12}, W_{12}^*, Z_{12}$ independent. Also,

$$\begin{aligned} Y_1 &= (1-\delta')Z_3 + [(1-\delta)I_\eta W_{12} + I_\delta Z_{12}], \\ Y_2 &= (1-\delta')Z_4 + [(1-\delta)I_\eta W_{12} + I_\delta Z_{12}], \quad (5.26) \end{aligned}$$

since $(1-\delta)I_\eta W_{12} + I_\delta Z_{12} \stackrel{d}{=} I_{\delta'} Z_{12}^*$ (this can be shown using moment generating functions). Using the representations (5.25) and (5.26), it follows that $(X_1, X_2) \prec_c (Y_1, Y_2)$ or that $F(\cdot;\delta) \prec_c F(\cdot;\delta')$.

5.9 For specific cases of (5.15) for the families M3, M4 and M6, show that (5.15) is not a proper copula if $\theta_1 > \theta_2$.

5.10 Study the two-parameter family of Archimedean copulas based on the family LTK in the Appendix.

5.11 Study the multivariate and negative dependent extensions of the family BB8 of bivariate copulas by extending the families LTJ, LTK of LTs to $\mathcal{L}_m$.

5.12 Verify the details below which show that the property of a TP_2 density does not hold for $\delta > 2$ for the family B2. Let $c(u, v; \delta)$ be the density of the family B2. Consider

$$c(u_1, v_1; \delta)\, c(u_2, v_2; \delta) - c(u_1, v_2; \delta)\, c(u_2, v_1; \delta), \qquad (5.27)$$

for $u_1 < u_2$, $v_1 < v_2$. This expression is negative when $\delta > 2$, u_1, u_2 are close to zero and v_1, v_2 are close to 1. Let $\overline{v} = 1-v$; expanding c to second order near $u = 0, v = 1$ leads to

$$c(u, v; \delta) \approx \delta^{-1}[1 + 2\rho(u + \overline{v}) + 3\rho^2(u^2 + \overline{v}^2) + \alpha u\overline{v}],$$

where $\rho = (\delta - 1)/\delta$ and $\alpha = 4\delta^{-2}(2\delta^2 - 5\delta + 3)$. Therefore (5.27) is equal in sign to $(4\rho^2 - \alpha)(u_1 - u_2)(v_1 - v_2) = 4\delta^{-2}(\delta - 1)(2 - \delta)(u_1 - u_2)(v_1 - v_2)$.

5.8 Unsolved problems

5.1 Verify or disprove the property of a TP_2 density for the families B7 and B8. Numerical checks seem to suggest that the property holds.

5.2 Numerical checks seem to suggest that the property of a TP_2 density holds for $1 < \delta \leq 2$ for the family B2. Also, numerically it appears that the property of a TP_2 cdf holds for the family B2. Can these properties be shown analytically?

5.3 Find other parametric families of Laplace transforms that can lead to interesting copulas.

5.4 For the families BB3, BB4 and BB5, show analytically the properties of the concordance ordering.

5.5 Find a parametric family of copulas of the form (4.4) that have positive density on $[0, 1]^m$ and more negative dependence than the families M3E and MB9E for $m \geq 3$.

5.6 This problem may be helpful to solve the preceding problem. Find the infimum of $\tau = 1 - 4\int_0^\infty s[\phi'(s)]^2 ds$ (the value of Kendall's tau from Theorem 4.3) subject to $\phi \in \mathcal{L}_m$.

CHAPTER 6

Multivariate extreme value distributions

This chapter is devoted to multivariate extreme value (MEV) models and their applications. A main goal is the construction of parametric families of MEV distributions with wide dependence structure. A typical example that provides some motivation for MEV models is given in Section 11.3.

A summary of the sections, including the highlights, is the following. Section 6.1 gives the brief background from univariate extreme theory that is relevant to statistical inference. Section 6.2 contains the background in MEV theory, including Pickand's representation and characterization for min-stable multivariate exponential (MSMVE) distributions, and shows the contrast between MEV limit theory and MVN limit theory. To relate to earlier chapters, we show that MEV distributions are one setting for naturally using copulas and exploiting the properties of copulas. Section 6.3 contains parametric families of MEV copulas in the form of MSMVE distributions; several families are derived from the extreme value limit of families of copulas that are given in Chapter 5. Section 6.4 is on the point process modelling approach for inference with multivariate extremes. Section 6.5 is on choice models and the use of MEV models in the psychology and econometrics literatures. Section 6.6 is devoted to mixtures of MEV distributions, which include max-geometric stable multivariate distributions.

6.1 Background: univariate extremes

Let $X_1, X_2, \ldots$ be iid rv's with continuous distribution function F. Let $S_n = X_1 + \cdots + X_n$, $M_n = \max\{X_1, \ldots, X_n\} = X_1 \vee \cdots \vee X_n$, and $L_n = \min\{X_1, \ldots, X_n\} = X_1 \wedge \cdots \wedge X_n$. Paralleling the central limit theorem and stable laws based on S_n, one can consider the possible limiting distributions for $(M_n - a_n)/b_n$ and $(L_n - c_n)/d_n$ as

$n \to \infty$ for suitably chosen sequences $\{a_n\}, \{b_n\}, \{c_n\}, \{d_n\}$. (The sequences need not be unique.) One needs only to study the case of maxima as the theory for minima is similar due to the identity $\min\{X_1, \ldots, X_n\} = -\max\{-X_1, \ldots, -X_n\}$.

The main well-known result or the 'three-types' theorem, with contributions from Fisher, Tippett, von Mises, Gnedenko and de Haan, is that the only possible limits are the location–scale families based on the cdfs:

1. $H_0(x) = \exp\{-e^{-x}\}$, $-\infty < x < \infty$ (Gumbel or extreme value distribution);
2. $H_1(x;\theta) = \exp\{-x^{-\theta}\}$, $x > 0, \theta > 0$ (Fréchet distribution);
3. $H_{-1}(x;\theta) = \exp\{-(-x)^{\theta}\}$, $x < 0, \theta > 0$ (Weibull distribution).

Necessary and sufficient conditions are given in books on extreme value theory, for example, Galambos (1987) and Resnick (1987). F is said to be in the **domain of attraction** of the Gumbel, Fréchet or Weibull distribution if one of these distributions is the extreme value limit.

By reflection, the possible limits for minima are the location–scale families based on the cdfs:

1. $H_0^*(x) = 1 - H_0(-x) = 1 - \exp\{-e^{x}\}$, $-\infty < x < \infty$;
2. $H_1^*(x;\theta) = 1 - H_1(-x;\theta) = 1 - \exp\{-(-x)^{-\theta}\}$, $x < 0, \theta > 0$;
3. $H_{-1}^*(x;\theta) = 1 - H_{-1}(-x;\theta) = 1 - \exp\{-x^{\theta}\}$, $x > 0, \theta > 0$.

After location/scale changes, the three types can be combined into the generalized extreme value (GEV) family:

$$H(x;\gamma) = \exp\{-(1+\gamma x)_+^{-1/\gamma}\}, \quad -\infty < x < \infty, \quad -\infty < \gamma < \infty,$$

where $(y)_+ = \max\{0, y\}$. $\gamma \to 0$ yields $H_0(x)$, $\gamma > 0$ yields $H_1(1 + \gamma x; 1/\gamma)$, and $\gamma < 0$ yields $H_{-1}(-1 - \gamma x; -1/\gamma)$.

Example 6.1 (Some special cases.) Note that $\Pr((M_n - a_n)/b_n \le z) = \Pr(M_n \le a_n + b_n z) = F^n(a_n + b_n z)$. For illustration, some examples are the following.

(a) Exponential. $F(x) = 1 - e^{-x}$, $x > 0$. Let $a_n = \log n = F^{-1}(1 - n^{-1})$, $b_n = 1$; then $F^n(a_n + b_n z) = (1 - n^{-1}e^{-z})^n \to \exp\{-e^{-z}\}$, $-\infty < z < \infty$, the Gumbel distribution.

(b) Pareto. $F(x) = 1 - x^{-1/\gamma}$, $x > 1$, $\gamma > 0$. The tail gets heavier as γ increases. Let $a_n = 0$, $b_n = n^{\gamma} = F^{-1}(1 - n^{-1})$; then $F^n(a_n + b_n z) = (1 - n^{-1}z^{-1/\gamma})^n \to \exp\{-z^{-1/\gamma}\}$, $z > 0$, the Fréchet distribution.

(c) Beta. $F(x) = 1 - (1-x)^{-1/\gamma}$, $0 < x < 1$, $\gamma < 0$. The tail gets heavier as γ increases or $-\gamma$ decreases and the upper end point of support is finite. Let $a_n = 1$, $b_n = n^\gamma = 1 - F^{-1}(1-n^{-1})$; then $F^n(a_n + b_n z) = [1 - n^{-1}(-z)^{-1/\gamma}]^n \to \exp\{-(-z)^{-1/\gamma}\}$, $z < 0$, the Weibull distribution.

(d) $F(x) = 1 - (\log x)^{-1}$, $x > e$. This has a heavier tail than the Pareto distributions and there is no extreme value limit based on linearity.

□

Since Pareto distributions are heavier-tailed than exponential distributions which in turn are heavier-tailed than distributions with finite upper end point of support, γ can be interpreted as a tail parameter that is larger for a heavier tail. For another example, the domain of attraction for a normal distribution is the Gumbel distribution but the sequences of constants are not simple in form (see the references cited earlier).

An application of univariate extreme value theory is as follows. For data $M_{1n}, \ldots, M_{kn}$ consisting of (approximately) iid maxima based on n observations each, the three-parameter GEV model $H((y-\mu)/\sigma; \gamma)$ is used as an approximation $F^n(y)$, assuming n is sufficiently large. From this, an approximation for the tail probability $\overline{F}(y)$, y large, is

$$\begin{aligned} 1 - F(y) &\approx 1 - H^{1/n}((y-\mu)/\sigma; \gamma) \\ &\approx -\log H^{1/n}((y-\mu)/\sigma; \gamma) \\ &= n^{-1}(1 + \gamma[y-\mu]/\sigma)_+^{-1/\gamma}, \end{aligned}$$

and for a large threshold T,

$$\frac{1 - F(x+T)}{1 - F(T)} \approx \left[\frac{1 + \gamma(x+T-\mu)/\sigma}{1 + \gamma(T-\mu)/\sigma}\right]_+^{-1/\gamma} = (1 + \gamma x/\sigma^*)_+^{-1/\gamma},$$

where $\sigma^* = 1 + \gamma(T-\mu)/\sigma$. This is the generalized Pareto approximation to the conditional tail distribution of a distribution that is in the domain of attraction of an extreme value distribution. The multivariate extension of this is given in Section 6.4.

A data analysis error is to fit a parametric family to data and extrapolate to extreme such as the 99th percentile. For these extreme value inferences, it is better to apply extreme value theory and the generalized Pareto distribution to the largest data values (say, less than the top 10%).

6.2 Multivariate extreme value theory

Let $(X_{i1}, \ldots, X_{im})$ be iid random vectors with distribution F, $i = 1, 2, \ldots$. Let $M_{jn} = \max_{1 \le i \le n} X_{ij}$, $j = 1, \ldots, m$, be the componentwise maxima. From a sample of vectors of maxima, one can make inferences about the upper tail of F using multivariate extreme value theory.

MEV distributions come from limits (in law) of $((M_{1n} - a_{1n})/b_{1n}, \ldots, (M_{mn} - a_{mn})/b_{mn})$. If a limiting distribution exists, then each univariate margin must be in the GEV family. The multivariate limiting distribution can then be written in the form

$$C(H(z_1; \gamma_1), \ldots, H(z_m; \gamma_m)),$$

where $H(z_j; \gamma_j)$ are GEV distributions and C is a multivariate copula. Further properties of MEV distributions and a method for constructing MEV distributions are given after the following univariate result on transforms and extremes.

Lemma 6.1 *Let $X_1, X_2, \ldots$ be iid with distribution function F and let $r(X_i) = X_i^*$, $i = 1, 2, \ldots$, be iid with distribution function F^*, where r is strictly increasing. Let $M_n = \max\{X_1, \ldots, X_n\}$, $M_n^* = \max\{X_1^*, \ldots, X_n^*\}$. Suppose $\Pr((M_n - a_n)/b_n \le z) \to H(z)$ and $\Pr((M_n^* - a_n^*)/b_n^* \le z^*) \to H^*(z^*)$, where H and H^* are extreme value distributions. Note that $\Pr((M_n^* - a_n^*)/b_n^* \le z^*) = \Pr(M_n \le r^{-1}(a_n^* + b_n^* z^*)) = \Pr((M_n - a_n)/b_n \le [r^{-1}(a_n^* + b_n^* z^*) - a_n]/b_n)$. Hence $\lim_n [r^{-1}(a_n^* + b_n^* z^*) - a_n]/b_n$ exists. Let $s(z^*)$ be the limit. Then $H^*(z^*) = H(s(z^*))$.*

Now, returning to the multivariate setting, let $(M_{1n}, \ldots, M_{mn})$ be a vector of componentwise maxima of the iid random vectors $(X_{i1}, \ldots, X_{im})$ from F, and suppose

$$\begin{aligned} G(\mathbf{z}) &= \lim_n F^n(a_{1n} + b_{1n} z_1, \ldots, a_{mn} + b_{mn} z_m) \\ &= \lim_n \Pr(M_{1n} \le a_{1n} + b_{1n} z_1, \ldots, M_{mn} \le a_{mn} + b_{mn} z_m) \\ &= C(H(z_1; \gamma_1), \ldots, H(z_m; \gamma_m)), \end{aligned} \tag{6.1}$$

where C is a copula. Let r_j be a strictly increasing transform of the X_{ij}, and let the transformed variables and maxima be X_{ij}^* and M_{jn}^*. Suppose the transforms are such that $(M_{jn}^* - a_{jn}^*)/b_{jn}^*$ converges in distribution as $n \to \infty$ for all j. Let

$$\begin{aligned} G^*(\mathbf{z}^*) &= \lim_n \Pr(M_{1n}^* \le a_{1n}^* + b_{1n}^* z_1^*, \ldots, M_{mn}^* \le a_{mn}^* + b_{mn}^* z_m^*) \\ &= C^*(H(z_1^*; \gamma_1^*), \ldots, H(z_m^*; \gamma_m^*)) \end{aligned}$$

for a copula C^*. Also, by Lemma 6.1,

$$\lim_n \Pr((M_{jn} - a_{jn})/b_{jn} \le [r_j^{-1}(a_{jn}^* + b_{jn}^* z_j^*) - a_{jn}]/b_{jn}, \forall j)$$
$$= G^*(\mathbf{z}^*) = C\big(H(s_1(z_1^*); \gamma_1), \ldots, H(s_m(z_m^*); \gamma_m)\big)$$

for some functions $s_1, \ldots, s_m$. Hence $H(s_j(z_j^*); \gamma_j) = H(z_j^*; \gamma_j^*)$, $j = 1, \ldots, m$, and $C = C^*$. The importance of this result is for obtaining MEV distributions through the taking of limits, since one can take univariate margins and constants a_{jn}, b_{jn} that lead to easier calculations (see Section 6.2.2).

The next result gives conditions that MEV copulas must satisfy and shows the relationship to MSMVE distributions, which are defined below.

Suppose (6.1) holds and let k be a positive integer. Then the sequences a_{jn}, b_{jn}, $j = 1, \ldots, m$, are such that

$$\lim_n \Pr(M_{1,kn} \le a_{1n} + b_{1n} z_1, \ldots, M_{m,kn} \le a_{mn} + b_{mn} z_m)$$
$$= G^k(z_1, \ldots, z_m) = C^k(H(z_1; \gamma_1), \ldots, H(z_m; \gamma_m)). \qquad (6.2)$$

Also

$$\Pr(M_{1,kn} \le a_{1n} + b_{1n} z_1, \ldots, M_{m,kn} \le a_{mn} + b_{mn} z_m)$$
$$= \Pr([M_{j,kn} - a_{j,kn}]/b_{j,kn} \le [a_{jn} + b_{jn} z_j - a_{j,kn}]/b_{j,kn}, 1 \le j \le m)$$
$$\to G([z_1 - \mu_1]/\sigma_1, \ldots, [z_m - \mu_m]/\sigma_m)$$
$$= C(H([z_1 - \mu_1]/\sigma_1; \gamma_1), \ldots, H([z_m - \mu_m]/\sigma_m; \gamma_m)) \qquad (6.3)$$

for some constants μ_j, σ_j, and $\Pr(M_{j,kn} \le a_{jn} + b_{jn} z_j) \to H^k(z_j, \gamma_j)$ for each j. Hence

$$G^k(\mathbf{z}) = C^*(H^k(z_1; \gamma_1), \ldots, H^k(z_k; \gamma_k)) \qquad (6.4)$$

for some copula C^*. From univariate extreme value theory, it follows that $H^k(z_j; \gamma_j) = H([z_j - \mu_j]/\sigma_j; \gamma_j)$ for each j, and hence $C = C^*$ from matching equations (6.2), (6.3) and (6.4).

Next, let $u_j = H(z_j, \gamma_j)$. Then from (6.2) and (6.4)

$$C^k(u_1, \ldots, u_m) = C(u_1^k, \ldots, u_m^k), \quad k = 1, 2, \ldots$$

or

$$C(u_1^{1/r}, \ldots, u_m^{1/r}) = C^{1/r}(\mathbf{u}), \quad r = 1, 2, \ldots.$$

Hence

$$C^{k/r}(\mathbf{u}) = C^k(u_1^{1/r}, \ldots, u_m^{1/r}) = C(u_1^{k/r}, \ldots, u_m^{k/r})$$

for all positive integers k, r. This can be extended to

$$C(u_1^t, \ldots, u_m^t) = C^t(\mathbf{u}) \quad \forall t > 0,$$

by continuity and approximation of a real number by a sequence of rational numbers. Let $D(\mathbf{y}) = C(e^{-y_1}, \ldots, e^{-y_m})$ be a multivariate distribution with unit exponential survival margins. Then

$$C(e^{-ty_1}, \ldots, e^{-ty_m}) = D(t\mathbf{y}) = C^t(e^{-y_1}, \ldots, e^{-y_m}) = D^t(\mathbf{y}).$$

Hence $A = -\log D$ satisfies $A(t\mathbf{y}) = tA(\mathbf{y})$ which implies that D is a MSMVE survival function as given in Theorem 6.2 below.

Definition. Let $\mathbf{X}$ be an m-dimensional random vector with survival function G. Suppose X_i is exponential with mean ν_i, $i = 1, \ldots, m$. $\mathbf{X}$ (or G) is **min-stable multivariate exponential** (MSMVE) if for all $\mathbf{w} \in (0,\infty)^m$, $\min\{X_1/w_1, \ldots, X_m/w_m\} = (X_1/w_1) \wedge \cdots \wedge (X_m/w_m)$ has an exponential distribution.

Note that the property of closure under weighted minima is analogous to the property of closure under linear combinations for the MVN family, with the operator $\wedge$ replacing the $+$ operator. The explanation of the term 'stable' is given later in this section. The next theorem shows that MSMVE distributions have survival functions G such that $-\log G$ is homogeneous of order 1.

Theorem 6.2 *Let G be a MSMVE survival function. Then $A = -\log G$ is homogeneous of order* 1, *i.e., $A(t\mathbf{x}) = tA(\mathbf{x})$ for all $t > 0$, $\mathbf{x} \in (0,\infty)^m$. Conversely, if G is a multivariate exponential survival function and $-\log G$ is homogeneous of order* 1, *then G is MSMVE.*

Proof. Let $\mathbf{X}$ be MSMVE with survival function G. From the definition, for $\mathbf{w} \in (0,\infty)^m$,

$$\Pr(\min_j X_j/w_j > t) = \Pr(X_j > tw_j, 1 \le j \le m) = G(t\mathbf{w}) = e^{-tA(\mathbf{w})}$$

for a constant A which depends on $\mathbf{w}$. Hence $A(\mathbf{w}) = -\log G(\mathbf{w})$ for all $\mathbf{w}$ (by letting $t = 1$). Therefore $A(t\mathbf{w}) = -\log G(t\mathbf{w}) = tA(\mathbf{w})$ for all $t > 0$ and $\mathbf{w} \in (0,\infty)^m$.

The converse is easy to prove, based on the preceding paragraph. □

Survival functions satisfying the homogeneity condition of Theorem 6.2 can be obtained from some families in Chapter 5, after substituting exponential survival margins into the copulas. The next two theorems show that the class of MSMVE distributions is infinite-dimensional (this is a contrast to the finite-dimensional MVN family that arises from limit theory of sums of random vectors with finite second moments). However for statistical inference, parametric inference is easier than nonparametric inference for the multi-dimensional situation, so that a goal is to find finite-dimensional parametric subfamilies that cover well the entire family represented by (6.5). Parametric families of MSMVE distributions, equivalently MEV copulas, are given in Section 6.3.

Theorem 6.3 *(The Pickands representation of a min-stable multivariate exponential distribution). Let $G(\mathbf{x})$ be a survival function with univariate exponential margins. G satisfies $-\log G(t\mathbf{x}) = -t\log G(\mathbf{x})$ for all $t > 0$ if and only if G has the representation*

$$-\log G(\mathbf{x}) = \int_{S_m} [\max_{1\le i\le m}(q_i x_i)]\, dU(\mathbf{q}), \quad x_i \ge 0, i = 1,\ldots,m, \tag{6.5}$$

where $S_m = \{\mathbf{q} : q_i \ge 0, i = 1,\ldots,m, \sum_i q_i = 1\}$ is the m-dimensional unit simplex and U is a finite measure on S_m.

Proof. This is a result of Pickands and an alternative statement of Theorem 5.4.5 of Galambos (1987); the proof is given in Galambos (1987). □

Remarks. If X has an exponential distribution with mean 1, then $Y = 1/X$ has a Fréchet distribution with cdf $\exp\{-y^{-1}\}$, $y > 0$. For maxima, sometimes it is more convenient to work with the representation for a **max-stable multivariate Fréchet** distribution. This has cdf:

$$F(\mathbf{y}) = \exp\Big\{-\int_{S_m} [\max_{1\le i\le m}(q_i y_i^{-1})]\, dU(\mathbf{q})\Big\}, \quad y_i \ge 0, i = 1,\ldots,m, \tag{6.6}$$

with S_m and U as defined above.

In the bivariate case, a simplification of the characterization is possible; the condition involves convexity.

Theorem 6.4 *Suppose B is a continuous non-negative function on $[0,1]$ with $B(0) = B(1) = 1$. Suppose that B has right and left derivatives up to second order except for at most a countable number of points. Then $G(x,y) = \exp\{-(x+y)B(x/(x+y))\}$ is a bivariate exponential survival function if and only if B is convex and $\max\{w, 1-w\} \le B(w) \le 1$ for $0 \le w \le 1$.*

Proof. The first-order derivatives are $-\partial G/\partial x = G(x,y)[B(w) + (1-w)B'(w)]$ and $-\partial G/\partial y = G(x,y)[B(w) - wB'(w)]$, with $w = x/(x+y)$. These equations hold for right and left derivatives. Because G must be decreasing in x, y in order to be a survival function, necessary conditions are: (i) $B(w) + (1-w)B'(w) \ge 0$ and $B(w) - wB'(w) \ge 0$, for all $w \in [0,1]$; and (ii) $B'(w+) \ge$

$B'(w-)$ if B' is discontinuous at w. Note that (ii) follows from $-\frac{\partial G}{\partial x}(x, y+) \le -\frac{\partial G}{\partial x}(x, y-)$ and $-\frac{\partial G}{\partial y}(x+, y) \le -\frac{\partial G}{\partial y}(x-, y)$. Hence $B'(0) \ge -1$ and $B'(1) \le 1$. The second-order mixed derivative (treat as right and left derivatives if B' is not continuous everywhere) is $G\{[B + (1-w)B'][B - wB'] + (1-w)wB''/(x+y)\}$. This must be non-negative in order for G to be a survival function. Letting $x, y \to 0$ with $x/(x+y) \to w \in (0,1)$, a necessary condition is $B''(w) \ge 0$ whenever B' is continuous at w. Putting everything together, B must be convex and $\max\{w, 1-w\} \le B(w) \le 1$.

Next for sufficiency, from convexity and the lower boundary constraint, $B'(w) \ge (B(w)-1)/w \ge (1-w-1)/w = -1$ and $B'(w) \le (1-B(w))/(1-w) \le (1-w)/(1-w) = 1$. Hence $B(w)+(1-w)B'(w) \ge (1-w)-(1-w) = 0$ and $B(w)-wB'(w) \ge w - w = 0$ and the first-order derivatives have the right signs. If B has a corner at point w_0 with $B'(w_0+) \ge B'(w_0-)$ (this must be the direction of the inequality from the convexity condition), then the first-order monotonicity is fine. Therefore if B' is continuous, the second-order mixed derivative is non-negative if $B'' \ge 0$. If B has some corners, so that G has a singular component, then the above second-order mixed derivative is non-negative and the absolutely continuous component is fine. Hence G is a survival function. □

Theorem 6.5 *Let G be given by the preceding theorem. If B has a corner point (or more than one), then G has a singular component.*

Proof. The survival function $G_{1|2}(x|y)$ is given by

$$e^y G(x,y)\left[B\left(\tfrac{x}{x+y}\right) - \tfrac{x}{x+y}B'\left(\tfrac{x}{x+y}\right)\right].$$

If B has a corner point at $w_0 \in (0,1)$ so that $B'(w_0+) > B'(w_0-)$, then $G_{1|2}(x|y)$ has a jump discontinuity at $x = w_0 y/(1-w_0)$, when $w_0 = x/(x+y)$. The conclusion now follows from Theorem 1.1. □

From earlier in this section, MSMVE distributions have MEV copulas. When the copulas take on univariate GEV cdfs, the results are MEV distributions for maxima (which can arise as extreme value limits). When the copulas take on GEV survival margins for minima, the results are MEV survival functions for minima.

The MEV distribution is max-stable (min-stable) for multivariate maxima (minima). (A multivariate distribution F is **max-stable** if for each $t > 0$, $F^t(\mathbf{x}) = F(a_{1t} + b_{1t}x_1, \ldots, a_{mt} + b_{mt}x_m)$ for some vectors $\mathbf{a}_t, \mathbf{b}_t$, and it is **min-stable** if for each $t > 0$, $\overline{F}^t(\mathbf{x}) = \overline{F}(a_{1t} + b_{1t}x_1, \ldots, a_{mt} + b_{mt}x_m)$ for some vectors $\mathbf{a}_t, \mathbf{b}_t$. With t being a positive integer n, this means that the vector of componentwise maxima (minima) has the same distribution up to location–scale changes.) If $G(\mathbf{x}) = e^{-A(\mathbf{x})}$ and A is homogeneous of order 1, then $G^t(\mathbf{x}) = G(t\mathbf{x})$ so that the min-stability property holds.

The final result in this section concerns the density of a MSMVE survival function. Note that maximum likelihood estimation for the MEV or MSMVE models requires the density.

If $G = e^{-A}$ is a survival function, then the density g, which is $(-1)^m$ times the mth-order mixed derivative, is

$$g(\mathbf{z}) = \exp\{-A(\mathbf{z})\} \left[\sum_{\{P_1,\dots,P_k\}\in\mathcal{P}} (-1)^{m-k} \prod_{i=1}^{k} \frac{\partial^{|P_i|} A}{\prod_{j\in P_i} \partial z_j}(\mathbf{z}) \right],$$

where $\mathcal{P}$ is the set of partitions of $\{1,\dots,m\}$. From this, we have the following theorem.

Theorem 6.6 *For $j = 1,\dots,m$, let a_j, b_j be reals with $\infty \le a_j < b_j \le \infty$. Let G be a function from $\times_{j=1}^m (a_j, b_j)$ to $[0,1]$, which is decreasing in each of the m arguments and satisfies (a) $\lim_{z_j\to b_j} G(\mathbf{z}) = 0$ and (b) $\lim_{z_j\to a_j \forall j} G(\mathbf{z}) = 1$. Let $A = -\log G$. Then G is a survival function if for every subset $S \in \mathcal{S}_m$,*

$$(-1)^{1+|S|} \frac{\partial^{|S|} A}{\prod_{i\in S} \partial z_i}(\mathbf{z}) \ge 0 \quad \forall \mathbf{z}.$$

6.2.1 Dependence properties

MSMVE distributions are max-id (since if G is MSMVE then $G^t(\mathbf{x}) = G(t\mathbf{x})$, $t > 0$), and hence so are MEV copulas. By Theorem 2.6, MEV copulas are positively dependent, being TP_2, for example, when $m = 2$. Furthermore, the stronger positive dependence property of association holds.

Theorem 6.7 *If C is an MEV copula, then C is associated. Hence any MEV distribution is associated.*

Proof. This is summarized from Proposition 5.1 of Marshall and Olkin (1983).

Because the dependence concept of association is preserved under strictly increasing or decreasing transformations of variables, it suffices to show that MSMVE distributions are associated. Because association is preserved under limits in distribution, and because of Pickand's representation (Theorem 6.3), it suffices to show association in the case where $U(\mathbf{q})$ in (6.5) consists of a finite number of masses. For this case, suppose $\mathbf{q}_k = (q_{k1}, \ldots, q_{km})$, $k = 1, \ldots, r$, are the support points with respective masses $\alpha_k > 0$. Then (6.5) becomes

$$A(\mathbf{x}) = \sum_{k=1}^{r} \alpha_k \max_{1 \le j \le m} q_{kj} x_j. \tag{6.7}$$

Hence e^{-A} is the distribution of $\mathbf{X}$ with $X_j = \min_{1 \le k \le r} a_{kj} Z_k$, where $a_{kj} = (q_{kj}\alpha_k)^{-1}$, and $Z_1, \ldots, Z_r$ are iid exponential rvs with mean 1. Note that

$$\begin{aligned} \Pr(X_j > x_j, j = 1, \ldots, m) &= \Pr\big(Z_k > \max_{1 \le j \le m} x_j / a_{kj}, k = 1, \ldots, r\big) \\ &= \exp\Big\{-\sum_{k=1}^{r} \max_{1 \le j \le m} x_j / a_{kj}\Big\}. \end{aligned}$$

Since $(Z_1, \ldots, Z_r)$ is associated by Theorem 2.4(d) and $\mathbf{X}$ consists of increasing functions of independent rvs, $\mathbf{X}$ is associated. □

The next result shows that for a bivariate copula with tail dependence, the extreme value limit has the same tail dependence parameter λ. Hence, the BEV limit is not the independence copula. In Section 5.1, the BEV limits are indicated for those families that have tail dependence. The result also applies to the bivariate margins of a multivariate copula.

Theorem 6.8 *Let C be a bivariate copula and let $F(x_1, x_2) = C(1 - e^{-x_1}, 1 - e^{-x_2})$. Suppose $\lim_{u \to 1} \overline{C}(u, u)/(1 - u) = \lambda$, where $\lambda \in (0, 1]$ and $\lim_{n \to \infty} F^n(x_1 + \log n, x_2 + \log n) = H(x_1, x_2) = \exp\{-\eta(x_1, x_2)\}$ with univariate margins $\exp\{-e^{-x_j}\}$, $j = 1, 2$. Let $C^*(u_1, u_2) = H(-\log[-\log u_1], -\log[-\log u_2])$. Then*

$$\lim_{u \to 1} \overline{C}^*(u, u)/(1 - u) = \lambda.$$

Proof. From (6.9) below,

$$\begin{aligned} \eta(x, x) &= \lim_{n \to \infty} n[1 - F(x + \log n, x + \log n)] \\ &= \lim_{n \to \infty} n[e^{-x - \log n} + e^{-x - \log n} - \overline{F}(x + \log n, x + \log n)] \\ &= 2e^{-x} - \lambda e^{-x}. \end{aligned}$$

Now, with $\hat{u} = -\log[-\log u]$,

$$\begin{aligned} \overline{C}^*(u, u)/(1 - u) &= [1 - 2u + \exp\{-\eta(\hat{u}, \hat{u})\}]/(1 - u) \\ &\sim [1 - 2u + u^{2 - \lambda}]/(1 - u) \to \lambda \end{aligned}$$

as $u \to 1$. □

6.2.2 Extreme value limit results

In this subsection, to avoid technicalities, limits are written under the assumption that they exist. In particular applications, one does have to check for their existence. General conditions for existence of limits can be found in Galambos (1987, Chapter 5), Resnick (1987, Chapter 5) and Marshall and Olkin (1983). Note that to check that a limit is a proper distribution, one needs to check the boundary conditions but not the rectangle condition of Section 1.4.2.

Theorem 6.9 *Let $(M_{1n}, \ldots, M_{mn})$ be a vector of maxima from iid random vectors $(X_{i1}, \ldots, X_{im})$ from F and suppose*

$$F^n(a_{1n} + b_{1n}z_1, \ldots, a_{mn} + b_{mn}z_m)$$
$$= \lim_n \Pr(M_{1n} \le a_{1n} + b_{1n}z_1, \ldots, M_{mn} \le a_{mn} + b_{mn}z_m)$$

converges weakly to $H(\mathbf{z})$. Suppose the limits

$$\lim_{n\to\infty} n\overline{F}_S(a_{jn} + b_{jn}z_j, j \in S) = r_S(z_j, j \in S)$$

are finite for all $S \in \mathcal{S}_m$. If

$$\exp\Big\{ \sum_{S\in\mathcal{S}_m} (-1)^{|S|} r_S(z_j, j \in S)\Big\}$$

is a non-degenerate distribution function, then it is equal to H.

Proof. This is part of Theorem 5.3.1 in Galambos (1987). An outline of the proof is as follows. We use the notation $\mathbf{a}_n + \mathbf{b}_n\mathbf{z}$ for the vector $(a_{1n}+b_{1n}z_1, \ldots, a_{mn}+b_{mn}z_m)$. If $\mathbf{z}$ is such that $F^n(\mathbf{a}_n+\mathbf{b}_n\mathbf{z})$ converges to a value in (0,1), then $F(\mathbf{a}_n + \mathbf{b}_n\mathbf{z}) \to 1$. Hence,

$$F^n(\mathbf{a}_n+\mathbf{b}_n\mathbf{z}) = \exp\{n \log F(\mathbf{a}_n+\mathbf{b}_n\mathbf{z})\} \sim \exp\{-n[1-F(\mathbf{a}_n+\mathbf{b}_n\mathbf{z})]\}.$$

The conclusion now follows with the use of equation (1.4). □

A general approach for deriving MEV distributions is based on a family of copulas. From earlier results in this section, the univariate margins can be conveniently taken to be exponential with mean 1 in order to derive the limiting MEV distribution, since the MEV copula that results does not depend on the univariate margins. Let the starting multivariate exponential distribution be denoted by F. The limiting MEV distribution is

$$\lim_{n\to\infty} F^n(x_1 + \log n, \ldots, x_m + \log n); \tag{6.8}$$

the linear transform $x_j + \log n$ comes from Example 6.1. This can be converted to a MSMVE survival distribution with the transforms $z_j = e^{-x_j}$, $j = 1, \ldots, m$, of the Gumbel univariate margins.

If F has bivariate tail dependence then the limit has dependent margins. If F does not have bivariate tail dependence such as for the MVN distribution or copula, then the limiting MEV distribution corresponds to the independence copula. If F is a member of a parametric family, then a parametric family of MEV distributions can result.

From the preceding theorem and its proof, (6.8) is the same as

$$\exp\{-\lim_{n\to\infty} n[1-F(x_1+\log n,\dots,x_m+\log n)]\}, \tag{6.9}$$

or

$$\exp\Big\{-\sum_{i=1}^{m} e^{-x_i} + \sum_{S\in\mathcal{S}_m,|S|\geq 2} (-1)^{|S|}\lim_n n\,\overline{F}_S(x_j+\log n, j\in S)\Big\}. \tag{6.10}$$

Let $\mathbf{X}\sim F$. Another form is based on using the identity

$$1-F(\mathbf{x}) = \sum_{i=1}^{m} \Pr(X_i > x_i)$$
$$-\textstyle\sum_{1\leq j<k\leq m} \Pr(X_j > x_j, X_k > x_k, X_i \leq x_i, j<i<k),$$

in (6.9). With $\lim n\exp\{-x_i - \log n\} = \exp\{-x_i\}$, $i=1,\dots,m$,

$$\zeta_{j,j+1}(x_j,x_{j+1}) \stackrel{\text{def}}{=} \lim_{n\to\infty} n[1-F_{j,j+1}(x_j+\log n, x_{j+1}+\log n)],$$

and

$$\zeta_{jk}(x_j,\dots,x_k) \stackrel{\text{def}}{=} \lim_{n\to\infty} n\Pr(X_j > x_j+\log n,$$
$$X_k > x_k+\log n, X_i\leq x_i+\log n,\ j<i<k),$$

for $j<k$, $k-j>1$, then

$$\lim_{n\to\infty} n[1-F(\mathbf{x}+\log n)] = \sum_i \exp\{-x_i\} - \sum_{j<k} \zeta_{jk}(x_j,\dots,x_k) \tag{6.11}$$

is the exponent in (6.9). Special cases are given below and in the next section.

For example, consider application to the limiting MEV distribution for (4.39) in Section 4.5. If $1-C_{jk}$ has simple form for all $j<k$, then for (4.39),

$$1-F_{1,\dots,m}(\mathbf{x}) = 1-F_{2,\dots,m-1}(x_2,\dots,x_{m-1}) \tag{6.12}$$
$$+\textstyle\int_0^{x_2}\cdots\int_0^{x_{m-1}} [1-C_{1m}(F_{1|2,\dots,m-1}, F_{m|2,\dots,m-1})]\,dF_{2,\dots,m-1},$$

so that a recursion formula is possible for the limit in the exponent of (6.8). Let $M=\log n$ and $\mathbf{z}=(z_2,\dots,z_{m-1})$. Then n times the integral in (6.12), with argument $\mathbf{x} + M$, leads to:

$$n \int_0^{x_2+M} \cdots \int_0^{x_{m-1}+M} \big[1 - C_{1m}(F_{1|2,\ldots,m-1}(x_1 + M|\mathbf{z}),$$
$$F_{m|2,\ldots,m-1}(x_m + M|\mathbf{z}))\big] \cdot f_{2,\ldots,m-1}(\mathbf{z})\, dz_2 \cdots dz_{m-1}$$
$$= \int_{-M}^{x_2} \cdots \int_{-M}^{x_{m-1}} \big[1 - C_{1m}(F_{1|2,\ldots,m-1}(x_1+M|v_2+M,\ldots,v_{m-1}+M),$$
$$F_{m|2,\ldots,m-1}(x_m + M|v_2 + M,\ldots,v_{m-1} + M))\big]$$
$$\cdot e^M f_{2,\ldots,m-1}(v_2 + M,\ldots,v_{m-1} + M)\, dv_2 \cdots dv_{m-1}$$
$$\to \int_{-\infty}^{x_2} \cdots \int_{-\infty}^{x_{m-1}} \big[1 - C_{1m}(1 - a_{1,2\ldots,m-1}(v_2 - x_1,\ldots,v_{m-1} - x_1),$$
$$1 - a_{m,2\ldots,m-1}(v_2 - x_m,\ldots,v_{m-1} - x_m))\big]$$
$$\cdot b_{2,\ldots,m-1}(v_2,\ldots,v_{m-1})\, dv_2 \cdots dv_{m-1} \qquad (6.13)$$

under conditions similar to those in Theorem 4.20 (note that the functions $a_{1,2\ldots,m-1}$, $a_{m,2\ldots,m-1}$ and $b_{2,\ldots,m-1}$ are defined in the proof of Theorem 4.20). For $1 \le j < k \le m$ with $k - j > 1$, let $\eta_{jk}(x_j,\ldots,x_k)$ be the limit similar to (6.13) starting with C_{jk} instead of C_{1m}, and for $1 \le j < m$, let

$$\eta_{j,j+1}(x_j, x_{j+1}) = \lim_{n\to\infty} n[1 - F_{j,j+1}(x_j + \log n, x_{j+1} + \log n)].$$

Then $\lim n[1 - F_{1,\ldots,m}(\mathbf{x} + \log n)]$, in the exponent of (6.9), is

$$\begin{cases} \sum_{i=1}^{m/2} \eta_{i,m+1-i}(x_i,\ldots,x_{m+1-i}) & m \text{ even}, \\ \sum_{i=1}^{(m-1)/2} \eta_{i,m+1-i}(x_i,\ldots,x_{m+1-i}) + \exp\{-x_{(m+1)/2}\} & m \text{ odd}. \end{cases} \qquad (6.14)$$

A simpler form of the limit may result from (6.11), with

$$\lim n \Pr(X_j > x_j + \log n, X_k > x_k + \log n, X_i \le x_i + \log n, j < i < k)$$
$$= \int_{-M}^{x_{j+1}} \cdots \int_{-M}^{x_{k-1}} \overline{C}_{jk}\big(F_{j|j+1,\ldots,k-1}(x_j+M|v_{j+1}+M,\ldots,v_{k-1}+M),$$
$$F_{k|j+1,\ldots,k-1}(x_k + M|v_{j+1} + M,\ldots,v_{k-1} + M)\big)$$
$$\cdot e^M f_{j+1,\ldots,k-1}(v_{j+1} + M,\ldots,v_{k-1} + M)\, dv_{j+1} \cdots dv_{k-1}$$
$$\to \int_{-\infty}^{x_{j+1}} \cdots \int_{-\infty}^{x_{k-1}} \overline{C}_{jk}\big(1 - a_{j,j+1,\ldots,k-1}(v_{j+1} - x_j,\ldots,v_{k-1} - x_j),$$
$$1 - a_{k,j+1,\ldots,k-1}(v_{j+1} - x_k,\ldots,v_{k-1} - x_k)\big)$$
$$\cdot b_{j+1,\ldots,k-1}(v_{j+1},\ldots,v_{k-1})\, dv_{j+1} \cdots dv_{k-1}$$
$$= \zeta_{jk}(x_j,\ldots,x_k). \qquad (6.15)$$

The functions $a_{j,j+1,\ldots,k-1}$, $a_{k,j+1,\ldots,k-1}$ and $b_{j+1,\ldots,k-1}$ are defined in the proof of Theorem 4.20.

6.3 Parametric families

A MEV distribution with univariate margins transformed to exponential survival functions becomes a MSMVE distribution. Let G be a min-stable m-dimensional exponential survival function. From Theorem 6.2, the exponent $A = -\log G$ satisfies

$$A(t\mathbf{z}) = tA(\mathbf{z}) \quad \forall t > 0. \tag{6.16}$$

Let G_S be a marginal survival function of G, with $S \in \mathcal{S}_m$. Then $G_S(\mathbf{z}_S) = \exp\{-A_S(\mathbf{z}_S)\}$, where A_S is obtained from A by setting $z_j = 0$ for $j \notin S$. The notation in this section is such that the arguments of A are shown through a subset S when more than one A is used.

In this section, in terms of the exponent A, we list some parametric families of MSMVE distributions that interpolate from independence to the Fréchet upper bound. As mentioned in Section 4.1, desirable properties are a wide range of dependence structure and closure under the taking of margins, etc. The families in Section 6.3.1 are such that each parameter has a dependence interpretation. Most of these families make use of the methods of construction in Chapter 4 and Section 6.2. Other parametric families are given in Section 6.3.2; for these the interpretation of parameters may not be as straightforward.

6.3.1 Dependence families

This subsection includes a listing of parameter families of MSMVE distributions with good dependence properties and is intended as a *reference* of useful parametric MSMVE families. We start with three one-parameter bivariate families and then go on to families that could be considered as multivariate extensions.

The three bivariate families B6, B7 and B8 in Section 5.1 are families of MEV copulas; in the min-stable exponential form, the exponents $-\log G$ are:

$$\begin{aligned}
A(z_1, z_2; \delta) &= (z_1^\delta + z_2^\delta)^{1/\delta}, \quad \delta \geq 1, && (6.17)\\
A(z_1, z_2; \delta) &= z_1 + z_2 - (z_1^{-\delta} + z_2^{-\delta})^{-1/\delta}, \quad \delta \geq 0, && (6.18)\\
A(z_1, z_2, \delta) &= z_1 \Phi(\delta^{-1} + \tfrac{1}{2}\delta[\log(z_1/z_2)]) && (6.19)\\
&\quad + z_2 \Phi(\delta^{-1} + \tfrac{1}{2}\delta[\log(z_2/z_1)]), \quad \delta \geq 0.
\end{aligned}$$

The independence case ($A(z_1, z_2) = z_1 + z_2$) obtains when $\delta = 1$, 0 and 0 in (6.17), (6.18) and (6.19), respectively; the Fréchet upper bound ($A(z_1, z_2) = \max\{z_1, z_2\}$) obtains when $\delta = \infty$ in all these cases.

Multivariate extensions of these bivariate models are given next together with some discussion and interpretations. Equation (6.19) derives from a (non-standard) extreme value limit of the BVN distribution and has a multivariate extension with a parameter for each bivariate margin. Equations (6.17) and (6.18) have two different parallel multivariate extensions, although the extension of having the bivariate family with a different parameter for each bivariate margin does not exist (or has not been constructed). The extensions that have been obtained satisfy different properties: (i) each bivariate margin is in the family (6.17) or (6.18), but there are only $m-1$ distinct bivariate parameters among the $m(m-1)/2$ bivariate margins (see Section 5.3); (ii) $m-1$ of the $m(m-1)/2$ bivariate margins are in the family (6.17) or (6.18), with different parameters possible, and the remaining bivariate margins are not of the same form, but overall there is a flexible dependence structure.

Family M8. The multivariate extension of (6.19) is closed under margins and has (6.19) with parameter $\delta_{ij} = \delta_{ji}$ for the (i, j) bivariate margin. The dependence structure is like that of the MVN (except that there is no negative dependence). Let $\alpha_{ij} = \delta_{ij}^{-1}, i \neq j$. Let $\rho_{kij} = [\alpha_{ik}^2 + \alpha_{jk}^2 - \alpha_{ij}^2]/[2\alpha_{ik}\alpha_{jk}]$ for i, j, k distinct. A symmetric form is given followed by a recursive (non-symmetric) form; the latter is more suitable for numerical computations in maximum likelihood estimation.

The symmetric form is:

$$A_{1\cdots m}(\mathbf{z}; \delta_{ij}, 1 \leq i < j \leq m) = z_1 + \ldots + z_m - \sum_{1 \leq i < j \leq m} \Big\{ z_i + z_j$$

$$- z_i \Phi(\delta_{ij}^{-1} + \tfrac{1}{2}\delta_{ij}[\log(z_i/z_j)]) - z_j \Phi(\delta_{ij}^{-1} + \tfrac{1}{2}\delta_{ij}[\log(z_j/z_i)]) \Big\}$$

$$+ \sum_{S:|S| \geq 3} (-1)^{|S|+1} r_S(z_i, i \in S; \delta_{ij}, i < j, i, j \in S),$$

where for $S = \{i_1, \ldots, i_k\}$ with $i_1 < \cdots < i_k$,

$$r_S(z_i, i \in S; \delta_{ij}, i < j, i, j \in S)$$
$$= \int_0^{z_{i_k}} \overline{\Phi}_{k-1}\big(\log(y/z_{i_j}) + 2\delta_{i_j, i_k}^{-2}, j = 1, \ldots, k-1; \Gamma\big)\, dy,$$

$\overline{\Phi}_{k-1}(\cdot;\Gamma)$ is the MVN survival function with covariance matrix Γ, and $\Gamma = \Gamma(\delta_{i_j,i_{j'}}, j, j' = 1,\ldots,k-1)$ is the $(k-1)\times(k-1)$ matrix with (j,j') element equal to $2(\delta^{-2}_{i_j,i_k} + \delta^{-2}_{i_{j'},i_k} - \delta^{-2}_{i_j,i_{j'}})$ for $1 \le j,j' \le k-1$, and δ^{-1}_{ii} is defined as zero for all i.

Proof. See Hüsler and Reiss (1989). The derivation was essentially based on Theorem 6.9 with the correlation parameters of a MVN distribution approaching 1. The sequences $a_{jn} = a_n$, $b_{jn} = b_n$ come from univariate extreme value theory. □

With parameters listed in lexicographic order, the exponent, in recursive form for $m \ge 3$, is:

$$A_{1\cdots m}(\mathbf{z};\delta_{12},\ldots,\delta_{m-1,m})$$
$$= A_{1,\cdots,m-1}(z_1,\ldots,z_{m-1};\delta_{12},\ldots,\delta_{m-2,m-1})+ \tag{6.20}$$
$$\int_0^{z_m} \Phi_{m-1}(\delta^{-1}_{i,m} + \tfrac{1}{2}\delta_{i,m}[\log(\tfrac{q}{z_i})], i \le m-1; (\rho_{mij})_{i<j\le m-1})\, dq.$$

By permuting the indices of (6.20), alternative representations are possible. The constraints on the δ_{ij} are that they are non-negative and the covariances matrices in all possible representations are positive definite.

Proof. The proof of the simpler form of the three-dimensional case is given here. The proofs for higher dimensions are similar.

For $m = 3$,

$$A_{123}(z_1,z_2,z_3;\delta_{12},\delta_{13},\delta_{23}) = z_1 + z_2 - B_2(z_1,z_2;\delta_{12}) + z_3$$
$$-B_2(z_1,z_3;\delta_{13}) - B_2(z_2,z_3;\delta_{23}) + B_3(z_1,z_2,z_3;\delta_{12},\delta_{13},\delta_{23}), \tag{6.21}$$

where

$$B_2(x_1,x_2;\delta) = \int_0^{x_2} \overline{\Phi}(\delta^{-1} + \tfrac{1}{2}\delta\log(q/x_1))\, dq$$

$$= x_1 + x_2 - x_1\Phi(\delta^{-1} + \tfrac{1}{2}\delta[\log(x_1/x_2)]) - x_2\Phi(\delta^{-1} + \tfrac{1}{2}\delta[\log(x_2/x_1)])$$

and

$$B_3(z_1,z_2,z_3;\delta_{12},\delta_{13},\delta_{23}) =$$
$$\int_0^{z_3} \overline{\Phi}_2(\delta^{-1}_{13} + \tfrac{1}{2}\delta_{13}[\log(q/z_1)], \delta^{-1}_{23} + \tfrac{1}{2}\delta_{23}[\log(q/z_2)]; \rho_{312})\, dq,$$

with ρ_{312} as defined above.

Now, $B_2(z_j,z_3;\delta_{j3})$ for $j = 1,2$ can be rewritten as

$$\int_0^{z_3} \overline{\Phi}_2(\delta^{-1}_{13} + \tfrac{1}{2}\delta_{13}[\log(q/z_1)], -\infty; \rho_{312})\, dq$$

and

$$\int_0^{z_3} \overline{\Phi}_2(-\infty, \delta_{23}^{-1} + \tfrac{1}{2}\delta_{23}[\log(q/z_2)]; \rho_{312})\, dq$$

respectively, and hence the last four terms of (6.21) simplify to the last term in (6.20) when $m = 3$. □

For the first generalization of (6.17) and (6.18), a derivation is given for a specific choice of clustering; the proof in general is the same, but there is no simple notation to cover all cases (see Section 5.3). We give one general form of clustering, followed by the second possibility for $m = 4$. The general form consists of root clusters $\{1, 2\}$ and $\{j+1\}, j = 2, \ldots, m-1$, and derived hierarchical clusters $\{1, \ldots, k\}$, $k = 3, \ldots, m$. In the notation of Section 5.3, the parameters are written as $\beta_{1,2}, \beta_{1,3}, \ldots, \beta_{1,m}$.

Family M6. (This is a continuation of family M6 in Section 5.3.) With A_{12} given by (6.17), the exponent, in recursive form, for $m \geq 3$, is:

$$A_{1\cdots m}(\mathbf{z}; \beta_{1,2}, \beta_{1,3}, \ldots, \beta_{1,m}) = \tag{6.22}$$

$$\left([A_{1,\ldots,m-1}(z_1, \ldots, z_{m-1}; \beta_{1,2}, \ldots, \beta_{1,m-1})]^{\beta_{1,m}} + z_m^{\beta_{1,m}}\right)^{1/\beta_{1,m}},$$

$$\beta_{1,2} \geq \beta_{1,3} \geq \cdots \geq \beta_{1,m} \geq 1.$$

The other case for $m = 4$ has root clusters $\{1, 2\}$ and $\{3, 4\}$, and the derived hierarchical cluster $\{1, 2, 3, 4\}$; the parameters are $\beta_{1,2}$, $\beta_{3,4}$, $\beta_{1,4}$. The exponent is

$$A_{1234}(\mathbf{z}; \beta_{1,2}, \beta_{3,4}, \beta_{1,4}) = \tag{6.23}$$

$$\left((z_1^{\beta_{1,2}} + z_2^{\beta_{1,2}})^{\beta_{1,4}/\beta_{1,2}} + (z_3^{\beta_{3,4}} + z_4^{\beta_{3,4}})^{\beta_{1,4}/\beta_{3,4}}\right)^{1/\beta_{1,4}},$$

$$1 \leq \beta_{1,4} \leq \beta_{1,2}, \beta_{3,4}.$$

Family M7. (A generalization of family B7.) The way to use a model that generalizes (6.17) to get one that generalizes (6.18) is as follows. The generalization of (6.18) is a sum of terms, with a term for each non-empty subset of $z_1, \ldots, z_m$. The sign of a term is $(-1)^{|S|+1}$ for a subset S. The last term with all m variables can be obtained from (6.22) by changing all β to $-\beta$. The term with variables $z_i, i \in S$, can be obtained from this last term by crossing out variables $z_{i'}$ with $i' \notin S$ and then simplifying. For example, with the right-hand side of (6.22) now denoted by $B_{1\cdots m}$, and with the same clusters as in (6.22), the generalization of (6.18) has exponent $A^*_{1\cdots m}$ given by

$$A^*_{1\cdots m}(\mathbf{z};\beta_{1,2},\ldots,\beta_{1,m}) = \sum_j z_j - \sum_{j_1<j_2} \left(z_{j_1}^{-\beta_{1,j_2}} + z_{j_2}^{-\beta_{1,j_2}}\right)^{-\frac{1}{\beta_{1,j_2}}}$$

$$+ \sum_{3\le k\le m} (-1)^{k+1} \sum_{j_1<\cdots<j_k} B_{1\cdots k}(z_{j_1},\ldots,z_{j_k};-\beta_{1,j_2},\ldots,-\beta_{1,j_k}),$$

$$\beta_{1,2}\ge\cdots\ge\beta_{1,m}\ge 0. \tag{6.24}$$

Similarly, there is an extension of (6.18) with dependence structure analagous to the model in (6.23).

The derivation of (6.24) and its extensions comes from the lower extreme value limit of the family M4 (or the upper extreme value limit applied to the associated copula of M4 when survival functions are used for the univariate arguments), using Theorem 6.9 and (6.10).

For $1 \le j_1 < j_2 \le m$, the bivariate margin of (6.22) (respectively model (6.24)) is family (6.17) ((6.18)) with parameter $\delta_{j_1 j_2} = \beta_{1,j_2}$. Hence for fixed $1 < k \le m$, the bivariate dependence between variables j and k is the same for all $j < k$. Also the bivariate dependence is decreasing with k. Hence to use (6.22), (6.23), (6.24) and their extensions, the variables have to be indexed in an appropriate order. All higher-order margins (for $m \ge 3$) of (6.22) and (6.24) are within the same family. For model (6.23), the (1,2), (3,4) bivariate margins are (6.17) with parameters $\beta_{1,2}$, $\beta_{3,4}$ respectively, and the remaining four bivariate margins have parameter $\beta_{1,4}$. The trivariate margins of (6.23) are in the family (6.22).

Next we turn to the second generalization of (6.17) and (6.18), which has $m(m-1)/2$ distinct bivariate dependence parameters, with $m-1$ of the $m(m-1)/2$ bivariate margins being (6.17) or (6.18). For some parameter vectors, the remaining bivariate margins are close to (6.17) or (6.18). The models that have these properties are given for $m = 3$ and 4 dimensions explicitly rather than in a general form because the form of the model is too complex. The general multivariate version can be obtained following the steps in Section 6.2.2, especially (6.14) and (6.11) respectively for generalizations of (6.17) and (6.18).

Family MM6. This family comes from the upper extreme value limit of the family MM5 (see Section 5.5) and some of the bivariate margins have the form of (6.17). With lexicographic ordering of the parameters, the exponents for $m = 3, 4$ are:

$$\begin{aligned} &A_{123}(z_1, z_2, z_3; \theta_{12}, \theta_{13}, \theta_{23}) \\ &\quad = z_2 + \int_{-\infty}^{-\log z_2} [\eta_{12}^{\theta_{13}} + \eta_{32}^{\theta_{13}} - \eta_{12}^{\theta_{13}}\eta_{32}^{\theta_{13}}]^{1/\theta_{13}} e^{-v_2}\, dv_2, \end{aligned} \tag{6.25}$$

where $\eta_{12} = \eta(z_1, e^{-v_2}, \theta_{12})$, $\eta_{32} = \eta(z_3, e^{-v_2}, \theta_{23})$ and $\eta(s, t, \theta) = 1 - [1 + (s/t)^{\theta}]^{-1+1/\theta}$; and

$$\begin{aligned} &A_{1234}(\mathbf{z}; \theta_{12}, \ldots, \theta_{34}) = (z_2^{\theta_{23}} + z_3^{\theta_{23}})^{1/\theta_{23}} \\ &\quad +(\theta_{23} - 1) \int_{-\infty}^{-\log z_2} \int_{-\infty}^{-\log z_3} [\eta_{123}^{\theta_{14}} + \eta_{432}^{\theta_{14}} - \eta_{123}^{\theta_{14}}\eta_{432}^{\theta_{14}}]^{1/\theta_{14}} \\ &\qquad \cdot (e^{-\theta_{23} v_2} + e^{-\theta_{23} v_3})^{-2+1/\theta_{23}} e^{-\theta_{23}(v_2+v_3)}\, dv_2 dv_3, \end{aligned} \tag{6.26}$$

where $\eta_{123} = \eta'(\eta_{12}, \eta_{32}^*, \theta_{13})$, $\eta_{432} = \eta'(\eta_{43}, \eta_{23}^*, \theta_{24})$, $\eta'(s, t, \theta) = 1 - [(s/t)^{\theta} + 1 - s^{\theta}]^{-1+1/\theta}(1 - s^{\theta})$, $\eta_{32}^* = \eta(e^{-v_3}, e^{-v_2}, \theta_{23})$, $\eta_{43} = \eta(z_4, e^{-v_3}, \theta_{34})$ and $\eta_{23}^* = \eta(e^{-v_2}, e^{-v_3}, \theta_{23})$.

In (6.25) and (6.26), the θ_{ij} are greater than or equal to 1. The models are such that the $(j, j+1)$ margins, $j = 1, \ldots, m-1$, are in the family (6.17) with parameters $\theta_{j,j+1}$. The remaining parameters have interpretations in terms of conditional dependence. The $(1, 2, 3)$ and $(2, 3, 4)$ margins of (6.26) are (6.25) with respective parameter vectors $(\theta_{12}, \theta_{13}, \theta_{23})$ and $(\theta_{23}, \theta_{24}, \theta_{34})$. The parameter θ_{13} measures the amount of conditional dependence of the first and third univariate margins given the second, and θ_{24} has a similar interpretation. A larger value of the parameter means more conditional dependence. Similarly, θ_{14} measures the amount of conditional dependence of the first and fourth univariate margins given the second and third margins. Numerical comparisons indicate that the (j, j') bivariate margins with $j' - j > 1$ become closer to (6.17) as the parameters, $\theta_{jj'}$, $j' - j > 1$, get closer to 1. This suggests that one way to assign variables to the indices $1, \ldots, m$ is such that the strengths of dependence for the resulting adjacent variables are greater.

Family MM7. This family comes from the upper extreme value limit of the family MM4 (see Section 5.5) and some of the bivariate margins have the form of (6.18). With lexicographic ordering of the parameters, the exponents for $m = 3, 4$ are:

$$\begin{aligned} &A_{123}(z_1, z_2, z_3; \theta_{12}, \theta_{13}, \theta_{23}) = z_1 + z_2 + z_3 - (z_1^{-\theta_{12}} + z_2^{-\theta_{12}})^{-1/\theta_{12}} \\ &-(z_2^{-\theta_{23}} + z_3^{-\theta_{23}})^{-1/\theta_{23}} - \int_{-\infty}^{-\log z_2} [\eta_{12}^{-\theta_{13}} + \eta_{32}^{-\theta_{13}} - 1]^{-1/\theta_{13}} e^{-v_2} dv_2, \end{aligned} \tag{6.27}$$

with $\eta_{12} = \eta(z_1, e^{-v_2}, \theta_{12})$, $\eta_{32} = \eta(z_3, e^{-v_2}, \theta_{23})$ and $\eta(s, t, \theta) =$

$[1+(t/s)^{\theta}]^{-1-1/\theta}$; and

$$A_{1234}(\mathbf{z};\theta_{12},\ldots,\theta_{34}) = A_{123}(z_1,z_2,z_3;\theta_{12},\theta_{13},\theta_{23}) + z_4$$

$$-(z_3^{-\theta_{34}}+z_4^{-\theta_{34}})^{-1/\theta_{34}} - \int_{-\infty}^{-\log z_3} [\eta_{23}^{-\theta_{24}}+\eta_{43}^{-\theta_{24}}-1]^{-1/\theta_{24}} e^{-v_3} dv_3$$

$$-(1+\theta_{23}) \int_{-\infty}^{-\log z_2} \int_{-\infty}^{-\log z_3} [\eta_{123}^{-\theta_{14}}+\eta_{432}^{-\theta_{14}}-1]^{-1/\theta_{14}}$$

$$\cdot(e^{\theta_{23}v_2}+e^{\theta_{23}v_3})^{-2-1/\theta_{23}} e^{\theta_{23}(v_2+v_3)}\, dv_2 dv_3, \qquad (6.28)$$

where $\eta_{123} = \eta'(\eta_{12},\eta_{32}^*,\theta_{13})$, $\eta_{432} = \eta'(\eta_{43},\eta_{23}^*,\theta_{24})$, $\eta'(s,t,\theta) = [(s/t)^{-\theta}+1-t^{\theta}]^{-1-1/\theta}$, $\eta_{32}^* = \eta(e^{-v_3},e^{-v_2},\theta_{23})$, $\eta_{43} = \eta(z_4,e^{-v_3},\theta_{34})$, $\eta_{23}^* = \eta(e^{-v_2},e^{-v_3},\theta_{23})$ and $\eta_{23} = \eta(z_2,e^{-v_3},\theta_{23})$.

In (6.27) and (6.28), the θ_{ij} are greater than or equal to 0. The interpretation of the parameters and the relations for the different dimensions are the same as for the models in (6.25) and (6.26).

Remarks. Equations (6.25) and (6.26) are of the form (6.14) and the family of copulas used for (4.37)–(4.39) in Section 4.5 is the family B5, i.e., $C(u_1,u_2;\theta) = 1-[(1-u_1)^{\theta}+(1-u_2)^{\theta}-(1-u_1)^{\theta}(1-u_2)^{\theta}]^{1/\theta}$, $\theta \geq 1$. Equations (6.27) and (6.28) are of the form (6.11) and (6.15) and the family of copulas used for (4.37)–(4.39) is the associated copulas of the family B4, i.e., $C(u_1,u_2,\theta) = u_1+u_2-1+[(1-u_1)^{-\theta}+(1-u_2)^{-\theta}-1]^{-1/\theta}$, $\theta \geq 0$. The details are mostly straightforward but care must be taken in dominating the integral to use the Lebesgue dominated convergence theorem before collecting terms and arriving at the terms in (6.25)–(6.28). The two families of copulas are chosen because in the bivariate case they have extreme value limits corresponding to (6.17) and (6.18), respectively.

Next, the families MM1 and MM3 from Section 5.5 are listed here in the form of the exponent of the MSMVE distribution. The family MM8 is also listed and derives from the lower extreme value limit of the family MM2 in Section 5.5.

Family MM1. In MSMVE form, the exponent is:

$$A_{1\cdots m}(\mathbf{z}) = \Big[\sum_{1\leq i<j\leq m} \big((p_i z_i^{\theta})^{\delta_{ij}}+(p_j z_j^{\theta})^{\delta_{ij}}\big)^{1/\delta_{ij}} + \sum_{i=1}^{m} \nu_i p_i z_i^{\theta}\Big]^{1/\theta}, \qquad (6.29)$$

where $p_i = (\nu_i+m-1)^{-1}$. The (i,j) bivariate margin is:

$$A_{ij}(z_i,z_j) = \big[((p_i z_i^{\theta})^{\delta_{ij}}+(p_j z_j^{\theta})^{\delta_{ij}})^{1/\delta_{ij}}$$

$$+(\nu_i+m-2)p_i z_i^{\theta}+(\nu_j+m-2)p_j z_j^{\theta}\big]^{1/\theta}.$$

Table 6.1. *Tail dependence parameters in a special trivariate case of family MM1.*

θ	$2-2^{1/\theta}$	$3-2(1.5)^{1/\theta}$
1	0	0
1.25	0.259	0.234
1.5	0.413	0.379
1.75	0.514	0.479
2	0.586	0.551
2.5	0.680	0.648
3	0.740	0.711
4	0.811	0.787
6	0.878	0.860
8	0.909	0.896
∞	1	1

With $m=2$, $p_i=p_j=1$ or $\nu_i=\nu_j=0$, this is the exponent in the family B6. The upper tail dependence parameter is $\lambda_{ij} = 2-[(p_i^{\delta_{ij}}+p_j^{\delta_{ij}})^{1/\delta_{ij}}+(\nu_i+m-2)p_i+(\nu_j+m-2)p_j]^{1/\theta}$; it increases as δ_{ij} or θ increases.

In a special case of $m=3$, we check for the range of dependence that is possible for the bivariate tail dependence parameters. With $\delta_{13}=1$, $\delta_{12}=\delta_{23}=\delta$, $\nu_1=\nu_3=-1$, $\nu_2=0$, and $p_1=p_3=1$, $p_2=\frac{1}{2}$, (6.29) and (4.25) become

$$A_{123}(z_1,z_2,z_3)=\left[(2^{-\delta}z_2^{\theta\delta}+z_1^{\theta\delta})^{1/\delta}+(2^{-\delta}z_2^{\theta\delta}+z_3^{\theta\delta})^{1/\delta}\right]^{1/\theta}. \quad (6.30)$$

The bivariate margins are $A_{j2}(z_j,z_2)=[(2^{-\delta}z_2^{\theta\delta}+z_j^{\theta\delta})^{1/\delta}+\frac{1}{2}z_2^{\theta}]^{1/\theta}$, $j=1,3$, and $A_{13}(z_1,z_3)=(z_1^{\theta}+z_3^{\theta})^{1/\theta}$. The bivariate tail dependence parameters are $\lambda_{12}=\lambda_{23}=2-[(2^{-\delta}+1)^{1/\delta}+\frac{1}{2}]^{1/\theta}$ and $\lambda_{13}=2-2^{1/\theta}$. As $\delta\to\infty$, $\lambda_{12}=\lambda_{23}\to 2-(1.5)^{1/\theta}$. A comparison of λ_{13} with the non-sharp lower bound $\max\{0,\lambda_{12}+\lambda_{23}-1\}=\max\{0,3-2(1.5)^{1/\theta}\}$ is given in Table 6.1. The table shows that there is a lot of flexibility in how small λ_{13} can get given $\lambda_{12}=\lambda_{23}$. Also from Theorem 4.17, (6.30) has a wide range for the triple of bivariate Kendall taus.

Family MM8. (Lower extreme value limit of family MM2.) The lower tail extreme value limit of the family MM2 is given below; it is analogous to (6.29) and generalizes the family B7. Let C_S

denote the margin of C in (5.20) with indices in $S \in \mathcal{S}_m$. Using Theorem 6.9, the function $r_S(z_i, i \in S)$ is defined as the limit of $nC_S(n^{-1}e^{-z_i}, i \in S)$ as $n \to \infty$, with $z_i > 0$, $i = 1, \ldots, m$. It is straightforward to verify that

$$r_S(z_i, i \in S) = \Big[\sum_{i \in S} z_i^{-\theta} - \sum_{i<j, i \in S, j \in S} (p_i^{-\delta_{ij}} z_i^{\theta\delta_{ij}} + p_j^{-\delta_{ij}} z_j^{\theta\delta_{ij}})^{-\frac{1}{\delta_{ij}}}\Big]^{-\frac{1}{\theta}}.$$

The limiting MEV copula has exponent

$$A_{1\cdots m}(\mathbf{z}) = z_1 + \cdots + z_m + \sum_{S \in \mathcal{S}, |S| \geq 2} (-1)^{|S|+1} r_S(z_i, i \in S). \quad (6.31)$$

Many of the dependence properties of this family are the same as for the family MM2.

Some special cases are given next. For $m = 2$, with $\nu_1 = \nu_2 = 0$, (and hence $p_1 = p_2 = 1$) and $\delta = \delta_{12}$, (6.31) becomes $A(z_1, z_2) = z_1 + z_2 - [z_1^{-\theta} + z_2^{-\theta} - (z_1^{\theta\delta} + z_2^{\theta\delta})^{-1/\delta}]^{-1/\theta}$ which appears from the BEV limit in the family BB4.

For $m = 3$, with $\delta_{13} \to 0$, $\delta_{12} = \delta_{23} = \delta$, $\nu_1 = \nu_3 = -1$, $\nu_2 = 0$, (6.31) becomes

$$\begin{aligned} A_{123}(\mathbf{z}) &= z_1 + z_2 + z_3 - (z_1^{-\theta} + z_3^{-\theta})^{-1/\theta} \\ &- [z_1^{-\theta} + z_2^{-\theta} - (z_1^{\theta\delta} + 2^{\delta} z_2^{\theta\delta})^{-1/\delta}]^{-1/\theta} \\ &- [z_3^{-\theta} + z_2^{-\theta} - (z_3^{\theta\delta} + 2^{\delta} z_2^{\theta\delta})^{-1/\delta}]^{-1/\theta} \\ &+ [z_1^{-\theta} + z_2^{-\theta} + z_3^{-\theta} - (z_1^{\theta\delta} + 2^{\delta} z_2^{\theta\delta})^{-1/\delta} \\ &\quad -(z_3^{\theta\delta} + 2^{\delta} z_2^{\theta\delta})^{-1/\delta}]^{-1/\theta}. \end{aligned}$$

The bivariate margins (by letting one of the z_i go to zero in turn) are $A_{j2}(z_j, z_2) = z_j + z_2 - [z_j^{-\theta} + z_2^{-\theta} - (z_j^{\theta\delta} + 2^{\delta} z_2^{\theta\delta})^{-1/\delta}]^{-1/\theta}$, $j = 1, 3$, and $A_{13}(z_1, z_3) = z_1 + z_3 - (z_1^{-\theta} + z_3^{-\theta})^{-1/\theta}$. The corresponding tail dependence parameters, $2 - A_{ik}(1, 1)$, are $\lambda_{j2} = [2 - (1 + 2^{\delta})^{-1/\delta}]^{-1/\theta}$, $j = 1, 3$, and $\lambda_{13} = 2^{-1/\theta}$. As $\delta \to \infty$, $\lambda_{12} = \lambda_{32} \to (1.5)^{-1/\theta}$. The calculations, given in Table 6.2, show that λ_{13} is close to the non-sharp lower bound $\max\{0, \lambda_{12} + \lambda_{23} - 1\} = \max\{0, 2(1.5)^{-1/\theta} - 1\} = L_\theta$, for part of the range of θ.

Family MM3. In MSMVE form, the exponent is

$$A_{1\cdots m}(\mathbf{z}) = \Big[\sum_{i=1}^{m} z_i^{\theta} - \sum_{1 \leq i < j \leq m} (p_i^{-\delta_{ij}} z_i^{-\theta\delta_{ij}} + p_j^{-\delta_{ij}} z_j^{-\theta\delta_{ij}})^{-1/\delta_{ij}}\Big]^{1/\theta}. \quad (6.32)$$

Note that the ν_i appear only implicitly in the p_i.

Table 6.2. *Tail dependence parameters in a special trivariate case of family MM8.*

θ	$2^{-1/\theta}$	L_θ
0	0	0
0.25	0.0625	0
0.5	0.250	0
0.75	0.397	0.165
1	0.500	0.333
1.5	0.630	0.526
2	0.707	0.633
3	0.794	0.747
4	0.841	0.807
5	0.871	0.844
∞	1	1

In a special trivariate case, the tail dependence analysis is similar to the family MM1, because from Section 5.5, $\lambda_{13} = 2 - 2^{1/\theta}$, and as $\delta \to \infty$, $\lambda_{12} = \lambda_{23} \to 2 - (1.5)^{1/\theta}$, and these are the same as for the family MM1.

6.3.2 Other parametric families

This subsection is devoted to existing parametric families where parameters are not all dependence parameters and individual parameters are harder to interpret.

An alternative representation, from de Haan (1984), for MSMVE distributions has the exponent $A = -\log G$ in the form:

$$A(\mathbf{z}) = \int_0^1 [\max_{1 \le j \le m} z_j g_j(v)]\, dv, \tag{6.33}$$

where the g_j are pdfs on $[0, 1]$. (This essentially follows from the definition of MSMVE and a limit on (6.7) as $r \to \infty$, with $\alpha_k q_{kj}$ being (multiples) of $g_j(v_k)$ for some points v_k in (0,1).) The families B6 and B7 (or (6.17) and (6.18)) can be unified with this form with $g_1(v) = (1 - \alpha)v^{-\alpha}$, $g_2(v) = (1 - \beta)(1 - v)^{-\beta}$, with $0 < \alpha = \beta < 1$ for (6.17) and $\alpha = \beta < 0$ for (6.18). With α, β both less than 1, a two-parameter extension of (6.17) and (6.18) obtains. But its interpretation is not as easy; the magnitude of the difference of α

and β measures asymmetry, and the average of α and β measures dependence. For $\alpha > 0$, $\beta > 0$, expansion of the integral in (6.33) leads to

$$A_{12}(z_1, z_2; \alpha, \beta) = (z_1 + z_2)B(z_2/(z_1 + z_2); \alpha, \beta), \quad 0 < \alpha, \beta < 1, \tag{6.34}$$

where $B(w; \alpha, \beta) = (1-w)u^{1-\alpha} + w(1-u)^{1-\beta}$, $0 \le w \le 1$, and $u = u(w; \alpha, \beta)$ is the root of the equation $(1-\alpha)(1-w)(1-u)^{\beta} - (1-\beta)wu^{\alpha} = 0$. For $\alpha = -\alpha_0 < 0$, $\beta = -\beta_0 < 0$, one obtains

$$A^*_{12}(z_1, z_2; \alpha_0, \beta_0) = (z_1 + z_2)B^*(z_2/(z_1 + z_2); \alpha_0, \beta_0), \quad \alpha_0, \beta_0 > 0, \tag{6.35}$$

where $B^*(w; \alpha_0, \beta_0) = 1 - w(1-u)^{1+\beta_0} - (1-w)u^{1+\alpha_0}$, $0 \le w \le 1$, and $u = u(w; \alpha_0, \beta_0)$ is the root of the equation $(1+\alpha_0)(1-w)u^{\alpha_0} - (1+\beta_0)w(1-u)^{\beta_0} = 0$. There is an obvious multivariate extension of (6.34) and (6.35), using more g_j of the same form in (6.33).

The distributions from (6.34) are increasing in concordance as α or β decreases for $\alpha, \beta > 0$; the Fréchet upper bound obtains in the limit as $\alpha = \beta \to 0$ and independence obtains as $\alpha = \beta \to 1$. The distributions from (6.35) are also increasing in concordance as α_0 or β_0 decreases; the Fréchet upper bound obtains in the limit as $\alpha_0 = \beta_0 \to 0$ and independence obtains as $\alpha_0 = \beta_0 \to \infty$. Independence obtains more generally as one of α, β (or α_0, β_0) is fixed and the other approaches 1 (∞). Different limits occur as one of the parameters approaches 0, for example, as $\beta \to 0$ in (6.34), or as $\beta_0 \to 0$ in (6.35).

Next we list the family of Marshall–Olkin (1967) multivariate exponential distributions and some of its extensions. Let S be an index variable over $\mathcal{S} = \mathcal{S}_m$. Let $\alpha_S > 0$ for each S and let $\nu_S = \sum_{T:S\subset T} \alpha_T$. For $|S| = 1$, simplifying notation such as α_i and ν_i will be used. For $S \in \mathcal{S}$, let Z_S be an exponential rv with rate parameter α_S (mean α_S^{-1}), and suppose $\{Z_S\}$ is a set of independent rvs. For $j = 1, \ldots, m$, let $X_j = \min\{Z_S : j \in S\}$.

As before, let the survival function be $G = e^{-A}$. The exponent A of the Marshall–Olkin distribution is

$$A(\mathbf{x}; \alpha_S, S \in \mathcal{S}) = \sum_{S\in\mathcal{S}} \alpha_S \max_{i\in S} x_i = \sum_{S\in\mathcal{S}} (-1)^{|S|+1} \nu_S \min_{i\in S} x_i. \tag{6.36}$$

The exponent in the last term in (6.36) can be rearranged to get

$$A(\mathbf{x}; \alpha_S, S \in \mathcal{S}) = \sum_{S\in\mathcal{S}} \alpha_S a_S(x_i, i \in S), \tag{6.37}$$

where

$$a_S(x_i, x \in S) = \sum_{T \subset S} (-1)^{|T|+1} [\min_{i \in T} x_i]. \tag{6.38}$$

The derivation from the stochastic representation is left as an exercise.

General families of MSMVE distributions can have the form of (6.37), where a_S is replaced by something other than (6.38) or $a_S = \max_{i \in S} x_i$. These include

$$A(\mathbf{x}; \alpha_S, S \in \mathcal{S}, \delta) = \sum_{S \in \mathcal{S}} \alpha_S \Big(\sum_{i \in S} x_i^{\delta}\Big)^{1/\delta}, \tag{6.39}$$

where $1 \le \delta \le \infty$, with $\delta = \infty$ meaning (6.39) is equivalent to (6.36), and

$$\begin{aligned} A(\mathbf{x}; \alpha_S, S \in \mathcal{S}, \delta) &= \sum_{S \in \mathcal{S}} (-1)^{|S|+1} \nu_S \big[\sum_{i \in S} x_i^{-\delta}\big]^{-1/\delta} \\ &= \sum_{S \in \mathcal{S}} \alpha_S a_S(x_i, i \in S; \delta), \end{aligned} \tag{6.40}$$

where

$$a_S(x_i, i \in S; \delta) = \sum_{T \subset S} (-1)^{|T|+1} \big[\sum_{i \in T} x_i^{-\delta}\big]^{-1/\delta}$$

generalizes (6.38) and $0 \le \delta \le \infty$, with $\delta = \infty$ meaning that (6.40) is equivalent to (6.36). Note that if $\delta = 1$ in (6.39) or if $\delta \to 0$ in (6.40), then G becomes a product of the univariate margins, i.e., $\exp\{-\sum_{i=1}^{m} \nu_i x_i\}$. Hence δ in both (6.39) and (6.40) is a global dependence parameter and α_S or ν_S for $|S| \ge 2$ are parameters indicating the strength of dependence for the variables in S. The univariate survival functions for (6.36), (6.39) and (6.40) are $G_j(x_j) = \exp\{-\nu_j x_j\}$, $j = 1, \ldots, m$, and each family has k-variate ($2 \le k < m$) marginal survival functions of the same form. By rescaling, one gets MSMVE distributions with unit means for the univariate margins. Special cases of (6.39) and (6.40) include the families B6 and B7 and their permutation-symmetric extensions (take $\alpha_S = 0$ if $S \ne \{1, \ldots, m\}$ and $\alpha_{\{1,\ldots,m\}} = 1$).

Another approach is to use Pickand's representation with parametric families of densities on the simplex S_m. An example is the Dirichlet distribution (or beta distribution for $m = 2$). After rescaling to get unit exponential margins, the exponent is:

$$A(\mathbf{z}) = \int_{S_m} \max_{1 \le j \le m} \{w_j z_j / \nu_j\} \Gamma(\alpha_+) \prod_{j=1}^{m} [w_j^{\alpha_j - 1} / \Gamma(\alpha_j)] \, d\mathbf{w},$$

where $\alpha_j > 0$, $j = 1, \ldots, m$, are the parameters of the Dirichlet distribution, $\alpha_+ = \alpha_1 + \cdots + \alpha_m$, and $\nu_j = \alpha_j/\alpha_+$, $j = 1, \ldots, m$. The independence and Fréchet upper bound copulas obtain when α_+ goes to 0 and ∞ respectively, with $\alpha_i/\alpha_+ \to \pi_i > 0$, $i = 1, \ldots, m$. For the bivariate case, there is the simplification to

$$A(z_1, z_2) = z_1[1 - B(y; \alpha_1 + 1, \alpha_2)] + z_2 B(y; \alpha_1, \alpha_2 + 1),$$

where $y = (z_2/\nu_2)/[z_1/\nu_1 + z_2/\nu_2]$ and B is the incomplete beta function. The interpretation of the parameters of the Dirichlet distribution in the resulting MSMVE distribution is not simple.

6.4 Point process approach *

Inference with multivariate extremes can be in the form of componentwise maxima (or minima), in which case the models in the previous section can be used directly. In addition, there is a point process approach for inference with multivariate tail probabilities; this may be more natural if data are not in the form of maxima. The background for this approach is given in this section.

Let $(X_{i1}, \ldots, X_{im})$ be iid random vectors from the distribution F, $i = 1, 2, \ldots$. Let $Z_{ij} = t_j(X_{ij})$ be (one-to-one) transforms such that $\Pr(Z_{ij} > z) \sim z^{-1}$, as $z \to \infty$. (This is essentially a transform to a tail that is like the Fréchet distribution with parameter of 1.) Let $M_{jn} = \max_{1 \le i \le n} Z_{ij}$, $j = 1, \ldots, m$, be the componentwise maxima.

Let P_n denote the point process in $\Re^n$ from $\{n^{-1}\mathbf{Z}_1, \ldots, n^{-1}\mathbf{Z}_n\}$. From multivariate extreme value theory, under some regularity conditions, P_n converges weakly to a non-homogeneous Poisson process on $[0, \infty)^m \backslash \{(0, \ldots, 0)\}$ as $n \to \infty$, with the intensity measure Λ of the limiting process satisfying $\Lambda(B/c) = c\Lambda(B)$ for all $c > 0$ and all measureable sets B that are bounded away from the origin. Let

$$V(\mathbf{z}) = \Lambda(\{[(0, z_1) \times \cdots \times (0, z_m)]^c\}),$$

so that V is homogeneous of order -1, i.e., $V(c\mathbf{z}) = c^{-1}V(\mathbf{z})$ for all $c > 0$, $\mathbf{z} \in (0, \infty)^m$.

If $A(\mathbf{w})$ is a possible exponent of a MSMVE distribution, then $V(\mathbf{z}) = A(z_1^{-1}, \ldots, z_m^{-1})$ is a possible intensity measure function.

Proof. Let $M'_{jn} = M_{jn}^{-1} = \min_{1 \le i \le n} Z_{ij}^{-1}$, $j = 1, \ldots, m$. Then for $x > 0$,

$$\Pr(M_{jn} \le nx) \sim (1 - [nx]^{-1})^n \to e^{-1/x}, \quad n \to \infty,$$

or for $w > 0$, $\Pr(M'_{jn} \geq w/n) \to e^{-w}$ as $n \to \infty$. If the sequence $n\,(M'_{1n}, \ldots, M'_{mn})$ converges in distribution as $n \to \infty$, then the limiting distribution is in the MSMVE class, say with exponent $A(\mathbf{w})$, and $n^{-1}(M_{1n}, \ldots, M_{mn})$ converges in distribution to a max-stable multivariate Fréchet distribution (cf. (6.6)), with exponent $A(z_1^{-1}, \ldots, z_m^{-1})$.

Next we look at the limit by taking a point process approach. Let $B = [(0, z_1) \times \cdots \times (0, z_m)]^c$. Then

$$\Pr(n^{-1}\mathbf{Z}_i \notin B, i = 1, \ldots, n) \to e^{-\Lambda(B)} = e^{-V(\mathbf{z})}, \qquad (6.41)$$

since the event on the left-hand side of (6.41) involves a count of 0 for the point process P_n and the limiting count is a Poisson rv with mean $\Lambda(B)$; this follows from the limiting non-homogeneous Poisson process result. However the left-hand side of (6.41) is also $\Pr(n^{-1}M_{jn} \leq z_j, j = 1, \ldots, m)$. Hence $V(\mathbf{z}) = A(z_1^{-1}, \ldots, z_m^{-1})$. □

The method of maximum likelihood can be used to obtain estimates of model parameters, including parameters of the functions t_j used to transform to the form of the assumed univariate tail.

Assuming that the point process does not have mass in lower-dimensional spaces (otherwise, see the references in Section 6.7), the likelihood given the transformed data is as follows. Let $\mathbf{x}_i = (x_{i1}, \ldots, x_{im})$, $i = 1, \ldots, n$, denote the original data and let $\mathbf{z}_i = (z_{i1}, \ldots, z_{im})$, $i = 1, \ldots, n$, denote the transformed data which have the required tail frequency distribution. The $\mathbf{x}_i$ are iid realizations of a random vector $\mathbf{X}$ and the $\mathbf{z}_i$ are treated as iid realizations of a random vector $\mathbf{Z}$. If the $n^{-1}\mathbf{z}_i$ which lie in a set B are a realization of a non-homogeneous Poisson point process with measure $\Lambda(\cdot; \boldsymbol{\theta})$ and intensity function $\lambda(\cdot; \boldsymbol{\theta})$, where $\boldsymbol{\theta}$ is a parameter vector, then the likelihood for $\boldsymbol{\theta}$ is

$$L(\boldsymbol{\theta}) = \Big[\prod_{i:\mathbf{z}_i/n \in B} \lambda(n^{-1}\mathbf{z}_i; \boldsymbol{\theta}) \Big] \exp\{-\Lambda(B; \boldsymbol{\theta})\}.$$

Note that $(-1)^m$ times the mixed derivative of V is the intensity function associated with the non-homogeneous Poisson process. Let $\boldsymbol{\eta}_1, \ldots, \boldsymbol{\eta}_m$ be vectors of univariate marginal parameters. Then we may write $\mathbf{z}_i = (t_1(x_{i1}; \boldsymbol{\eta}_1), \ldots, t_m(x_{im}; \boldsymbol{\eta}_m))$, $i = 1, \ldots, n$, where $t_1, \ldots, t_m$ are strictly increasing transformations. The full likelihood for $\boldsymbol{\theta}, \boldsymbol{\eta}_1, \ldots, \boldsymbol{\eta}_m$ (with asymptotic approximations) is then

$$L(\boldsymbol{\theta}, \boldsymbol{\eta}_1, \ldots, \boldsymbol{\eta}_m) = \exp\{-\Lambda(B; \boldsymbol{\theta})\} \qquad (6.42)$$
$$\cdot \prod_{\mathbf{x}_i:\mathbf{z}_i/n \in B} \Big[\lambda(n^{-1}\mathbf{z}_i; \boldsymbol{\theta}) \prod_{j=1}^m n^{-1} \tfrac{\partial t_j}{\partial x_j}\, (x_{ij}; \boldsymbol{\eta}_j)\Big].$$

A weak assumption is that univariate distributions are in the domain of attraction of a GEV distribution. Then the transformations of the variables come from assuming that the upper tails are generalized Pareto with unknown parameters, with the remainder of the distributions being arbitrary but known or estimated with an empirical distribution (alternatively rank approximations can be used for the remainder).

This leads to the following transformations for the t_j. Assume that the sample size n is large. For $j = 1, \ldots, m$, assume that the cdf of X_j is known below the threshold u_j and conditionally generalized Pareto with unknown parameters $\boldsymbol{\eta}_j = (\gamma_j, \sigma_j)$ above the thresholds. That is, the jth cdf is

$$F_j(x_j; \boldsymbol{\eta}_j) = \begin{cases} F_j(x_j), & x_j \le u_j, \\ 1-(1-F_j(u_j))(1+\gamma_j[x_j - u_j]/\sigma_j)_+^{-1/\gamma_j}, & x_j \ge u_j, \end{cases}$$

where $y_+ = \max\{0, y\}$. The transformation Z_j of X_j is such that Z_j has the Fréchet distribution $G(z) = \exp\{-1/z\}$, $z > 0$. Note that $G(z) \sim 1 - 1/z$ for z large, as required from the point process derivation. Therefore,

$$Z_j = t_j(X_j) = [-\log F_j(X_j; \boldsymbol{\eta}_j)]^{-1}, \quad X_j = F_j^{-1}(\exp\{-1/Z_j\}; \boldsymbol{\eta}_j)$$

and, for $X_j \ge u_j$, $Z_j = [-\log\{1 - (1 - F_j(u_j))(1 + \gamma_j[(X_j - u_j)/\sigma_j])_+^{-1/\gamma_j}\}]^{-1}$. For $x_j > u_j$, $\partial t_j/\partial x_j$ evaluated at $z = t_j(x_j)$ becomes $\sigma_j^{-1} z^2 (1 - e^{-1/z})^{1+\gamma_j} (1 - F_j(u_j))^{-\gamma_j} e^{1/z}$.

For (6.42), there is a simplification if $B = (\times_{j=1}^m [0, n^{-1} t_j(u_j)])^c$, since in this case t_j depends on $\boldsymbol{\eta}_j$ only if $x_{ij} > u_j$.

The intensity function is $\lambda(\mathbf{z}) = (-1)^m \partial^m V / \prod \partial z_j$. For (6.29) and (6.32), V is of the form D^ς, with D in turn taking the form $\sum_{i=1}^m D_i + \sum_{i<j} D_{ij}$. For (6.31), λ comes from the last term $(-1)^{m+1} r_{\{1,\ldots,m\}}$ and $r_{\{1,\ldots,m\}}$ has the form D^ς of the preceding sentence. The mixed derivative of V can then be obtained from the derivatives of the D_i and D_{ij} and λ obtains in a reasonable form. Let $D_{ji} = D_{ij}$ for $i < j$, and let

$$W_i = \frac{\partial D}{\partial z_i} = \frac{\partial D_i}{\partial z_i} + \sum_{j \ne i} \frac{\partial D_{ij}}{\partial z_i} \quad \forall i$$

and

$$W_{ij} = \frac{\partial^2 D}{\partial z_i \partial z_j} = \frac{\partial^2 D_{ij}}{\partial z_i \partial z_j} \quad \forall i \ne j.$$

The mth-order mixed derivative of D^ζ is

$$\frac{\partial^m V}{\partial z_1 \cdots \partial z_m} = \zeta(\zeta-1)\cdots(\zeta-m+1)D^{\zeta-m}\prod_{i=1}^m W_i \tag{6.43}$$
$$+ \quad \zeta(\zeta-1)\cdots(\zeta-m+2)D^{\zeta-m+1}\Big(\sum_{i<j} W_{ij}\prod_{k\neq i,j} W_k\Big)$$
$$+ \quad \zeta\cdots(\zeta-m+3)D^{\zeta-m+2}\Big(\sum\nolimits^{*} W_{i_1,j_1}W_{i_2,j_2}\prod_{k\neq i_1,i_2,j_1,j_2} W_k\Big),$$

where $\sum^*$ is over the set $\{i_1<j_1, i_2<j_2, i_1<i_2\}$. Note that the third term on the right-hand side of (6.43) is null if $m<4$.

6.5 Choice models

In this section, we illustrate the use of MEV distributions for choice models. This is just one of many possible families for choice models that has appeared in the econometrics and mathematical psychology literature. One of its advantages is that closed-form choice probabilities obtain if a closed-form parametric family of MEV distributions is used.

In general, a **choice model** for m items consists of an absolutely continuous multivariate distribution for rvs $X_1,\ldots,X_m$, where X_j is a utility or merit rv associated with the jth option or item. (The assumption of absolute continuity is needed here to eliminate the possibility of a positive probability of ties among weighted maxima or minima.) A **choice probability** has the form $\pi_{j,S} = \Pr(X_j > X_i, i\in S, i\neq j)$, where $S\in\mathcal{S}_m$, $|S|\geq 2$; $\pi_{j,S}$ is interpreted as the probability that option j is the preferred or chosen option among the choice set S of options.

Additional notation and background ideas are needed before we get to the calculation of choice probabilities for MEV distributions.

Let $G(\mathbf{x}) = e^{-A(\mathbf{x})}$ be an absolutely continuous min-stable m-variate exponential survival function and let $\mathbf{X}\sim G$. For $j=1,\ldots,m$, let A_j be the partial derivative of A with respect to x_j. (We retain the notation of Section 6.3, with A_S denoting the S-margin of A for $|S|\geq 2$, so that $A_{S,j}$ for $j\in S$ refers to the jth partial derivative of A_S.) Then $-\partial G/\partial x_j = GA_j$. By writing $A(\mathbf{x}) = x_1A(1, x_2/x_1,\ldots,x_m/x_1)$ (cf. condition (6.16)), $-\partial G/\partial x_1$ is also equal to

$$G\Big[A(1,x_2/x_1,\ldots,x_m/x_1) - \sum_{i=2}^m (x_i/x_1)A_i(1,x_2/x_1,\ldots,x_m/x_1)\Big].$$

Therefore

$$A_1(\mathbf{x}) = A(\mathbf{x}/x_1) - \sum_{i=2}^{m}(x_i/x_1)A_i(\mathbf{x}/x_1) = A_1(\mathbf{x}/x_1).$$

That is, $A_1(\mathbf{x})$ depends only on the ratios x_i/x_1 and is homogeneous of order 0. Similarly, for $j = 2, \ldots, m$, $A_j(\mathbf{x})$ depends only on the ratios x_i/x_j, $i = 1, \ldots, m$.

Conditional survival functions are $\Pr(X_i > x_i, i \neq j | X_j = x_j) = -e^{x_j}\partial G/\partial x_1 = e^{x_j} G A_j$, $j = 1, \ldots, m$. Let $X^* = \min\{X_1, \ldots, X_m\}$. Then

$$\Pr(X^* > t, X^* = X_j) = \int_t^\infty e^{-A(x\mathbf{1}_m)} A_j(x\mathbf{1}_m)\, dx \qquad (6.44)$$
$$= \int_t^\infty e^{-xA(\mathbf{1}_m)} A_j(\mathbf{1}_m)\, dx = [A_j(\mathbf{1}_m)/A(\mathbf{1}_m)] e^{-tA(\mathbf{1}_m)}.$$

(This shows that $\sum_{j=1}^m A_j(\mathbf{1}_m) = A(\mathbf{1}_m)$.) But the right-hand side of (6.44) is also equal to $\Pr(X^* = X_j)\Pr(X^* > t)$, so that the events $\{X^* = X_j\}$ and $\{X^* > t\}$ are independent for all j. However, this independence criterion does not characterize MSMVE distributions among the class of multivariate exponential distributions. If $(X_1, \ldots, X_m)$ is a vector of exchangeable exponential rvs, then $\Pr(X^* > t, X^* = X_j) = m^{-1}\Pr(X^* > t) = \Pr(X^* = X_j)\Pr(X^* > t)$ for all j by symmetry.

Now suppose that e^{-A} is a MSMVE survival function and $\mathbf{V}$ has the MEV distribution $F(\mathbf{x}) = \exp\{-A(e^{-x_1}, \ldots, e^{-x_m})\}$. Suppose also that there are location parameters μ_j, which are merit parameters, associated with the jth option, $j = 1, \ldots, m$, and option j is preferred to the other $m-1$ options if $V_j + \mu_j \geq V_i + \mu_i$ for all $i \neq j$. Let π_j, $j = 1, \ldots, m$, be the probability that the jth option is preferred (or chosen). These are the **choice probabilities** with a MEV distribution as a choice model when the choice set is $S = \{1, \ldots, m\}$. Let $w_j = e^{\mu_j}$ and $Z_j = e^{-V_j}$, $j = 1, \ldots, m$, and let $\mu_{ij} = \mu_i - \mu_j$, $w_{ij} = e^{\mu_{ij}} = w_i/w_j$ for $i, j = 1, \ldots, m$. Then

$$\begin{aligned}\pi_1 &= \pi_{1,\{1,\ldots,m\}} = \Pr(V_1 > V_i + \mu_{i1}, i = 2, \ldots, m) \\ &= \Pr(V_i < V_1 - \mu_{i1}, 2 \leq i \leq m) = \Pr(Z_i > w_{i1}Z_1, 2 \leq i \leq m).\end{aligned}$$

Using properties of MSMVE distributions,

$$\begin{aligned}\pi_1 &= \int_0^\infty e^{-A(z, w_{21}z, \ldots, w_{m1}z)} A_1(1, w_{21}, \ldots, w_{m1})\, dz \\ &= A_1(1, w_{21}, \ldots, w_{m1})/A(1, w_{21}, \ldots, w_{m1}) \\ &= A_1(\mathbf{w})/[w_1^{-1}A(\mathbf{w})] = w_1 A_1(\mathbf{w})/A(\mathbf{w}).\end{aligned}$$

Similarly,

$$\pi_j = \pi_{j,\{1,\ldots,m\}} = w_j A_j(\mathbf{w})/A(\mathbf{w}), \tag{6.45}$$

$j = 2, \ldots, m$. For other $S \in \mathcal{S}_m$, probabilities $\pi_{j,S}$ have a similar form with $A_{S,j}, A_S$ replacing A_j, A in (6.45), since the class of MSMVE distributions is closed under margins.

Logit-type probabilities result when the permutation-symmetric copula M6 is used (see case 1 in Example 6.2 below). Other choice models occur from other MEV copulas. Note that the models deriving from max-stable MEV distributions are convenient in that closed-form expressions are obtained for the choice probabilities $\pi_{j,S}$. The parameters μ_j can be functions of covariates when the options have a factorial or regression structure.

Example 6.2 Some cases, using MSMVE distributions from Section 6.3.1, are:

1. $A(\mathbf{z}) = (z_1^\theta + \cdots + z_m^\theta)^{1/\theta}$, $\theta \geq 1$. Then
$$A_j(\mathbf{z}) = (z_1^\theta + \cdots + z_m^\theta)^{-1+1/\theta} z_j^{\theta-1},$$
and, from (6.45),
$$\pi_j = w_j^\theta/(w_1^\theta + \cdots + w_m^\theta) = e^{\theta\mu_j} \Big/ \sum_{i=1}^m e^{\theta\mu_i}, \quad j = 1, \ldots, m.$$

2. $A(z_1, z_2) = z_1 + z_2 - (z_1^{-\theta} + z_2^{-\theta})^{-1/\theta}$, $\theta > 0$. Then, for $j = 1, 2$,
$$A_j(z_1, z_2) = 1 - z_j^{-\theta-1}(z_1^{-\theta} + z_2^{-\theta})^{-1-1/\theta}$$
and
$$\pi_j = [w_j - w_j^{-\theta}(w_1^{-\theta} + w_2^{-\theta})^{-1-1/\theta}]/[w_1 + w_2 - (w_1^{-\theta} + w_2^{-\theta})^{-1/\theta}].$$

3. $A(z_1, z_2, z_3) = [(z_1^\theta + z_2^\theta)^{\delta/\theta} + z_3^\delta]^{1/\delta}$, with $1 \leq \delta \leq \theta$. Then,
$$\pi_j = \{w_j^\theta/(w_1^\theta + w_2^\theta)\} \cdot \{(w_1^\theta + w_2^\theta)^{\delta/\theta}/[(w_1^\theta + w_2^\theta)^{\delta/\theta} + w_3^\delta]\},$$
$j = 1, 2$, and $\pi_3 = w_3^\delta/[(w_1^\theta + w_2^\theta)^{\delta/\theta} + w_3^\delta]$.

□

In the remainder of this section, we look further at a generalized 'independence' criterion. Let $c_1, \ldots, c_m$ be positive constants and let $X^*(\mathbf{c}) = \min\{X_i/c_i : i = 1, \ldots, m\}$. This is equivalent to the maximum of the shifted extreme value rvs given above. If $\mathbf{X}$ is MSMVE with survival function G, then the events $\{X^*(\mathbf{c}) > t\}$ and $\{X^*(\mathbf{c}) = X_j/c_j\}$ are independent. Since $G = e^{-A}$, where A

is homogeneous of order 1, it is straightforward to obtain:

$$\begin{aligned}
&\Pr(X^*(\mathbf{c}) > t, X^*(\mathbf{c}) = X_j/c_j) \\
&= \int_{c_j t}^{\infty} \exp\{-A(\tfrac{c_1 x}{c_j}, \ldots, \tfrac{c_m x}{c_j})\} A_j(\tfrac{c_1 x}{c_j}, \ldots, \tfrac{c_m x}{c_j})\, dx \\
&= [A_j(\tfrac{c_1}{c_j}, \ldots, \tfrac{c_m}{c_j})/A(\tfrac{c_1}{c_j}, \ldots, \tfrac{c_m}{c_j})] \exp\{-c_j t A(\tfrac{c_1}{c_j}, \ldots, \tfrac{c_m}{c_j})\} \\
&= c_j [A_j(\mathbf{c})/A(\mathbf{c})] \exp\{-tA(\mathbf{c})\} \\
&= \Pr(X^*(\mathbf{c}) = X_j/c_j) \Pr(X^*(\mathbf{c}) > t),
\end{aligned}$$

$j = 1, \ldots, m$. (Hence $\sum_{i=1}^m c_i A_i(\mathbf{c}) = A(\mathbf{c})$ for all $\mathbf{c} \in (0, \infty)^m$.)

The generalized 'independence' criterion need not hold for other exchangeable exponential rvs. This can be verified directly for some specific cases (e.g., with the copula B10).

The theorem below shows that the above is a characterization of MSMVE distributions.

Theorem 6.10 *If* $\mathbf{X}$ *is (absolutely continuous) multivariate exponential and* $\{\min_i X_i/c_i > t\}$ *is independent of* $\{\min_i X_i/c_i = X_j/c_j\}$ *for all* $t > 0$, $j \in \{1, \ldots, m\}$, $\mathbf{c} \in (0, \infty)^m$, *then* $\mathbf{X}$ *must be min-stable multivariate exponential.*

Proof. The proof follows that in Robertson and Strauss (1981) for the most part.

Let $G(\mathbf{x})$ be the survival function of $\mathbf{X}$, let $G_j = -\partial G/\partial x_j$, $j = 1, \ldots, m$, and let $X^*(\mathbf{c}) = \{\min_i X_i/c_i\}$. Then

$$\Pr(X^*(\mathbf{c}) > t, X^*(\mathbf{c}) = X_j/c_j) = \int_{c_j t}^{\infty} G_j(c_1 x/c_j, \ldots, c_m x/c_j)\, dx$$

$$= c_j \int_t^{\infty} G_j(c_1 x, \ldots, c_m x)\, dx, \quad j = 1, \ldots, m, \tag{6.46}$$

$$\Pr(X^*(\mathbf{c}) = X_j/c_j) = c_j \int_0^{\infty} G_j(c_1 x, \ldots, c_m x)\, dx, \quad j = 1, \ldots, m.$$

If

$$\Pr(X^*(\mathbf{c}) > t, X^*(\mathbf{c}) = X_j/c_j) = \Pr(X^*(\mathbf{c}) = X_j/c_j) \Pr(X^*(\mathbf{c}) > t)$$

for all t, j, then for $i \neq j$,

$$\Pr(X^*(\mathbf{c}) > t, X^*(\mathbf{c}) = X_i/c_i) / \Pr(X^*(\mathbf{c}) > t, X^*(\mathbf{c}) = X_j/c_j)$$

$$= \Pr(X^*(\mathbf{c}) = X_i/c_i) / \Pr(X^*(\mathbf{c}) = X_j/c_j) \tag{6.47}$$

does not depend on t. Let the right-hand side of (6.47) be denoted by $p_{ij}(\mathbf{c})$. Then p_{ij} is increasing as c_i increases and decreasing as

c_j increases. Differentiating ratios of the right-hand side of (6.46) implies

$$[c_i G_i(c_1 t, \ldots, c_m t)]/[c_j G_j(c_1 t, \ldots, c_m t)] = p_{ij}(\mathbf{c}), \quad \forall t. \tag{6.48}$$

Rewrite (6.48) as

$$c_i G_i(c_1 t, \ldots, c_m t) = c_j G_j(c_1 t, \ldots, c_m t)\, p_{ij}(\mathbf{c}), \quad \forall t. \tag{6.49}$$

Let $G_{\ell k} = \partial G_\ell / \partial x_k = -\partial^2 G / \partial x_\ell \partial x_k$, for $1 \leq \ell, k \leq m$. Then differentiation of (6.49) with respect to t yields $c_i \sum_{k=1}^m c_k G_{ik} = p_{ij}(\mathbf{c})\, c_j \sum_{k=1}^m c_k G_{jk}$ or

$$G_j \sum_{k=1}^m c_k G_{ik} = G_i \sum_{k=1}^m c_k G_{jk}. \tag{6.50}$$

Let $t = 1$; then (6.50) holds for all arguments $\mathbf{c} \in (0, \infty)^m$. This implies $\sum_k c_k G_{ik}(\mathbf{c}) / G_i(\mathbf{c}) = \eta(\mathbf{c})$ for some function η that does not depend on the index i, or $\sum_k c_k G_{ik}(\mathbf{c}) = \eta(\mathbf{c}) G_i(\mathbf{c})$, $i = 1, \ldots, m$. This has a solution only if $\eta(\mathbf{c}) = \phi'(G(\mathbf{c}))$ for some function ϕ with first derivative ϕ', and then by integration

$$\sum_k c_k G_k(\mathbf{c}) = \phi(G(\mathbf{c})). \tag{6.51}$$

(The proof is as follows: Let $L(\mathbf{c}) = \sum_k c_k G_k(\mathbf{c})$; then $L_i(\mathbf{c}) = \sum_k c_k G_{ik}(\mathbf{c}) = \eta(\mathbf{c}) G_i(\mathbf{c})$, $i = 1, \ldots, m$, so that $L_i G_j - G_i L_j = 0$ for all $i \neq j$. Hence the Jacobian $\partial(G, L)/\partial(c_i, c_j)$ vanishes for all $i \neq j$. For $m = 2$, this means that L and G are functionally related. For $m \geq 3$, let $\mathbf{c}_{-ij}$ or $\mathbf{x}_{-ij}$ be $\mathbf{c}$ or $\mathbf{x}$ respectively without the ith and jth components. Then there are functions $\phi_{ij}(u, \mathbf{x}_{-ij})$, $i < j$, such that $L(\mathbf{c}) = \phi_{ij}(G(\mathbf{c}), \mathbf{c}_{-ij})$ for all $i < j$ and $L_k = \frac{\partial \phi_{ij}}{\partial u} \cdot G_k + \frac{\partial \phi_{ij}}{\partial x_k} \cdot I(k \neq i, j)$. Hence for $k \neq i, j$,

$$\frac{L_k}{L_i} = \frac{\frac{\partial \phi_{ij}}{\partial u} \cdot G_k + \frac{\partial \phi_{ij}}{\partial x_k}}{\frac{\partial \phi_{ij}}{\partial u} \cdot G_i} = \frac{G_k}{G_i}$$

only if $\frac{\partial \phi_{ij}}{\partial x_k} = 0$. This implies that $\phi_{ij} = \phi$ for all $i < j$ and that G and L are functionally related.) Since $G_{ik} \leq 0$, $G_i > 0$, $i, k = 1, \ldots, m$, then $\phi' \leq 0$ and $\phi \geq 0$.

Replace c_k now by x_k. The first-order partial differential equation (6.51) can be solved by the method of characteristics (Bluman and Cole 1974) to get the only possible solutions. This means first solving

$$\frac{dx_1}{x_1} = \cdots = \frac{dx_m}{x_m} = \frac{dG}{\phi(G)},$$

to get independent solutions $u_1(\mathbf{x}) = a_1, \ldots, u_m(\mathbf{x}) = a_m$, where $a_1, \ldots, a_m$ are constants, so that the solution to (6.51) has the form $\beta(u_1, \ldots, u_m) = 0$ or $u_m = \alpha(u_1, \ldots, u_{m-1})$ for some functions α, β. From solving $dx_1/x_1 = dx_{i+1}/x_{i+1}$, one has $u_i = x_{i+1}/x_1 = a_i$, $i = 1, \ldots, m-1$. From solving $dx_1/x_1 = dG/\phi(G)$, one gets $u_m = \zeta(G)/x_1 = a_m$, where $\log \zeta(G)$ is the anti-derivative of $[\phi(G)]^{-1}$. Since $\phi' \le 0$, then ζ is a decreasing function. Therefore $\zeta(G(\mathbf{x}))/x_1 = \alpha(x_2/x_1, \ldots, x_m/x_1)$, or

$$\zeta(G(\mathbf{x})) = x_1 \alpha(x_2/x_1, \ldots, x_m/x_1) = A(\mathbf{x})$$

where A is a homogeneous function of order 1, or

$$G(\mathbf{x}) = \zeta^{-1} \circ A(\mathbf{x}) = \psi(A(\mathbf{x})) \tag{6.52}$$

with $\psi = \zeta^{-1}$. The homogeneity condition means that A has margin $A(0, \ldots, 0, x_i, 0, \ldots, 0) = b_i x_i$ for a constant b_i, $i = 1, \ldots, m$. To get exponential survival distributions as margins, $\zeta^{-1}(t) = e^{-t}$ or $\zeta(s) = -\log s$, and hence G is a MSMVE distribution. □

Other solutions G to (6.51) (or other survival functions with the generalized independence property) have the form of (6.52) with some other decreasing function ψ. Sufficient conditions are: (i) ψ is a LT (see Section 4.3); (ii) $-\log \psi \in \mathcal{L}_m^*$ (see Section 4.4); and (iii) $\psi \in \mathcal{L}_m$ (see (1.1) in Section 1.3) if A satisfies

$$(-1)^{1+|S|} \frac{\partial^{|S|} A}{\prod_{i \in S} \partial x_i}(\mathbf{x}) \ge 0 \quad \forall S \in \mathcal{S}_m.$$

We return to this in the next section.

6.6 Mixtures of MEV distributions *

In this section, we study the class of mixtures of powers of a MEV distribution. This is a special case of the larger class of mixtures of powers of a max-id distribution, as given in Section 4.3. This class has some closure properties that the larger class need not have. It is these closure properties and other properties that make the class interesting.

The three types of univariate extreme value margins are Gumbel, Weibull and Fréchet (see Section 6.1). Since maxima can be transformed into minima and vice versa, we will consider Weibull survival margins with minima, and Fréchet and Gumbel margins with maxima so that we can work on either $[0, \infty)$ or $(-\infty, \infty)$ for each univariate margin. Without loss of generality, we assume that the univariate margins are identical and standardized. As shown in Section 6.2.1, a property of a MEV distribution G is that all positive powers of G are also distributions.

Let G be a min-stable m-variate exponential survival function with unit exponential margins and let $A = -\log G$. Since A is homogeneous of order 1, $A(x_1, \ldots, x_m) = x_j$ if all arguments are zero except x_j, and $G^t(\mathbf{x}) = \exp\{-tA(\mathbf{x})\} = \exp\{-A(t\mathbf{x})\}$ is a survival function for all $t > 0$.

By making the transformations $x_j \to x_j^{\alpha}$, with $\alpha > 0$, the resulting min-stable m-variate Weibull survival function is

$$G_1(\mathbf{x}; \alpha) = \exp\{-A(x_1^{\alpha}, \ldots, x_m^{\alpha})\}. \tag{6.53}$$

If $\mathbf{X}$ has the distribution in (6.53), then

$$\begin{aligned}\Pr(\min\{X_1/c_1, \ldots, X_m/c_m\} > t) &= \exp\{-A((tc_1)^{\alpha}, \ldots, (tc_m)^{\alpha})\} \\ &= \exp\{-t^{\alpha} A(c_1^{\alpha}, \ldots, c_m^{\alpha})\}, \quad t > 0,\end{aligned}$$

for all $\mathbf{c} \in (0, \infty)^m$. That is, $\min\{X_1/c_1, \ldots, X_m/c_m\}$ has a scaled Weibull distribution for all $\mathbf{c} \in (0, \infty)^m$.

Similar results hold for transforms to other extreme value margins. By making the transformations $x_j \to x_j^{-\beta}$, with $\beta > 0$, the resulting max-stable m-variate Fréchet distribution function is

$$G_2(\mathbf{x}; \beta) = \exp\{-A(x_1^{-\beta}, \ldots, x_m^{-\beta})\}. \tag{6.54}$$

If $\mathbf{X}$ has the distribution in (6.54), then

$$\begin{aligned}\Pr\Big(\max\Big\{\frac{X_1}{c_1}, \ldots, \frac{X_m}{c_m}\Big\} \le t\Big) &= \exp\{-A((tc_1)^{-\beta}, \ldots, (tc_m)^{-\beta})\} \\ &= \exp\{-t^{-\beta} A(c_1^{-\beta}, \ldots, c_m^{-\beta})\}, \quad t > 0,\end{aligned}$$

for all $\mathbf{c} \in (0, \infty)^m$. That is, $\max\{X_1/c_1, \ldots, X_m/c_m\}$ has a scaled Fréchet distribution for all $\mathbf{c} \in (0, \infty)^m$.

By making the transformations $x_j \to e^{-x_j}$, the resulting max-stable m-variate Gumbel distribution function is

$$G_3(\mathbf{x}) = \exp\{-A(e^{-x_1}, \ldots, e^{-x_m})\}. \tag{6.55}$$

If $\mathbf{X}$ has the distribution in (6.55), then

$$\Pr(\max\{X_1 - c_1, \ldots, X_m - c_m\} \le t) = \exp\{-e^{-t} A(e^{-c_1}, \ldots, e^{-c_m})\},$$

$-\infty < t < \infty$, for all $\mathbf{c} \in (0, \infty)^m$. That is, $\max\{X_1 - c_1, \ldots, X_m - c_m\}$, a maximum of shifted random variables, has a location-shifted Gumbel distribution for all $\mathbf{c} \in (0, \infty)^m$. The weighting is done with additive rather than multiplicative constants in this case.

Note that a positive power of (6.53), (6.54) or (6.55) is a survival or distribution function since either a scale or location shift occurs. By taking mixtures of powers of one of these MEV distributions, we get distributions with other univariate margins which satisfy the closure property of weighted minima or maxima in the same scale or location family. Let M be the distribution function of a positive rv and let its LT be ψ. The mixtures of powers of (6.53), (6.54) and (6.55) lead to:

$$\int_0^\infty \exp\{-\gamma A(x_1^\alpha, \ldots, x_m^\alpha)\} dM(\gamma) = \psi\big(A(x_1^\alpha, \ldots, x_m^\alpha)\big), \quad (6.56)$$

$$\int_0^\infty \exp\{-\gamma A(x_1^{-\beta}, \ldots, x_m^{-\beta})\} dM(\gamma) = \psi\big(A(x_1^{-\beta}, \ldots, x_m^{-\beta})\big) \quad (6.57)$$

and

$$\int_0^\infty \exp\{-\gamma A(e^{-x_1}, \ldots, e^{-x_m})\} dM(\gamma) = \psi\big(A(e^{-x_1}, \ldots, e^{-x_m})\big). \quad (6.58)$$

The univariate survival margins in (6.56) are $\psi(x_j^\alpha)$ and the univariate cdfs in (6.57), (6.58) are respectively $\psi(x_j^{-\beta})$ and $\psi(e^{-x_j})$. If $\mathbf{X}$ has the distribution in (6.56), $\min_{1\le i\le m}\{X_i/c_i\}$ has the survival function $\psi([t/\sigma(\mathbf{c})]^\alpha)$, $t > 0$, with $\sigma(\mathbf{c}) = [A(c_1^\alpha, \ldots, c_m^\alpha)]^{-1/\alpha}$. If $\mathbf{X}$ has the distribution in (6.57), $\max_{1\le i\le m}\{X_i/c_i\}$ has the distribution $\psi([t/\eta(\mathbf{c})]^{-\beta})$, $t > 0$, with $\eta(\mathbf{c}) = [A(c_1^{-\beta}, \ldots, c_m^{-\beta})]^{1/\beta}$, and if $\mathbf{Y}$ has the distribution in (6.58), $\max_{1\le i\le m}\{Y_i - c_i\}$ has the distribution $\psi(\exp\{-[t-\mu(\mathbf{c})]\})$, $-\infty < t < \infty$, with $\mu(\mathbf{c}) = \log A(e^{-c_1}, \ldots, e^{-c_m})$.

A special case of (6.58) or (6.57) arises when $\psi(s) = (1+s)^{-1}$. For $H = e^{-A}$ being a general max-stable distribution, $F = \int H^\gamma dM(\gamma)$ is a max-geometric stable distribution. A multivariate distribution F is **max-geometric infinitely divisible** if for $\mathbf{X} \sim F$, then for any $0 < p < 1$, there exist iid random vectors $\mathbf{X}_{p,i}$, independent of N_p, such that $\mathbf{X} \stackrel{d}{=} \max_{i\le N_p} \mathbf{X}_{p,i}$ (with componentwise maxima), where N_p is a geometric rv with parameter p ($\Pr(N_p = k) = p(1-p)^{k-1}$, $k = 1, 2, \ldots$), and F is **max-geometric stable** (a stronger property) if for all $0 < p < 1$, $\mathbf{X}_{p,1}$ is in the location–scale family generated by F. Hence as a consequence of this definition, $F/[p+(1-p)F]$ must be in the location–scale family generated by F for any $0 < p < 1$. It is easy to show that this property is satisfied if $F(\mathbf{x}) = [1+ \quad {}^{-x_1}, \ldots, e^{-x_m})]^{-1}$ where A(is homogeneous of order 1. With H being the Gumbel distribution and $\psi(s) = (1+s)^{-1}$, the univariate margins of (6.58) become the logistic distribution $(1+e^{-x_j})^{-1}$ and F is a max-geometric stable multivariate logistic distribution.

In the remainder of this section, some dependence properties and characterizations are given.

Theorem 6.11 *Let* $F = \int_0^\infty H^\gamma dM(\gamma)$ *be the survival function or distribution function given in (6.56), (6.57) or (6.58). Then* F *is the distribution of associated rvs.*

Proof. The proof is similar to that of Theorem 4.5 and is left as an exercise. □

We mentioned above that (6.58) could result in multivariate distributions with univariate logistic marginals with the choice of the LT $\psi(s) = (1+s)^{-1}$. However, for logistic margins, only strictly positively dependent multivariate distributions can result; it is easily checked that the multivariate distribution with independent univariate logistic margins does not satisfy the property of closure under weighted maxima. This division of independence versus positive dependence is true in general. We show below that multivariate distributions with the independence copula can arise from (6.56), (6.57) and (6.58) only if the univariate margins are Weibull, Fréchet and Gumbel, respectively.

Theorem 6.12 *Suppose that* $\psi(A(x_1^\alpha, \ldots, x_m^\alpha))$ *in (6.56) is equal to* $\prod_{j=1}^m \psi(x_j^\alpha)$ *for all* $\mathbf{x} \in (0,\infty)^m$. *Then all possible solutions are covered by taking* $\psi(s) = \exp\{-\lambda s^{1/\sigma}\}$ *for some positive constants* λ *and* σ.

Proof. Let $X_1, \ldots, X_m$ be iid with survival function $\overline{F}(x) = \psi(x^\alpha)$. Then $\overline{F}^2(t) = \Pr(X_1 > t, X_2 > t) = \psi((ta)^\alpha) = \overline{F}(ta)$ for all $t > 0$, where $a^\alpha = A(1,1,0,\ldots,0)$ is a constant (exceeding 1). Let $r(t) = -\log \overline{F}(t)$. Then $r(0) = 0$, $r(\infty) = \infty$, r is increasing and $2r(t) = r(at)$ for all $t > 0$. Let σ be a constant satisfying $a = 2^\sigma$ so that $2^\sigma r^\sigma(t) = ar^\sigma(t) = r^\sigma(at)$. Next, let $\eta(t) = r^\sigma(t)$ so that $a\eta(t) = \eta(at)$ for all $t > 0$. Since the LT ψ is differentiable, η is differentiable and $a\eta'(t) = a\eta'(at)$ for all $t > 0$. The conditions on r and η then imply that η' is a constant and η is linearly increasing. Since $\eta(0) = 0$, $\eta(t) = \lambda^\sigma t$ for a positive constant λ. Hence $r(t) = \lambda t^{1/\sigma}$ and $\sigma = \log a / \log 2$, or $\overline{F}(t) = \exp\{-\lambda t^{1/\sigma}\}$ for some positive constants σ, λ. □

Theorem 6.13 *Suppose* $\psi(A(x_1^{-\beta}, \ldots, x_m^{-\beta}))$ *in (6.57) is equal to* $\prod_{j=1}^m \psi(x_j^{-\beta})$ *for all* $\mathbf{x} \in (0,\infty)^m$. *Then all possible solutions are* *covered by taking* $\psi(s) = \exp\{-\lambda s^{1/\sigma}\}$ *for some positive constants* λ *and* σ.

Proof. The proof is accomplished by a similar argument to that of the above theorem. We omit the details. □

Theorem 6.14 *Suppose* $\psi(A(e^{-x_1}, \ldots, e^{-x_m}))$ *in (6.58) is equal to* $\prod_{j=1}^m \psi(e^{-x_j})$ *for all* $\mathbf{x} \in \Re^m$. *Then all possible solutions are covered by taking* $\psi(s) = \exp\{-\lambda s^{1/\sigma}\}$ *for some positive constants* λ *and* σ.

Proof. Let $\mathbf{X}$ have the distribution $\psi(A(e^{-x_1}, \ldots, e^{-x_m}))$. Set $Y_j = \exp\{-X_j/\alpha\}$, $j = 1, \ldots, m$, where $\alpha > 0$. Then $(Y_1, \ldots, Y_m)$ has the survival function

$$\begin{aligned}\Pr(Y_j > y_j, 1 \le j \le m) &= \Pr(X_1 < -\alpha \log y_1, \ldots, X_m < -\alpha \log y_m) \\ &= \psi(A(y_1^\alpha, \ldots, y_m^\alpha))\end{aligned}$$

for all $\mathbf{y} \in (0, \infty)^m$. If $X_1, X_2, \ldots, X_m$ are independent with survival function $\psi(e^{-x})$, then $Y_1, Y_2, \ldots, Y_m$ are independent with survival function $\psi(y^\alpha)$. From the proof of Theorem 6.12, we get that $\psi(s) = \exp\{-\lambda s^{1/\sigma}\}$ for some positive constants λ, σ. This completes the proof. □

Choice probability properties from Section 6.5 hold for (6.56) to (6.58). The independence criterion also holds for scale mixtures of MSMVE distributions, i.e., survival functions of the form $\overline{F}(\mathbf{x}) = \int_0^\infty e^{-\gamma A(\mathbf{x})} dM(\gamma)$ where M is the distribution of a non-negative rv Γ. If $\mathbf{X} \sim F$ and $X^* = \min\{X_1, \ldots, X_m\}$, then for $j = 1, \ldots, m$,

$$\Pr(X^* > t, X^* = X_j) = \int_0^\infty \Pr(X^* > t, X^* = X_j \mid \Gamma = \gamma)\, dM(\gamma)$$

$$= \frac{A_j(\mathbf{1}_m)}{A(\mathbf{1}_m)} \int_0^\infty e^{-\gamma t A(\mathbf{1}_m)}\, dM(\gamma) = \Pr(X^* = X_j) \Pr(X^* > t).$$

Also the same probability as (6.45) results if $(V_1, \ldots, V_m) \sim F$ with F given by $\int_0^\infty \exp\{-\gamma A(e^{-x_1}, \ldots, e^{-x_m})\} dM(\gamma)$.

6.7 Bibliographic notes

Books on extreme value theory include Galambos (1987) and Resnick (1987). Representations of multivariate extreme value distributions are given in Pickands (1981), Deheuvels (1983) and de Haan (1984). For copulas and multivariate extreme value distributions, see also Deheuvels (1978). The definition of min-stable multivariate exponential is from Pickands (1981).

References for the BEV and MEV families in Section 6.3.1 are Hüsler and Reiss (1989), Tawn (1988; 1990), Joe (1994; 1996a). The family B6 (and its extension to M6) is called a logistic model in Tiago de Oliveira (1980) and Tawn (1988; 1990). The adjective 'logistic' comes from the fact that the difference of dependent Gumbel or extreme value rvs with the copula family B6 has the logistic distribution; this also relates to the choice model results in Section 6.5. The models are not called logistic here because logistic regression comes up later in the book.

References for the families in Section 6.3.2 are Smith (1990), Joe (1990a; 1993) and Coles and Tawn (1991). The models of Joe (1990a) have not been successful in fitting multivariate extreme value data, partly because the parameters α_B of (6.39) and (6.40) are not interpretable as solely dependence parameters.

For the point process approach, see Joe, Smith and Weissman (1992) for the bivariate case and Coles and Tawn (1991) for the multivariate case. The presentation in Section 6.4 is a little bit different from that in Coles and Tawn, where the densities of the point process measure are emphasized more.

References for choice models are McFadden (1974; 1981). See the comment in McFadden (1974, p. 108) on the general difficulty of specifying a joint distribution in order to get closed-form choice probabilities. For a property of coverage of choice probabilities from MEV choice models, see Dagsvik (1995).

The results from Section 6.6 are mainly from Joe and Hu (1996). See Rachev and Resnick (1991) for max-geometric multivariate distributions and Arnold (1996) for max-geometric multivariate logistic distributions.

6.8 Exercises

6.1 Find the parameter of the extreme value distribution in which the t distribution with ν degrees of freedom is in the domain of attraction. The t distribution has density of the form: $f(x) = c_\nu(1 + x^2/\nu)^{-(\nu+1)/2}$. [Hint: approximate the tail of the density and hence the survival distribution, and then compare with Example 6.1.]

6.2 If $e^{-A(\mathbf{z})}$ and $e^{-B(\mathbf{z})}$ are MSMVE survival functions, show that $e^{-\alpha A(\mathbf{z}) - \beta B(\mathbf{z})}$ is also MSMVE for all $\alpha, \beta > 0$.

6.3 Derive the survival function for the Marshall–Olkin multivariate exponential distribution from the stochastic representation in Section 6.3.2.

6.4 If $\mathbf{X} = (X_1, \ldots, X_m)$ is MSMVE, then show that $\mathbf{Z}$ with $Z_j = \min_{1\le i\le m} a_{ij}X_i$, $a_{kj} \ge 0$, $k, j = 1, \ldots, m$, is MSMVE.

6.5 Consider the min-stable trivariate exponential survival function $G = e^{-A}$ with $A(z_1, z_2, z_3) = [(z_1^\theta + z_2^\theta)^{\delta/\theta} + z_3^\delta]^{1/\delta}$. Show that G is not a proper survival function if $\delta > \theta \ge 1$.

6.6 Take the extreme value limit of the bivariate copulas in the families B2 to B6 and B9 to B11 in Section 5.1.

6.7 Prove Theorem 6.11.

6.8 Verify that the extreme value limits for the families MM4, MM5 and MM2 are respectively the families MM6, MM7 and MM8.

6.9 Suppose $(V_1, \ldots, V_m, T)$ has an absolutely continuous MEV distribution,

$$F(x_1, \ldots, x_m, x_{m+1}) = \exp\{-A(e^{-x_1}, \ldots, e^{-x_m}, e^{-x_{m+1}})\},$$

where T is a threshold random variable and the V_j have a similar interpretation to that in Section 6.5. Let S be a non-empty subset of $\{1, \ldots, m\}$. Show that

$$\Pr\Big(\min_{j\in S}\{V_j + \mu_j\} > T > \max_{i\in S^c}\{V_i + \mu_i\}\Big)$$

has closed form based on A and its margins. This is an example of a subset choice probability. [Hint: start with $m = 3, 4$ to see the pattern.]

(A.J.J. Marley, personal communication, 1996)

6.9 Unsolved problems

6.1 Find other approaches to deriving parametric families of MEV copulas with better dependence and closure properties.

6.2 Conjecture: For a multivariate distribution of the form (6.58), the only possible symmetric univariate distributions that can result over all LTs ψ are the scaled logistic distributions. This is equivalent to showing that

$$\psi(s) = 1 - \psi(s^{-1}), \quad \forall s > 0,$$

holds only for the LTs $\psi(s) = (1 + s^{1/\delta})^{-1}$, $\delta \ge 1$.

CHAPTER 7

Multivariate discrete distributions

This chapter is devoted to multivariate discrete distributions for binary, count, ordinal categorical and nominal categorical responses (Sections 7.1 to 7.4, respectively), and extensions to models that include covariates. Included are models that could be considered as multivariate logit models for multivariate binary and ordinal response data. Dependence structures that are covered are exchangeable and general dependence, with both positive and negative dependence. The time series dependence structure is discussed in Chapter 8; it can be obtained from some models as a special case of the general dependence structure. The data analyses in Sections 11.1 and 11.2 make use of the theory in this chapter.

Approaches for multivariate models include:

(a) mixtures over Bernoulli or Poisson parameters for multivariate binary or count data;

(b) latent variable models from copulas with discrete probability distributions for the univariate margins;

(c) conditional independence models and random effects models.

The types of models that can be constructed depend on the response type. For some types of models, it is not possible to separate out the dependence from the univariate parameters, i.e., the range of dependence depends on the univariate margins. Because of the discreteness, this separation may not be as critical.

7.1 Multivariate binary

An exchangeable multivariate binary model may be reasonable for some familial or cluster data, where the same binary variable is measured for each member of a family or cluster. If there is no reason to assume exchangeable dependence, then one should use a

Table 7.1. *Bivariate Bernoulli distribution*

$Y_1\backslash Y_2$	0	1	
0	p_{00}	p_{01}	p_{0+}
1	p_{10}	p_{11}	p_{1+}
	p_{+0}	p_{+1}	1

model that can cover as general a dependence structure as possible. An example is the measurement of a vector of different binary outcomes (at the same time) on each individual in a study. In this section, we start with simple models for multivariate binary response and proceed to more complex models.

7.1.1 *Bivariate Bernoulli and binomial* °

Table 7.1 shows the natural bivariate Bernoulli or binary distribution. In it, $\Pr(Y_1 = 1, Y_2 = 1) = p_{11}$, $\Pr(Y_1 = 1) = p_{1+} = \pi_1$, $\Pr(Y_2 = 1) = p_{+1} = \pi_2$, etc. The bivariate Bernoulli distribution can be parametrized to have two univariate parameters π_1, π_2, and one bivariate parameter p_{11} (or the correlation $\rho = (p_{11} - \pi_1\pi_2)/\sqrt{\pi_1(1-\pi_1)\pi_2(1-\pi_2)}$). From the Fréchet bound inequalities in Section 3.1,

$$\max\{0, \pi_1 + \pi_2 - 1\} \le p_{11} \le \min\{\pi_1, \pi_2\},$$

so that

$$\max\left\{-\sqrt{\frac{\pi_1\pi_2}{\overline{\pi}_1\overline{\pi}_2}}, -\sqrt{\frac{\overline{\pi}_1\overline{\pi}_2}{\pi_1\pi_2}}\right\} \le \rho \le \sqrt{\frac{\pi_{\min}(1-\pi_{\max})}{\pi_{\max}(1-\pi_{\min})}}, \tag{7.1}$$

where $\overline{\pi}_j = 1 - \pi_j$, $j = 1, 2$, $\pi_{\min} = \min\{\pi_1, \pi_2\}$, and $\pi_{\max} = \max\{\pi_1, \pi_2\}$. Because the range depends on the univariate margin parameters, the correlation is not a good dependence measure to use except possibly for the case $\pi_1 = \pi_2$.

For the bivariate binomial, let (Y_{i1}, Y_{i2}), $i = 1, \ldots, n$, be iid with the distribution in Table 7.1. Then $(S_1, S_2) \stackrel{\text{def}}{=} (\sum_{i=1}^n Y_{i1}, \sum_{i=1}^n Y_{i2})$ has a bivariate binomial distribution with pmf

$$\sum_a \binom{n}{a, s_1 - a, s_2 - a, n - s_1 - s_2 + a} p_{11}^a p_{10}^{s_1-a} p_{01}^{s_2-a} p_{00}^{n-s_1-s_2+a},$$

$s_1, s_2 = 0, 1, \ldots, n$.

7.1.2 General multivariate Bernoulli °

Let $(Y_1, \ldots, Y_m)$ be a multivariate binary vector with the distribution $\Pr(Y_j = y_j, j = 1, \ldots, m) = p(\mathbf{y})$, $y_j = 0$ or 1, $j = 1, \ldots, m$. This has $2^m - 1$ parameters and generalizes the bivariate Bernoulli distribution; it is the most general possible. Note that $Y_j \sim$ Bernoulli (π_j), where

$$\pi_j = \sum_{y_1=0}^{1} \cdots \sum_{y_{j-1}=0}^{1} \sum_{y_{j+1}=0}^{1} \cdots \sum_{y_m=0}^{1} p(\mathbf{y}),$$

with $y_j = 1$ for each $\mathbf{y}$ in the sum. The sum $Y_1 + \cdots + Y_m$ is a correlated binomial rv.

This model can be considered as a special case of a multinomial distribution. It has too many parameters for applications, unless one has a sufficiently large sample (the sense of 'large' depends on m). Hence it is useful to obtain parametric subfamilies which can cover different types of dependence pattern.

7.1.3 Exchangeable mixture model °

We start with a fairly general model for exchangeable binary rvs. This comes from a conditional independence or mixture model in which the Bernoulli parameter p is random with some density, and given p, the m binary rvs Y_j are conditionally iid. The model is

$$f(\mathbf{y}) = \Pr(Y_j = y_j, j = 1, \ldots, m) = \int_0^1 p^k (1-p)^{m-k} dG(p), \quad (7.2)$$

where $k = \sum_{j=1}^{m} y_j$, each y_j is 0 or 1, and G is a distribution with support on [0,1].

Properties of (7.2) are the following.

1. Marginal probabilities include $\pi = \Pr(Y_j = 1) = \int_0^1 p\, dG(p)$, $j = 1, \ldots, m$, and $\eta = \Pr(Y_j = 1, Y_{j'} = 1) = \int_0^1 p^2 dG(p)$, $j \neq j'$. Then for $j \neq j'$, $\mathrm{Cov}(Y_j, Y_{j'}) = \eta - \pi^2 = \mathrm{Var}(P) \geq 0$, where P is a rv with cdf G. That is, only positive dependence is possible. The case of zero covariance can occur only if G is degenerate, in which case $Y_1, \ldots, Y_m$ are independent.

2. The pairwise correlation is $(\eta - \pi^2)/(\pi - \pi^2)$. It can vary from 0 to 1. The correlation of 1 (or Fréchet upper bound) is achieved only if G has support on the points 0 and 1, since this is the only case in which $\eta = \pi$.

If G is a Beta(α,β) distribution, with density

$$g(p) = [B(\alpha,\beta)]^{-1} p^{\alpha-1}(1-p)^{\beta-1}, \quad 0 < p < 1,$$

then (7.2) becomes

$$f(\mathbf{y};\alpha,\beta,m) = B(\alpha+k,\beta+m-k)/B(\alpha,\beta), \qquad (7.3)$$

a two-parameter family (reparametrization to one parameter for marginal probability and one for equicorrelation is possible). Furthermore, $\pi = \alpha/(\alpha+\beta)$, $\eta = \alpha(\alpha+1)/[(\alpha+\beta)(\alpha+\beta+1)]$, and the correlation is $\rho = (\alpha+\beta+1)^{-1}$. Hence correlation of 1 or the Fréchet upper bound is achieved if $\alpha,\beta \to 0$ such that $\alpha/(\alpha+\beta) \to \pi$. If g is a Beta(α,β) density, then $\sum_j Y_j$ has a Beta-binomial(α,β) distribution.

In some cases, such as when G is the beta distribution, the functional form, but not the mixture representation, of (7.2) can be extended to include negative dependence.

With the parameters π for the probability of occurrence of a 1 and ρ for correlation, the extension of (7.3) is given in Theorem 7.1. It is shown in Theorem 7.2 that $\rho = (1+\gamma^{-1})^{-1}$, where γ is the parameter in (7.4) below.

Theorem 7.1 *The function*

$$f(\mathbf{y};\pi,\gamma,m) = \frac{\prod_{i=0}^{k-1}[\pi + i\gamma]\prod_{i=0}^{m-k-1}[(1-\pi)+i\gamma]}{\prod_{i=0}^{m-1}[1+i\gamma]}, \qquad (7.4)$$

$$k = 0,\ldots,m, \; y_1 + \cdots y_m = k, \; y_j = 0,1 \;\forall j,$$

with $0 < \pi < 1$, $-\infty < \gamma < \infty$, *is the pmf of* m *exchangeable binary rvs, if* $\pi + (m-1)\gamma \geq 0$ *and* $(1-\pi) + (m-1)\gamma \geq 0$ *(or* $\gamma \geq -(m-1)^{-1}\min\{\pi, 1-\pi\}$*).*

Proof. The conditions on π,γ,m are clearly necessary in order for (7.4) to be non-negative for $k = 0,\ldots,m$. Let f_S be equal to $\binom{m}{k}$ times the right-hand side of (7.4); we will show that f_S is the pmf of a rv S which is a count made from m draws in an urn model. Expression (7.4) then results if, given $S = k$, binary variables $Y_1,\ldots,Y_m$ are defined so that k of the m Y_j are randomly chosen to have unit values and the remainder are given zero values.

Consider an urn with two types of matter, called type 1 and type 2. There are m draws and each draw results in either type 1 or type 2. Then f_S is the probability distribution of the total number of draws of type 1 in the m draws, when the following sequential scheme is used. Start with π units of type 1 and $1-\pi$ units of type 2. For $r = 1,\ldots,m-1$, after the rth draw, γ units of type j are added if the rth draw was type j, $j = 1, 2$. (If $\gamma < 0$, this means that a negative amount is added or something is subtracted.) Let I_r be the number of draws of type 1 in the first r draws. The probability of type 1 is π on the first draw. The probability of type 1 on draw $r+1$ ($r = 1, \ldots, m-1$) conditional on I_r, is $(\pi + I_r\gamma)/(1 + r\gamma)$, and the probability of type 2 is hence $[(1-\pi)+(r-I_r)\gamma]/(1+r\gamma)$. Putting everything together, the probability of exactly k draws of type 1 in m draws is

$$\binom{m}{k}\frac{\pi(\pi+\gamma)\cdots(\pi+[k-1]\gamma)(1-\pi)\cdots(1-\pi+[m-k-1]\gamma)}{1(1+\gamma)\cdots(1+[m-1]\gamma)}, \tag{7.5}$$

and this is the same as f_S. Note that all sequences with exactly k draws of type 1 have the same probability. □

We show four special cases of f_S, which in the above proof is defined to be $\binom{m}{k}$ times the right-hand side of (7.4). These are the beta-binomial, hypergeometric, binomial and Pólya–Eggenberger distributions.

From (7.3), the beta-binomial pmf is $f_S(k) = \binom{m}{k}B(\alpha+k, \beta+m-k)/B(\alpha, \beta)$, $k = 0, \ldots, m$, $\alpha, \beta > 0$. This can be expanded as:

$$\binom{m}{k}\frac{\prod_{i=0}^{k-1}[\pi(\rho^{-1}-1)+i]\prod_{i=0}^{m-k-1}[(1-\pi)(\rho^{-1}-1)+i]}{\prod_{i=0}^{m-1}[\rho^{-1}-1+i]}$$

$$= \binom{m}{k}\frac{\prod_{i=0}^{k-1}[\pi+i\gamma]\prod_{i=0}^{m-k-1}[(1-\pi)+i\gamma]}{\prod_{i=0}^{m-1}[1+i\gamma]},$$

where $\pi = \alpha/\theta$, $\theta = \alpha+\beta = \gamma^{-1}$, $\gamma = (\rho^{-1}-1)^{-1} > 0$ (or $\alpha = \pi(\rho^{-1}-1) = \pi\gamma^{-1}$ and $\beta = (1-\pi)\gamma^{-1}$). In this case, $\rho = (1+\gamma^{-1})^{-1} = (1+\theta)^{-1}$ is the pairwise equicorrelation parameter for the m binary rvs that lead to the beta-binomial distribution.

We continue with the same definitions of θ, α, β. If γ is negative, then ρ is negative and the hypergeometric distribution becomes a special case of f_S if α and β are positive integers. The hypergeometric distribution, with α items of type 1, β items of type 2 and m draws without replacement, has pmf

$$\binom{m}{k}\frac{\alpha!}{(\alpha-k)!}\frac{\beta!}{(\beta-m+k)!}\frac{(\theta-m)!}{\theta!}, \quad k = 0, 1, \ldots, m.$$

This can be rewritten as:

$$\binom{m}{k}\frac{\prod_{i=0}^{k-1}[\pi-i/\theta]\prod_{i=0}^{m-k-1}[1-\pi-i/\theta]}{\prod_{i=0}^{m-1}[1-i/\theta]},$$

and this is the same as (7.5) with $\gamma = -1/\theta$. The Bernoulli variables, $Y_1, \ldots, Y_m$, with Y_j being the indicator variable for whether the jth draw is of type 1, have pairwise equicorrelation parameter $\rho = -(\theta - 1)^{-1} = (1 + \gamma^{-1})^{-1}$.

The binomial distribution is a special case when $\gamma = 0$. This is well known to be at the boundary of the families of beta-binomial and hypergeometric distributions.

The Pólya–Eggenberger distribution is

$$\binom{m}{k} \frac{b(b+s)\cdots(b+[k-1]s)w(w+s)\cdots(w+[m-k-1]s)}{(b+w)(b+w+s)\cdots(b+w+[m-1]s)},$$

$k = 0, \ldots, m$, where b, w are the starting number of black and white balls in the urn, and s is the number of additional balls added, of the same colour as the rth draw, after the rth draw. Letting $b, w, s \to \infty$ such that $b/(b+w) \to \pi$ and $s/(b+w) \to \gamma$, then (7.5) results.

Theorem 7.2 *For (7.4), the univariate marginal distributions are Bernoulli(π), and the equicorrelation parameter is* $(1 + \gamma^{-1})^{-1}$.

Proof. Let $(Y_1, \ldots, Y_m)$ have the pmf in (7.4) and let $S = \sum_j Y_j$ have the pmf, denoted by $f_S(\cdot; \pi, \gamma, m)$, in (7.5). Then $\mathrm{E}(Y_j) = m^{-1}\mathrm{E}(S)$, $j = 1, \ldots, m$, and

$$\mathrm{Cov}(Y_j, Y_{j'}) = [\mathrm{Var}(S) - m\mathrm{Var}(Y_1)]/[m(m-1)], \quad j \neq j',$$

by exchangeability. Using the standard method,

$$\mathrm{E}(S) = m\pi \sum_{k=1}^{m} f_S\left(k-1; \frac{\pi+\gamma}{1+\gamma}, \frac{\gamma}{1+\gamma}, m-1\right) = m\pi,$$

and

$$\mathrm{E}[S(S-1)] = m(m-1)\pi(\pi+\gamma)(1+\gamma)^{-1}$$

$$= m(m-1)\pi(\pi+\gamma)(1+\gamma)^{-1} \sum_{k=2}^{m} f_S\left(k-2; \frac{\pi+2\gamma}{1+2\gamma}, \frac{\gamma}{1+2\gamma}, m-2\right).$$

Therefore, $\mathrm{E}(Y_j) = \pi$ and Y_j is Bernoulli(π) since it is binary. Next, after some elementary algebra, $\mathrm{Var}(S) = m\pi(1-\pi)(1+m\gamma)/(1+\gamma)$ and $\mathrm{Cov}(Y_1, Y_2) = \pi(1-\pi)\gamma/(1+\gamma)$. Finally, the correlation ρ of Y_j, Y_j' for $j \neq j'$ is $\gamma/(1+\gamma) = (1+\gamma^{-1})^{-1}$. □

Theorem 7.3 *The family in (7.4) is closed under the taking of margins.*

Proof. This result is obvious for the cases of $\gamma = 0$ (independent case) and $\gamma > 0$ (from the mixture construction). For the general proof, let (7.5) be denoted by $f_S(\cdot; \pi, \gamma, m)$. Without loss of generality, we show that if $(Y_1, \ldots, Y_m)$ has a pmf of the form (7.4), then so does $(Y_1, \ldots, Y_{m-1})$. The pmf of $(Y_1, \ldots, Y_{m-1})$ at $(y_1, \ldots, y_{m-1})$ when $k = y_1 + \ldots + y_{m-1}$ is

$$\binom{m}{k}^{-1} f_S(k; \pi, \gamma, m) + \binom{m}{k+1}^{-1} f_S(k+1; \pi, \gamma, m)$$

$$= \binom{m-1}{k}^{-1} f_S(k; \pi, \gamma, m-1) \frac{[(1-\pi) + (m-k-1)\gamma] + [\pi + k\gamma]}{1 + (m-1)\gamma}$$

$$= \binom{m-1}{k}^{-1} f_S(k; \pi, \gamma, m-1).$$

□

How does the family (7.4) do in terms of range of negative dependence? Some comparisons are made with other multivariate exchangeable Bernoulli distributions in Section 7.1.10.

7.1.4 Extensions to include covariates

We consider extensions of models in Section 7.1.3 to include covariates. Exchangeable dependence may be reasonable for familial data such as when $(Y_1, \ldots, Y_m)$ are responses from the same family or cluster. If there is a covariate $\mathbf{x}$ common to the family, then model parameters can be functions of $\mathbf{x}$. If covariates $\mathbf{x}_j$ exist at the individual level within a family, then one can consider models that are modifications of models with exchangeable dependence.

For notational convenience in the following models, $\mathbf{x}$s are column vectors and $\boldsymbol{\beta}$s are row vectors.

If covariates are at the cluster level, an extension of (7.3) is:

$$\int_0^1 \prod_{j=1}^m p^{y_j}(1-p)^{1-y_j} \cdot [B(\alpha_1(\mathbf{x}), \alpha_2(\mathbf{x}))]^{-1} p^{\alpha_1(\mathbf{x})-1}(1-p)^{\alpha_2(\mathbf{x})-1} dp$$

$$= B(\alpha_1(\mathbf{x}) + y_+, \alpha_2(\mathbf{x}) + m - y_+)/B(\alpha_1(\mathbf{x}), \alpha_2(\mathbf{x})),$$

for some choices of $\alpha_1(\mathbf{x}), \alpha_2(\mathbf{x})$, where $y_+ = \sum_{i=1}^m y_j$. For example, if $\alpha_1(\mathbf{x}) = \theta_1 e^{\boldsymbol{\beta}_1 \mathbf{x}}$, $\alpha_2(\mathbf{x}) = \theta_2 e^{\boldsymbol{\beta}_2 \mathbf{x}}$, then $\Pr(Y_j = 1|\mathbf{x}) = \pi(\mathbf{x}) = \alpha_1(\mathbf{x})/[\alpha_1(\mathbf{x}) + \alpha_2(\mathbf{x})]$,

$$\log[\pi(\mathbf{x})/(1-\pi(\mathbf{x}))] = \log \alpha_1(\mathbf{x}) - \log \alpha_2(\mathbf{x}) = \log[\theta_1/\theta_2] + (\boldsymbol{\beta}_1 - \boldsymbol{\beta}_2)\mathbf{x}$$

and

$$\rho(\mathbf{x}) = (\alpha_1(\mathbf{x}) + \alpha_2(\mathbf{x}) + 1)^{-1} = (\theta_1 e^{\boldsymbol{\beta}_1 \mathbf{x}} + \theta_2 e^{\boldsymbol{\beta}_2 \mathbf{x}} + 1)^{-1}.$$

If covariates are at the individual level, there could be many modifications or extensions, one of which is given below.

With covariate $\mathbf{x}_j$ for the jth individual, $j = 1, \ldots, m$, an extension of (7.3) is:

$$\int_0^1 \prod_{j=1}^{m} [h(\mathbf{x}_j, p)]^{y_j} [1 - h(\mathbf{x}_j, p)]^{1-y_j} \cdot [B(\alpha_1, \alpha_2)]^{-1} p^{\alpha_1 - 1} (1-p)^{\alpha_2 - 1} dp, \tag{7.6}$$

for some function h with range in [0,1]. An example is $h(\mathbf{x}, p) = p^{\exp\{-\boldsymbol{\beta}\mathbf{x}\}}$. Since only one-dimensional integrations are required, model (7.6) is fairly easy to work with computationally. A larger family of functions is $h(\mathbf{x}, p) = F(F^{-1}(p) + \boldsymbol{\beta}\mathbf{x})$, where F is a univariate cdf. Note that if $\boldsymbol{\beta} = \mathbf{0}$ or $\mathbf{x} = \mathbf{0}$, then (7.3) obtains. If $F(z) = \exp\{-e^{-z}\}$, then $F^{-1}(p) = -\log[-\log p]$ and $h(\mathbf{x}, p) = \exp\{-e^{-\boldsymbol{\beta}\mathbf{x}}(-\log p)\} = p^{\exp\{-\boldsymbol{\beta}\mathbf{x}\}}$. If $F(z) = (1 + e^{-z})^{-1}$ is logistic, then $h(\mathbf{x}, p) = (1 + ce^{-\boldsymbol{\beta}\mathbf{x}})^{-1}$, where $c = (1-p)/p$.

For the case of $h(\mathbf{x}, p) = p^{\exp\{-\boldsymbol{\beta}\mathbf{x}\}}$, expected values and covariances of the Y_j are easily obtained. Calculations are:

(a) $\mathrm{E}(Y_j) = B(\alpha_1 + e^{-\boldsymbol{\beta}\mathbf{x}_j}, \alpha_2)/B(\alpha_1, \alpha_2) = \mathrm{E}(Y_j^2)$;

(b) $\mathrm{E}(Y_j Y_k) = B(\alpha_1 + e^{-\boldsymbol{\beta}\mathbf{x}_j} + e^{-\boldsymbol{\beta}\mathbf{x}_k}, \alpha_2)/B(\alpha_1, \alpha_2)$, $j \neq k$;

(c) $\mathrm{Cov}(Y_j, Y_k) = [B(\alpha_1 + e^{-\boldsymbol{\beta}\mathbf{x}_j} + e^{-\boldsymbol{\beta}\mathbf{x}_k}, \alpha_2) B(\alpha_1, \alpha_2) - B(\alpha_1 + e^{-\boldsymbol{\beta}\mathbf{x}_j}, \alpha_2) B(\alpha_1 + e^{-\boldsymbol{\beta}\mathbf{x}_k}, \alpha_2)]/B^2(\alpha_1, \alpha_2)$.

Note that

$$\mathrm{E}(Y_j) = [\Gamma(\alpha_1 + e^{-\boldsymbol{\beta}\mathbf{x}_j})/\Gamma(\alpha_1 + \alpha_2 + e^{-\boldsymbol{\beta}\mathbf{x}_j})][\Gamma(\alpha_1 + \alpha_2)/\Gamma(\alpha_1)]$$

and $\Gamma(\alpha_1 + t)/\Gamma(\alpha_1 + \alpha_2 + t)$ is decreasing in $t > 0$, so that $\mathrm{E}(Y_j)$ is increasing in $\boldsymbol{\beta}\mathbf{x}_j$. Furthermore, $\mathrm{E}(Y_j) \to 0$ as $\boldsymbol{\beta}\mathbf{x}_j \to -\infty$ and $\mathrm{E}(Y_j) \to 1$ as $\boldsymbol{\beta}\mathbf{x}_j \to \infty$.

7.1.5 Other exchangeable models

There are many other possible exchangeable models for multivariate binary data, including those which are special cases of the general dependence models in Section 7.1.7. Simple exponential family models also exist but they have some undesirable properties. We analyse an exponential family model here and illustrate the typical problem of non-closure that exists in general for non-normal multivariate exponential family models.

Let

$$f(\mathbf{y};\gamma_1,\gamma_2) = [c(\gamma_1,\gamma_2)]^{-1} \exp\Big\{\gamma_1 \sum_{j=1}^{m} y_j + \gamma_2 \sum_{j<j'} y_j y_{j'}\Big\}, \quad (7.7)$$

for $\mathbf{y} \in \{0,1\}^m$, where c is a normalizing constant and $-\infty < \gamma_1, \gamma_2 < \infty$. Let $\mathbf{Y}$ be a random vector with this pmf. The marginal pmfs are not of the same form and the parameter $\pi = \Pr(Y_j = 1)$ depends on both γ_1 and γ_2. For example,

$$\Pr(Y_j = y_j, 1 \le j \le m-1) = \Big[1 + \exp\Big\{\gamma_1 + \gamma_2 \sum_{j=1}^{m-1} y_j\Big\}\Big]$$

$$\cdot [c(\gamma_1,\gamma_2)]^{-1} \exp\Big\{\gamma_1 \sum_{j=1}^{m-1} y_j + \gamma_2 \sum_{j<j'<m} y_j y_{j'}\Big\}.$$

This non-closure property may make this model and other exponential family models harder to use. Because π does not have simple form, the extension to include covariates is not easy. However the parameters γ_1, γ_2 can be given interpretations:

$$\begin{aligned} e^{\gamma_2} &= \frac{\Pr(Y_j = 1, Y_{j'} = 1,\ Y_k = y_k, k \ne j, j')}{\Pr(Y_j = 1, Y_{j'} = 0,\ Y_k = y_k, k \ne j, j')} \\ &\quad \cdot \frac{\Pr(Y_j = 0, Y_{j'} = 0,\ Y_k = y_k, k \ne j, j')}{\Pr(Y_j = 0, Y_{j'} = 1,\ Y_k = y_k, k \ne j, j')}, \end{aligned}$$

for $j \ne j'$, and

$$e^{\gamma_1} = \frac{\Pr(Y_j = 1,\ Y_k = 0, k \ne j)}{\Pr(Y_j = 0,\ Y_k = 0, k \ne j)}.$$

Next we study some dependence properties including the range of dependence. For $i = 0, \ldots, m$, let $a_i = a_i(\gamma_1, \gamma_2) = \exp\{i\gamma_1 + i(i-1)\gamma_2/2\}$ and let $p_i = c^{-1}a_i = f(\mathbf{y};\gamma_1,\gamma_2)$ when $i = \sum_{j=1}^m y_j$. With this notation, $c(\gamma_1,\gamma_2) = \sum_{i=0}^m \binom{m}{i} a_i$. Then $\pi_i = \Pr(Y_j = 1) = \sum_{i=1}^m \binom{m-1}{i-1} p_i$ and $\pi_{12} = \Pr(Y_1 = 1, Y_2 = 1) = \sum_{i=2}^m \binom{m-2}{i-2} p_i$. Let $h(\gamma_1,\gamma_2) = \sum_{i=1}^m \binom{m-1}{i-1} p_i = [\sum_{i=1}^m \binom{m-1}{i-1} a_i(\gamma_1,\gamma_2)]/c(\gamma_1,\gamma_2)$.

We show below that h is strictly increasing in γ_1, γ_2. For fixed γ_2, $h \to 0$ as $\gamma_1 \to -\infty$ (since $p_1, \ldots, p_m \to 0$, $p_0 \to 1$) and $h \to 1$ as $\gamma_1 \to \infty$ (since $p_m \to 1$ and $p_0, \ldots, p_{m-1} \to 0$). Hence for a fixed π and γ_2, there exists a unique $t(\gamma_2, \pi)$ such that $\pi = h(t(\gamma_2,\pi), \gamma_2)$, and t is decreasing in γ_2. The model (7.7) could be reparametrized in terms of π and γ_2. With this new parametrization γ_2 can be considered a dependence parameter since $\pi_{12} = \pi_{12}(\pi, \gamma_2)$ is increasing in γ_2. Furthermore, the Fréchet upper bound obtains in the limit as $\gamma_2 \to \infty$ and the most negative exchangeable multivariate binary distribution obtains in the limit as $\gamma_2 \to -\infty$. The former result is shown in the next paragraph and the latter result is proved in Section 7.1.10 (in order to compare with other exchangeable multivariate binary distributions). Also when $\gamma_2 = 0$, the distribution has independence of the univariate margins.

For $\gamma_2 \to \infty$, let $\gamma_2 = N$ and $\gamma_1 \sim -\frac{1}{2}(m-1)N + \epsilon$ with $N \to \infty$ and ϵ a fixed real; then $i\gamma_1 + i(i-1)\gamma_2/2 = i\epsilon + \frac{1}{2}i(i-m)N$, $a_0 = 1$, $a_i \sim \exp\{i\epsilon + \frac{1}{2}i(i-m)N\}$, $i = 1, \ldots, m$. Hence $a_m \sim e^{m\epsilon}$ and $a_i \to 0$, $i = 1, \ldots, m-1$. Therefore $p_0 \to 1/(1+e^{m\epsilon})$, $p_i \to 0$, $i = 1, \ldots, m-1$ and $p_m \to e^{m\epsilon}/(1+e^{m\epsilon})$. By choosing ϵ appropriately $h(\gamma_1, \gamma_2) \sim p_m$ can have a limiting value of any $\pi \in (0,1)$.

Now we prove that h is increasing in γ_1, γ_2. Since $\partial a_i/\partial\gamma_1 = ia_i$ and $\partial a_i/\partial\gamma_2 = i(i-1)a_i/2$, then

$$\begin{aligned} c^2\frac{\partial h}{\partial\gamma_1} &= \sum_{i=1}^{m}\binom{m-1}{i-1} ia_i \cdot \sum_{j=0}^{m}\binom{m}{j} a_j \\ &- \sum_{i=1}^{m}\binom{m-1}{i-1} a_i \cdot \sum_{j=0}^{m}\binom{m}{j} ja_j \end{aligned} \tag{7.8}$$

and

$$\begin{aligned} 2c^2\frac{\partial h}{\partial\gamma_2} &= \sum_{i=1}^{m}\binom{m-1}{i-1} i(i-1)a_i \cdot \sum_{j=0}^{m}\binom{m}{j} a_j \\ &- \sum_{i=1}^{m}\binom{m-1}{i-1} a_i \cdot \sum_{j=0}^{m}\binom{m}{j} j(j-1)a_j. \end{aligned} \tag{7.9}$$

It is straightforward to show that both (7.8) and (7.9) are positive. Finally, we outline the proof that $\pi_{12}(\pi, \gamma_2)$ is increasing in γ_2; the details of this result are more tedious. Let

$$h^*(\gamma_1, \gamma_2) = \sum_{i=2}^{m}\binom{m-2}{i-2} p_i = \Big[\sum_{i=2}^{m}\binom{m-2}{i-2} a_i(\gamma_1, \gamma_2)\Big]\Big/ c(\gamma_1, \gamma_2),$$

so that $\pi_{12}(\pi, \gamma_2) = h^*(t(\gamma_2, \pi), \gamma_2)$. Then

$$\begin{aligned} \partial\pi_{12}/\partial\gamma_2 &= [\partial h^*/\partial\gamma_1][\partial t/\partial\gamma_2] + [\partial h^*/\partial\gamma_2] \qquad (7.10) \\ &= [\partial h^*/\partial\gamma_1][-\partial h/\partial\gamma_2]/[\partial h/\partial\gamma_1] + [\partial h^*/\partial\gamma_2] \\ &\overset{\text{sgn}}{=} [\partial h/\partial\gamma_1][\partial h^*/\partial\gamma_2] - [\partial h/\partial\gamma_2][\partial h^*/\partial\gamma_1]. \end{aligned}$$

With the help of symbolic manipulation software, this can be shown to be non-negative.

7.1.6 Mixture models °

A mixture model that generalizes (7.2) to a general dependence structure is:

$$f(\mathbf{y}) = \int_{[0,1]^m} \prod_{j=1}^{m} p_j^{y_j}(1-p_j)^{1-y_j}\, G(d\mathbf{p}), \qquad (7.11)$$

where G is a cdf with support in $[0,1]^m$. Compared with (7.2), this generalization has m distinct probability parameters that are mixed instead of one.

To get a parametric family with flexible dependence structure, one choice for G is the multivariate logit-normal family; $\mathbf{P}$ is **multivariate logit-normal** with parameters $\boldsymbol{\mu}$ and $\Sigma = (\sigma_{jk})$, if

$$(\log[P_1/(1-P_1)], \ldots, \log[P_m/(1-P_m)]) \sim N_m(\boldsymbol{\mu}, \Sigma). \qquad (7.12)$$

The univariate logit-normal density with parameters μ, σ^2 is:

$$\phi(\{\log[p/(1-p)] - \mu\}/\sigma) \cdot [\sigma p(1-p)]^{-1}, \quad 0 < p < 1,$$

where ϕ is the standard normal pdf. This univariate family has the approximate shapes of the family of beta densities; the density is unimodal if σ is small and U-shaped if σ is sufficiently large (except the density approaches 0 at the end points of 0 and 1). The log of the density is a constant plus $-\frac{1}{2}\sigma^{-2}\{\log[p/(1-p)] - \mu\}^2 - \log\sigma - \log[p(1-p)]$ and this has derivative equal in sign to $-\{\log[p/(1-p)] - \mu\} + \sigma^2(2p-1)$. Hence the above description of the density follows.

If $\mathbf{P}$ has cdf G and $\mathbf{Y}$ has the pmf in (7.11), then $\mathrm{E}(Y_j) = \mathrm{E}(P_j)$, $\mathrm{Var}(Y_j) = \mathrm{Var}(P_j) + \mathrm{E}[P_j(1-P_j)] = \mathrm{E}(P_j)[1-\mathrm{E}(P_j)]$, $j = 1, \ldots, m$, and $\mathrm{Cov}(Y_j, Y_k) = \mathrm{Cov}(P_j, P_k)$, $j \neq k$. Hence the model is not identifiable, unless some univariate parameters are fixed, e.g., σ_{jj}^2. For further analysis of this model, we assume that $\sigma_{jj} = \sigma^2$ for all j. In this case, μ_j is the parameter for the jth univariate margin and the correlation $\rho_{jk} = \sigma_{jk}/\sigma^2$ is the parameter for the (j,k) bivariate margin for $j \neq k$.

Let $\Sigma = \sigma^2 R$, where R is a correlation matrix, and let $\mathbf{Z} \sim N_m(\mathbf{0}, R)$. Assuming (7.12), a stochastic representation for $\mathbf{P}$ is $P_j = [1 + \exp\{-(\sigma Z_j + \mu_j)\}]^{-1}$, $j = 1, \ldots, m$. The resulting multivariate binary distribution in (7.11) has univariate parameters

$\pi_j = \mathrm{E}(P_j) = \Pr(Y_j = 1)$, and

$$\Pr(Y_j = y_j, j = 1, \ldots, m) = \mathrm{E}\left[\prod_{j=1}^{m} P_j^{y_j}(1 - P_j)^{1-y_j}\right]$$

$$= \mathrm{E}\left\{\prod_{j=1}^{m}[1 + \exp\{(1 - 2y_j)(\sigma Z_j + \mu_j)\}]^{-1}\right\}. \qquad (7.13)$$

Marginal distributions of (7.11) have a similar form. For $1 \leq j \leq m$, π_j is an increasing function of μ_j with σ^2 fixed. As $\sigma^2 \to \infty$ with μ_j fixed, $\pi_j \to \frac{1}{2}$, and as $\sigma^2 \to 0$ with μ_j fixed, $\pi_j \to (1+e^{-\mu_j})^{-1}$. The (j, k) correlation of (7.11) is $\mathrm{Corr}(Y_j, Y_k) = [\Pr(Y_j = y_j, Y_k = y_k) - \pi_j\pi_k]/\sqrt{\pi_j(1 - \pi_j)\pi_k(1 - \pi_k)}$. For fixed μ_i and π_i, $i = 1, \ldots, m$, $\mathrm{Corr}(Y_j, Y_k)$ increases as ρ_{jk} increases. A wider range exists for $\mathrm{Corr}(Y_j, Y_k)$ as σ^2 increases. For example, as $\sigma^2 \to 0$, $\mathrm{Corr}(Y_j, Y_k)$ goes to 0 for all $j \neq k$. Hence for application of model (7.11), it might be best to fix σ^2 at a large enough value in order to allow a wide range of dependence.

The multivariate probit model (see the next subsection) is included as a special limiting case of (7.11). Let $\mu_j = \nu_j\sigma$, $j = 1, \ldots, m$. Then as $\sigma^2 \to \infty$, the limit of (7.13) is

$$\Pr((-1)^{y_j} Z_j < -(-1)^{y_j}\nu_j, j = 1, \ldots, m).$$

A given π_j is achieved in the limit if ν_j is chosen to be $\Phi^{-1}(\pi_j)$ where Φ is the standard normal cdf.

A special case of (7.11), which compares with (7.6) and the model in Exercise 4.1, is

$$\int \prod_{j=1}^{m}\left\{[p_j(\alpha)]^{y_j}[1 - p_j(\alpha)]^{1-y_j}\right\} dM(\alpha), \qquad (7.14)$$

where M is the distribution of a rv A. If $\mathbf{Y}$ has the distribution in (7.14), then $Y_1, \ldots, Y_m$ are conditionally independent given A. If the functions $p_j(\cdot)$ belong to the same parametric family $p(\cdot; \theta)$, then (7.14) can be written as

$$\int \prod_{j=1}^{m}\left\{[p(\alpha; \theta_j)]^{y_j}[1 - p(\alpha; \theta_j)]^{1-y_j}\right\} dM(\alpha), \qquad (7.15)$$

Conaway (1990) has a model of form (7.15), with M being a gamma distribution and $-\log[-\log p(\alpha; \theta)] = \alpha + \theta$.

7.1.7 Latent variable models °

A general approach is to assume that there is a continuous latent vector, or equivalently the **latent variable model** comes from the discretization of a continuous m-variate family $F(\cdot;\boldsymbol{\theta}) \in \mathcal{F}(F_0,\ldots,F_0)$. We use a stochastic representation to present the model. Let $\mathbf{Z} \sim F$, with each $Z_j \sim F_0$. Define a binary random vector $\mathbf{Y}$ with $Y_j = I(Z_j \le \alpha_j)$, $j = 1,\ldots,m$. This corresponds to a discretization of F or $\mathbf{Z}$, and $\mathbf{Z}$ is a latent vector. (Alternatively, one could define $Y_j = I(Z_j > \alpha_j)$, but the former usage corresponds to that in univariate logistic and probit regression.) There are m univariate parameters: α_j for cutoff points or $\pi_j = F_0(\alpha_j)$ for binary probability parameters. The number of dependence parameters is equal to the dimension of $\boldsymbol{\theta}$. If $F(\cdot;\boldsymbol{\theta}_U)$ is the Fréchet upper bound, then the distribution of $\mathbf{Y}$ is the Fréchet upper bound with univariate Bernoulli(π_j) margins. If there is a covariate vector $\mathbf{x}$, then the parameters α_j and $\boldsymbol{\theta}$ can depend on $\mathbf{x}$.

To generalize the probit model for a binary response to a **multivariate probit** model, F_0 is the standard normal cdf and F is a MVSN cdf with correlation matrix $R = \boldsymbol{\theta}$ (with $m(m-1)/2$ parameters). The multivariate binary probabilities are:

$$\begin{aligned}&\Pr(Y_j = y_j, j = 1,\ldots,m)\\ &\quad = \Pr((-1)^{1-y_j} Z_j \le (-1)^{1-y_j}\alpha_j,\ j = 1,\ldots,m),\end{aligned} \tag{7.16}$$

for $\mathbf{y} \in \{0,1\}^m$, where $\mathbf{Z} \sim N_m(\mathbf{0}, R)$.

Usually, in the multivariate probit model, α_j is linear in the covariates and $\boldsymbol{\theta}$ is constant over the covariates. Extensions such that the correlation matrix is a function of the covariates are not easy. An example is given in the next subsection, and other possible extensions are also left as an unsolved problem.

To generalize the logit model or logistic regression to a **multivariate logit** model, F_0 is the (standard) logistic distribution, $F_0(z) = (1+e^{-z})^{-1}$, and a family of multivariate logistic distributions is needed for F. There is no obvious or natural choice, but candidates are $F = C(F_0,\ldots,F_0;\boldsymbol{\theta})$ for families of copulas $C(\cdot;\boldsymbol{\theta})$ with a wide range of dependence. Additional desirable properties to yield a multivariate logit model are: (i) closed form for the copulas (since one reason for the popularity of the logit model in comparison to the probit model is the closed form of the former); and (ii) reflection symmetry.

For the bivariate case, perhaps the families B2 and B3 of copulas are better because of the reflection symmetry property. These families of copulas C have the possibly desirable property of $C(u_1, u_2; \theta) = u_1 + u_2 - 1 + C(1 - u_1, 1 - u_2; \theta)$, which means that the latent variable model does not depend on the orientation of the discretization, i.e., $Y_j = I(Z_j \le \alpha_j)$ and $Y_j = I(Z_j > \alpha_j)$ are equivalent in use. Also they extend to negative dependence, and if $(U, V) \sim C(\cdot; \theta)$, then $(U, 1 - V)$ has a copula within the same family, since $u - C(u, 1 - v; \theta) = C(u, v; \theta^{-1})$ for the family B2 and $u - C(u, 1 - v; \theta) = C(u, v; -\theta)$ for the family B3. There are no known extensions of these families that have similar invariant properties in higher dimensions. Hence possible copulas in the multivariate case are those from Section 4.3 or 5.5 that have a wide range of dependence, or from the construction in Section 4.8.

To get a model with copulas with the exchangeable type of dependence, one can consider Archimedean copulas. With $\psi(s) = -\theta^{-1} \log(1 - [1 - e^{-\theta}]e^{-s})$, $\theta > 0$, the permutation-symmetric extension of the family B3 is

$$C(\mathbf{u}; \theta) = \psi\Big(\sum_{j=1}^{m} \psi^{-1}(u_j)\Big) = -\theta^{-1} \log\Big(1 - \frac{\prod_j (1 - e^{-\theta u_j})}{(1 - e^{-\theta})^{m-1}}\Big). \tag{7.17}$$

This extension does not have the property of **reflection symmetry**, which would hold if $\mathbf{U} \sim C$ and $\mathbf{U} \stackrel{d}{=} 1 - \mathbf{U}$. We show the lack of the property for $m = 3$, and this then follows also for $m > 3$.

We now refer to the copula in (7.17) as C_m to show explicitly the dimension m. We will show that

$$\overline{C}_3(1 - u_1, 1 - u_2, 1 - u_3) = u_1 + u_2 + u_3 - 2 + C_2(1 - u_1, 1 - u_2)$$
$$+C_2(1 - u_1, 1 - u_3) + C_2(1 - u_2, 1 - u_3) - C_3(1 - u_1, 1 - u_2, 1 - u_3) \tag{7.18}$$

is not the same as (7.17). Let $\alpha = 1 - e^{-\theta}$. Expanding (7.17) and (7.18) about $u = 0$ when $u_1 = u_2 = u_3 = u$, leads to:

- $C_3(u, u, u) \sim -\theta^{-1} \log[1 - \theta^3 u^3/\alpha^2] \sim \theta^2 u^3/\alpha^2$;
- $1 - e^{-\theta(1-u)} \sim \alpha - (1 - \alpha)\theta u[1 + \theta u/2 + \theta^2 u^2/6]$;
- $1 - (1 - e^{-\theta(1-u)})^2/\alpha \sim e^{-\theta}[1 + 2\theta u + \theta^2(2\alpha - 1)\alpha^{-1}u^2 + (3\alpha)^{-1}\theta^3(4\alpha - 3)u^3]$;
- $C_2(1 - u, 1 - u) \sim 1 - 2u + \theta\alpha^{-1}u^2 - \theta^2\alpha^{-1}u^3$;
- $1 - (1 - e^{-\theta(1-u)})^3/\alpha^2 \sim e^{-\theta}[1 + 3\theta u + 3\theta^2\alpha^{-1}(-2 + 3\alpha)u^2/2 + \theta^3\alpha^{-2}(2 - 10\alpha + 9\alpha^2)u^3/2]$;
- $C_3(1 - u, 1 - u, 1 - u) \sim 1 - 3u + 3\theta\alpha^{-1}u^2 - \theta^2\alpha^{-2}(4\alpha + 1)u^3$;

- $\overline{C}_3(1-u,1-u,1-u) \sim 3u-2+3[1-2u+\theta\alpha^{-1}u^2-\theta^2\alpha^{-1}u^3]-[1-3u+3\theta\alpha^{-1}u^2-\theta^2\alpha^{-2}(4\alpha+1)u^3] = \theta^2\alpha^{-2}u^3(1+\alpha) = \theta^2\alpha^{-2}u^3(2-e^{-\theta})$.

Since $2-e^{-\theta}>1$ for $\theta>0$, $\overline{C}_3(1-u,1-u,1-u)>C_3(u,u,u)$ for $\theta>0$ and u near 0. The inequality is reversed for the range of negative θ for which C_3 is a negatively dependent copula (see the family M3E in Section 5.4). For the expansions, a result that is used is

$$\log(1+a_1x+a_2x^2+a_3x^3) \sim b_1x+b_2x^2+b_3x^3, \quad x \to 0$$

with $b_1=a_1$, $b_2=a_2-a_1^2/2$, $b_3=a_3-a_1a_2+a_1^3/3$.

The generalization of (7.17) to allow for a more flexible dependence structure is obtained by substituting a family of max-id copulas for the K_{ij} in (4.25) or (4.31) with this ψ. The result is

$$C(\mathbf{u}) = -\theta^{-1}\log\Big[1-(1-e^{-\theta})\prod_{1\le i<j\le m} K_{ij}(\hat{u}_i,\hat{u}_j)\prod_{i=1}^{m}\hat{u}_i^{\nu_i}\Big], \quad (7.19)$$

where $\hat{u}_j=[(1-e^{-\theta u_j})/(1-e^{-\theta})]^{p_j}$, $p_j=(\nu_j+m-1)^{-1}$, $j=1,\ldots,m$. Also other families of LTs, such as LTA, LTB LTC in the Appendix, can be used in (4.25) or (4.31), so that many combinations of families for ψ and K_{ij} exist.

Another choice for the distribution of the latent random vector is the multivariate copula with general dependence structure from the Molenberghs and Lesaffre construction in Section 4.8 with bivariate copulas in the family B2 or B3. Actually the multivariate objects in Section 4.8 have not been proved to be proper multivariate copulas, but they can be used for the parameter range that leads to positive orthant probabilities for the resulting probabilities for the multivariate binary vector. With the choice of the bivariate copula family B2 or B3, and with parameters $\psi_S=1$ for $|S|\ge 3$, the property of reflection symmetry holds.

For all of these models, for the extensions to include covariates, the univariate cutoff parameters can be made linear in the covariate as in logistic regression, and the dependence parameters of the multivariate copulas could be taken as constant or as functions of the covariates (as in the multivariate probit model, the latter may not be easy or obvious to specify).

For applications of the models in this section, see Sections 11.1 and 11.2.

7.1.8 Random effects models

This subsection combines some of the ideas in the preceding two subsections, by introducing **random effects** in which parameters of a simpler model are assumed random. The resulting models could be considered as mixture models as well as random effects models. This type of model is reasonable if subjects each have outcomes that follow the simple model but with different parameters.

Let $\mathbf{Y}$ be an m-variate binary vector. Suppose $Y_j = I(Z_j \le \alpha_j)$ where the α_j are random, the Z_j have distribution F_0, and the Y_j are conditionally independent given the α_j. Hence

$$\Pr(Y_j = 1, \forall j) = \int_{\Re^m} F_0(\alpha_1) \cdots F_0(\alpha_m)\, G(d\alpha_1, \ldots, d\alpha_m),$$

where G is the cdf of $(\alpha_1, \ldots, \alpha_m)$.

For the case of a covariate column vector $\mathbf{x}$, write $Y_j = I(Z_j \le \alpha_j + \boldsymbol{\beta}_j \mathbf{x})$, where the α_j and $\boldsymbol{\beta}_j$ are random and the Y_j are conditionally independent given the α_j and $\boldsymbol{\beta}_j$. (The $\boldsymbol{\beta}_j$ are row vectors.) Then

$$\Pr(Y_1 = 1, \ldots, Y_m = 1) = \int F(\alpha_1 + \boldsymbol{\beta}_1 \mathbf{x}) \cdots F(\alpha_m + \boldsymbol{\beta}_m \mathbf{x}) \cdot G(d\alpha_1, \ldots, d\alpha_m, d\boldsymbol{\beta}_1, \ldots, d\boldsymbol{\beta}_m).$$

In the remainder of this subsection, we specialize to the case of a MVN latent vector. First consider the case of no covariates. If $Z_1, \ldots, Z_m$ are iid $N(0,1)$ and $(\alpha_1, \ldots, \alpha_m) \sim N(\boldsymbol{\mu}, \Omega)$ with $\Omega = (\omega_{ij})$, then $Y_j = I(Z_j' \le \mu_j)$, where $(Z_1', \ldots, Z_m')$ is MVN with zero mean vector and covariance matrix $\Sigma = (\sigma_{ij})$, $\sigma_{jj} = 1 + \omega_{jj}$, $j = 1, \ldots, m$, and $\sigma_{ij} = \omega_{ij}$, $i \ne j$. As the parameters $\boldsymbol{\mu}, \Omega$ vary, all MVN distributions are possible for the latent vector $(Z_1', \ldots, Z_m')$ (by letting μ_j, ω_{ij} be arbitrarily large).

Next consider the case of a scalar covariate and $m \ge 2$. Suppose $\mathbf{Z} \sim N_m(\mathbf{0}, R)$ where $R = (\rho_{ij})$. Let $Y_j = I(Z_j \le \alpha_j + \beta_j x)$, $j = 1, \ldots, m$, with

$$(\boldsymbol{\alpha} \quad \boldsymbol{\beta}) \sim N_{2m}\left((\boldsymbol{\mu} \quad \boldsymbol{\nu}), \begin{pmatrix} \Omega & \Gamma \\ \Gamma^T & \Omega^\circ \end{pmatrix} \right),$$

independently of $\mathbf{Z}$, where $\boldsymbol{\alpha} = (\alpha_1, \ldots, \alpha_m)$, $\boldsymbol{\beta} = (\beta_1, \ldots, \beta_m)$, and $\Omega = (\omega_{ij})$, $\Omega^\circ = (\omega_{ij}^\circ)$, $\Gamma = (\gamma_{ij})$ are $m \times m$ matrices. Then for $j = 1, \ldots, m$, $Y_j = I(Z_j' \le \mu_j + \nu_j x)$, with

$$\mathrm{Var}\,(Z_j') = \sigma_{jj}(x) = 1 + \omega_{jj} + x^2 \omega_{jj}^\circ + 2x\gamma_{jj}.$$

The covariance of Z_i', Z_j' for $i \neq j$ is

$$\mathrm{Cov}\,(Z_i', Z_j') = \sigma_{ij}(x) = \rho_{ij} + \omega_{ij} + x^2\omega_{ij}^{\circ} + x(\gamma_{ij} + \gamma_{ji}).$$

This model is equivalent to the stochastic representation $Y_j = I(Z_j'' \leq (\mu_j + \nu_j x)/\sqrt{\sigma_{jj}(x)})$, where Z_j'' have variances of 1 and the correlation of (Z_i'', Z_j'') is $\sigma_{ij}(x)/\sqrt{\sigma_{ii}(x)\sigma_{jj}(x)}$. The special case of $\Omega = \Omega^{\circ} = \Gamma = 0$ leads to the usual multivariate probit model. The univariate margins are $\Phi((\mu_j + \nu_j x)/\sqrt{\sigma_{jj}(x)})$ so that the cutoff points are non-linear functions of x and do not necessarily have limits of 0,1 as $x \to \pm\infty$. This model has a lot of parameters, but some of the latent covariances can be set to 0 for simplification.

Note that the derivations given above lead to 'natural' forms for correlation matrices that are functions of the covariates, but also to univariate margins that are not probit models. However, one could still consider a multivariate probit model with covariates with a correlation matrix of the form in the preceding paragraph.

7.1.9 *Other general dependence models*

There are many other possible models for multivariate binary response. For example, the exponential family models of Section 7.1.5 can be generalized to accommodate more general dependence, but still have the undesirable property of non-closure under the taking of margins. An exponential family for a multivariate binary response with covariates is derived in Section 9.2.3 from conditional logistic regressions. Another approach comes from a representation of the multivariate Bernoulli distribution; this is given below with the bivariate case first.

Let $p(y_1, y_2) = \Pr(Y_1 = y_1, Y_2 = y_2)$ and let $p_1(y_1) = \Pr(Y_1 = y_1)$, $p_2(y_2) = \Pr(Y_2 = y_2)$, $y_1, y_2 = 0, 1$. If the correlation is fixed as ρ, then for $y_1, y_2 = 0, 1$,

$$p(y_1, y_2; \rho) = p_1(y_1)p_2(y_2)\left\{1 + \rho\left[\frac{y_1 - p_1(1)}{\sqrt{p_1(1)p_1(0)}}\right]\left[\frac{y_2 - p_2(1)}{\sqrt{p_2(1)p_2(0)}}\right]\right\}. \tag{7.20}$$

This parametrization may not be very desirable because from (7.1)

$$\max\{-\sqrt{p_1(1)p_2(1)/[p_1(0)p_2(0)]}, -\sqrt{p_1(0)p_2(0)/[p_1(1)p_2(1)]}\} \leq \rho$$
$$\leq \min\{\sqrt{p_1(1)p_2(0)/[p_1(0)p_2(1)]}, \sqrt{p_1(0)p_2(1)/[p_1(1)p_2(0)]}\}.$$

There are further modelling problems if $p_j(y_j; \mathbf{x})$, $j = 1, 2$, are functions of the covariate vector $\mathbf{x}$ (either ρ depends on $\mathbf{x}$ or it is constant over $\mathbf{x}$ with further constraints).

Next let $p(\mathbf{y}) = \Pr(Y_j = y_j, j = 1, \ldots, m)$, and let $p_j(y_j) = \Pr(Y_j = y_j)$, $\pi_j = p_j(1)$ and $z_j = (y_j - \pi_j)/\sqrt{\pi_j(1-\pi_j)}$, $j = 1, \ldots, m$. A multivariate extension of (7.20) is

$$p(\mathbf{y}; m, \rho_S, S \in \mathcal{S}_m) = \prod_{j=1}^{m} \pi_j^{y_j}(1-\pi_j)^{1-y_j}\Big\{1 + \sum_{S:|S|\geq 2} \rho_S \prod_{k\in S} z_k\Big\}. \tag{7.21}$$

There are constraints on the parameters ρ_S in order that (7.21) is non-negative for all $\mathbf{y}$, but all multivariate Bernoulli distributions have this representation. It is straightforward to show that (7.21) is closed under margins with no change to the parameters $\{\rho_S\}$. For example,

$$p(y_1, \ldots, y_{m-1}, 0; m, \rho_S, S \in \mathcal{S}_m) + p(y_1, \ldots, y_{m-1}, 1; m, \rho_S, S \in \mathcal{S}_m)$$
$$= p(y_1, \ldots, y_{m-1}; m-1, \rho_S, S \in \mathcal{S}_{m-1}).$$

Let $Z_j = (Y_j - \pi_j)/\sqrt{\pi_j(1-\pi_j)}$, $j = 1, \ldots, m$. From the closure property, it follows that for S with $|S| \geq 2$,

$$\mathrm{E}\Big[\prod_{k\in S} Z_k\Big] = \mathrm{E}^*\Big[\prod_{k\in S} Z_k\Big\{1 + \sum_T \rho_T \prod_{r\in T} Z_r\Big\}\Big] = \rho_S,$$

where E^* is an expectation assuming the Y_j are independent Bernoulli rvs with respective parameters π_j (and hence $Z_1, \ldots, Z_m$ are independent rvs with mean 0 and variance 1). In particular, the parameter ρ_S is the correlation of $Y_j, Y_{j'}$ if $S = \{j, j'\}$.

There are too many parameters for (7.21) to be useful as a model for multivariate binary data, unless m is small, such as 2 or 3. Also it may not be a convenient form for extensions to include covariates. However, its closure property makes it better than an exponential family model with high-order moment terms. If (7.21) is truncated after the bivariate or trivariate terms (i.e., $|S| \leq 2$ or $|S| \leq 3$), the result may not cover much range of dependence. For example, for the Fréchet upper bound when the univariate margins all have a mean of π, the representation (7.21) has $\rho_S = [(1-\pi)^{|S|-1} + \pi^{|S|-1}(-1)^{|S|}]/[\pi(1-\pi)]^{|S|/2-1}$ (the details are left as an exercise).

7.1.10 Comparisons

In this subsection, we make some comparisons of the multivariate binary distributions in the preceding subsections. The first is a comparison of the most negatively dependent distribution among exchangeable families of multivariate Bernoulli distributions. For this comparison, we derive the most negatively dependent distribution among all multivariate exchangeable Bernoulli distributions with marginal probability π (of getting a 1). A second comparison is for the trivariate case and considers the range of one bivariate margin given the other two bivariate margins are fixed; this uses the bounds of the Fréchet class $\mathcal{F}(F_{12}, F_{23})$ in Section 3.2. A third comparison for the trivariate case considers the range of the trivariate margin given the three bivariate margins; this uses the bounds of the Fréchet class $\mathcal{F}(F_{12}, F_{13}, F_{23})$ in Section 3.4. For both the second and third comparisons, summaries are given for the trivariate probit model. Other (numerical) comparisons of these types can be done for the models in Section 7.1, although we do not do so because of space considerations.

For the most negatively dependent multivariate exchangeable Bernoulli distribution, first suppose that $m = 3$. We give more details for the trivariate case, as the solution in this case led to the conjecture of the solution in the multivariate case. Let $\overline{\pi} = 1 - \pi$. Consider the three-way table, such that the (1,2), (1,3) and (2,3) bivariate margins are $\begin{bmatrix} \overline{\pi}^2 + \theta & \pi\overline{\pi} - \theta \\ \pi\overline{\pi} - \theta & \pi^2 + \theta \end{bmatrix}$. Then for the trivariate distribution with $f_{ijk} = \Pr(Y_1 = i, Y_2 = j, Y_3 = k)$, the bivariate constraints lead to $f_{000} = x$, $f_{001} = f_{010} = f_{100} = \overline{\pi}^2 + \theta - x$, $f_{011} = f_{101} = f_{110} = \pi\overline{\pi} - \overline{\pi}^2 - 2\theta + x$, $f_{111} = \pi^2 + \overline{\pi}^2 - \pi\overline{\pi} + 3\theta - x$. The non-negativity of each term implies $x \geq 0$, $x - \theta \leq \overline{\pi}^2$, $x - 2\theta \geq \overline{\pi}^2 - \pi\overline{\pi}$ and $x - 3\theta \leq \pi^2 + \overline{\pi}^2 - \pi\overline{\pi}$. An analysis of the inequalities in an (x, θ) graph leads to a minimum of $\theta = (\pi\overline{\pi} - \pi^2 - \overline{\pi}^2)/3$ $(x = 0)$, a correlation of $\frac{1}{3}(1 - \pi/\overline{\pi} - \overline{\pi}/\pi)$, if $1/3 \leq \overline{\pi} \leq 2/3$; to $\theta = -\pi^2$ $(x = 3\overline{\pi} - 2)$, a correlation of $-\pi/\overline{\pi}$ if $\overline{\pi} > 2/3$; and to $\theta = -\overline{\pi}^2$ $(x = 0)$, a correlation of $-\overline{\pi}/\pi$ if $\overline{\pi} < 1/3$.

Hence for $\overline{\pi} \geq 2/3$ or $\pi \leq 1/3$, the non-zero probabilities are $f_{000} = 3\overline{\pi} - 2 = 1 - 3\pi$, $f_{001} = f_{010} = f_{100} = \pi$. This is a Fréchet lower bound distribution, and also if X is the number of 1s among Y_1, Y_2, Y_3, then X has a (generalized) hypergeometric distribution. If $\pi = n^{-1}$ where $n \geq 3$ is an integer, then the hypergeometric distribution is $\binom{1}{k}\binom{n-1}{3-k}/\binom{n}{3}$, $k = 0, 1$, or respectively, $(n-3)/n = 1 - 3\pi$ and $3/n = 3\pi$. If π is not the reciprocal of an integer, then

$$\binom{1}{k}\frac{(n-1)\cdots(n-3+k)}{(3-k)!}\Big/\frac{n(n-1)(n-2)}{3!}, \quad k = 0, 1,$$

leads to the same distribution as in the preceding sentence. Similarly, if $\pi \geq 2/3$, the non-zero probabilities are $f_{111} = 3\pi - 2$, $f_{011} = f_{101} = f_{110} = 1 - \pi$. If $1/3 < \pi < 2/3$, the non-zero probabilities are $f_{001} = f_{010} = f_{100} = 2/3 - \pi$ and $f_{011} = f_{101} = f_{110} = \pi - 1/3$. There is no Fréchet lower bound distribution for π in this range (see Example 3.1). Also, there is no relation to a hypergeometric distribution except when $\pi = \frac{1}{2}$.

For the general multivariate situation, the most negatively dependent m-variate exchangeable binary distribution has pmf

$$f_{i_1 i_2 \cdots i_m} = \begin{cases} (r+1-m\pi)/\binom{m}{r}, & \text{if } \sum i_j = r, \\ (m\pi - r)/\binom{m}{r+1}, & \text{if } \sum i_j = r+1, \\ 0, & \text{otherwise,} \end{cases} \tag{7.22}$$

when $r \leq m\pi < r+1$. Let $\mathbf{Y}$ be such that $f_{i_1 \cdots i_m} = P(Y_1 = i_1, \cdots, Y_m = i_m)$. Then $\mathrm{E}(Y_1 Y_2) = r(2m\pi - r - 1)/[m(m-1)]$ and the pairwise correlation of rvs is

$$\rho = \left[\frac{r}{m(m-1)}(2m\pi - r - 1) - \pi^2\right] \Big/ [\pi(1-\pi)].$$

An outline of the proof is as follows. To obtain f corresponding to a most negatively dependent distribution (in the multivariate concordance ordering), one should minimize $f_{0\cdots0}$ followed by $f_{0\cdots01}$, etc., as well as minimize $f_{1\cdots1}$ followed by $f_{1\cdots10}$, etc. Hence $f_{i_1 i_2 \cdots i_m}$ is non-zero for at most two distinct values of $\sum i_j$; the distinct values are unique for a given π.

Next we show that (7.22) is at the boundary of the exponential family model (7.7) in Section 7.1.5. Using the notation there, (7.22) obtains when $\gamma_2 \to -\infty$. The proof is divided into cases with $\pi \in [r/m, (r+1)/m)$, $r = 0, 1, \ldots, m-1$. Let ϵ be a (fixed) real.

(a) First suppose $\gamma_2 = -N$ and $\gamma_1 \to \epsilon$, with $N \to \infty$. Then $a_i \to 0$ for $i = 2, \ldots, m$, $a_0 = 1$ and $a_1 \to e^\epsilon$. Hence $c \to 1 + me^\epsilon$, $p_0 \to [1 + me^\epsilon]^{-1}$, $p_1 \to e^\epsilon/[1 + me^\epsilon]$ and $h(\gamma_1, \gamma_2) \sim p_1 \to e^\epsilon/[1 + me^\epsilon] = \pi$. As ϵ varies from $-\infty$ to ∞, π can be in the range $(0, m^{-1})$ for this case (of $r = 0$).

(b) For $r > 0$, suppose $\gamma_2 = -N$ and $\gamma_1 \sim rN + \epsilon$, with $N \to \infty$. Then $a_i \sim \exp\{i\epsilon + N[ri - i(i-1)/2]\}$, $i = 0, \ldots, m$. Since $ri - i(i-1)/2$ is maximized at $r(r+1)/2$ for $i = r$ and $r+1$, then $p_i \to 0$ for $i \neq r, r+1$, $p_r \to [\binom{m}{r} + \binom{m}{r+1} e^\epsilon]^{-1}$, $p_{r+1} \to e^\epsilon [\binom{m}{r} + \binom{m}{r+1} e^\epsilon]^{-1}$ and

$$h(\gamma_1, \gamma_2) \sim \binom{m-1}{r-1} p_r + \binom{m-1}{r} p_{r+1} \sim \frac{\binom{m-1}{r-1} + \binom{m-1}{r} e^\epsilon}{\binom{m}{r} + \binom{m}{r+1} e^\epsilon} = \pi. \tag{7.23}$$

Table 7.2. *Correlations of most negatively dependent multivariate exchangeable Bernoulli distributions within some parametric families.*

m	π	(7.22)	(7.4)	(7.16)
3	0.1	−0.111	−0.053	−0.103
3	0.2	−0.250	−0.111	−0.197
3	0.3	−0.429	−0.176	−0.271
3	0.4	−0.389	−0.250	−0.317
3	0.5	−0.333	−0.333	−0.333
4	0.1	−0.111	−0.034	−0.083
4	0.2	−0.250	−0.071	−0.143
4	0.3	−0.270	−0.111	−0.184
4	0.4	−0.250	−0.154	−0.208
4	0.5	−0.333	−0.200	−0.216
5	0.1	−0.111	−0.026	−0.068
5	0.2	−0.250	−0.053	−0.111
5	0.3	−0.190	−0.081	−0.139
5	0.4	−0.250	−0.111	−0.155
5	0.5	−0.200	−0.143	−0.161

The right-hand side of (7.23) is an increasing function of ϵ so that π can be in the range $[r/m, (r+1)/m]$.

Hence the form of the limiting cases with $\gamma_2 \to -\infty$ is the same as (7.22).

We next compare the negative dependence of other models to (7.22). Table 7.2 lists the correlation of the most negatively dependent multivariate exchangeable Bernoulli distribution for certain values of m and π, as well as the correlation of the most negatively dependent distribution in the family (7.4) and in the multivariate probit model (7.16). For the probit model, the correlation in Table 7.2 comes from using $-(m-1)^{-1}$ for the latent equicorrelation parameter. There is symmetry about 0.5, in that the correlations for π are the same as those for $1-\pi$, so Table 7.2 has only π in the range of 0 to 0.5. The table shows that the multivariate probit model attains a greater range of negative dependence than (7.4).

A second comparison is made in the trivariate case for the range of the third bivariate margin given the other two bivariate margins. Some calculations are done with the trivariate probit model. Let

Table 7.3. *Bounds for* $\pi_{13} = p_{13}(1,1)$ *given* $\pi_1, \pi_2, \pi_3, \pi_{12}, \pi_{23}$*: nonparametric versus trivariate probit.*

					Nonpar.		Probit	
π_1	π_2	π_3	π_{12}	π_{23}	π_{13}^L	π_{13}^U	π_{13}^{L*}	π_{13}^{U*}
0.3	0.3	0.3	0.090	0.090	0.000	0.300	0.000	0.300
0.3	0.3	0.3	0.157	0.157	0.014	0.300	0.033	0.300
0.3	0.3	0.3	0.115	0.211	0.026	0.204	0.041	0.200
0.3	0.3	0.3	0.033	0.033	0.000	0.300	0.033	0.300
0.3	0.3	0.3	0.066	0.005	0.000	0.239	0.042	0.200
0.3	0.3	0.3	0.033	0.157	0.000	0.176	0.000	0.157
0.5	0.5	0.5	0.250	0.250	0.000	0.500	0.000	0.500
0.5	0.5	0.5	0.333	0.333	0.166	0.500	0.166	0.500
0.5	0.5	0.5	0.282	0.398	0.180	0.384	0.180	0.384
0.5	0.5	0.5	0.167	0.167	0.166	0.500	0.166	0.500
0.5	0.5	0.5	0.218	0.102	0.384	0.500	0.180	0.384
0.5	0.5	0.5	0.167	0.333	0.000	0.334	0.000	0.334
0.1	0.5	0.7	0.050	0.350	0.000	0.100	0.000	0.100
0.1	0.5	0.7	0.084	0.422	0.006	0.100	0.035	0.100
0.1	0.5	0.7	0.064	0.471	0.035	0.100	0.040	0.100
0.1	0.5	0.7	0.016	0.278	0.006	0.100	0.035	0.100
0.1	0.5	0.7	0.036	0.229	0.035	0.100	0.040	0.100
0.1	0.5	0.7	0.010	0.422	0.000	0.100	0.000	0.094

(Y_1, Y_2, Y_3) be a trivariate binary random vector, and, for $j \neq k$, let $p_{jk}(y_j, y_k) = \Pr(Y_j = y_j, Y_k = y_k)$ and $\pi_{jk} = p_{jk}(1,1)$. Given $\pi_j = \Pr(Y_j = 1)$, $j = 1,2,3$, and π_{12}, π_{23}, we compare the maximum and minimum value of π_{13} for the probit model with the nonparametric bounds. Let $p_{j|2}(y_j|y_2) = \Pr(Y_j = y_j|Y_2 = y_2)$ for $j = 1,3$; these can be written in terms of $\pi_1, \pi_2, \pi_3, \pi_{12}, \pi_{23}$. Using Theorem 3.10, the nonparametric bounds are:

$$\begin{aligned}\pi_{13}^L &\stackrel{\text{def}}{=} \sum_{y=0}^{1} [p_{1|2}(1|y) + p_{3|2}(1|y) - 1]_+ \Pr(Y_2 = y) \le \pi_{13} \\ &\le \sum_{y=0}^{1} \min\{p_{1|2}(1|y), p_{3|2}(1|y)\} \Pr(Y_2 = y) \stackrel{\text{def}}{=} \pi_{13}^U.\end{aligned}$$

Table 7.4. *Bounds for* $p_{123}(1,1,1)$ *given* $\pi_1, \pi_2, \pi_3,\ \pi_{12},\ \pi_{13},\ \pi_{23}$*: non-parametric versus trivariate probit.*

π_1	π_2	π_3	π_{12}	π_{13}	π_{23}	π_{123}^L	π_{123}	π_{123}^U
0.3	0.3	0.3	0.090	0.090	0.090	0.000	0.027	0.090
0.3	0.3	0.3	0.157	0.157	0.157	0.014	0.101	0.157
0.3	0.3	0.3	0.115	0.100	0.211	0.026	0.078	0.100
0.3	0.3	0.3	0.033	0.100	0.033	0.000	0.002	0.033
0.3	0.3	0.3	0.066	0.100	0.005	0.000	0.001	0.005
0.3	0.3	0.3	0.033	0.100	0.157	0.000	0.025	0.033
0.5	0.5	0.5	0.250	0.250	0.250	0.000	0.125	0.250
0.5	0.5	0.5	0.333	0.333	0.333	0.166	0.250	0.333
0.5	0.5	0.5	0.282	0.282	0.398	0.180	0.231	0.282
0.5	0.5	0.5	0.167	0.333	0.167	0.000	0.083	0.167
0.5	0.5	0.5	0.218	0.282	0.102	0.000	0.051	0.102
0.5	0.5	0.5	0.167	0.167	0.333	0.000	0.083	0.167
0.1	0.5	0.7	0.050	0.070	0.350	0.020	0.035	0.050
0.1	0.5	0.7	0.084	0.080	0.422	0.064	0.071	0.080
0.1	0.5	0.7	0.064	0.070	0.471	0.035	0.058	0.064
0.1	0.5	0.7	0.016	0.080	0.278	0.000	0.009	0.016
0.1	0.5	0.7	0.036	0.070	0.229	0.006	0.012	0.035
0.1	0.5	0.7	0.016	0.050	0.422	0.000	0.012	0.016

The bounds π_{13}^{L*} and π_{13}^{U*} within the trivariate probit distributions come from using the bounds for ρ_{13} in the inequality:

$$\rho_{12}\rho_{32} - \sqrt{1-\rho_{12}^2}\sqrt{1-\rho_{32}^2} \leq \rho_{13} \leq \rho_{12}\rho_{32} + \sqrt{1-\rho_{12}^2}\sqrt{1-\rho_{32}^2},$$

where $\alpha_j = \Phi^{-1}(\pi_j)$, $j = 1,2,3$, and ρ_{j2} is the unique root of $\Phi_2(\alpha_j, \alpha_2; \rho_{j2}) = \pi_{j2}$, $j = 1,3$, with Φ_2 being the BVSN cdf. The bounds for π_{13} are given in Table 7.3 for selected values of $\pi_1, \pi_2, \pi_3,\ \pi_{12},\ \pi_{23}$. The values suggest that the trivariate probit model achieves a wide range of (bivariate) dependence among trivariate Bernoulli distributions.

A third comparison is made for the range of the trivariate distribution given the three bivariate margins. We again use the trivariate probit model to illustrate the comparisons and use the same notation as before. Let $p_{123}(y_1, y_2, y_3) = \Pr(Y_1 = y_1, Y_2 = y_2, Y_3 = y_3)$. Given $\pi_j = \Pr(Y_j = 1)$, $j = 1,2,3$, and compatible probabilities $\pi_{12} = p_{12}(1,1)$, $\pi_{13} = p_{13}(1,1)$, and $\pi_{23} = p_{23}(1,1)$, we compare the value of $\pi_{123} = p_{123}(1,1,1)$ for the probit model with the nonparametric bounds. From the proof of Theorem 3.11, sharp bounds on π_{123} are

$$\pi_{123}^L \stackrel{\text{def}}{=} \max\{0, \pi_{12} + \pi_{13} - \pi_1, \pi_{12} + \pi_{23} - \pi_2, \pi_{13} + \pi_{23} - \pi_3\} \leq$$

$$\pi_{123} \leq \min\{\pi_{12}, \pi_{13}, \pi_{23}, 1 - \pi_1 - \pi_2 - \pi_3 + \pi_{12} + \pi_{13} + \pi_{23}\} \stackrel{\text{def}}{=} \pi_{123}^U.$$

However, for the trivariate probit model, there is a unique value of π_{123} given π_1, π_2, π_3, π_{12}, π_{13}, π_{23}, since the given quantities uniquely determine the parameters $\alpha_1, \alpha_2, \alpha_3$, ρ_{12}, ρ_{13} and ρ_{23} of the trivariate probit distribution. Hence the trivariate (and the general multivariate) probit model does not allow a range of third- and higher-order dependence. The bounds for π_{123} are given in Table 7.4 for selected values of π_1, π_2, π_3, π_{12}, π_{13}, π_{23}.

7.2 Multivariate count

Models for univariate count data are the Poisson distributions and larger families that include the Poisson distributions. One such family combines the negative binomial, Poisson and two-parameter binomial distributions into a two-parameter family. For distributions for count data, an important summary is the **index of dispersion** or variance to mean ratio. This is 1 for Poisson distributions; if it is larger (less) than 1, then the distribution is said to be **overdispersed (underdispersed)** relative to Poisson.

Models for multivariate count data include mixture models and copula-based models.

7.2.1 Background for univariate count data °

Let the negative binomial, Poisson and two-parameter binomial be parametrized by the mean μ and ν, where $D = \nu + 1$ is the index of dispersion. Then the **negative binomial** distribution has pmf:

$$f(k;\mu,\nu) = \frac{\Gamma(k+\mu/\nu)}{k!\,\Gamma(\mu/\nu)}\nu^k(1+\nu)^{-k-\mu/\nu}, \quad k = 0,1,\ldots;$$

it has mean $\mu = \theta q/p$, variance $\sigma^2 = \theta q/p^2$ and $\nu = p^{-1} - 1$. The two-parameter binomial distribution has mean $\mu = np$, variance $\sigma^2 = npq$ and $\nu = -p$; its pmf is

$$f(k;\mu,\nu) = \frac{\Gamma(1-\mu/\nu)}{k!\,\Gamma(1-k-\mu/\nu)}(-\nu)^k(1+\nu)^{-k-\mu/\nu},$$

$k = 0, 1, \ldots, -\mu/\nu$. These can be combined together as one family (including the Poisson distribution when $\nu = 0$), if written as

$$f(k; \mu, \nu) = (k!)^{-1}\mu(\mu + \nu) \cdots (\mu + [k-1]\nu)(1+\nu)^{-k-\mu/\nu},$$

$k = 0, 1, \ldots, \min\{\infty, -\mu/\nu\}$.

The negative binomial distribution obtains as a Gamma$(\mu/\nu, \nu)$ mixture of Poisson distributions. That is, let $Y \sim$ Poisson (α) given $A = \alpha$ and $A \sim$ Gamma$(\mu/\nu, \nu)$. Then for $y = 0, 1, \ldots,$

$$\begin{aligned} \Pr(Y = y) &= \int_0^\infty \frac{\alpha^{\mu/\nu - 1}}{\Gamma(\mu/\nu)} \nu^{-\mu/\nu} e^{-\alpha/\nu} \frac{e^{-\alpha}\alpha^y}{y!} d\alpha \\ &= \frac{\Gamma(y + \mu/\nu)}{y!\,\Gamma(\mu/\nu)} \nu^y (1+\nu)^{-y-\mu/\nu}. \end{aligned}$$

Other mixing distributions for overdispersed Poisson models include the lognormal and inverse Gaussian distributions. More generally, consider $f(y) = \int_0^\infty [y!]^{-1}\alpha^y e^{-\alpha} dM(\alpha)$, $y = 0, 1, \ldots,$ where M is a distribution function on $(0, \infty)$. Let $A \sim M$ and let Y have the pmf f. Then $\mathrm{E}(Y) = \mathrm{E}(A)$ and $\mathrm{Var}(Y) = \mathrm{E}[\mathrm{Var}(Y|A)] + \mathrm{Var}[\mathrm{E}(Y|A)] = \mathrm{E}(A) + \mathrm{Var}(A)$. Hence $D = \mathrm{Var}(Y)/\mathrm{E}(Y) \geq 1$, with equality only when A has a degenerate distribution. Therefore general mixtures of Poisson distributions are overdispersed relative to Poisson.

Another model with overdispersion is the generalized Poisson family (see Problem 7.2).

7.2.2 Multivariate Poisson °

A natural bivariate Poisson distribution has the stochastic representation $(Y_1, Y_2) \stackrel{d}{=} (Z_1 + Z_{12}, Z_2 + Z_{12})$, where Z_1, Z_2, Z_{12} are independent Poisson rvs with parameters $\theta_1, \theta_2, \theta_{12}$, respectively. This, together with its m-variate extension to a construction based on $2^m - 1$ independent Poisson rvs, is a special case of the multivariate models given in Section 4.6. A Markov chain time series model based on this multivariate distribution is given in Section 8.4, with a data analysis example in Section 11.5.

Other multivariate Poisson distributions obtain from copulas with univariate Poisson margins. These may not as interpretable, but copula-based models with appropriate families of parametric copulas can cover a wide range of dependence, including negative dependence, whereas the bivariate and multivariate Poisson distributions from Section 4.6 have positive dependence only.

7.2.3 Mixture models and overdispersed Poisson °

In this subsection, we parallel the development in Section 7.1, with mixtures of independent Poisson rvs instead of Bernoulli rvs to obtain multivariate distributions. We start with exchangeable mixtures and then go on to general mixtures.

An exchangeable mixture model is:

$$f(\mathbf{y}) = \int_0^\infty \prod_{j=1}^m \left[e^{-\alpha}\frac{\alpha^{y_j}}{y_j!}\right] dM(\alpha). \tag{7.24}$$

The main drawback with this model is that the univariate margins cannot be separated from the dependence. (For the mixture of Bernoulli distributions, the problem did not occur because a mixture of Bernoulli distributions is still Bernoulli.) If $A \sim M$, then for (7.24), $\mathrm{E}(Y_j) = \mathrm{E}(A)$, $\mathrm{Var}(Y_j) = \mathrm{E}(A) + \mathrm{Var}(A)$, $j = 1, \ldots, m$, and $\mathrm{Cov}(Y_j, Y_{j'}) = \mathrm{Var}(A)$, $j \neq j'$. The equicorrelation parameter is $\rho = \mathrm{Var}(A)/[\mathrm{E}(A) + \mathrm{Var}(A)] = D/(1 + D)$, where $D = \mathrm{Var}(Y_j)/\mathrm{E}(Y_j)$. Hence the correlation is increasing as the index of dispersion D increases.

For example, if M is the Gamma(θ, σ) distribution, (7.24) is a multivariate negative binomial distribution with pmf

$$f(\mathbf{y}) = \frac{\Gamma(\theta + y_+)}{\Gamma(\theta)\prod y_j!}\sigma^{-\theta}(\sigma^{-1} + m)^{-\theta - y_+}, \quad y_j = 0, 1, \ldots,$$

where $y_+ = \sum_{j=1}^m y_j$. The expectations, variances and covariances are: $\mathrm{E}(Y_j) = \theta\sigma$, $\mathrm{Var}(Y_j) = \theta\sigma(\sigma + 1)$, $\mathrm{Cov}(Y_j, Y_{j'}) = \mathrm{Var}(A) = \theta\sigma^2$. Therefore $\rho = \mathrm{Corr}(Y_j, Y_{j'}) = \sigma/(\sigma + 1)$, $j \neq j'$, and both $D = \sigma + 1$ and ρ are increasing as σ increases.

One can also take gamma mixtures of an exchangeable multivariate Poisson distribution in order to get an exchangeable negative binomial multivariate distribution. Take $\mathbf{Y}$ given $A = \alpha$ to have the stochastic representation $\mathbf{Z} + Z_0$, where Z_j are independent Poisson$(\alpha\theta_j)$ rvs, $j = 0, 1, \ldots, m$, $\theta_1 = \ldots = \theta_m = \theta$ and $A \sim$ Gamma$(\mu/\nu, \nu)$. Then the pmf of $\mathbf{Y}$ is:

$$f(\mathbf{y}) = \sum_{i=0}^{y_{\min}} \frac{\nu^{-\mu/\nu}\Gamma(y_+ - [m-1]i + \mu/\nu)}{(m\theta + \theta_0 + 1/\nu)^{y_+ - (m-1)i + \mu/\nu}\Gamma(\mu/\nu)} \frac{\theta_0^i}{i!} \prod_{j=1}^m \frac{\theta^{y_j - i}}{(y_j - i)!}, \tag{7.25}$$

where $y_+ = \sum_{j=1}^m y_j$, $y_{\min} = \min\{y_1, \ldots, y_m\}$.

With (7.25), $\mathrm{E}(Y_j) = (\theta + \theta_0)\mu$, $\mathrm{Var}(Y_j) = (\theta + \theta_0)^2\mu\nu + (\theta + \theta_0)\mu$ and $\mathrm{Cov}(Y_j, Y_{j'}) = (\theta + \theta_0)^2\mu\nu + \theta_0\mu$, $j \neq j'$. The sum $\theta + \theta_0$ can be taken to be 1 without loss of generality. Then $D = \nu + 1$ and

$\rho = \text{Corr}(Y_j, Y_{j'}) = (\nu + \theta_0)/(\nu + 1)$, $j \neq j'$, where $0 \leq \theta_0 \leq 1$. This family is still quite restricted in the range of dependence; the confounding between the index of dispersion and the range of dependence has not been eliminated.

A mixture model, which includes (7.24) and allows the univariate margins to be different, is:

$$\int_0^\infty \prod_{j=1}^m \Big\{\exp\{-\gamma(\alpha, \theta_j)\}[\gamma(\alpha, \theta_j)]^{y_j}/y_j!\Big\} dM(\alpha), \tag{7.26}$$

where $\gamma(\alpha, \theta) \geq 0$ and M is the mixing distribution of a rv A. If $\mathbf{Y}$ has the distribution in (7.26), then $Y_1, \ldots, Y_m$ are conditionally independent given A; compare (7.15) and Exercise 4.1.

To get a model with greater flexibility in dependence structure and indices of dispersion, consider using an m-variate mixing distribution. This leads to the pmf

$$f(\mathbf{y}) = \int_{[0,\infty)^m} \prod_{j=1}^m \left[e^{-\alpha_j} \frac{\alpha_j^{y_j}}{y_j!}\right] G(d\alpha_1, \ldots, d\alpha_m). \tag{7.27}$$

A concordance study is not possible but through conditional expectations one can study correlations. Let $(A_1, \ldots, A_m) \sim G$, and suppose $(Y_1, \ldots, Y_m)$ has pmf f. Then $\text{E}(Y_j) = \text{E}(A_j)$, $\text{Var}(Y_j) = \text{Var}(A_j) + \text{E}(A_j)$, $j = 1, \ldots, m$, and $\text{Cov}(Y_j, Y_k) = \text{Cov}(A_j, A_k)$, $j \neq k$. Therefore the correlation of Y_j, Y_k is

$$\rho_{jk} = \frac{\text{Cov}(A_j, A_k)}{[\text{Var}(A_j) + \text{E}(A_j)]^{1/2}[\text{Var}(A_k) + \text{E}(A_k)]^{1/2}}.$$

Hence negatively correlated A_j imply negatively correlated Y_j but to a lesser extent. The upper limit $\rho_{jk} \to 1$ is reached if $A_j = A_k$ and $\text{Var}(A_j)/\text{E}(A_j) \to \infty$.

The choice of multivariate lognormal mixing distribution for G is a special case, with a wide range of dependence. This leads to the multivariate Poisson-lognormal distribution. Let

$$\begin{aligned} g(\boldsymbol{\theta}; \boldsymbol{\mu}, \Sigma) &= (2\pi)^{-m/2}(\theta_1 \cdots \theta_m)^{-1}|\Sigma|^{-1/2} \\ &\cdot \exp\{-\tfrac{1}{2}(\log\boldsymbol{\theta} - \boldsymbol{\mu})\Sigma^{-1}(\log\boldsymbol{\theta} - \boldsymbol{\mu})^T\}, \ \theta_j > 0, \ 1 \leq j \leq m, \end{aligned}$$

be the m-variate lognormal density, with mean vector $\boldsymbol{\mu}$ and covariance matrix $\Sigma = (\sigma_{ij})$, where $\log\boldsymbol{\theta} = (\log\theta_1, \ldots, \log\theta_m)$. The **multivariate Poisson-lognormal distribution** is

$$\begin{aligned} \Pr(Y_1 = y_1, \ldots, Y_m = y_m) &= f(\mathbf{y}; \boldsymbol{\mu}, \Sigma) \\ &= \textstyle\int_{[0,\infty)^m} \prod_{j=1}^m p(y_j; \alpha_j) \cdot g(\boldsymbol{\alpha}; \boldsymbol{\mu}, \Sigma)\, d\boldsymbol{\alpha}, \end{aligned} \tag{7.28}$$

$y_j = 0, 1, 2, \ldots, j = 1, \ldots, m$, where $p(y; \alpha) = e^{-\alpha}\alpha^y/y!$. There is no simpler form, but moments have closed forms:

$$\mathrm{E}(Y_j) = \exp\{\mu_j + \tfrac{1}{2}\sigma_{jj}\} \stackrel{\text{def}}{=} \beta_j, \quad \mathrm{Var}(Y_j) = \beta_j + \beta_j^2[\exp(\sigma_{jj}) - 1],$$

$$\mathrm{Cov}(Y_j, Y_k) = \beta_j \beta_k[\exp(\sigma_{jk}) - 1], \quad j \neq k.$$

The (j, k) correlation ρ_{jk} is

$$\rho_{jk} = \frac{\sqrt{\beta_j \beta_k}\,(e^{\sigma_{jk}} - 1)}{\sqrt{[1 + \beta_j(e^{\sigma_{jj}} - 1)][1 + \beta_k(e^{\sigma_{kk}} - 1)]}}.$$

The range of correlations depends on the univariate parameters $\mu_j, \sigma_{jj}, j = 1, \ldots, m$. For example, if $\mu_j = \mu_k = \mu$, $\sigma_{jj} = \sigma_{kk} = \sigma^2$, $\sigma_{jk} = \omega\sigma^2$, $\beta_j = \beta_k = \beta$, then

$$\frac{\beta[e^{-\sigma^2} - 1]}{1 + \beta(e^{\sigma^2} - 1)} \leq \rho_{jk} \leq \frac{\beta[e^{\sigma^2} - 1]}{1 + \beta(e^{\sigma^2} - 1)}.$$

A multivariate gamma distribution would be needed to get a multivariate distribution in (7.27) with negative binomial margins, but there is no known multivariate gamma distribution with convenient form for the pdf or cdf that leads to a simple form for (7.27).

For addition of a covariate vector $\mathbf{x}$, the parameters of the mixing distribution G can depend on $\mathbf{x}$. For example, for the multivariate Poisson-lognormal distribution, $\boldsymbol{\mu}$ can depend on $\mathbf{x}$, say through a linear function. As for the multivariate probit model, the dependence of the covariance matrix Σ on $\mathbf{x}$ is harder to specify.

7.2.4 Other models

The mixture models in the preceding subsection have univariate margins and an amount of dependence that depends on the mixing distribution. Although perhaps not as interpretable, copula models with dependence separated from the univariate margins (say, negative binomial or generalized Poisson) could be used as models for multivariate count data. The parametric families of copulas in Chapter 5 could be used.

7.3 Multivariate models for ordinal responses °

Latent variable models from Section 7.1.7 generalize with more cutoff points and so can also be used for ordinal categorical response variables. These models also are physically meaningful because one can usually assume that there is a continuous latent variable associated with an ordinal categorical variable. A data analysis example with latent variable models for a multivariate ordinal response is given in Section 11.2.

The other models for a multivariate binary response vector do not extend to a multivariate ordinal random vector. One could mix multinomial distributions, but the ordinal feature of the categorical variable would not be used; such models are given in the next section on multivariate models for nominal responses.

For a **latent variable** model, we discretize a continuous m-variate family $F(\cdot;\boldsymbol{\theta}) \in \mathcal{F}(F_0,\ldots,F_0)$. We first use a stochastic representation to define the model. Let $\mathbf{Z} \sim F$, with each $Z_j \sim F_0$. Define a random vector $\mathbf{Y}$ of ordinal categorical components with $Y_j = k$ if $\alpha_{j,k-1} < Z_j \le \alpha_{j,k}$, $k = 1,\ldots,r_j$, where r_j is the number of categories of the jth variable, $j = 1,\ldots,m$. (Without loss of generality, assume $\alpha_{j,0} = -\infty$ and $\alpha_{j,r_j} = \infty$ for all j.) From this definition,

$$\begin{aligned}&\Pr(Y_1 = k_1,\ldots,Y_m = k_m)\\&\quad= \textstyle\sum_{i_1=k_1-1}^{k_1}\cdots\sum_{i_m=k_m-1}^{k_m}(-1)^{\Sigma_j(k_j-i_j)}\,F(\alpha_{1i_1},\ldots,\alpha_{mi_m};\boldsymbol{\theta}).\end{aligned}$$

There are $\sum_{j=1}^{m}(r_j-1)$ univariate parameters or cutoff points. Univariate probability parameters are $\pi_{jk} = F_0(\alpha_{j,k}) - F_0(\alpha_{j,k-1})$, $k = 1,\ldots,r_j-1$, $j = 1,\ldots,m$. The number of dependence parameters is the dimension of $\boldsymbol{\theta}$.

If there is a covariate vector $\mathbf{x}$, then the parameters $\alpha_{j,k}$ and $\boldsymbol{\theta}$ can depend on $\mathbf{x}$, with the constraint that $\alpha_{j,k-1}(\mathbf{x}) < \alpha_{j,k}(\mathbf{x})$.

To generalize the probit model to a **multivariate probit** model, F_0 is the standard normal cdf and F is a MVSN cdf with correlation matrix $R = \boldsymbol{\theta}$ (with $m(m-1)/2$ parameters). Usually, in the multivariate probit model, $\alpha_{j,k}$ is linear in the covariates (this is acceptable with different regression coefficients for different cutoff points of the same variable if the range of $\mathbf{x}$ is not too large) and $\boldsymbol{\theta}$ is constant over the covariates. A **multivariate logit** model obtains if F_0 is the logistic cdf.

7.4 Multivariate models for nominal responses

Classes of models that extend those in Section 7.1 are mixtures of multinomial distributions. A multinomial distribution could include something like a bivariate Bernoulli distribution, since the categories, which could be labelled as (0,0), (0,1), (1,0), (1,1), are not ordered. Consider first the exchangeable case in which the nominal rvs $Y_1, \ldots, Y_m$ take values of $1, \ldots, r$ for categories 1 to r (with $r > 2$). Let $I_{jk} = I(y_j = k)$, $j = 1, \ldots, m$, $k = 1, \ldots, r$. Then a multivariate model is

$$f(\mathbf{y}) = \int \prod_{j=1}^{m} \prod_{k=1}^{r} p_k^{I_{jk}}\, G(d\mathbf{p}),$$

where G is a distribution on the simplex $S_r = \{\mathbf{p} : \sum_{i=1}^r p_i = 1, p_i \geq 0\}$.

A possible mixing distribution G is the Dirichlet distribution with density $\{\Gamma(\alpha_1 + \cdots + \alpha_r)/[\Gamma(\alpha_1) \cdots \Gamma(\alpha_r)]\} \prod_{k=1}^r p_k^{\alpha_k - 1}$. This leads to

$$f(\mathbf{y}; \alpha_1, \ldots, \alpha_r, m) = \frac{\prod_{k=1}^r \prod_{\ell=0}^{s_k - 1} (\alpha_k + \ell)}{\prod_{\ell=0}^{m-1} (\alpha_+ + \ell)}, \quad y_j \in \{1, \ldots, r\},$$

where $s_k = \sum_{j=1}^m I_{jk}$ is the number of occurrences of category k among the m responses, $\alpha_+ = \sum_{k=1}^r \alpha_k$ and a null product is defined to be 1. Analogous to (7.4) and with a similar proof, this extends to

$$f(\mathbf{y}; \pi_1, \ldots, \pi_r, \gamma, m) = \frac{\prod_{k=1}^r \prod_{\ell=0}^{s_k - 1} (\pi_k + \ell\gamma)}{(1+\gamma) \cdots (1 + [m-1]\gamma)}, \tag{7.29}$$

where the parameters satisfy $\pi_k \geq 0$ and $\pi_k + (m-1)\gamma \geq 0$, $k = 1, \ldots, r$, and $\sum_{k=1}^r \pi_k = 1$.

With the coefficient $\binom{m}{s_1, \ldots, s_r}$ in front of the right-hand side of (7.29), the Dirichlet-multinomial distribution obtains for $\gamma > 0$, the multinomial distribution obtains for $\gamma = 0$ and some multivariate hypergeometric and multivariate Pólya–Eggenberger distributions obtain for $\gamma < 0$.

A model for nominal rvs $Y_1, \ldots, Y_m$ with the same r categories, but without necessarily the same univariate margins, is

$$f(\mathbf{y}) = \int \prod_{j=1}^{m} \prod_{k=1}^{r} [p_k(\alpha, \theta_j)]^{I_{jk}}\, dM(\alpha), \tag{7.30}$$

where $\sum_{k=1}^r p_k(\alpha, \theta) = 1$ for all α and θ. This model generalizes (7.15) and has conditional independence. If A has distribution M and $\mathbf{Y}$ has the distribution in (7.30), then (7.30) is equivalent to

$$\Pr(Y_1 = y_1, \ldots, Y_m = y_m) = \int \prod_{j=1}^{m} \Pr(Y_j = y_j \mid A = \alpha)\, dM(\alpha),$$

where $\Pr(Y_j = k \mid A = \alpha) = p_k(\alpha, \theta_j)$.

For a general dependence model, with r_j categories for the nominal variable Y_j, one can use a more general mixing distribution. Let $I_{jk} = I(y_j = k)$, $j = 1, \ldots, m$, $k = 1, \ldots, r_j$. Then a multivariate model is

$$f(\mathbf{y}) = \int \prod_{j=1}^{m} \prod_{k=1}^{r_j} p_{jk}^{I_{jk}} \, G(d\mathbf{p}_1, \ldots, d\mathbf{p}_m),$$

where $\mathbf{p}_j = (p_{j1}, \ldots, p_{jr_j})$ and G is a distribution on the product of simplices $S_{r_1} \times \cdots \times S_{r_m}$.

7.5 Bibliographic notes

An early paper on the analysis of multivariate binary data is Cox (1972). Prentice (1986) has the extension of the beta-binomial distribution without mention of the hypergeometric and Pólya–Eggenberger distributions; see Johnson and Kotz (1977) for the latter. The representation in Section 7.1.9 is given in Bahadur (1961). The most negatively dependent exchangeable multivariate Bernoulli distribution in Section 7.1.10 was obtained by T. Hu.

References for the multivariate probit model are Ashford and Sowden (1970) and Lesaffre and Molenberghs (1991) for binary responses, and Anderson and Pemberton (1985) for ordinal responses. Meester and MacKay (1994) make use of the exchangeble extension of the copula family B3 as a latent variable distribution for clustered correlated ordinal response. More on multivariate logistic distributions can be found in Arnold (1991; 1996).

Connolly and Liang (1988) generalize (7.3) to include covariates for the case of cluster or familial data, with the exponential family type model: $A \exp\{\log f(\mathbf{y}; \alpha_1, \alpha_2, m) + \sum_i (\boldsymbol{\beta}_i \mathbf{x}_i) y_i\}$, where A is a normalizing constant and f is given by (7.3).

The gamma mixture of bivariate Poisson distributions is given in Kocherlakota and Kocherlakota (1992). See Aitchison and Ho (1989) for the multivariate Poisson-lognormal distribution. A unification of mixture and random effects models for count and binary data is given in Xu (1996).

Section 7.2 has multivariate negative binomial distributions that arise from the generalization of gamma mixtures of Poisson distributions. Other multivariate negative binomial distributions come from multivariate waiting times that generalize the waiting time to the rth success; see Marshall and Olkin (1985) and Kocherlakota and Kocherlakota (1992). Also there are multivariate negative binomial distributions that come from generalizing the univariate probability generating function; see for example Doss (1979).

For $m = 2$, McCloud and Darroch (1995) have the model (7.30) in Section 7.4 with a special form for $p_k(\alpha, \theta_j)$.

There are other classes of models for multivariate binary and discrete data in the statistics literature. They are not mentioned here because they do not fit within the framework of the approach in this book. One class that may overlap is that in Glonek and McCullagh (1995). In the bivariate case, their model is equivalent to the use of the copula B2 in Section 5.1 with the dependence parameter linear in covariates. Their multivariate extension appears to overlap with that of Molenberghs and Lesaffre (1994), but the approach is different from that in Section 7.1.7. The framework of generalized linear models (McCullagh and Nelder 1989) is not used, as it is tied to exponential families and does not seem to lead to a unified approach for models for multivariate responses.

7.6 Exercises

7.1 Show that any bivariate Bernoulli distribution obtains as a latent variable model with the family B3 of copulas (with extension to negative dependence), but that not all trivariate Bernoulli distributions obtain as latent variable models from the trivariate family M3 of copulas in Section 5.3.

7.2 Consider a non-homogeneous Poisson process N in $\Re^2$ with intensity function λ and mean value function μ. Let R_1, R_2 be regions in $\Re^2$ with non-empty intersection. For $j = 1, 2$, let $Y_j = N(R_j)$ be the count of the process in region R_j. What is the joint distribution of (Y_1, Y_2)?

7.3 Verify the derivation of the first- and second-order moments for the multivariate Poisson-lognormal distribution.

7.4 Study the range of dependence for the multivariate Poisson-lognormal distribution.

7.5 Consider a bivariate Poisson distribution with the BVN copula with correlation ρ and univariate Poisson(μ_j) margins, $j = 1, 2$. Let (Y_1, Y_2) be a random pair with this distribution. Investigate (analytically and computationally) how the correlation of Y_1, Y_2 varies as ρ, μ_1, μ_2 vary. For $\rho = -1$, how does the correlation of Y_1, Y_2 vary with μ_1, μ_2 and what is the minimum correlation?

7.6 Verify the special cases of the family in (7.29).

7.7 Show that (7.21) is closed under margins.

7.8 For the Fréchet upper bound when the univariate margins are Bernoulli(π), show that the representation (7.21) has

$$\rho_S = [(1-\pi)^{|S|-1} + \pi^{|S|-1}(-1)^{|S|}]/[\pi(1-\pi)]^{|S|/2-1},$$

for $S \in \mathcal{S}_m$, $|S| \geq 2$.

7.9 Compare different choices of the bivariate family K_{ij} in (7.19). Are there other choices of LTs and bivariate copulas in (4.25) or (4.31) that would be lead to a suitable multivariate copula for the latent random vector for binary rvs?

7.10 Show that (7.8), (7.9) and (7.10) are non-negative.

7.11 Do a dependence analysis of latent variable multivariate binary models that come from using the Molenberghs–Lesaffre construction with the bivariate family B2 or B3; obtain a summary comparable to that in Table 7.3.

7.12 (Construction of a negatively dependent bivariate Poisson distribution.) Let $f(y_1, y_2; n_1, n_2, p_{00}, p_{01}, p_{10}, p_{11})$ be a bivariate binomial pmf with univariate Binomial(n_j, π_j) margins, $j = 1, 2$, where $\pi_1 = p_{10} + p_{11}$, $\pi_2 = p_{01} + p_{11}$. (Assume that a Binomial$(0,\pi)$ distribution means a degenerate distribution at 0.) Let N_1, N_2 be random variables taking values on the non-negative integers. Suppose that $\Pr(Y_1 = y_1, Y_2 = y_2 \mid N_1 = n_1, N_2 = n_2) = f(y_1, y_2; n_1, n_2, p_{00}, p_{01}, p_{10}, p_{11})$.

(a) Show that if $N_j \sim$ Poisson(θ_j), $j = 1, 2$, with N_1 independent of N_2, then $Y_j \sim$ Poisson$(\theta\pi_j)$, $j = 1, 2$.

(b) Show that if $p_{00}, p_{01}, p_{10}, p_{11}$ are such that the bivariate binomial distribution is negatively dependent for all n_1, n_2, then the unconditional distribution of (Y_1, Y_2) is negatively dependent (say, as measured by the covariance or correlation).

(c) One possible bivariate binomial distribution has the stochastic representation:

$$(S_1, S_2) = \sum_{i=1}^{n_1 \wedge n_2} (X_{i1}, X_{i2}) + I(n_1 > n_2)\Big(\sum_{i=n_2+1}^{n_1} X_{i1}, 0\Big)$$

$$+ I(n_1 < n_2)\Big(0, \sum_{i=n_1+1}^{n_2} X_{i2}\Big),$$

where $(X_{i1}, X_{i2}), i = 1, 2, \ldots$, are iid bivariate Bernoulli random pairs with parameters $p_{00}, p_{01}, p_{10}, p_{11}$ (see Section 7.1.1). Obtain the pmf for this distribution.

(d) Show that if $p_{11} - \pi_1\pi_2 < 0$, then the distribution in (c) has negative correlation.

7.13 Study the dependence structure of the models in (7.14) or (7.15).

7.14 Study the dependence structure of the model in (7.26).

7.7 Unsolved problems

7.1 The **generalized Poisson** distribution (see Consul 1989, for details) has pmf:

$$f(x) = \theta(\theta + \eta x)^{x-1} e^{-\theta-\eta x}/x!, \quad x = 0, 1, \ldots, \theta > 0, \eta \geq 0.$$

This is based on the identity $e^{\theta} = \sum_{i=0}^{\infty} (i!)^{-1}\theta(\theta + i\eta)^{i-1}e^{-i\eta}$. An unknown property is whether this family of distributions is a mixture of Poisson distributions (when $\eta > 0, \theta > 0$).

7.2 Consider a multivariate probit model in which the correlation matrix depends on the covariates. What are choices of functions for the correlation matrix? Or, what are functional forms that guarantee a positive definite correlation matrix in the range of the covariates? [Note that one example is given in Section 7.1.8.]

7.3 Let $Y_j = I(Z_j \leq \alpha_j), j = 1, 2$, with (Z_1, Z_2) being BVSN with correlation ρ. Let $\rho_B(\alpha_1, \alpha_2; \rho)$ be the correlation of Y_1, Y_2. Numerically the behaviour of ρ_B is as follows:

(i) for $\rho > 0$, $\rho_B(\alpha, \alpha, \rho)$ decreases as $|\alpha|$ increases;

(ii) for $\rho < 0$, $\rho_B < 0$ and $|\rho_B(\alpha, \alpha, \rho)|$ decreases as $|\alpha|$ increases;

(iii) for $\rho > 0$, $\rho_B(\alpha_1, \alpha_2, \rho)$ is unimodal in α_2 with α_1 fixed and the mode is between 0 and α_1;

(iv) for $\rho < 0$, $|\rho_B(\alpha_1, \alpha_2, \rho)|$ is unimodal in α_2 with α_1 fixed and the mode is between 0 and $-\alpha_1$.

Establish these properties analytically. Note that these results imply that for $\rho > 0$, $\rho_B(\alpha_1, \alpha_2, \rho)$ is maximized at $\alpha_1 = \alpha_2 = 0$ with a value of $\frac{2}{\pi}\arcsin(\rho)$ (see Exercise 2.14).

CHAPTER 8

Multivariate models with serial dependence

In this chapter, we present some (multivariate) models for time series, longitudinal or repeated measures (over time) data when the response variable can be discrete, continuous or categorical. The multivariate dependence structure is time series dependence or dependence decreasing with lag. Stationary time series models that allow arbitrary univariate margins are first studied and then generalized to the non-stationary case, in which there are time-dependent or time-independent covariates or time trends. It is considerations of having univariate margins in given families that make the models here different from the approach of much of the research in the time series literature.

For time series with normal rvs, standard models are autoregressive (AR) and moving average (MA) models. These can be generalized for the convolution-closed infinitely divisible class (see Section 8.4). In allowing for time series models with arbitrary univariate margins, *autoregressive* is replaced by *Markov* and *moving average* is replaced by *k-dependent* (only rvs that are separated by a lag of k or less are dependent). These are studied in Sections 8.1 and 8.2, respectively. In particular, the case of Markov of order 1 as a replacement for autoregressive of order 1 is a simple starting point, and these types of models can be constructed from families of bivariate copulas. For these Markov models, general results on the decrease in dependence with lags are given in Section 8.5.

In Section 8.3, latent variable models, mainly based on the MVN distribution with correlation matrix of the form of stationary autoregressive moving average (ARMA) time series, are considered.

The models in this chapter are applied in the data analysis examples in Sections 11.4, 11.5 and 11.6.

The following is a summary of the main ideas of the chapter. Let $\{Y_t : t = 1, 2, \ldots\}$ denote a stationary time series. For Markov models, let $\{\epsilon_t\}$ be an iid sequence of rvs such that ϵ_t is independent of $\{Y_{t-1}, Y_{t-2}, \ldots\}$. A classification is:

(a) Markov of order 1: $Y_t = g(Y_{t-1}, \epsilon_t)$ for some real-valued function g;

(b) AR(1): $Y_t = \alpha Y_{t-1} + \epsilon_t$, where α is a scalar;

(c) convolution-closed infinitely divisible univariate margin: $Y_t = A_t(Y_{t-1}) + \epsilon_t$, where A_t are independent realizations of a stochastic operator.

For 1-dependent models, let $\{\epsilon_t\}$ be an iid sequence of rvs. A classification is:

(a) 1-dependent: $Y_t = h(\epsilon_t, \epsilon_{t-1})$ for some real-valued function h;

(b) MA(1): $Y_t = \epsilon_t + \beta\epsilon_{t-1}$, where β is an appropriate scalar;

(c) convolution-closed infinitely divisible univariate margin: $Y_t = \epsilon_t + A_t(\epsilon_{t-1})$; where A_t are independent realizations of a stochastic operator.

Obviously, given $Y_t \sim F$, the possible choices of ϵ_t depend on F.

8.1 Markov chain models

For time series with non-normal response variables, one possible class of models consists of Markov chains, with the simple case being those of order 1. If the order of the Markov chain is not mentioned, then it may be taken to be 1.

8.1.1 Stationary time series based on copulas °

A (stationary) Markov chain of first order with any given univariate margin can be constructed from a bivariate copula. (This is an important application of copulas.) This is a generalization of the normal AR(1) time series to those which admit any possible univariate margin, since the normal AR(1) time series arises as a special case with the bivariate normal copula and a univariate normal margin. A Markov chain of second order, with any given univariate margin, can be constructed from a trivariate copula which has the property that the (1,2) and (2,3) bivariate margins are the same. This generalizes the AR(2) normal time series. Extensions to Markov chains of higher order require multivariate copulas with the obvious constraints on the margins.

The description of the stationary Markov chain time series based on a (twice differentiable) bivariate copula $C(u, v)$ is given next, separately for the absolutely continuous case and the discrete case of non-negative integers. Let the time series be denoted by Y_1,Y_2,... or $\{Y_t : t = 1, 2, \ldots\}$.

(a) Absolutely continuous case. Suppose $Y_t \sim F$, where F is a continuous univariate cdf with density f. Then $F_{12}(x, y) = C(F(x), F(y))$ is a bivariate distribution with univariate margins both equal to F. Let $C_{2|1}(v|u) = (\partial C/\partial u)(u, v)$ denote the conditional distribution of the copula. The transition distribution of $\{Y_t\}$ is

$$H(y_t|y_{t-1}) = \Pr(Y_t \le y_t|Y_{t-1} = y_{t-1}) = C_{2|1}(F(y_t)|F(y_{t-1})).$$

(b) Discrete case. Suppose Y_t takes values on the non-negative integers. Let F and f be the cdf and pmf, respectively. As above, $F_{12}(x, y) = C(F(x), F(y))$ is a bivariate distribution with univariate margins both equal to F. The transition distribution of $\{Y_t\}$ is

$$\begin{aligned} H(y_t|y_{t-1}) &= \Pr(Y_t \le y_t \mid Y_{t-1} = y_{t-1}) \\ &= \left[C(F(y_{t-1}), F(y_t)) - C(F(y_{t-1} - 1), F(y_t))\right]/f(y_{t-1}). \end{aligned}$$

If a bivariate distribution in $F_{12} \in \mathcal{F}(F, F)$ is not conveniently specified through a copula, then the equivalent of (a) or (b) can still be obtained directly from F_{12}. This is done in Section 8.4 for multivariate distributions with univariate margins in a convolution-closed infinitely divisible class.

If a parametric family of copulas, such as one of those in Section 5.1, that interpolates between independence and the Fréchet upper bound is chosen, then one has a parametric family of time series models, which includes an iid sequence at one boundary and a perfectly dependent (or persistent) sequence at the other boundary.

Stationary Markov chains of order $m - 1$ can be constructed from an m-variate copula C that satisfies the following conditions: (i) the bivariate margins C_{ij} are such that $C_{i,i+\ell} = C_{1,1+\ell}$, $\ell = 1, \ldots, m-2$, $i = 2, \ldots, m-\ell$; (ii) the higher-dimensional margins are such that $C_{i_1,\ldots,i_k} = C_{1,i_2-i_1+1,\ldots,i_k-i_1+1}$ for $1 \le i_1 < \cdots < i_k \le m$, $3 \le k \le m-1$; and (iii) C is differentiable in its first $m-1$ arguments.

For the trivariate case, these conditions become $C_{12} = C_{23}$. Candidates for families of trivariate copulas with these conditions are in Sections 4.3 and 4.5.

If $F_{1\cdots m} = C(F, \ldots, F)$ is an m-variate cdf, such that F is absolutely continuous and C is a copula with the above properties, then the transition cdf of the stationary Markov chain is:

$$H(y_t|y_{t-m+1},\ldots,y_{t-1}) = \frac{a(F(y_{t-m+1}),\ldots,F(y_t))}{b(F(y_{t-m+1}),\ldots,F(y_{t-1}))},$$

where

$$a(u_1,\ldots,u_m) = \frac{\partial^{m-1}C}{\partial u_1\cdots\partial u_{m-1}}(\mathbf{u})$$

and

$$b(u_1,\ldots,u_{m-1}) = \frac{\partial^{m-1}C_{1\cdots m-1}}{\partial u_1\cdots\partial u_{m-1}}(u_1,\ldots,u_{m-1}),$$

with $C_{1\cdots m-1}$ being an $(m-1)$-dimensional marginal of C.

8.1.2 Binary time series

We look at simple Markov chains of order 1 for binary time series, and consider several ways to extend these to include covariates. Of particular interest is how logistic regression can be incorporated. The models that can be constructed depend on the nature of the data, such as whether: (i) there is one time series, or time series for many different subjects; (ii) the time series are short, moderate-length or long, and equal or unequal in length, if there are many different subjects; and (iii) covariates are time-varying or time-independent.

We start with the simple stationary case with the marginal probability of 1 being $p = 1 - q$. Consider the Markov chain based on the bivariate distribution

$$P = \begin{bmatrix} p_{00} & p_{01} \\ p_{10} & p_{11} \end{bmatrix} = \begin{bmatrix} q^2+\theta & pq-\theta \\ pq-\theta & p^2+\theta \end{bmatrix},$$

where $-\min\{p^2,q^2\} \le \theta \le pq$ (θ is the covariance). The transition matrix is

$$H = \begin{bmatrix} p_{0|0} & p_{1|0} \\ p_{0|1} & p_{1|1} \end{bmatrix} = \begin{bmatrix} q+\theta/q & p-\theta/q \\ q-\theta/p & p+\theta/p \end{bmatrix}.$$

If $\{Y_t\}$ is a stationary Markov chain with transition matrix H, the joint distribution of (Y_1, Y_j) is $P_j = PH^{j-2} = P_{j-1}H$, $j \ge 3$, with $P_2 = P$. Suppose P_{j-1} has the form $\begin{bmatrix} q^2+\theta_{j-1} & pq-\theta_{j-1} \\ pq-\theta_{j-1} & p^2+\theta_{j-1} \end{bmatrix}$; then

$$P_{j-1}H = \begin{bmatrix} q^2+\theta_{j-1}\theta/(pq) & pq-\theta_{j-1}\theta/(pq) \\ pq-\theta_{j-1}\theta/(pq) & p^2+\theta_{j-1}\theta/(pq) \end{bmatrix},$$

so that P_j has the same form as P_{j-1} with $\theta_j = \theta_{j-1}\theta/(pq)$. Since $\theta_2 = \theta$, then $\theta_j = \theta^{j-1}/(pq)^{j-2}$, $j \ge 3$. Note that if $0 < \theta < pq$ then the dependence on P_j is decreasing as j increases. More specifically, $\theta_j/(pq)$ is the correlation and its absolute value is decreasing geometrically. In the form of a fixed lag 1 correlation ρ, the transition matrix is $H = \begin{bmatrix} 1-p(1-\rho) & p(1-\rho) \\ (1-\rho)q & p+\rho q \end{bmatrix}$.

To get a parametric family of models, one can take P to come from a family of discretized bivariate copulas, i.e., $p_{00} = C(\alpha, \alpha; \delta)$ or $p_{11} = C(\alpha, \alpha; \delta)$ for some α for a family $C(\cdot; \delta)$. For this model, the correlation ρ of two binary variables will not be constant over α (or marginal probability p) unless C is the family B11 in Section 5.1 with $C(u, v; \rho) = \rho \min\{u, v\} + (1-\rho)uv$.

If the Markov chain depends on a time-independent covariate vector $\mathbf{x}$, then one can have $p = p(\mathbf{x})$ and $\rho = \rho(\mathbf{x})$. If ρ is constant over $\mathbf{x}$, then $p_{1|1} = (1-\rho)p(\mathbf{x}) + \rho$ is increasing as $p(\mathbf{x})$ increases. For example, with a logistic regression margin, $p(\mathbf{x}) = e^{\alpha+\boldsymbol{\beta}\mathbf{x}}/(1 + e^{\alpha+\boldsymbol{\beta}\mathbf{x}})$, where $\mathbf{x}$ is a column vector and $\boldsymbol{\beta}$ is a row vector,

$$\Pr(Y_t = 1 \mid Y_{t-1} = 0, \mathbf{x}) = (1-\rho)e^{\alpha+\boldsymbol{\beta}\mathbf{x}}/(1 + e^{\alpha+\boldsymbol{\beta}\mathbf{x}})$$

and

$$\Pr(Y_t = 1 \mid Y_{t-1} = 1, \mathbf{x}) = (\rho + e^{\alpha+\boldsymbol{\beta}\mathbf{x}})/(1 + e^{\alpha+\boldsymbol{\beta}\mathbf{x}}).$$

The constraint on ρ is that $\rho \geq -e^{\alpha+\boldsymbol{\beta}\mathbf{x}}$ for all $\mathbf{x}$. Note that the conditional probabilities are not logits.

For the situation in which there are time-dependent covariates or there is non-stationarity, we can look at transition matrices that take $(1, e^{\boldsymbol{\beta}\mathbf{x}_1})/(1 + e^{\boldsymbol{\beta}\mathbf{x}_1})$ to $(1, e^{\boldsymbol{\beta}\mathbf{x}_2})/(1 + e^{\boldsymbol{\beta}\mathbf{x}_2})$, or (q_1, p_1) to (q_2, p_2). With the correlation ρ fixed, the transition matrix has the form $\begin{bmatrix} 1-a & a \\ 1-b & b \end{bmatrix}$, with $a = p_2 - \rho\sqrt{p_1 p_2 q_2/q_1}$ and $b = p_2 + \rho\sqrt{q_1 q_2 p_2/p_1}$. Therefore if p_1, p_2 depend on time-varying covariates, the transition probabilities depend on the covariates at the current and next time points. This may be reasonable, but the assumption of the correlation being fixed over all possible marginal probabilities is unlikely to be so.

Next consider the situation of specified conditional probabilities with covariates that are not time-varying. If one requires the conditional probabilities to be logits, then what is the marginal stationary probability $p(\mathbf{x}) = \Pr(Y_t = 1|\mathbf{x})$? Let $p_{j|i}(\mathbf{x}) = \Pr(Y_t = j \mid Y_{t-1} = i, \mathbf{x})$, $i, j = 0, 1$. Suppose $\log[p_{1|y}(\mathbf{x})/p_{0|y}(\mathbf{x})] = \alpha + \boldsymbol{\beta}\mathbf{x} + \gamma y$. (This form is chosen so that $p_{1|1}(\mathbf{x}) - p_{1|0}(\mathbf{x})$ has the same sign for all $\mathbf{x}$.) The transition matrix $(p_{i|j}(\mathbf{x}))$ is

$$\begin{bmatrix} 1/(1+e^{\alpha+\boldsymbol{\beta}\mathbf{x}}) & e^{\alpha+\boldsymbol{\beta}\mathbf{x}}/(1+e^{\alpha+\boldsymbol{\beta}\mathbf{x}}) \\ 1/(1+e^{\alpha+\gamma+\boldsymbol{\beta}\mathbf{x}}) & e^{\alpha+\gamma+\boldsymbol{\beta}\mathbf{x}}/(1+e^{\alpha+\gamma+\boldsymbol{\beta}\mathbf{x}}) \end{bmatrix}.$$

The condition $\gamma > 0$ corresponds to positive serial dependence and $\gamma < 0$ corresponds to negative serial dependence of lag 1. The stationary probability $p(\mathbf{x})$ satisfies

$$(1-p(\mathbf{x}))\frac{e^{\alpha+\boldsymbol{\beta}\mathbf{x}}}{1+e^{\alpha+\boldsymbol{\beta}\mathbf{x}}} + p(\mathbf{x})\frac{e^{\alpha+\gamma+\boldsymbol{\beta}\mathbf{x}}}{1+e^{\alpha+\gamma+\boldsymbol{\beta}\mathbf{x}}} = p(\mathbf{x}),$$

so that

$$\begin{aligned} p(\mathbf{x}) &= p_{1|0}(\mathbf{x})/[p_{1|0}(\mathbf{x}) + p_{0|1}(\mathbf{x})] \\ &= e^{\alpha+\boldsymbol{\beta}\mathbf{x}}(1+e^{\alpha+\gamma+\boldsymbol{\beta}\mathbf{x}})/(1+2e^{\alpha+\boldsymbol{\beta}\mathbf{x}}+e^{2\alpha+\gamma+2\boldsymbol{\beta}\mathbf{x}}). \end{aligned}$$

Hence the marginal probability is not logit or close to logit. Let $b = e^{\alpha+\boldsymbol{\beta}\mathbf{x}}$ and $c = e^{\gamma}$. Then $p(\mathbf{x})$ increases as $\boldsymbol{\beta}\mathbf{x}$ increases since $p_{1|0}(\mathbf{x})/p_{0|1}(\mathbf{x}) = b(1+bc)/(1+b)$ is the product of two terms that are increasing in b.

Note that in general, any two of the functions $p(\mathbf{x})$, $p_{1|0}(\mathbf{x})$, $p_{1|1}(\mathbf{x})$ determine the third for a stationary binary Markov chain with time-independent covariate (assuming compatibility). Alternatively, $p(\mathbf{x})$ and the correlation (or another dependence measure) $\rho(\mathbf{x})$ determine $p_{1|0}(\mathbf{x})$, $p_{1|1}(\mathbf{x})$.

We next consider what might be done if there are observed binary time series on many subjects. The type of modelling that is possible might depend on the lengths of the series (see the beginning of Chapter 10). If one has a long stationary binary time series for each subject, one could estimate the parameters, such as p and ρ, by subject, and then regress these on the covariate vector $\mathbf{x}$ before combining everything. Another possibility is a Markov chain random effects model, with parameters that vary with subjects. If there are many subjects and short series, it might be harder to use a random effects model, and one possibility for an initial data analysis is to aggregate subjects by clustering of covariates and estimate parameters within each group (cluster).

8.1.3 Categorical response

If the response is a categorical variable, either nominal or ordinal, then the ideas in the preceding subsections can be extended to a Markov chain with more than two states. That is, one can start with a stationary Markov chain and then make modifications to include covariates. If there are r categories for the categorical response, then the transition matrix has potentially $r(r-1)$ parameters if there are no covariates, and more parameters if there are covariates. If the response is ordinal, it may be possible to reduce the number of dependence parameters by using a copula-based Markov chain model, as given in Section 8.1.1.

In the statistical literature, Markov models have been commonly used, but not usually with considerations of both the univariate marginal distributions and the transitional probabilities. The specific details of the modelling depend on the nature of the data and the types of inferences of interest. Factors include: (i) whether the data are aggregated at each time point or consist of individual time series; and (ii) whether a stationarity assumption can be made or whether there is non-stationarity such as a progression through a sequence of states.

8.1.4 Extreme value behaviour

This subsection is concerned with extreme value dependence behaviour for stationary Markov chains with state space $\Re$. This behaviour is relevant for extreme value inference from time series.

If the bivariate copula has (upper) tail dependence (see Section 2.1.10), then the time series $\{Y_t\}$ with continuous univariate margin F has clustering of observations above high thresholds (extreme value dependence). The property of upper tail dependence implies that the extremal index β of the time series is less than 1 (in fact, less than $1-\lambda$, where λ is the upper tail dependence parameter). The extremal index can be interpreted roughly as the reciprocal of the mean length of clusters of consecutive values above high thresholds. More rigorously, for a stationary dependent sequence, the **extremal index** is defined as

$$\beta \stackrel{\text{def}}{=} \lim_{n\to\infty} -\log F_n(y_n)/[-n\log F(y_n)]$$

with $y_n \uparrow \infty$ at an appropriate rate, where F is the univariate marginal cdf and F_n is the distribution of $\max\{Y_t,\ldots,Y_{t+n-1}\}$ With this definition, $F_n(y) \approx F^{n\beta}(y)$ for large y and n. For ARMA normal time series, the extremal index is $\beta = 1$. See the two simulated time series in Figure 8.1 for a comparison of an AR(1) normal time series and a time series with a normal univariate margin based on the copula $C'(u,v;\delta) = u+v-1+C(1-u,1-v;\delta)$, where C is in the family B4 (C' has upper tail dependence) with parameter chosen so that the Kendall tau value is approximately that of the BVN distribution with a correlation of 0.75. The property of tail dependence shows itself clearly. Markov chain models based on bivariate copulas with tail dependence can be used to model a variety of extreme value (clustering) behaviour.

For the Markov chains based on bivariate copulas, we will use the function

$$\beta_n(y) = -\log F_n(y)/[-n \log F(y)]$$

and the limit $\beta = \lim_{n\to\infty} \lim_{y\to\infty} \beta_n(y)$ as the measures of (upper) extreme value dependence or clustering. Properties of $\beta_n(y)$, including inequalities, bounds, limits and monotonicities, which relate to (serial) dependence in the stationary sequence, are studied as they help in determining what patterns are possible for $\beta_n(y)$.

For a stationary dependent sequence $\{Y_t\}$ with univariate margin F, let $\alpha_1(y) = F(y)$ and let

$$\alpha_i(y) = \Pr(Y_i \le y \mid Y_1 \le y, \ldots, Y_{i-1} \le y) \tag{8.1}$$

for $i = 2, 3, \ldots$. Then $F_n(y) = \prod_{i=1}^n \alpha_i(y)$ and

$$\beta_n(y) = n^{-1} \sum_{i=1}^{n} [-\log \alpha_i(y)]/[-\log \alpha_1(y)].$$

It should be intuitive that $\alpha_i(y)$ converges as $i \to \infty$, especially if the sequence $\{Y_t\}$ does not have long-range dependence. If $\alpha_i(y)$ converges to $\alpha_\infty(y)$, then

$$\beta_n(y) \to [-\log \alpha_\infty(y)]/[-\log \alpha_1(y)] \stackrel{\text{def}}{=} \beta_\infty(y), \quad n \to \infty.$$

A dependence condition implying the monotonicity (and hence convergence) of $\alpha_i(y)$ in i is given below.

A rough connection between the extremal index and reciprocal cluster size of large exceedances comes from the following results. We begin with the inequality:

$$\begin{aligned} \beta_r(y) &> [1 - F_r(y)]/[r(1 - F(y))] \qquad (8.2) \\ &= \frac{(1 - F(y)) + (F(y) - F_2(y)) + \ldots + (F_{r-1}(y) - F_r(y))}{r(1 - F(y))} \\ &> (F_{r-1}(y) - F_r(y))/(1 - F(y)). \end{aligned}$$

The first inequality comes from $[-\log a]/[-\log b] > (1-a)/(1-b)$ if $0 < a < b < 1$; the difference of the two quantities in this inequality gets smaller as a, b increase towards 1. The second inequality comes from the sequence $F_{i-1}(y) - F_i(y) = \Pr(Y_1 > y, Y_2 \le y, \ldots, Y_i \le y)$ decreasing in i. For fixed r and y, with r large, Leadbetter (1983)

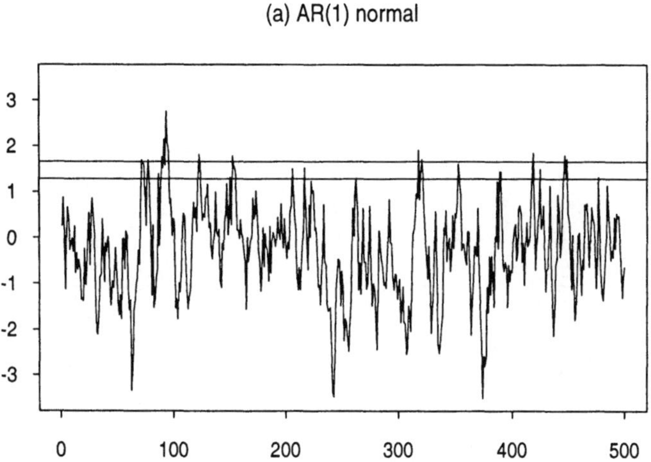

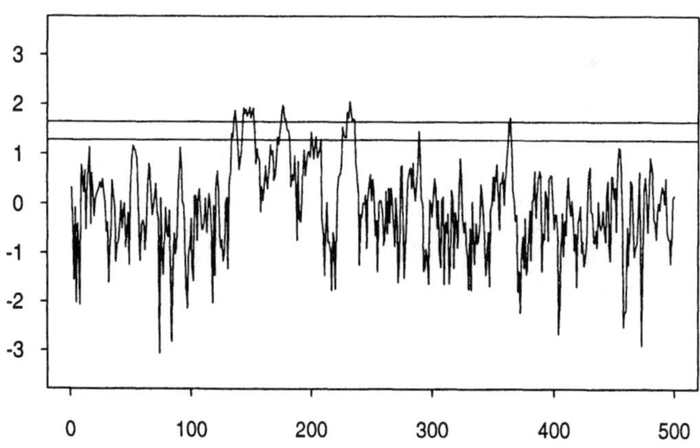

Figure 8.1. *Comparison of extreme value properties for Markov chains based on different bivariate copulas; parameters are (a)* $\rho = 0.75$ *and (b)* $\delta = 2.34$.

interprets the reciprocal of the second term in (8.2) as the expected number of exceedances above y in a block of r consecutive Y_i given that at least one of the Y_i exceeds y. That is,

$$\frac{r(1-F(y))}{1-F_r(y)} = \frac{\mathrm{E}\sum_{i=j}^{j+r-1} I(Y_i > y)}{\Pr(A)} = \sum_{i=j}^{j+r-1} \Pr(Y_i > y \mid A)$$

$$= \frac{\mathrm{E}\sum_{i=j}^{j+r-1} I(\{Y_i > y\} \cap A)}{\Pr(A)},$$

where A is the event $\{\max\{Y_j, \ldots, Y_{j+r-1}\} > y\}$. The exceedances in a block can be considered as a cluster.

Properties of $\beta_n(y)$:

(a) (Bound based on a positive or negative dependence condition.) If the positive dependence condition $F_n \geq F^n$ holds for all n, then $\beta_n(y) \leq 1$. Similarly, $\beta_n(y) \geq 1$ if $\{Y_t\}$ exhibits enough negative dependence such that $F_n \leq F^n$.

(b) (Bounds on $\beta_n(y)$: $\beta_n(y)$ can be larger than 1 but not β.) From the Fréchet upper and lower bounds for a multivariate distribution with given univariate margins, $\max\{0, nF(y) - (n-1)\} \leq F_n(y) \leq F(y)$. The upper bound results when $Y_i = Y_1$ for all i. The lower bound is probably quite crude when considered as a bound for multivariate distributions that come from stationary dependence sequences. From these bounds, $\beta_n(y) \geq n^{-1}$, $\beta_n(y) \leq -\log(np-n+1)/[-n\log p]$ if y is such that $F(y) = p > 1-n^{-1}$ (and there is no upper bound if $F(y) \leq 1-n^{-1}$). Since $\lim_{p\to 1} -\log(np-n+1)/[-n\log p] = 1$, then $\limsup_{y\to\infty} \beta_n(y) \leq 1$.

(c) (Monotonicity of $\alpha_i(y)$ in i.) If $F_{n+1}(y) \geq F_n(y)F(y)$ for all n, then $\alpha_n(y) \geq \alpha_1(y)$ for all n (and hence $\beta_n(y) \leq 1$ for all n). The preceding statement is also valid with all inequalities reversed. Under some stronger positive dependence assumptions, $\alpha_i(y)$ is increasing in i (which implies that $\beta_n(y)$ is decreasing in n). Similarly, there are negative dependence assumptions for which $\alpha_i(y)$ is decreasing in i. Sufficient conditions for monotonicity of $\alpha_i(y)$ are given in Glaz and Johnson (1984). By Theorem 2.3 of Glaz and Johnson, $\alpha_i(y)$ is increasing in i for all y, if for all n the density f_n of $(Y_1, \ldots, Y_n)$ is MTP_2.

Next are two examples where $\alpha_i(y)$ is monotone in i.

Example 8.1 If $(Y_1, \ldots, Y_n)$ is MVN with covariance matrix Σ_n and inverse covariance matrix $A_n = \Sigma_n^{-1} = (a_{ij})$, then a necessary and sufficient condition for the density of $(Y_1, \ldots, Y_n)$ to be MTP_2 is that all off-diagonal elements of A_n are non-positive. For $\{Y_t\}$ being a stationary AR(1) normal sequence with lag 1 autocorrelation $\rho > 0$, A_n has the form $a_{11} = a_{nn} = (1-\rho^2)^{-1}$, $a_{ii} = (1+\rho^2)/(1-\rho^2)$ for $2 \le i \le n-1$, $a_{i,i+1} = a_{i+1,i} = -\rho/(1-\rho^2)$, $i = 1, \ldots, n-1$, and $a_{ij} = 0$ for $|i-j| \ge 2$. Hence the MTP_2 condition holds and $\alpha_i(y)$ is increasing in i for all y. □

Example 8.2 Let $\{Y_t\}$ be a Markov chain of order 1 with continuous marginal distribution F. Let $h(y_{t-1}, y_t) = p(y_t|y_{t-1})$ be the transition pdf. From Proposition 3.10 in Karlin and Rinott (1980a), the density of $Y_1, \ldots, Y_n$ is MTP_2 for all n if h is TP_2. Let C be the bivariate copula associated with (Y_{t-1}, Y_t). Then h is TP_2 if the density c of C is TP_2 since $h(y_{t-1}, y_t) = c(F(y_{t-1}), F(y_t))f(y_t)$. The property of TP_2 density holds for a number of the families of bivariate copulas in Section 5.1.

8.2 k-dependent time series models

MA(k) normal time series models are examples of k-dependent sequences. In this section, we look at k-dependent sequences, based on copulas, that allow for arbitrary univariate margins. These models are probably less useful for applications compared with Markov models; however, they are included for theoretical completeness.

8.2.1 1-dependent series associated with copulas

Let $C(u, v)$ be a bivariate copula with conditional distribution $C_{2|1}(v|u) = \partial C(u, v)/\partial u$. The inverse conditional distribution is denoted by $C_{2|1}^{-1}(s|u)$. Let F be a (continuous) univariate cdf and let $\epsilon_0, \epsilon_1, \ldots$ be a sequence of iid $U(0,1)$ rvs. (The development could be extended to discrete distributions F.) A 1-dependent sequence with stationary distribution F is:

$$Y_t = h(\epsilon_{t-1}, \epsilon_t), \quad t \ge 1, \tag{8.3}$$

where

$$h(u, v) = F^{-1}[C_{2|1}^{-1}(v|u)].$$

The marginal distribution is:

$$\begin{aligned}\Pr\big(F^{-1}[C_{2|1}^{-1}(\epsilon_t|\epsilon_{t-1})] \le y\big) &= \Pr\big(\epsilon_t \le C_{2|1}(F(y)|\epsilon_{t-1})\big) \\ = \textstyle\int_0^1 \Pr\big(\epsilon_t \le C_{2|1}(F(y)|u)\big)\, du &= \textstyle\int_0^1 C_{2|1}(F(y)|u)\, du = F(y).\end{aligned}$$

The joint distribution of (Y_{t-1}, Y_t), $\Pr(Y_{t-1} \le x, Y_t \le y)$, becomes

$$\begin{aligned}
&\Pr\big(\epsilon_{t-1} \le C_{2|1}(F(x)|\epsilon_{t-2}),\, \epsilon_t \le C_{2|1}(F(y)|\epsilon_{t-1})\big) \\
&= \int_0^1 \int_0^1 \Pr\big(u_2 \le C_{2|1}(F(x)|u_1),\, \epsilon_t \le C_{2|1}(F(y)|u_2)\big)\, du_2 du_1 \\
&= \int_0^1 \int_0^{C_{2|1}(F(x)|u_1)} C_{2|1}(F(y)|u_2)\, du_2 du_1 \\
&= \int_0^1 C\big(C_{2|1}(F(x)|u), F(y)\big)\, du. \qquad (8.4)
\end{aligned}$$

For the independence copula, note that $Y_t = F^{-1}(\epsilon_t)$, $t \ge 1$, is an iid sequence. For a BVN copula $C(u,v) = \Phi_\rho(\Phi^{-1}(u), \Phi^{-1}(v))$, $C_{2|1}(v|u) = \Phi([\Phi^{-1}(v) - \rho\Phi^{-1}(u)]/\sqrt{1-\rho^2})$, and $C_{2|1}^{-1}(s|u) = \Phi(\rho\Phi^{-1}(u) + \sqrt{1-\rho^2}\,\Phi^{-1}(s))$. If $F = \Phi$, the sequence becomes

$$Y_t = \rho\, a_{t-1} + \sqrt{1-\rho^2}\, a_t,$$

where a_t are iid $N(0,1)$ rvs. The lag 1 correlation is $\rho\sqrt{1-\rho^2}$, which reaches a maximum of $\frac{1}{2}$ when $\rho = \sqrt{\frac{1}{2}}$ and a minimum of $-\frac{1}{2}$ when $\rho = -\sqrt{\frac{1}{2}}$. The joint distribution from (8.4) is

$$\begin{aligned}
&\int_0^1 \Phi_\rho\big([x - \rho\Phi^{-1}(u)]/\sqrt{1-\rho^2}, y\big)\, du \\
&\quad = \int_{-\infty}^{\infty} \Phi_\rho\big((x-\rho z)/\sqrt{1-\rho^2}, y\big)\, d\Phi(z) \\
&\quad = \Pr\big(\sqrt{1-\rho^2}\, X + \rho Z \le x, Y \le y\big), \qquad (8.5)
\end{aligned}$$

where (X,Y,Z) is normal with covariance matrix $\begin{bmatrix} 1 & \rho & 0 \\ \rho & 1 & 0 \\ 0 & 0 & 1 \end{bmatrix}$. Hence the bivariate distribution in (8.5) is a BVSN distribution with correlation $\rho\sqrt{1-\rho^2}$.

For the upper Fréchet bound copula, $C_{2|1}(v|u) = 0$ if $v < u$ and 1 if $v \ge u$. Hence $C_{2|1}^{-1}(s|u) = u$ and the sequence reduces to $Y_t = F^{-1}(\epsilon_{t-1})$, an iid sequence. For the lower Fréchet bound copula, $C_{2|1}(v|u) = 0$ if $v < 1-u$ and 1 if $v \ge 1-u$. Hence $C_{2|1}^{-1}(s|u) = 1-u$ and the sequence reduces to $Y_t = F^{-1}(1-\epsilon_{t-1})$, another iid sequence.

Not all 1-dependent stationary sequences have a copula representation. One example consists of the MA sequences in Section 8.4.2 with univariate margins in the convolution-closed infinitely divisible class. Another example is that if $\xi_1, \xi_2, \dots$ are iid, then the sequence $X_t = \max\{\xi_t, \xi_{t+1}\}$, $t = 1, 2, \dots$, does not have the form of (8.3).

Next we state some properties of the distribution in (8.4). If the copula C is PQD, then the distribution in (8.4) pointwise exceeds $\int_0^1 C_{2|1}(F(x)|u)F(y)du = F(x)F(y)$ so that (8.4) is also PQD. Similarly, (8.4) is NQD if C is NQD. A comparison of the copula in (8.4) and C is as follows. Let $C^*(x,y) = \int_0^1 C(C_{2|1}(x|u), y)\, du$ be the copula in (8.4). Notice that $\int_0^1 C_{2|1}(x|u)\, du = x$ so that if $C(\cdot; y)$ is concave for all y, then $C^*(x,y) \le C(x,y)$ for all $0 \le x, y \le 1$. This follows because the condition of concavity is the same as $C_{2|1}(y|u)$ decreasing in u for all y, or equivalently $C_{2|1}(\cdot|u)$ is SI as u increases. Similarly, if $C_{2|1}(\cdot|u)$ is stochastically decreasing as u increases, then $C^*(x,y) \ge C(x,y)$ for all $0 \le x, y \le 1$.

8.2.2 Higher-order copulas

Let C be a trivariate copula that is differentiable with respect to the first two arguments. For stationary 2-dependent sequences with univariate margin F, based on a trivariate copula C, the generalization of (8.4) is:

$$Y_t = h(\epsilon_{t-2}, \epsilon_{t-1}, \epsilon_t), \quad t = 1, 2, \dots, \tag{8.6}$$

where

$$h(u_1, u_2, u_3) = F^{-1}\big(C_{3|12}^{-1}(u_3|u_1, C_{2|1}^{-1}(u_2|u_1))\big).$$

It is easily checked that the special case of the independence copula leads to an iid sequence.

Let $\Sigma = (\rho_{ij})$ be a trivariate correlation matrix. For the trivariate normal copula, $C(\mathbf{u}) = \Phi_\Sigma(\Phi^{-1}(u_1), \Phi^{-1}(u_2), \Phi^{-1}(u_3))$,

$$C_{3|12}(v_3|v_1, v_2) = \Phi\big(\big[\Phi^{-1}(v_3) - a_1\Phi^{-1}(v_1) - a_2\Phi^{-1}(v_2)\big]/a_3\big),$$

where $a_1 = (\rho_{13} - \rho_{12}\rho_{23})/(1-\rho_{12}^2)$, $a_2 = (\rho_{23} - \rho_{12}\rho_{13})/(1-\rho_{12}^2)$, $a_3 = [(1-\rho_{12}^2-\rho_{13}^2-\rho_{23}^2+2\rho_{12}\rho_{13}\rho_{23})/(1-\rho_{12}^2)]^{1/2}$. The model (8.6) becomes equivalent to $Y_t = (a_1 + a_2\rho_{12})Z_{t-2} + a_2\sqrt{1-\rho_{12}^2}\, Z_{t-1} + a_3 Z_t$, where Z_t are iid $N(0,1)$ rvs. If $\rho_{12} = \rho_{13} = \rho_{23} = \rho \to 1$, then $a_1 = a_2 \to \frac{1}{2}$, $a_3 \to 0$ and $Y_t = Z_{t-2}$.

The generalization of (8.6) to k-dependent sequences based on a $(k+1)$-dimensional copula C that is differentiable with respect to the first k arguments is obvious.

Table 8.1. *Correlations for binary discretization of extreme 1-dependent normal sequences.*

p or q	0.1	0.2	0.25	0.3	0.4	0.5
$\delta = 0.5$	0.249	0.295	0.308	0.318	0.330	0.333
$\delta = -0.5$	−0.103	−0.197	−0.237	−0.271	−0.317	−0.333

8.2.3 1-dependent binary series

In this subsection, we study some special forms for 1-dependent stationary binary time series $\{Y_t\}$, in order to obtain bounds on the maximum and minimum lag 1 correlations, as a function of $p = \Pr(Y_t = 1) = 1 - q$. This then provides a comparison for studying the range of dependence of 1-dependent binary time series.

From consideration of the non-negative definite correlation matrix for an infinite 1-dependent stationary sequence, the maximum correlation is less than or equal to $\frac{1}{2}$ and the minimum correlation is greater than or equal to $-\frac{1}{2}$. We have the exact maximum and minimum only for some p values and leave the other cases as an unsolved problem. The extension to k-dependent stationary binary series is also left as an unsolved problem; for the generalization, one has to consider an appropriate quantification of 'most dependent' and 'least dependent' over lags.

Candidates to consider for obtaining bounds are the discretizations of the extreme 1-dependent normal sequences $\{Z_t\}$ that have lag 1 correlations $\delta = \pm\frac{1}{2}$; i.e., $Y_t = I(Z_t > \Phi^{-1}(1-p))$, with $0 < p < 1$. For Bernoulli (p) margins, the lag 1 correlation of $\{Y_t\}$ is $[\Phi_\delta(\Phi^{-1}(q), \Phi^{-1}(q)) - q^2]/(pq)$ with $q = 1 - p$ and $\delta = \frac{1}{2}$ or $-\frac{1}{2}$. This leads to the correlation values in Table 8.1 (there is symmetry about $p = q = \frac{1}{2}$).

Next we consider some 1-dependent sequences given by:

(a) $Y_t = I(\max\{\xi_t, \xi_{t-1}\} > s)$;

(b) $Y_t = I(\min\{\xi_t, \xi_{t-1}\} > s)$;

(c) $Y_t = I(\max\{1 - \xi_{t-1}, \xi_t\} > s)$;

(d) $Y_t = I(\min\{1 - \xi_{t-1}, \xi_t\} > s)$.

In (a) and (b), ξ_t are iid continuous rvs, and in (c) and (d), ξ_t are iid $U(0,1)$ rvs. In each case, with $0 < p < 1$ fixed, s can be chosen so that Y_t is Bernoulli (p). Let $p_2 = \Pr(Y_t = Y_{t+1} = 1)$, so that the correlation is $\rho = (p_2 - p^2)/(p - p^2)$.

Table 8.2. *Maximum correlations over the 1-dependent binary sequences in (a) and (b).*

p or q	0.1	0.2	0.25	0.3	0.4	0.5
ρ	0.487	0.472	0.464	0.456	0.436	0.414

Table 8.3. *Minimum correlations over the 1-dependent binary sequences in (c) and (d).*

p or q	0.1	0.2	0.25	0.3	0.4	0.5
ρ	−0.111	−0.250	−0.333	−0.292	−0.225	−0.172

For (a), $p_2 = 1-2\gamma^2+\gamma^3$ with $\gamma = (1-p)^{1/2}$. For (b), $p_2 = p^{3/2}$. For (a) with $p \le \frac{1}{2}$ and (b) with $p \ge \frac{1}{2}$, the correlations exceed those in Table 8.1; these are given in Table 8.2 (there is symmetry in the maximum correlations about $p = q = \frac{1}{2}$).

For (c), $p = 1 - s^2$ if $s = (1-p)^{1/2}$, $p_2 = 1 - 2s^2$ if $s \le \frac{1}{2}$ and $p_2 = 1 - 2s^2 + s^2(2s-1)$ if $s \ge \frac{1}{2}$. For $s \le \frac{1}{2}$ or $p \ge 0.75$, this leads to $\rho = -s^2/(1-s^2) = -(1-p)/p$. For (d), $p = (1-s)^2$ if $s = 1 - p^{1/2}$, $p_2 = 0$ if $s \ge \frac{1}{2}$ and $p_2 = (1-s)^2(1-2s)$ if $s \le \frac{1}{2}$. For $s \ge \frac{1}{2}$ or $p \le 0.25$, this leads to $\rho = -p/(1-p)$. Table 8.3 has the minimum correlations that are possible from (c) and (d) (there is symmetry in the minimum correlations about $p = q = \frac{1}{2}$).

An upper bound on the maximum correlation and a lower bound on the minimum correlation for a 1-dependent stationary binary sequence can be obtained by considering three-way tables, four-way tables, etc., with appropriate margins. Consider the three-way table such that the (1,2) and (2,3) bivariate margins are $\begin{bmatrix} q^2+\theta & pq-\theta \\ pq-\theta & p^2+\theta \end{bmatrix}$ and the (1,3) bivariate margin is $\begin{bmatrix} q^2 & pq \\ pq & p^2 \end{bmatrix}$. Then for the trivariate distribution with $p_{ijk} = \Pr(Y_t = i, Y_{t+1} = j, Y_{t+2} = k)$, the bivariate constraints lead to $p_{000} = x$, $p_{001} = p_{100} = q^2 + \theta - x$, $p_{010} = q^2 - x$, $p_{011} = p_{110} = pq - q^2 - \theta + x$, $p_{101} = pq - q^2 - 2\theta + x$, $p_{111} = p^2 + q^2 - pq + 2\theta - x$. The non-negativity of each term implies $0 \le x \le q^2$, $q^2 - pq \le x - \theta \le q^2$, $q^2 - pq \le x - 2\theta \le q^2 - pq + p^2$. The maximum and minimum of θ are now reduced to linear programming problems. The inequalities can be drawn in the (x, θ) plane. The maximum is $\theta = \frac{1}{2}pq$

Table 8.4. *Bounds on extreme correlations for 1-dependent binary sequences.*

p or q	0.1	0.2	0.25	0.3	0.4	0.5
UB (max)	0.5	0.5	0.5	0.5	0.5	0.5
LB (max)	0.487	0.472	0.464	0.456	0.436	0.414
UB (min)	−0.111	−0.250	−0.333	−0.292	−0.317	−0.333
LB (min)	−0.111	−0.250	−0.333	−0.429	−0.5	−0.5

(when $x = q^2$ on the line $x - 2\theta = q^2 - pq$), and the maximum correlation is $\frac{1}{2}$. The graph for the case $q < \frac{1}{2}$ and $q \le p^2$ leads to the minimum θ value of $-q^2$ (correlation equal to $-q/p$), from the intersection of $x = 0$ and $x - \theta = q^2$. The graph for the case $q < \frac{1}{2}$ and $q > p^2$ leads to the minimum θ value of $\frac{1}{2}(-q^2 + pq - p^2)$ (correlation equal to $\frac{1}{2}(1 - q/p - p/q)$), from the intersection of $x = 0$ and $x - 2\theta = q^2 - pq + p^2$. The graph for the case $q \ge \frac{1}{2}$ and $p \le q^2$ leads to the minimum θ value of $-p^2$ (correlation equal to $-p/q$), from the intersection of $x - 2\theta = q^2 - pq + p^2$ and $x - \theta = q^2 - pq$ ($x = q^2 + pq - p^2$). The graph for the case $q \ge \frac{1}{2}$ and $p > q^2$ leads to the minimum θ value of $\frac{1}{2}(-q^2 + pq - p^2)$ (correlation equal to $\frac{1}{2}(1 - q/p - p/q)$), from the intersection of $x = 0$ and $x - 2\theta = q^2 - pq + p^2$. The constraint $p = q^2$ corresponds to $q = (-1 + \sqrt{5})/2 = 0.618$ and $p = (3 - \sqrt{5})/2$. The bound for the minimum correlation is $-\frac{1}{2}$ at $q = \frac{1}{2}$ and dips below $-\frac{1}{2}$ for q from 1/3 to 2/3.

A summary for all of the bounds is given in the Table 8.4. Note that the sequences from (c) with $p \ge 0.75$ and for (d) with $p \le 0.25$ attain the lower bound on the lag 1 correlation. The derivation of sharp bounds in the remaining cases is left as an unsolved problem.

8.3 Latent variable models

Models from Chapter 7 that have the MVN distribution, as a mixing distribution or for latent variables, can be used for longitudinal data, if the correlation or covariance matrices have patterns of correlations depending on lags, i.e., $\rho_{ij} = \gamma_{|j-i|}$ for $i \ne j$, for a sequence γ_k. Examples are the multivariate Poisson-lognormal distribution, the multivariate logit-normal distribution, and the discretization of ARMA normal time series for binary and ordinal response (equivalently, multivariate probit model with patterned covariance matrix). Simple patterned matrices that could be used for initial modelling are the AR(1) and AR(2) correlation structures. For AR(1), $\rho_{ij} = \rho^{|j-i|}$ for some $-1 < \rho < 1$. For AR(2), $\rho_{ij} = \rho_{|j-i|}$, with ρ_k being the autocorrelation of lag k; the autocorrelations satisfy $\rho_k = \phi_1\rho_{k-1} + \phi_2\rho_{|k-2|}$, $k \geq 3$, where $\phi_1 = \rho_1(1-\rho_2)/(1-\rho_1^2)$, $\phi_2 = (\rho_2 - \rho_1^2)/(1-\rho_1^2)$, and are determined from ρ_1, ρ_2. Note that if $Y_t = I(Z_t \leq \alpha)$, where $\{Z_t\}$ is a dependent AR sequence, then $\{Y_t\}$ is not a Markov chain.

8.4 Convolution-closed infinitely divisible class

A unified approach for time series models with non-negative serial dependence can be obtained for the case where the response variable has a distribution in the convolution-closed infinitely divisible class. The class includes Poisson, negative binomial (with fixed probability parameter), gamma (with fixed scale parameter), generalized Poisson (with one fixed parameter), inverse Gaussian (with one fixed parameter) and normal. The models are the same as or have similar form to the autoregressive moving average (ARMA) time series models in the case of a normal univariate margin, but only a subclass of the ARMA normal models is obtained. Following the usage in the statistical literature, we will refer to the models here as ARMA models for non-normal distributions. A Poisson, negative binomial or generalized Poisson margin can be used for count data and a gamma or inverse Gaussian margin can be used for a positive response variable.

Stationary first-order Markov or AR(1) time series models are stochastically described in Section 8.4.1, along with some properties and interesting special cases. Extensions to AR(p), MA(q) and ARMA models are covered in subsequent subsections. A non-stationary extension is mentioned briefly in Section 8.4.4.

The ideas in this section do not seem to extend to models with negative dependence for lags.

8.4.1 Stationary AR(1) time series $^\circ$

The theory here is a special case of the stationary Markov chain time series in Section 8.1.1; the joint distribution of a consecutive pair of observations has one of the bivariate distributions in Section 4.6. However, the time series models are best presented through stochastic representations rather than through transition probabilities.

Let F_θ, $\theta > 0$, be a convolution-closed infinitely divisible parametric family such that $F_{\theta_1} * F_{\theta_2} = F_{\theta_1+\theta_2}$, where $*$ is the convolution operator. It is assumed that F_0 corresponds to the degenerate distribution at 0. For $Z_j \sim F_{\theta_j}$, $j = 1, 2$, with Z_1, Z_2 independent, let $G_{\theta_1,\theta_2,z}$ be the distribution of Z_1 given that $Z_1 + Z_2 = z$. Let A be a **random operator** such that $A(Y)$ given $Y = y$ has distribution $G_{\alpha\theta,(1-\alpha)\theta,y}$, and $A(Y) \sim F_{\alpha\theta}$ when $Y \sim F_\theta$. (In later subsections, the operator is denoted by $A(\cdot;\alpha)$ to show the dependence on α.) A stationary time series with margin F_θ and autocorrelation $0 < \alpha < 1$ (of lag 1) can now be constructed as

$$Y_t = A_t(Y_{t-1}) + \epsilon_t, \tag{8.7}$$

where the innovations ϵ_t are iid with distribution $F_{(1-\alpha)\theta}$, $Y_t \sim F_\theta$ for all t, and $\{A_t : t \geq 1\}$ are independent replications of the operator A. (The term **innovation** is used because ϵ_t need not have a mean of 0; rather ϵ_t is new or innovative at time t.)

Here is the intuition reasoning behind the operator $A(\cdot)$. A consecutive pair (Y_{t-1}, Y_t) has a common latent or unobserved component X_{12} through the representation:

$$Y_{t-1} = X_{12} + X_1, \quad Y_t = X_{12} + X_2,$$

where X_{12}, X_1, X_2 are independent rvs with distributions $F_{\alpha\theta}$, $F_{(1-\alpha)\theta}$, $F_{(1-\alpha)\theta}$, respectively. The operator $A(Y_{t-1})$ 'recovers' the unobserved X_{12}; hence the distribution of $A(y)$ given $Y_{t-1} = y$ must be the same as the distribution of X_{12} given $X_{12} + X_1 = y$.

Interesting examples are the following.

(a) If F_θ is Gamma(θ, ξ) with ξ fixed, then $G_{\alpha\theta,(1-\alpha)\theta,y}$ is the distribution of y times a Beta$(\alpha\theta, (1-\alpha)\theta)$ rv. That is, the model could be represented as

$$Y_t = A_t Y_{t-1} + \epsilon_t, \tag{8.8}$$

where the A_t are iid Beta$(\alpha\theta, (1-\alpha)\theta)$ rvs and the ϵ_t are iid Gamma$((1-\alpha)\theta, \xi)$ rvs.

(b) If F_θ is $N(0, \theta)$, then $G_{\alpha\theta,(1-\alpha)\theta,y}$ is $N(\alpha y, \alpha(1-\alpha)\theta)$, and the usual normal AR(1) model results, since $A(Y) \sim N(0, \alpha\theta)$.

(c) If F_θ is Poisson(θ), then $G_{\alpha\theta,(1-\alpha)\theta,y}$ is Binomial(y, α).

(d) If F_θ is Negative Binomial(θ, p), as given in (8.10) below, with p fixed, then $G_{\alpha\theta,(1-\alpha)\theta,y}$ is Beta-binomial$(y, \alpha\theta, (1-\alpha)\theta)$ (with pmf given in (8.11) below).

(e) If F_θ is inverse Gaussian with parameters θ, λ (λ fixed), mean θ, variance $\theta\lambda^2$, and density of the form

$$f_\theta(y) = \frac{\theta}{(2\pi y^3)^{1/2}\lambda} \exp\{-\theta^2/(2y\lambda^2) - y/(2\lambda^2) + \theta/\lambda^2\},\ y > 0, \tag{8.9}$$

then the inverse Gaussian subfamily is infinitely divisible. The density of $G_{\alpha\theta,(1-\alpha)\theta,y}$ is given in (8.12) below.

(f) If F_θ is generalized Poisson with parameters θ, η ($\eta \geq 0$ fixed) and density of the form

$$f_\theta(y) = \theta(\theta + \eta y)^{y-1} e^{-\theta-\eta y}/y!, \quad y = 0, 1, \ldots,$$

then the generalized Poisson subfamily is infinitely divisible, and $G_{\alpha\theta,(1-\alpha)\theta,y}$ is a quasi-binomial distribution (given in (8.13) below).

By relaxing the condition of infinite divisibility, one gets the following additional interesting example.

(g) If F_θ is Binomial(θ, p) and α is restricted to a multiple of θ^{-1}, then $G_{\alpha\theta,(1-\alpha)\theta,y}$ is Hypergeometric$(\alpha\theta, (1-\alpha)\theta, y)$ (the pmf is given in (8.14) below).

The models in (c), (d), (f) and (g) could be used for count data, with (d) and (f) for overdispersed counts relative to Poisson, and (g) for underdispersed counts relative to Poisson. The model in (g) might be useful for inferences when θ is large and unknown.

Some details for the specific examples of interest are given next.

(b) (Normal.) Let $Z_j \sim N(0, \theta_j)$ independently, $j = 1, 2$. Let ϕ denote the standard normal density. Then the density of Z_1, given that $Z_1 + Z_2 = y$, is

$$\begin{aligned} g(w; \theta_1, \theta_2, y) &= \frac{(\theta_1\theta_2)^{-1/2}\phi(w/\sqrt{\theta_1})\,\phi((y-w)/\sqrt{\theta_2})}{(\theta_1+\theta_2)^{-1/2}\phi(y/\sqrt{\theta_1+\theta_2})} \\ &= [\theta/(\theta_1\theta_2)]^{1/2}\phi((w-\alpha y)/\sqrt{\alpha(1-\alpha)\theta}), \end{aligned}$$

with $\theta = \theta_1 + \theta_2$, $\alpha = \theta_1/\theta$. Equivalently, $Z_1 \mid Z_1 + Z_2 = y$ is $N(\alpha y, \alpha(1-\alpha)\theta)$. Hence, for (8.7), a stochastic representation is $Y_t = \alpha Y_{t-1} + \omega_t + \epsilon_t$, where ω_t are iid $N(0, \alpha(1-\alpha)\theta)$ independently of the $\{\epsilon_t\}$, and $\omega_t + \epsilon_t$ are iid $N(0, (1-\alpha^2)\theta)$.

(d) (Negative binomial and beta-binomial.) For the negative binomial (NB) distribution with parameters θ, p, we mean the distribution with pmf

$$f(k;\theta,p) = \{\Gamma(\theta+k)/[k!\,\Gamma(\theta)]\}p^\theta q^k,\ k=0,1,\ldots,\ q=1-p. \tag{8.10}$$

If $Z_1 \sim \mathrm{NB}(\theta_1,p)$, $Z_2 \sim \mathrm{NB}(\theta_2,p)$, and Z_1, Z_2 are independent rvs, then

$$\Pr(Z_1 = k \mid Z_1+Z_2=y) = \frac{f(k;\theta_1,p)f(y-k;\theta_2,p)}{f(y;\theta_1+\theta_2,p)}$$

$$= \binom{y}{k}\frac{B(\theta_1+k,\theta_2+y-k)}{B(\theta_1,\theta_2)},\quad k=0,1,\ldots,y. \tag{8.11}$$

This is the beta-binomial pmf, which is a Beta (θ_1,θ_2) mixture of Binomial (y,p) distributions.

(e) (Inverse Gaussian.) Let Z_1, Z_2 be independent inverse Gaussian rvs with respective parameters θ_1,θ_2. Using (8.9), the conditional density of Z_1 given $Z_1+Z_2=y$, $f_{\theta_1}(w)f_{\theta_2}(y-w)/f_{\theta_1+\theta_2}(y)$, simplifies to

$$(2\pi)^{-1/2}\left[\frac{y}{w(y-w)}\right]^{3/2}\frac{\theta_1\theta_2}{\lambda(\theta_1+\theta_2)}$$
$$\cdot\exp\left\{-\tfrac{1}{2\lambda^2}\left[\tfrac{\theta_1^2}{w}+\tfrac{\theta_2^2}{y-w}-\tfrac{(\theta_1+\theta_2)^2}{y}\right]\right\}, \tag{8.12}$$

for $0<w<y$. From this, the conditional density for $Z_1/(Z_1+Z_2)$ given $Z_1+Z_2=y$ depends on y, so that there is not a simpler stochastic representation for (8.7), as for the case of gamma margins.

(f) (Generalized Poisson and quasi-binomial.) Let Z_1, Z_2 be independent generalized Poisson rvs with respective parameters θ_1,θ_2. The conditional density of Z_1 given $Z_1+Z_2=y$ is the quasi-binomial distribution with pmf given by

$$g_k = \binom{y}{k}\frac{p(1-p)}{1+\zeta y}\left[\frac{p+\zeta k}{1+\zeta y}\right]^{k-1}\left[\frac{1-p+\zeta(y-k)}{1+\zeta y}\right]^{y-k-1}, \tag{8.13}$$

$k=0,1,\ldots,y$, where $p=\theta_1/(\theta_1+\theta_2)$, $\zeta=\eta/(\theta_1+\theta_2)$.

(g) (Binomial and hypergeometric.) If $Z_1 \sim \mathrm{Binomial}(\theta_1,p)$ and $Z_2 \sim \mathrm{Binomial}(\theta_2,p)$, and Z_1,Z_2 are independent, then Z_1 given $Z_1+Z_2=y$ is hypergeometric with pmf

$$g_k = \binom{\theta_1}{k}\binom{\theta_2}{y-k}\Big/\binom{\theta}{y},\quad k=0,1,\ldots,y,$$

where $\theta = \theta_1 + \theta_2$. To make this appear similar to the beta-binomial distribution, the pmf can be rewritten as

$$g_k = \binom{y}{k} \frac{\theta_1!}{(\theta_1 - k)!} \frac{\theta_2!}{(\theta_2 - y + k)!} \frac{(\theta - y)!}{\theta!}. \tag{8.14}$$

Theorem 8.1 *Properties of the process (8.7) are the following.*

(a) The process is Markov of order 1 and is time-reversible.

(b) If F_θ has moments of second order, then the autocorrelation of lag j is α^j, $j = 1, 2, \ldots$.

(c) An iid sequence is obtained if $\alpha \to 0$ and a perfectly dependent sequence is obtained if $\alpha \to 1$.

Proof. The proofs are left as exercises. □

The final result of this subsection concerns the bivariate margins for the AR(1) time series. For the Poisson margin, the bivariate distribution of (Y_1, Y_{j+1}) is the standard (and most natural) bivariate Poisson distribution (see Section 7.2.2).

Theorem 8.2 *For the AR(1) Poisson time series of the form (8.7), the bivariate distribution of (Y_1, Y_{j+1}) is bivariate Poisson with parameters $\theta, \theta, \alpha^j\theta$,* i.e., *the pmf is*

$$\sum_{k=0}^{y\wedge z} \frac{e^{-\lambda_{12}}\lambda_{12}^k}{k!} \frac{e^{-\lambda_1+\lambda_{12}}(\lambda_1 - \lambda_{12})^{y-k}}{(y-k)!} \frac{e^{-\lambda_2+\lambda_{12}}(\lambda_2 - \lambda_{12})^{z-k}}{(z-k)!},$$

$y, z = 0, 1, \ldots,$ *with* $\lambda_1 = \lambda_2 = \theta$, $\lambda_{12} = \alpha^j\theta$.

Proof. The proof is based on a stochastic representation. Let ϵ_t, $t \geq 2$, be iid Poisson $((1-\alpha)\theta)$ rvs, and let δ_{jki} be iid Bernoulli (α) rvs which are independent of the ϵ_t and Y_1. Then the Poisson AR(1) series of the form (8.7) can be represented as:

- $Y_2 = \sum_{i=1}^{Y_1} \delta_{21i} + \epsilon_2$,
- $Y_3 = \sum_{i=1}^{Y_1} \delta_{21i}\delta_{31i} + \sum_{i=1}^{\epsilon_2} \delta_{32i} + \epsilon_3$,
- $\ldots$,
- $Y_{j+1} = \sum_{i=1}^{Y_1} \delta_{21i} \cdots \delta_{j+1,1i} + \sum_{k=2}^{j} \sum_{i=1}^{\epsilon_k} \delta_{k+1,ki} \cdots \delta_{j+1,ki} + \epsilon_{j+1}$.

Since the rvs $Y_1, \epsilon_2, \ldots, \epsilon_{j+1}$ are independent, Y_{j+1} is stochastically equal to $A_{j+1}^*(Y_1) + \epsilon$, where ϵ is independent of Y_1 and $A_{j+1}^*(y)$ has the Binomial(y, α^j) distribution. Hence (Y_1, Y_{j+1}) has distribution similar in form to the distribution of (Y_1, Y_2) with α replaced by α^j. □

Remark. There is not a similar result for the negative binomial or gamma distributions. For example, using (8.8) as the stochastic representation for the gamma AR(1) time series, $Y_3 = A_3A_2Y_1 + A_3\epsilon_2 + \epsilon_3$, where A_2, A_3 are iid Beta $(\alpha\theta, (1-\alpha)\theta)$ rvs, independent of $Y_1, \epsilon_2, \epsilon_3$. $A_3A_2Y_1$ is stochastically equal to A^*Y_1 with A^* having a Beta $(\alpha^2\theta, (1-\alpha^2)\theta)$ distribution, and $A_3\epsilon_2 + \epsilon_3$ is stochastically equal to ϵ^*, a Gamma$(1-\alpha^2, \xi)$ rv. The pair $(Y_1, A^*Y_1 + \epsilon^*)$ has a form stochastically equal to that in (4.49) in Section 4.6 (assuming Y_1, A^* and ϵ^* are independent), but (Y_1, Y_3) does not since $A_3A_2Y_1$ and $A_3\epsilon_2 + \epsilon_3$ are dependent.

8.4.2 Moving average models

There are versions of the models in preceding subsection for stationary moving average (MA) models. For the MA(1) models, if $Y \sim F_\theta$ and $0 < \beta < 1$, let $A(Y;\beta)$ denote a rv that, given $Y = y$, has distribution $G_{\beta\theta,(1-\beta)\theta,y}$. Let ϵ_t, $t = 0, 1, 2, \ldots$, be iid rvs with distribution F_η. An MA(1) time series with marginal distribution F_θ has the form

$$Y_t = \epsilon_t + A_t(\epsilon_{t-1}; \alpha), \quad t = 1, 2, \ldots, \tag{8.15}$$

where $0 \le \alpha \le 1$ and $\eta = \theta/(\alpha+1)$. As before, A_t, $t \ge 1$, are independent operators. Assuming that F_θ has finite second moment $\theta\sigma^2$, $\mathrm{Cov}\,(Y_t, Y_{t+1}) = \mathrm{Cov}\,(\mathrm{E}\,[\epsilon_t|\epsilon_t], \mathrm{E}\,[A_{t+1}(\epsilon_t;\alpha)|\epsilon_t]) = \mathrm{Cov}\,(\epsilon_t, \alpha\epsilon_t) = \alpha\eta\sigma^2 = \alpha\theta\sigma^2/(\alpha+1)$, and the correlation is $\alpha/(\alpha+1)$ which is bounded above by $\frac{1}{2}$ (when $\alpha = 1$, $\eta = \theta/2$). The lower bound is 0 when $\alpha = 0$.

The MA(q) model, with $Y_t \sim F_\theta$, has the form

$$Y_t = \sum_{j=0}^{q} A_{t,j}(\epsilon_{t-j}; \alpha_j),$$

where $\alpha_0 = 1$, $0 \le \alpha_j \le 1$, $j = 1, \ldots, q$, ϵ_t are iid with distribution F_η and $\theta = \eta \sum_0^q \alpha_j$. The operators $A_{t,j}$ are independent over t and j.

For $1 \le k \le q$, the autocovariances and autocorrelations are

$$\mathrm{Cov}\,(Y_t, Y_{t+k}) = \mathrm{Cov}\,\Big(\sum_{j=0}^{q} A_{t,j}(\epsilon_{t-j}; \alpha_j),$$

$$\sum_{j'=-k}^{q-k} A_{t+k,j'+k}(\epsilon_{t-j'}; \alpha_{j'+k})\Big) = \sum_{j=0}^{q-k} \alpha_j\alpha_{j+k}\eta\sigma^2$$

and $\sum_{j=0}^{q-k} \alpha_j \alpha_{j+k} / \sum_{j=0}^{q} \alpha_j$, respectively.

Combining the ideas for the AR(1) and MA(q) models, one can get an ARMA(1,q) model. This has the form

$$\begin{aligned} Y_t &= W_{t-q} + \sum_{j=1}^{q} A_{t,j}(\epsilon_{t+1-j}; \alpha_j), \quad t \geq 1, \\ W_t &= A_t(W_{t-1}; \beta) + \epsilon_t, \quad t \geq 1-q, \end{aligned} \tag{8.16}$$

where $\{W_t\}$ is the autoregressive component, $W_t \sim F_\gamma$, $0 \leq \beta \leq 1$, $0 \leq \alpha_j \leq 1$, $j = 1, \ldots, q$, ϵ_t are iid with distribution F_η, ϵ_t is independent of $W_{t-1}, W_{t-2}, \ldots$, $A_{t,j}, A_t$ are independent operators over different t, j, $\gamma = \eta/(1-\beta)$ and $\eta = \theta/[(1-\beta)^{-1} + \sum_1^q \alpha_j]$. Note that this form of the ARMA model is not the same as the usual one for an ARMA time series with normal rvs. In general, convolution of dependent rvs, each with distribution in the family F_θ, need not result in a rv in the same family. In (8.16), the autoregressive component W_{t-q} is independent of $\epsilon_t, \ldots, \epsilon_{t-q+1}$ in order that Y_t is the sum of independent rvs and $Y_t \sim F_\theta$.

The form of the autocorrelation function for (8.16) has a simple form, like that for the usual ARMA(1,q) normal model, only in cases where the operators have an additive property, as given below. Let ρ_j be the autocorrelation of lag j. If the decomposition

$$\begin{aligned} &\operatorname{Cov}\left(A(X_1 + X_2; \beta_1), A^*(Y; \beta_2)\right) \\ &\quad = \operatorname{Cov}\left(A(X_1; \beta_1), A^*(Y; \beta_2)\right) + \operatorname{Cov}\left(A(X_2; \beta_1), A^*(Y; \beta_2)\right) \end{aligned}$$

makes sense for independent operators A, A^* and arbitrary $\beta_1, \beta_2 \in (0,1)$, then a formula (with the proof left as an exercise) is:

$$\rho_k = \Big[\beta^k \gamma + \eta \sum_{i=1}^{q-k} \alpha_i \alpha_{i+k} + \eta \sum_{i=q-k+1}^{q} \beta^{i-q+k-1} \alpha_i\Big] \Big/ \theta, \; k = 1, \ldots, q,$$

$$\rho_{q+k} = \beta \rho_{q+k-1}, \quad k = 1, 2, \ldots.$$

8.4.3 Higher-order autoregressive models

The generalization of the AR(1) models to AR(p), $p > 1$, is not as straightforward. There is the possibility of more than one generalization. Our extension is based on the multivariate generalization of a family of univariate distributions in the convolution-closed infinitely divisible class. That is, the joint distribution of $p+1$ consecutive observations has one of the multivariate distributions in Section 4.6. These higher-order AR models do not have conditional linearity except for the special case of the normal distribution. We next state the general AR(2) model that extends (8.7). From this, the extension to AR(p), $p > 2$, is straightforward conceptually, although the notation is a bit cumbersome. The extension to ARMA(p, q) models then combines the ideas in (8.16) and (8.17). Note again that the AR models are examples of the stationary Markov chain time series models in Section 8.1.1; stochastic representations can be given for the transition probabilities.

For the AR(2) model, we make use of the distribution of $Z_{13} + Z_{23} + Z_{123}$ given $Z_1 + Z_{12} + Z_{13} + Z_{123} = y_1$, $Z_2 + Z_{12} + Z_{23} + Z_{123} = y_2$, where $Z_1, Z_2, Z_{12}, Z_{13}, Z_{23}, Z_{123}$ have distributions in the family F with respective parameters $\theta - \theta_1 - \theta_2 - \theta_3$, $\theta - 2\theta_1 - \theta_3$, θ_1, θ_2, θ_1, θ_3 (θ is defined so that the first two parameters are non-negative). Let this distribution be denoted by $G_{\theta_1,\theta_2,\theta_3,\theta,y_1,y_2}$ and let a rv with this distribution be denoted by $A(y_1, y_2)$. Let ϵ_t be iid rvs with distribution $F_{\theta-\theta_1-\theta_2-\theta_3}$. The AR(2) time series is defined as

$$Y_t = A_t(Y_{t-2}, Y_{t-1}) + \epsilon_t, \quad t = 3, 4, \ldots, \tag{8.17}$$

where A_t, $t \geq 3$, are independent replicates of the operator A. To get a stationary series, let $Y_1 \sim F_\theta$ and $Y_2 = A_2^*(Y_1) + \epsilon_2$, where $A_2^*(y)$ has distribution $G_{\theta_1+\theta_3,\theta-\theta_1-\theta_3,y}$ (from Section 8.4.1) and $\epsilon_2 \sim F_{\theta-\theta_1-\theta_3}$. By construction $Y_t \sim F_\theta$.

Here is some of the intuition reasoning behind the operator $A(\cdot)$. A consecutive triple (Y_{t-2}, Y_{t-1}, Y_t) has common latent or unobserved components X_{123}, X_{12}, X_{13}, X_{23} through the representation:

$$\begin{aligned}
Y_{t-2} &= X_{123} + X_{12} + X_{13} + X_1, \\
Y_{t-1} &= X_{123} + X_{12} + X_{23} + X_2, \\
Y_t &= X_{123} + X_{13} + X_{23} + X_3,
\end{aligned}$$

where $X_{123}, X_{12}, X_{13}, X_{23}, X_1, X_2, X_3$ are independent with distributions corresponding to the Z_S defined above. The operator $A(Y_{t-2}, Y_{t-1})$ 'recovers' the unobserved sum $X_{123} + X_{13} + X_{23}$; hence the distribution of $A(y_1, y_2)$ given $Y_{t-2} = y_1$ and $Y_{t-1} = y_2$ must be the same as the distribution of $X_{123} + X_{13} + X_{23}$ given $X_{123} + X_{12} + X_{13} + X_1 = y_1$ and $X_{123} + X_{12} + X_{23} + X_2 = y_2$.

By comparison with the stochastic representation of (Y_{t-2}, Y_{t-1}, Y_t) in the AR(1) model of Section 8.4.1, the AR(1) model is a special case of (8.17) when $(\theta_1, \theta_2, \theta_3)$ has the form $(\theta\alpha[1-\alpha], 0, \theta\alpha^2)$, where $0 < \alpha < 1$. The details are left as an exercise.

Let $\zeta_1 = \theta - \theta_1 - \theta_2 - \theta_3$, $\zeta_2 = \theta - 2\theta_1 - \theta_3$. Let f_θ be the density of F_θ with respect to the measure ν. The density g associated with $G_{\theta_1,\theta_2,\theta_3,\theta,y_1,y_2}$ is:

$$g(x) = \int_u \int_v \int_w f_{\theta_3}(u) f_{\theta_1}(v) f_{\theta_1}(w) f_{\theta_2}(x-u-v) f_{\zeta_1}(y_1 - x + v - w)$$
$$\cdot f_{\zeta_2}(y_2 - u - v - w)\, d\nu d\nu d\nu \,/ h(y_1, y_2),$$

where h is the joint density of $(Z_1 + Z_{12} + Z_{13} + Z_{123}, Z_2 + Z_{12} + Z_{23} + Z_{123})$. This density does not simplify unless F_θ is the normal family (with parameter θ for the variance).

In the special case of the normal margin, let $X = Z_{13}+Z_{23}+Z_{123}$, $Y_1 = Z_1 + Z_{12} + Z_{13} + Z_{123}$, $Y_2 = Z_2 + Z_{12} + Z_{23} + Z_{123}$. The joint distribution of (Y_1, Y_2, X) is trivariate normal with zero mean vector and covariance matrix $\begin{bmatrix} \theta & \delta_1 & \delta_2 \\ \delta_1 & \theta & \delta_1 \\ \delta_2 & \delta_1 & \delta_1 + \theta_2 \end{bmatrix}$, where $\delta_1 = \theta_1 + \theta_3$, $\delta_2 = \theta_2 + \theta_3$. The conditional distribution of X given $Y_1 = y_1$, $Y_2 = y_2$ is normal with mean $c_1 y_1 + c_2 y_2$ and variance $V = \delta_1 + \theta_2 - c_1\delta_2 - c_2\delta_1$, where $c_1 = (\theta\delta_2 - \delta_1^2)/(\theta^2 - \delta_1^2)$, $c_2 = \delta_1(\theta - \delta_2)/(\theta^2 - \delta_1^2)$. Therefore, $A(Y_1, Y_2) \stackrel{d}{=} c_1 Y_1 + c_2 Y_2 + W$, where $W \sim N(0, V)$, and $\mathrm{Var}\,(A(Y_1, Y_2)) = (c_1^2 + c_2^2)\theta + 2c_1c_2\delta_1 + V = \delta_1 + \theta_2$. Also one can verify that $\mathrm{Cov}\,(Y_3, Y_2) = \mathrm{Cov}\,(c_1 Y_1 + c_2 Y_2, Y_2) = c_1\delta_1 + c_2\theta = \delta_1$ and $\mathrm{Cov}\,(Y_3, Y_1) = c_1\theta + c_2\delta_1 = \delta_2$.

Properties that are discussed briefly are: (i) multivariate cumulants; (ii) time reversibility; and (iii) no general form for the autocorrelation function.

(i) For the model (8.17), the parameters are related to multivariate cumulants (see Section 4.6). There are constants γ_2, γ_3 such that $\kappa_{12} = \gamma_2(\theta_1 + \theta_3)$, $\kappa_{13} = \gamma_2(\theta_2 + \theta_3)$, $\kappa_{123} = \gamma_3\theta_3$, where $\kappa_{12}, \kappa_{13}, \kappa_{123}$ are respectively the mixed cumulants of (Y_t, Y_{t+1}), (Y_t, Y_{t+2}) and (Y_t, Y_{t+1}, Y_{t+2}).

(ii) By construction, the joint distribution of (Y_t, Y_{t+1}, Y_{t+2}) is the same as that of $(Z_1 + Z_{12} + Z_{13} + Z_{123}, Z_2 + Z_{12} + Z_{23} + Z_{123}, Z_3 + Z_{13} + Z_{23} + Z_{123})$, where $Z_1, Z_2, Z_3, Z_{12}, Z_{13}, Z_{23}, Z_{123}$ have distributions with respective parameters $\zeta_1, \zeta_2, \zeta_1, \theta_1, \theta_2, \theta_1, \theta_3$. The joint density of (Y_t, Y_{t+1}, Y_{t+2}) is

$$f_{123}(y_1, y_2, y_3) = \int_u \int_v \int_w \int_x f_{\theta_3}(u) f_{\theta_1}(v) f_{\theta_1}(w) f_{\theta_2}(x)$$
$$\cdot f_{\zeta_1}(y_1 - u - v - x) f_{\zeta_2}(y_2 - u - v - w) f_{\zeta_1}(y_3 - u - w - x)$$
$$d\nu(x)\, d\nu(w)\, d\nu(v)\, d\nu(u).$$

This density is symmetric in y_1, y_3 so that the process $\{Y_t\}$ is time-reversible.

(iii) The approach for the AR(1) and MA(q) models of using conditional expectations to obtain the autocorrelation function does not work for (8.17) because there is no general result for the conditional expectation $\mathrm{E}(Y_{t+2}|Y_t, Y_{t+1})$. This conditional expectation is non-linear in general and would have to be evaluated separately for different families $\{F_\theta\}$.

8.4.4 Models for longitudinal data

In this subsection, we first indicate one approach to obtaining a non-stationary extension of the AR(1) model in (8.7). In some applications, the parameter θ may depend on time. There could be a time trend, or the time series may consist of repeated measures or longitudinal data with the parameter θ depending on (time-varying) covariates. In the former case, we could have the model $\theta_t = g(t)$ for a positive-valued function g, and in the latter case, we could have the model $\theta_t = g(\mathbf{x}_t)$, where g is positive-valued and $\mathbf{x}_t$ is the covariate (column) vector at time t. For covariates, a convenient choice of g to ensure that θ_t is non-negative is $g(\mathbf{x}_t) = \exp\{\beta_0 + \boldsymbol{\beta}\mathbf{x}_t\}$ for a constant β_0 and a row vector $\boldsymbol{\beta}$.

Because Y_t, $A_t(Y_{t-1};\alpha)$, ϵ_t in (8.7) are in the same convolution-closed family, (8.7) can be adapted to a changing θ, as follows:

$$Y_t = A_t(Y_{t-1};\alpha) + \epsilon_t, \tag{8.18}$$

with $Y_t \sim F_{\theta_t}$, $\epsilon_t \sim F_{\eta_t}$, $\eta_t = \theta_t - \alpha\theta_{t-1} \geq 0$. In particular, if F corresponds to the Poisson, generalized Poisson, negative binomial or binomial distribution, (8.18) is a potential model for longitudinal count data with covariates.

Similarly, one can extend (8.17) as follows. We use the distribution of $Z_{13} + Z_{23} + Z_{123}$ given $Z_1 + Z_{12} + Z_{13} + Z_{123} = y_1$, $Z_2 + Z_{12} + Z_{23} + Z_{123} = y_2$, where $Z_1, Z_2, Z_{12}, Z_{13}, Z_{23}, Z_{123}$ have distributions in the family F with respective parameters $\theta - \theta_1 - \theta_2 - \theta_3$, $\theta' - \theta_1 - \theta_1' - \theta_3$, θ_1, θ_2, θ_1', θ_3 (θ, θ' are defined so that the first two parameters are non-negative). Let this distribution be denoted by $G_{\theta_1,\theta_1',\theta_2,\theta_3,\theta,\theta',y_1,y_2}$ and let a rv with this distribution be denoted by $A(y_1, y_2; \Theta)$ with $\Theta = (\theta_1, \theta_1', \theta_2, \theta_3, \theta, \theta')$. Let $\alpha_1, \alpha_2, \alpha_3$ be non-negative constants such that $0 \leq \alpha_1 + \alpha_2 + \alpha_3 < 1$; α_3 is a measure of dependence in three consecutive observations Y_{t-2}, Y_{t-1}, Y_t, while α_1 is a measure of extra dependence in two consecutive observations Y_{t-1}, Y_t, and α_2 is a measure of extra dependence in two observations lagged by 2, Y_{t-2}, Y_t. For $t \geq 1$, let $Y_t \sim F_{\theta_t}$, and for $t \geq 3$, let ϵ_t have distribution $F_{\theta_t-\theta_{1t}-\theta_{2t}-\theta_{3t}}$, with $\theta_{1t} = \alpha_1\theta_{t-1}$, $\theta_{2t} = \alpha_2\theta_{t-2}$, $\theta_{3t} = \alpha_3 \min\{\theta_{t-1}, \theta_{t-2}\}$, and $\theta_t - \theta_{1t} - \theta_{2t} - \theta_{3t} \geq 0$. The AR(2) non-stationary time series is defined as

$$Y_t = A_t(Y_{t-2}, Y_{t-1}; \Theta_t) + \epsilon_t, \quad t = 3, 4, \ldots, \tag{8.19}$$

where A_t, $t \geq 3$, are independent operators with parameter vectors

$$\Theta_t = (\alpha_1\theta_{t-2},\ \alpha_1\theta_{t-1},\ \alpha_2\theta_{t-2},\ \alpha_3 \min\{\theta_{t-1}, \theta_{t-2}\},\ \theta_{t-2},\ \theta_{t-1}).$$

The start of the time series for Y_1, Y_2 is the same as (8.18).

The models presented here are used in the data analysis example in Section 11.5. Alternative ways of making the parameters depend on covariates are possible, particularly when univariate marginal family, such as the negative binomial or generalized Poisson, has another parameter besides θ, which can depend on the covariate. For example, for count data, a statistical modelling consideration is the variance to mean relationship.

8.4.5 Other non-normal time series models

There are other classes of stationary non-normal time series models that exist in the probability and statistics literature, but they do not necessarily have nice extensions to non-stationary time series (because of unusual innovation rvs) or to higher-order autoregression (stationary univariate margin does not have simple form).

One general approach is for univariate distributions in the class of self-decomposable distributions, a subset of infinitely divisible continuous distributions. The main drawback of these models is that the distribution of the innovation term in the time series can have a mass at 0, leading to a singularity that is usually not reasonable for modelling time series of continuous response variables.

A rv Y is in the **self-decomposable** class if for every $0 < \alpha < 1$, there exists a rv $\epsilon = \epsilon(\alpha)$, independent of Y, such that $\alpha Y + \epsilon(\alpha)$ is equal in distribution to Y.

If Y_t, $t \geq 1$, have the same distribution as Y, the time series is

$$Y_t = \alpha Y_{t-1} + \epsilon_t, \quad t = 1, 2, \ldots,$$

where ϵ_t are iid with the appropriate distribution. $\{Y_t\}$ is iid if $\alpha \to 0$ and perfectly dependent if $\alpha \to 1$.

Some special cases are the following.

(a) If Y is Exponential (1), then, with probability α, ϵ_t is zero, and with probability $1 - \alpha$, ϵ_t is an Exponential (1) rv.

(b) If Y is Gamma$(\theta, 1)$, and θ is a positive integer, then with probability α^θ, ϵ_t is zero, and with probability p_j $(1 \le j \le \theta)$, ϵ_t is a Gamma$(j, 1)$ rv, where the p_j are probabilities from a Binomial$(\theta, 1-\alpha)$ distribution.

(c) If Y is Gamma$(\theta, 1)$, and θ is not a integer, then with probability α^θ, ϵ_t is zero, and with probability p_j $(j \ge 1)$, ϵ_t is a Gamma(j, α) rv, where the p_j are probabilities from a NB(θ, α) distribution $(p_j = \{\Gamma(\theta + j)/[j!\,\Gamma(\theta)]\}\alpha^\theta(1-\alpha)^j)$.

(d) If Y is $N(0, \sigma^2)$, the usual normal AR(1) time series results.

Note that, for examples (a) to (c), $Y_t = \alpha Y_{t-1}$ with positive probability. This is an undesirable property as one does not expect this behaviour for time series encountered in practice. This shows that linearity (Y_t linear in Y_{t-1}) is not an appropriate assumption to use in general for non-normal rvs.

There are time series models for discrete response variables, based on operators different from those in Section 8.4.1. For example, there is an operator for the negative binomial distribution in the following model. Let $\{Y_t\}$ be a sequence of NB(θ, p) rvs. An autoregressive-like sequence satisfies

$$Y_t = \alpha * Y_{t-1} + \epsilon_t, \quad t = 1, 2, \ldots,$$

where $\alpha * Y$ given $Y = y$ is a Binomial(y, α) rv and ϵ_t has the probability generating function $g(1-s) = [\alpha + (1-\alpha)\lambda/(\lambda + s)]^\theta$, where $\lambda = p/(1-p)$. If θ is a positive integer, then ϵ_t is a Binomial$(\theta, 1-\alpha)$ mixture of NB(j, p) distributions $(j = 0, \ldots, \theta)$ and ϵ_t is zero with positive probability.

8.5 Markov chains: dependence properties *

It may be intuitive that the dependence decreases with lag for stationary Markov chains, i.e., for a stationary Markov chain, $\{Y_t\}$, (Y_i, Y_{i+j}) has less dependence as j increases. In this section, conditions for which this is true for first-order Markov chains are obtained. Different notions of dependence are considered, some of which depend on the form of the state space of the Markov chain.

In applications to Markov chain models it is useful to know the conditions needed for the behaviour of decrease in dependence with lag for different notions of dependence. This behaviour then roughly holds for Markov chains not starting in a stationary distribution, if the convergence to the stationary distribution is fast.

The following notation will be used. The distribution of Y_t is F.

If the state space is the real line or the integers and the corresponding measure is Lebesgue or counting measure, then the density is denoted by f if it exists. For $m \geq 2$, the bivariate distribution for Y_1, Y_m is denoted by F_{1m} or F_{Y_1,Y_m} and its density is denoted by f_{1m} if it exists. In addition, let $F^{(2)}(y_1, y_2) = F(y_1)F(y_2)$; this is the limit of F_{1m} as $m \to \infty$. The transition distribution, which is $F_{2|1}$, is also denoted by $H(\cdot|\cdot)$; its density, if it exists, is denoted by $h(\cdot|\cdot)$.

We now prove a sequence of dependence results. The first result involves the bivariate concordance ordering, i.e., (Y_1, Y_m) decreasing in concordance as m increases. This then implies that measures of association for Y_1, Y_m, such as Spearman's correlation and Kendall's tau, decrease as m increases (see Exercise 2.10). Since the concordance ordering is a positive dependence ordering, a positive dependence requirement is required on the transition distribution. A sufficient condition is SI, and from counterexamples it can be shown that it cannot be weakened to PQD, LTD or RTI (the details are left as an exercise).

Theorem 8.3 *Let $Y_1, Y_2, \ldots$ be a stationary Markov chain with state space in $\Re$. If H is SI, then (a) $Y_j \uparrow_{\rm st} Y_1$, $j = 2, 3, \ldots$, and (b) $F^{(2)} \prec_{\rm c} \cdots \prec_{\rm c} F_{1m} \prec_{\rm c} \cdots \prec_{\rm c} F_{13} \prec_{\rm c} F_{12}$.*

Proof. (a) By Lemma 5.4.8 of Barlow and Proschan (1981), since H is SI, there is an increasing function $g(\cdot,\cdot)$ such that

$$(Y_{j-1}, Y_j) \stackrel{d}{=} (Y_{j-1}, g(Y_{j-1}, U_j)), \quad j > 1,$$

where $U_2, U_3, \ldots$ is a sequence of independent rvs with U_j independent of Y_{j-1}. Let $g_1(y, u_2) = g(y, u_2)$ and recursively define

$$g_k(y, u_2, \ldots, u_{k+1}) = g(g_{k-1}(y, u_2, \ldots, u_k), u_{k+1}), \quad k > 1.$$

By induction, g_k is increasing in each of its arguments. From the structure of the Markov chain,

$$Y_j \stackrel{d}{=} g_{j-1}(Y_1, U_2, \ldots, U_j), \quad j \geq 2.$$

Hence $Y_j \uparrow_{\rm st} Y_1$ for all $j \geq 2$.

(b) For a first-order Markov chain, Y_1 and Y_m are conditionally independent given Y_{m-1}. Hence,

$$\begin{aligned} F_{1,m}(u, v) &= \Pr(Y_1 \leq u, Y_m \leq v) \\ &= \int_{y_1=-\infty}^{u} \int_{-\infty}^{\infty} F_{Y_m|Y_{m-1}}(v|z)\, F_{Y_1,Y_{m-1}}(dy_1, dz) \end{aligned}$$

$$\begin{aligned}
&= \int_{-\infty}^{u}\int_{-\infty}^{\infty} H(v|z)\, F_{1,m-1}(dy_1, dz) \\
&\le \int_{-\infty}^{u}\int_{-\infty}^{\infty} H(v|z)\, F_{1,m-2}(dy_1, dz) \\
&= \Pr(Y_1 \le u, Y_{m-1} \le v) = F_{1,m-1}(u, v),
\end{aligned}$$

if $(Y_1, Y_{m-1}) \prec_c (Y_1, Y_{m-2})$, since H is SI and the function $\phi(y_1, z) = H(v|z) I_{(-\infty,u]}(y_1)$ satisfies the condition of Theorem 2.8. To start the induction, we need $F_{13}(u,v) \le F_{12}(u,v)$ for all u, v. But

$$\begin{aligned}
F_{13}(u,v) &= \int_{y_1=-\infty}^{u} \int_{-\infty}^{\infty} H(v|z)\, F_{12}(dy_1, dz) \\
&= \mathrm{E}\,[H(v|Y_2) I_{(-\infty,u]}(Y_1)] \le \mathrm{E}\,[H(v|Y_1) I_{(-\infty,u]}(Y_1)] = F_{12}(u,v),
\end{aligned}$$

since $(Y_1, Y_2) \prec_c (Y_1, Y_1)$. □

If the transition distribution is negatively dependent, then intuitively one may have the property that (Y_1, Y_m) is negatively dependent if m is even and positively dependent if m is odd with the overall dependence decreasing in m. A sufficient condition for this is given in next theorem.

Theorem 8.4 *Let $Y_1, Y_2, \ldots$ be a stationary Markov chain with state space in $\Re$. If H is stochastically decreasing then*

(a) $Y_{2n} \downarrow_{\mathrm{st}} Y_1$ and $Y_{2n+1} \uparrow_{\mathrm{st}} Y_1$, $n = 1, 2, \ldots$, and

(b) $F_{12} \prec_c F_{14} \prec_c \cdots \prec_c F^{(2)} \prec_c \cdots \prec_c F_{15} \prec_c F_{13}$.

Proof. Note that $Y_1, Y_3, Y_5, \ldots$ is a Markov chain with transition kernel $H^*(y|x) = \Pr(Y_3 \le y | Y_1 = x)$. The second parts of (a) and (b) now follow from parts (a) and (b) of Theorem 8.3 if H^* is SI. A stochastic representation argument that shows that H^* is SI is as follows. Similar to Lemma 5.4.8 of Barlow and Proschan (1981), H stochastically decreasing implies that there exists a function g such that $Y_2 \stackrel{d}{=} g(Y_1, U_2)$, $Y_3 \stackrel{d}{=} g(Y_2, U_3)$, where (i) U_2, U_3 are independent rvs and are respectively independent of Y_1, Y_2, and (ii) $g(u, v)$ is decreasing in u and increasing in v. Hence $Y_3 \stackrel{d}{=} g(g(Y_1, U_2), U_3)$ and $Y_3 \uparrow_{\mathrm{st}} Y_1$.

Similarly, $Y_2, Y_4, \ldots$ is a Markov chain with transition kernel $H^*(y|x)$. Hence $Y_{2n} \uparrow_{\mathrm{st}} Y_2$ from part (a) of Theorem 8.3. From Lemma 5.4.8 of Barlow and Proschan (1981), there is an increasing function g_{2n} in two real arguments such that $Y_{2n} \stackrel{d}{=} g_{2n}(Y_2, U_{2n})$. Hence $Y_{2n} \stackrel{d}{=} g_{2n}(g(Y_1, U_2), U_{2n})$ and $Y_{2n} \downarrow_{\mathrm{st}} Y_1$ and the first half of (a) is established.

Finally, the proof of the first half of (b) is similar to the proof of Theorem 8.3. Since two rvs in a first-order Markov chain are conditionally independent given intermediate rvs in the sequence, Y_1 and Y_{2n} are conditionally independent given Y_{2n-2}. Now

$$\begin{aligned} F_{1,2n}(u,v) &= \Pr(Y_1 \le u, Y_{2n} \le v) \\ &= \int_{y_1=-\infty}^{u} \int_{-\infty}^{\infty} F_{Y_{2n}|Y_{2n-2}}(v|z)\, F_{Y_1,Y_{2n-2}}(dy_1, dz) \\ &= \int_{-\infty}^{u} \int_{\infty}^{-\infty} H^*(v|z)\, F_{1,2n-2}(dy_1, dz) \\ &\ge \int_{-\infty}^{u} \int_{\infty}^{-\infty} H^*(v|z)\, F_{1,2n-4}(dy_1, dz) \\ &= \Pr(Y_1 \le u, Y_{2n-2} \le v) = F_{1,2n-2}(u,v) \end{aligned}$$

if $(Y_1, Y_{2n-4}) \prec_c (Y_1, Y_{2n-2})$ since H^* is SI and the function $\phi(y_1, z) = H^*(v|z) I_{(-\infty,u]}(y_1)$ satisfies the condition of Theorem 2.8. To start the induction, we need $F_{14} \ge F_{12}$. Now

$$F_{14}(u,v) = \int_{y_1=-\infty}^{u} \int_{-\infty}^{\infty} H(v|z)\, F_{13}(dy_1, dz)$$

$$= \mathrm{E}\,[H(v|Y_3)\, I_{(-\infty,u]}(Y_1)] \ge \mathrm{E}\,[H(v|Y_1)\, I_{(-\infty,u]}(Y_1)] = F_{12}(u,v),$$

since $(Y_1, Y_3) \prec_c (Y_1, Y_1)$ and $-H(v|z)\, I_{(-\infty,u]}(y_1)$ is lattice superadditive in y_1, z (see Tchen 1980, for this condition).

Finally, $F_{1,2n} \prec_c F^{(2)} \prec_c F_{1,2n+1}$ follows from (a) and Theorem 2.3. □

The next result is valid for arbitrary state spaces whenever all densities exist (including marginal and transition densities). The notion of dependence used is a measure of dependence based on **directed divergence**. Let p, q be probability densities on a space $\mathcal{H}$ with measure ν, and let ψ be a convex function on $[0, \infty)$, strictly convex at 1 and satisfying $\psi(1) = 0$. Then the **ψ-divergence** of p from q is

$$I_\psi(p,q) = \int_{\mathcal{H}} q\, \psi(p/q)\, d\nu. \tag{8.20}$$

A measure of bivariate dependence is obtained when q is the product of univariate marginal densities and p has these univariate margins. The special case $\psi(u) = u \log u$ leads to the relative entropy measure of dependence. The next theorem concerns the decrease in lag with dependence for measures of dependence based on (8.20).

Theorem 8.5 *Let $Y_1, Y_2, \ldots$ be a stationary Markov chain on a state space $\mathcal{H}$ with measure ν. Let f be the density of Y_1 and let f_{1m} be the density of Y_1, Y_m. For a fixed convex function ψ on $[0, \infty)$, strictly convex at 1 and satisfying $\psi(1) = 0$, let*

$$\delta_{1m} = \int_{\mathcal{H}} f(x)f(y)\,\psi\big(f_{1m}(x,y)/[f(x)f(y)]\big)\,d\nu(x)d\nu(y)$$

for $m = 2, 3, \ldots$. Then δ_{1m} decreases as m increases.

Proof. With the conditional density $h^{(m-1)}(y|z) = f_{1m}(z,y)/f(z)$,

$$\delta_{1m} = \int_{\mathcal{H}} f(z) \int_{\mathcal{H}} f(y)\,\psi\big(h^{(m-1)}(y|z)/f(y)\big)\,d\nu(y)d\nu(z).$$

Hence $\delta_{1m} \geq \delta_{1,m+1}$ if for all $z \in \mathcal{H}$, $m \geq 2$,

$$\int_{\mathcal{H}} f(y)\,\psi\big(h^{(m-1)}(y|z)/f(y)\big)\,d\nu(y) \geq \int_{\mathcal{H}} f(y)\,\psi\big(h^{(m)}(y|z)/f(y)\big)\,d\nu(y). \tag{8.21}$$

This inequality obtains from results in Joe (1990b) as follows.

The function $k(x,y) = h(y|x)$ on $\mathcal{H} \times \mathcal{H}$ satisfies $k(x,y) \geq 0$, $\int k(x,y)\,d\nu(y) = 1$ for all x, $\int f(x)k(x,y)\,d\nu(x) = f(y)$ for all y, and $h^{(m)}(y|z) = \int h^{(m-1)}(x|z)k(x,y)\,d\nu(x)$ for all y, z. Hence $h^{(m)}(\cdot|z)$ is r-majorized by $h^{(m-1)}(\cdot|z)$ with respect to f (the interpretation is that $h^{(m)}(\cdot|z)$ is closer to f than $h^{(m-1)}(\cdot|z)$) and (8.21) holds. □

The next result is for a state space that is discrete and finite. The dependence measure is the Goodman–Kruskal λ and no other conditions are needed for the decrease in dependence with lag.

Theorem 8.6 *Let $Y_1, Y_2, \ldots$ be a stationary Markov chain with a finite discrete state space $\mathcal{H}$ (of unordered states). Let f_{1m} denote the joint pmf for Y_1, Y_m and let f denote the pmf for each Y_i. Let λ_{1m} be the Goodman–Kruskal λ for Y_1, Y_m, i.e.,*

$$\lambda_{1m} = \frac{\sum_{i \in \mathcal{H}} f_{1m}(i,*) - f(*)}{1 - f(*)}, \quad m = 2, 3, \ldots,$$

where $f(*) = \max_{j \in \mathcal{H}} f(j)$ and $f_{1m}(i,*) = \max_{j \in \mathcal{H}} f_{1m}(i,j)$. Then λ_{1m} is decreasing in m.

Proof. This is straightforward and left as an exercise. □

The next two results strengthen the conclusion of Theorems 8.3 and 8.4 to the $\prec_{\mathrm{SI}}$ ordering. The stronger dependence conditions that are needed are the TP_2 and RR_2 conditions, respectively.

Theorem 8.7 *Let $\{Y_i : i = 1, 2, \ldots\}$ be a stationary Markov chain with state space $\Re$ and let f_{12} denote the density of (Y_1, Y_2). If $f_{12}(x_1, x_2)$ is TP_2 in x_1, x_2, then for $n \geq 2$,*

$$F^{(2)} \prec_{\mathrm{SI}} F_{1,n+1} \prec_{\mathrm{SI}} F_{1n} \prec_{\mathrm{SI}} \cdots \prec_{\mathrm{SI}} F_{12}.$$

Proof. Let $F_{m|j}, f_{m|j}$ denote the conditional distribution and pdf of Y_m given Y_j. For any real y, y',

$$F_{3|1}(y|x) - F_{2|1}(y'|x) = \int_{-\infty}^{\infty} \left[F_{2|1}(y|u) - I_{(-\infty,y']}(u)\right] f_{2|1}(u|x) du.$$

Since $f_{12}(y_1, y_2)$ is TP_2 in y_1, y_2, $f_{2|1}(u|x)$ is TP_2 in u and x. By the variation diminishing property (Karlin 1968), the one sign change of $F_{2|1}(y|u) - I_{(-\infty,y']}(u)$ in u implies that the number of sign changes of $F_{3|1}(y|x) - F_{2|1}(y'|x)$ in x is at most one, and if there is a sign change it is from $-$ to $+$ as x goes from $-\infty$ to ∞. By Theorem 2.10, $F_{13} \prec_{\mathrm{SI}} F_{12}$.

Now we proceed by induction. Let $n \geq 3$ and suppose that $F_{1n} \prec_{\mathrm{SI}} F_{1,n-1} \prec_{\mathrm{SI}} \cdots \prec_{\mathrm{SI}} F_{12}$. Then

$$F_{n+1|1}(y|x) - F_{n|1}(y'|x) = \int_{-\infty}^{\infty} [F_{n|1}(y|u) - F_{n-1|1}(y'|u)] f_{2|1}(u|x) du.$$

By induction and using Theorem 2.10, $F_{n|1}(y|u) - F_{n-1|1}(y'|u)$ has at most one sign change in u. Using the same argument as before, $F_{n+1|1}(y|x) - F_{n|1}(y'|x)$ has at most one sign change in x (from $-$ to $+$) as x goes from $-\infty$ to ∞, and $F_{1,n+1} \prec_{\mathrm{SI}} F_{1n}$.

Finally, the proof of $F^{(2)} \prec_{\mathrm{SI}} F_{1,n+1}$ follows from Theorems 8.3, 2.3(a) and 2.11. □

Theorem 8.8 *Let $\{Y_i : i = 1, 2, \ldots\}$ be a stationary Markov chain with state space $\Re$ and let f_{12} denote the density of (Y_1, Y_2). If $f_{12}(y_1, y_2)$ is RR_2 in y_1, y_2, then for $n \geq 1$,*

$$F_{1,2n} \prec_{\mathrm{SI}} F_{1,2n+2} \prec_{\mathrm{SI}} F^{(2)} \prec_{\mathrm{SI}} F_{1,2n+3} \prec_{\mathrm{SI}} F_{1,2n+1}.$$

Proof. Let $F_{m|j}, f_{m|j}$ be as defined in the proof of Theorem 8.7. Let $f_{13}(y_1, y_3) = \int_{-\infty}^{\infty} f_{1|2}(y_1|y_2) f_{3|2}(y_3|y_2)\, dF(y_2)$. By the basic composition theorem of Karlin (1968), $f_{13}(y_1, y_3)$ is TP_2 in y_1, y_3. Since $\{Y_{2n+1} : n = 0, 1, 2, \ldots\}$ is a Markov chain (based on the bivariate density f_{13}), the second half of the conclusion follows from Theorem 8.7. The proof of the first half is similar to the proof of Theorem 8.7. We have

$$
\begin{aligned}
&F_{2n|1}(y|x) - F_{2(n+1)|1}(y'|x) \\
&\quad = \int_{-\infty}^{\infty} \left[F_{2n-1|1}(y|u) - F_{2n+1|1}(y'|u)\right] f_{2|1}(u|x)\, du \\
&\quad = \int_{-\infty}^{\infty} \left[F_{2n-1|1}(y|-u) - F_{2n+1|1}(y'|-u)\right] f_{2|1}(-u|x)\, du.
\end{aligned}
$$

By assumption, $f_{2|1}(-u|x)$ is TP_2 in u and x. Applying the variation diminishing theorem of Karlin (1968) and Theorem 2.10 completes the proof. □

Next we compare two Markov chains with different transition probabilities.

Theorem 8.9 *Let F, F' be two bivariate distributions in $\mathcal{F}(F_1, F_1)$ and suppose that $F \prec_c F'$. Suppose that the conditional distributions $F_{2|1}, F'_{2|1}, F_{1|2}, F'_{1|2}$ are all SI. Let $Y_1, Y_2, \ldots$ be a Markov chain with transition distribution $F_{2|1}$ and let $Y_1', Y_2', \ldots$ be a Markov chain with transition distribution $F'_{2|1}$ (both with state space in $\Re$). Then $(Y_1, Y_j) \prec_c (Y_1', Y_j')$ for $j = 2, 3, \ldots$.*

Proof. Let $\overline{F}_{2|1} = 1 - F_{2|1}$, $\overline{F}'_{2|1} = 1 - F'_{2|1}$, etc..

First consider $j = 3$. Fix u_1, u_3. Then

$$
\begin{aligned}
&\Pr(Y_1' > u_1, Y_3' > u_3) - \Pr(Y_1 > u_1, Y_3 > u_3) \\
&\quad = \int_{-\infty}^{\infty} \left[\overline{F}'_{1|2}(u_1|v)\overline{F}'_{2|1}(u_3|v) - \overline{F}_{1|2}(u_1|v)\overline{F}_{2|1}(u_3|v)\right] dF_1(v) \\
&\quad = \int_{-\infty}^{\infty} \overline{F}'_{1|2}(u_1|v)\left[\overline{F}'_{2|1}(u_3|v) - \overline{F}_{2|1}(u_3|v)\right] dF_1(v) \\
&\qquad + \int_{-\infty}^{\infty} \left[\overline{F}'_{1|2}(u_1|v) - \overline{F}_{1|2}(u_1|v)\right]\overline{F}_{2|1}(u_3|v)\, dF_1(v).
\end{aligned}
$$

The last two summands are similar so we only show that the first is non-negative. Using Exercise 2.23, suppose that there are $2r-1$ sign changes ($r \geq 1$) in $F_{2|1}(y|\cdot) - F'_{2|1}(y|\cdot)$. Let the locations of the changes be denoted by b_i, $i = 1, \ldots, 2r-1$. Let $b_0 = -\infty$, $b_{2r} = \infty$. Then

$$
\begin{aligned}
&\int_{-\infty}^{\infty} \overline{F}'_{1|2}(u_1|v)\left[\overline{F}'_{2|1}(u_3|v) - \overline{F}_{2|1}(u_3|v)\right] dF_1(v) \\
&\quad = \sum_{i=1}^{2r} \int_{b_{i-1}}^{b_i} \overline{F}'_{1|2}(u_1|v)\left[\overline{F}'_{2|1}(u_3|v) - \overline{F}_{2|1}(u_3|v)\right] dF_1(v) \\
&\quad \geq \sum_{i=1}^{r} \overline{F}'_{1|2}(u_1|b_{2i-1}) \int_{b_{2i-2}}^{b_{2i}} \left[\overline{F}'_{2|1}(u_3|v) - \overline{F}_{2|1}(u_3|v)\right] dF_1(v).
\end{aligned}
$$

This is 0 if $r = 1$, and for $r \geq 2$ it is

$$\sum_{i=2}^{r} \Big\{ [\overline{F}'_{1|2}(u_1|b_{2i-1}) - \overline{F}'_{1|2}(u_1|b_{2i-3})] \cdot \int_{b_{2i-2}}^{\infty} [\overline{F}'_{2|1}(u_3|v) - \overline{F}_{2|1}(u_3|v)]\, dF_1(v) \Big\} \geq 0$$

since $\overline{F}'_{1|2}(u_1|b_{2i-1}) - \overline{F}'_{1|2}(u_1|b_{2i-3}) \geq 0$ from the SI assumption and the integral is non-negative from the concordance assumption.

Now we proceed by induction. Let $j \geq 4$. Suppose we have $(Y_2, Y_m) \prec_c (Y'_2, Y'_m)$ for $m \leq j$. Let $F_{j|2}, F'_{j|2}$ denote the conditional distributions of Y_j given Y_2 and Y'_j given Y'_2, respectively. Then

$$\begin{aligned} &\Pr(Y'_1 > u_1, Y'_j > u_j) - \Pr(Y_1 > u_1, Y_j > u_j) \\ &\quad = \textstyle\int_{-\infty}^{\infty} [\overline{F}'_{1|2}(u_1|v)\overline{F}'_{j|2}(u_j|v) - \overline{F}_{1|2}(u_1|v)\overline{F}_{j|2}(u_j|v)]\, dF_1(v). \end{aligned}$$

The above argument can be applied to conclude that this difference of probabilities is non-negative, since $\overline{F}'_{j|2}$ and $\overline{F}_{j|2}$ are also SI, $(Y_2, Y_j) \prec_c (Y'_2, Y'_j)$ by induction, and Exercise 2.23 can be applied to $\overline{F}'_{j|2} - \overline{F}_{j|2}$. □

Finally, we have a result involving the PFD concept (see Section 2.1.6) that is relevant to likelihood inference for dependent sequences. Mainly, we show that (asymptotic) standard errors computed based on an assumption of time independence for a stationary time series are too small when in fact there is time dependence (in the form of Markov chains which satisfy certain conditions). An example with data that illustrates this is given in Section 11.6.

Given (differentiable) parametric families of bivariate copulas $C(\cdot;\boldsymbol{\delta})$ and univariate cdfs $F(\cdot;\boldsymbol{\theta})$,

$$G(x, y; \boldsymbol{\delta}, \boldsymbol{\theta}) = C(F(x;\boldsymbol{\theta}), F(y;\boldsymbol{\theta}); \boldsymbol{\delta})$$

is a bivariate family with univariate margins F. For the Markov chain based on G, the Markov transition density is

$$\begin{aligned} h(y|x;\boldsymbol{\theta},\boldsymbol{\delta}) &= \frac{\partial^2 G(x, y; \boldsymbol{\delta}, \boldsymbol{\theta})}{\partial x \partial y} \Big/ \frac{\partial F(x;\boldsymbol{\theta})}{\partial x} \\ &= c(F(x;\boldsymbol{\theta}), F(y;\boldsymbol{\theta}); \boldsymbol{\delta})\, f(y;\boldsymbol{\theta}), \end{aligned}$$

where $c(u, v; \boldsymbol{\delta}) = \partial^2 C(u, v; \boldsymbol{\delta})/\partial u \partial v$ and $f(y;\boldsymbol{\theta}) = \partial F(y;\boldsymbol{\theta})/\partial y$. The likelihood for observations $y_1, \ldots, y_n$ based on this transition density is $f(y_1;\boldsymbol{\theta})\prod_{t=2}^n h(y_t|y_{t-1};\boldsymbol{\theta},\boldsymbol{\delta})$. The log-likelihood is

$$L(\boldsymbol{\delta},\boldsymbol{\theta}) = \sum_{t=1}^{n} \log f(y_t;\boldsymbol{\theta}) + \sum_{t=2}^{n} \log c(F(y_{t-1};\boldsymbol{\theta}), F(y_t;\boldsymbol{\theta});\boldsymbol{\delta}). \quad (8.22)$$

Assuming that the standard regularity conditions hold, then from Billingsley (1961) (see also Section 10.4), the maximum likelihood estimate (MLE) of $(\boldsymbol{\delta},\boldsymbol{\theta})$ from (8.22) is asymptotically normal and the inverse Hessian matrix (matrix of second-order derivatives) for L evaluated at the MLE $(\hat{\boldsymbol{\delta}},\hat{\boldsymbol{\theta}})$ is the inverse Fisher information, and it can be used for SEs for functions of $\boldsymbol{\theta}$ such as quantiles.

For comparison, we also consider the MLE of $\boldsymbol{\theta}$ and its asymptotic covariance matrix based on the log-likelihood

$$L(\boldsymbol{\theta}) = \sum_{t=1}^{n} \log f(y_t;\boldsymbol{\theta}), \quad (8.23)$$

i.e., the log-likelihood assuming data are iid from the density $f(\cdot;\boldsymbol{\theta})$. Theoretically, if the true dependence structure for the sequence is Markov with any of the families B1, B3, B4, B5, B6, B7 of copulas in Section 5.1, then the use of (8.23) for maximum likelihood estimation leads to SEs that are too small, although the MLE of $\tilde{\boldsymbol{\theta}}$ from (8.23) is consistent. An outline of the proof of this result is given below. We conjecture this result to be true for the family B2 and for other models for the time dependence. It is similar to the well-known property for the variance of a sample mean — the variance is larger with positive dependence of the rvs than with independence because of the additional positive covariance terms.

Proof. Let $Y_1,\ldots,Y_n$ be a stationary dependent sequence. Suppose the marginal density is in the parametric family $f(y;\boldsymbol{\theta})$, where $\boldsymbol{\theta}$ is a column vector, and let $F(y;\boldsymbol{\theta})$ be the family of cdfs. Assume that the usual regularity conditions of asymptotic likelihood theory hold (see Serfling, 1980). Let $\ell(\boldsymbol{\theta};y) = \log f(y;\boldsymbol{\theta})$, $S(\boldsymbol{\theta};y) = \partial\ell(\boldsymbol{\theta};y)/\partial\boldsymbol{\theta}$ (the score vector) and $S_i = S(\boldsymbol{\theta};Y_i)$. Let $\mathcal{I}(\boldsymbol{\theta}) = -\mathrm{E}\,[\partial^2\ell(\boldsymbol{\theta};Y_i)/\partial\boldsymbol{\theta}\partial\boldsymbol{\theta}^T] = \mathrm{Var}\,(S_i)$ be the Fisher information matrix.

The asymptotic normality for $\tilde{\boldsymbol{\theta}}$ comes from the approximation

$$n^{1/2}(\tilde{\boldsymbol{\theta}}-\boldsymbol{\theta}) \approx \mathcal{I}^{-1}(\boldsymbol{\theta})\left[n^{-1/2}\sum S_i\right],$$

where $\mathcal{I}^{-1}(\boldsymbol{\theta})$ is the matrix inverse of $\mathcal{I}(\boldsymbol{\theta})$. Let $g(\boldsymbol{\theta})$ be a real-valued function. From the delta (Taylor expansion) method for a function of the parameter,

$$n^{1/2}(g(\tilde{\boldsymbol{\theta}}) - g(\boldsymbol{\theta})) \approx (\partial g/\partial \boldsymbol{\theta})^T \mathcal{I}^{-1}(\boldsymbol{\theta}) \big[n^{-1/2} \sum S_i\big].$$

The asymptotic variance of $n^{1/2}(g(\tilde{\boldsymbol{\theta}}) - g(\boldsymbol{\theta}))$ for an independent sequence is $(\partial g/\partial \boldsymbol{\theta})^T \mathcal{I}^{-1}(\boldsymbol{\theta})(\partial g/\partial \boldsymbol{\theta})$. The asymptotic variance of $n^{1/2}(g(\tilde{\boldsymbol{\theta}}) - g(\boldsymbol{\theta}))$ for a dependent sequence is

$$(\partial g/\partial \boldsymbol{\theta})^T \mathcal{I}^{-1}(\boldsymbol{\theta})(\partial g/\partial \boldsymbol{\theta}) + n^{-1} \sum_{i \neq i'} \mathrm{Cov}\,(B_i, B_{i'}),$$

where $B_i = (\partial g/\partial \boldsymbol{\theta})^T \mathcal{I}^{-1}(\boldsymbol{\theta})\, S_i = (\partial g/\partial \boldsymbol{\theta})^T \mathcal{I}^{-1}(\boldsymbol{\theta})\, S(\boldsymbol{\theta}; Y_i)$, $i = 1, \ldots, n$. Therefore, for all real-valued g, from the PFD condition, the asymptotic variance of $g(\tilde{\boldsymbol{\theta}})$ is greater than under independence if, for all $i \neq i'$,

$$\mathrm{Cov}\,(h(Y_i), h(Y_{i'})) \geq 0, \quad \forall \text{ real-valued } h \tag{8.24}$$

such that the covariances exist (compare Gleser and Moore, 1983).

Now suppose $\{Y_i\}$ is a reversible Markov chain based on a copula C that satisfies $C(u, v) = C(v, u)$ for all u, v. The joint distribution of Y_i, Y_{i+1} is $C(F(y_i; \boldsymbol{\theta}), F(y_{i+1}; \boldsymbol{\theta}))$. If (8.24) is satisfied for Y_i, Y_{i+1}, then it is satisfied for Y_i, Y_{i+j} for all $j \geq 2$. The proof is by induction on the lag j:

$$\begin{aligned} \mathrm{Cov}\,\big(h(Y_i), h(Y_{i+j})\big) &= \mathrm{E}\big\{\mathrm{Cov}\,\big(h(Y_i), h(Y_{i+j}) \mid Y_{i+1}, \ldots, Y_{i+j-1}\big)\big\} \\ &+ \mathrm{Cov}\,\big(\mathrm{E}\,[h(Y_i) \mid Y_{i+1}, \ldots, Y_{i+j-1}], \mathrm{E}\,[h(Y_{i+j}) \mid Y_{i+1}, \ldots, Y_{i+j-1}]\big) \\ &= 0 + \mathrm{Cov}\,\big(a(Y_{i+1}), a(Y_{i+j-1})\big) \geq 0, \end{aligned} \tag{8.25}$$

where $a(y) = \mathrm{E}\,[h(Y_i) \mid Y_{i-1} = y] = \mathrm{E}\,[h(Y_i) \mid Y_{i+1} = y]$. The first term in (8.25) is 0 since two rvs in a Markov chain are conditionally independent given intermediate rvs in the sequence. The second term follows from the Markov property and reversibility.

For the families B3–B6, (8.24) follows from Theorem 4.6. For the family B1, the positive dependence by mixture condition in Section 2.1.6 can be used to obtain (8.24) (Exercise 8.3). For the family B7, (8.24) can be shown via an extreme value limit (Exercise 8.13).
□

In conclusion, for some Markov time series based on bivariate copulas, the estimation of the univariate parameter $\boldsymbol{\theta}$ from the likelihood, assuming independence, leads to a consistent estimate of $\boldsymbol{\theta}$ but the SEs of functions of $\boldsymbol{\theta}$ are too small. The main result used is the positive dependence condition in (8.24). Since (8.24) may hold even for non-Markov models or Markov models of order more than 1, we can expect more generally that SEs are too small when there is positive dependence in the time series and the likelihood assuming independence is used for estimation.

8.6 Bibliographic notes

References for Markov chains and binary time series include Cox and Snell (1989) and Muenz and Rubinstein (1985). Binary Markov chains with a time-independent covariate vector $\mathbf{x}$ are studied in Darlington and Farewell (1992), but in one model, there is an inconsistency in the definition of $\rho = \rho(\mathbf{x})$. Markov chain random effects models are considered in Gardner (1990). A more complicated model for random effects is given in Stiratelli, Laird and Ware (1984). The results in Section 8.2 for 1-dependent and k-dependent sequences are new. Latent variable discrete time series models based on the MVN distribution have been mentioned in the econometrics literature; see, for example, Heckman (1981).

References for Markov chain transitional models for categorical data include Diggle, Liang and Zeger (1994), Albert (1994), Follmann (1994) and Gottschau (1994). For longitudinal data, another class of Markov-like models uses previous responses as covariates. Some references for this approach are Bonney (1987) and Fahrmeir and Kaufman (1987). These models do not consider the joint distribution of subsets of $\{Y_t\}$.

References for the extremal index relevant to the material in this chapter are Leadbetter, Lindgren and Rootzén (1983), O'Brien (1987), Smith (1992) and Smith and Weissman (1994).

Some references for non-normal AR and ARMA models are Joe (1996b), McKenzie (1988), Lewis, McKenzie and Hugus (1989), Al-Osh and Alzaid (1993; 1994) and Alzaid and Al-Osh (1993). In Section 8.4.4, with a negative binomial margin, the dependence of the parameters on the covariates is different from that in Lawless (1987), as can be seen in the variance to mean relationship. References for Section 8.4.5 are Bernier (1970), Gaver and Lewis (1980) and McKenzie (1986).

Further results on the decrease in dependence with lag for stationary Markov chains are given in Fang, Hu and Joe (1994) and Hu and Joe (1995). Theorems 8.7 and 8.8 are due to T. Hu; he also assisted in the proof of Theorem 8.9, which is new.

8.7 Exercises

8.1 Obtain the transition density for a stationary Markov chain associated with the trivariate copula (4.29) in Section 4.3, when K is in the family B6 and ψ is in the LT family LTA.

8.2 Let $Y_1, Y_2, \ldots$ be the Markov chain based on the copula family B10. Obtain the copula for (Y_1, Y_m), $m > 1$. [Hint: use induction and show that it is within the same family B10.]

8.3 Show that the BVN distribution with positive correlation has a representation of the form $\int P(u;\alpha)P(v;\alpha)\,dM(\alpha)$ for appropriately chosen P, M.

8.4 In (8.4), consider a 1-dependent series based on the copula family B10 in Section 5.1. What is the copula for (Y_1, Y_2)?

8.5 Study the tail dependence properties of the 1-dependent series associated with copulas in Section 8.2.1.

8.6 Prove the properties in Theorem 8.1 for AR(1) time series with univariate margin in the convolution-closed infinitely divisible class. (Joe 1996b)

8.7 Show that the AR(1) model in Section 8.4.1 is a special case of the AR(2) model in Section 8.4.3 when $(\theta_1, \theta_2, \theta_3)$ has the form $(\theta\alpha[1-\alpha], 0, \theta\alpha^2)$, where $0 < \alpha < 1$.

8.8 Relating to Section 8.1.4, let $\{Y_t\}$ be a stationary AR(d) ($d \geq 2$) normal sequence. Let ρ_k be the autocorrelation coefficient of lag k, let Σ_{11} be the $d \times d$ correlation matrix with $\rho_{|i-j|}$ in the (i,j) position for $i \neq j$, and let $\Sigma_{12} = (\rho_d, \rho_{d-1}, \ldots, \rho_1)^T$ be a column vector of length d. The coefficients of $\Sigma_{11}^{-1}\Sigma_{12} = (\phi_d, \ldots, \phi_2, \phi_1)^T$ are the coefficients in the linear representation $Y_t = \sum_{i=1}^d \phi_i Y_{t-i} + \epsilon_t$. Show that a necessary and sufficient condition for the density of $(Y_1, \ldots, Y_n)$ to be MTP$_2$ for all $n > d$ is that $\phi_i \geq 0$, $i = 1, \ldots, d$ (which is equivalent to the partial autocorrelations $\pi_1, \ldots, \pi_d$ of $\{Y_t\}$ being non-negative), and $\phi_i - \sum_{j=1}^{d-i} \phi_j \phi_{j+i} \geq 0$, $i = 1, \ldots, d-1$. [Hint: let A_n be the inverse correlation matrix for $Y_1, \ldots, Y_n$. Obtain conditions for all of the off-diagonal elements of A_n to be non-positive.]

8.9 Prove Theorem 8.6. (Fang, Hu and Joe 1994)

8.10 For stationary AR(p) normal sequences, obtain conditions for the autocorrelations to be decreasing with lag.

(Fang, Hu and Joe 1994)

8.11 If the SI condition of Theorem 8.3 is weakened to PQD, LTD or RTI, then the conclusion need not hold. Show this through counterexamples. (Fang, Hu and Joe 1994)

8.12 Construct a Poisson time series with negative autocorrelation of lag 1 based on the bivariate Poisson distribution in Exercise 7.12.

8.13 Prove the PFD property for the Markov chain based on the copula family B7 (a result due to T. Hu). [Hint: from Section 8.5, it suffices to prove the PFD property for the family B7. The idea is to use a stochastic representation from the extreme value limit of the copula family B4, since the PFD property is closed under weak convergence.]

8.8 Unsolved problems

8.1 Conjecture: Assuming that $\{Y_t\}$ is a stationary sequence with marginal distribution F, the sequence $\alpha_i(y)$ defined by (8.1) converges as $i \to \infty$, under minimal conditions.

8.2 Find the maximum and minimum lag 1 correlation for 1-dependent stationary binary sequences with marginal probability p for the occurrence of a 1. (See Section 8.2.3.)

8.3 For $k \geq 1$, among k-dependent stationary binary sequences with marginal probability p for the occurrence of a 1, find the most and least dependent sequences.

8.4 Can the approach for AR models with univariate margins in the convolution-closed infinitely divisible class be extended to allow for negative autocorrelations?

8.5 Prove (or disprove) the PFD property for the Markov chain based on the copula family B2 (see Section 8.5).

8.6 Extend some of the results in Section 8.5 on decrease in dependence with lag to higher-order stationary Markov chains.

8.7 Prove some results on the $\prec_{\text{pfd}}$ ordering for (Y_1, Y_j) associated with a Markov chain from a bivariate copula, i.e., stronger results than PFD.

8.8 Develop some theory for spatial processes with given univariate margins.

CHAPTER 9

Models from given conditional distributions

This chapter complements and supplements Chapters 3 and 4, in that we study the construction of multivariate models from families of conditional distributions (with compatibility conditions on the conditional distributions). This approach has been considered in the statistical literature partly because of the difficulty of construction of families of multivariate distributions with given margins. The type of dependence that can arise is surprising in some cases.

Examples that are illustrated in some detail are: (i) conditional distributions in exponential families; and (ii) multivariate binary response with conditional logistic regressions. The examples are in Section 9.2, following some theory in Section 9.1. The model of type (ii) is applied in the data analysis example in Section 11.1.

9.1 Conditional specifications and compatibility conditions

Throughout this section, we assume that all densities exist with respect to appropriate measure spaces.

Consider first the bivariate case. Suppose the conditional densities $f_{1|2}(\cdot|y)$ are given for all y and $f_{2|1}(\cdot|x_0)$ is given for a particular value x_0. Assume that $f_{1|2}(x_0|y) > 0$ for all y. Then for the joint density f_{12},

$$f_{12}(x, y) \propto \frac{f_{1|2}(x|y)f_{2|1}(y|x_0)}{f_{1|2}(x_0|y)}, \tag{9.1}$$

with the proportionality constant equal to

$$f_1(x_0) = \left\{\int\int [f_{1|2}(x|y)f_{2|1}(y|x_0)/f_{1|2}(x_0|y)]\, d\nu(x)\, d\nu(y)\right\}^{-1},$$

where ν is the appropriate measure. That is, for two rvs, the set of conditional densities given one variable, plus one conditional density given the other variable, determines the joint bivariate distribution.

Now if the conditional densities $f_{1|2}(\cdot|y)$ are given for all y and $f_{2|1}(\cdot|x)$ are given for all x, then from (9.1) a condition for compatibility is that

$$\frac{f_{1|2}(x|y)f_{2|1}(y|x_1)}{f_{1|2}(x_1|y)} \Big/ \frac{f_{1|2}(x|y)f_{2|1}(y|x_2)}{f_{1|2}(x_2|y)} = \frac{f_{1|2}(x_2|y)f_{2|1}(y|x_1)}{f_{1|2}(x_1|y)f_{2|1}(y|x_2)} \tag{9.2}$$

does not depend on x, y for all choices of $x_1 \neq x_2$. (Another compatibility condition is that there exist non-negative functions $a(x)$, $b(y)$ such that $a(x)f_{2|1}(y|x) = b(y)f_{1|2}(x|y)$ for all x, y.)

As an example, consider exponential conditional distributions: $f_{1|2}(x|y) = \lambda_1(y)\,e^{-\lambda_1(y)x}$, $f_{2|1}(y|x) = \lambda_2(x)\,e^{-\lambda_2(x)y}$, $x, y > 0$, $\lambda_1(y), \lambda_2(x) \geq 0$. Condition (9.2) becomes

$$[\lambda_2(x_1)/\lambda_2(x_2)]\,e^{-\lambda_1(y)[x_2-x_1]}\,e^{-y[\lambda_2(x_1)-\lambda_2(x_2)]},$$

and this is independent of x, y if and only if λ_1, λ_2 are linearly nondecreasing with a common slope ($\lambda_2(x) = \alpha + \gamma x$, $\lambda_1(y) = \beta + \gamma y$, $\alpha, \beta, \gamma \geq 0$).

The trivariate extension is as follows. Suppose the conditional densities $f_{1|23}(\cdot|y, z)$ are given for all y, z, $f_{2|13}(\cdot|x_0, z)$ are given for all z and a fixed x_0, and $f_{3|12}(\cdot|x_0, y_0)$ is given for some fixed y_0 (and the same x_0 as for $f_{2|13}$). Assume that $f_{1|23}(x_0|y, z) > 0$ for all y, z and $f_{2|13}(y_0|x_0, z) > 0$ for all z. Then for the joint density f_{123},

$$f_{123}(x, y, z) \propto \frac{f_{1|23}(x|y, z)f_{2|13}(y|x_0, z)f_{3|12}(z|x_0, y_0)}{f_{1|23}(x_0|y, z)f_{2|13}(y_0|x_0, z)},$$

with the proportionality constant equal to $f_{12}(x_0, y_0)$. If the conditional densities $f_{1|23}(\cdot|y, z)$ are given for all y, z, $f_{2|13}(\cdot|x, z)$ are given for all x, z, and $f_{3|12}(\cdot|x, y)$ are given for all x, y, then a condition for compatibility is that

$$\frac{f_{1|23}(x_2|y, z)f_{2|13}(y_2|x_2, z)f_{2|13}(y|x_1, z)f_{3|12}(z|x_1, y_1)}{f_{1|23}(x_1|y, z)f_{2|13}(y_1|x_1, z)f_{2|13}(y|x_2, z)f_{3|12}(z|x_2, y_2)}$$

does not depend on x, y, z for all choices of $(x_1, y_1) \neq (x_2, y_2)$.

The ideas in the preceding paragraph clearly extend to higher dimensions, when one has the set of conditional distributions of each variable given the remainder. The result is as follows and the proof is left as an exercise. Let $f_{1\cdots m}$ be the joint density of rvs $Y_1, \ldots, Y_m$ and let $f_{i|r}$ be the conditional density of Y_i given the remaining variables Y_j, $j \neq i$. Then

$$f_{1\cdots m}(\mathbf{y}) \propto \frac{\prod_{i=1}^m f_{i|r}(y_i|y_1^0, \ldots, y_{i-1}^0, y_{i+1}, \ldots, y_m)}{\prod_{i=1}^m f_{i|r}(y_i^0|y_1^0, \ldots, y_{i-1}^0, y_{i+1}, \ldots, y_m)} \tag{9.3}$$

for a given $\mathbf{y}^0$ for which all of the conditional densities in the above expression are positive. The compatibility condition for given sets of conditional densities $f_{i|r}$ is that

$$\frac{\prod_{i=1}^m f_{i|r}(y_i|y_1^0, \ldots, y_{i-1}^0, y_{i+1}, \ldots, y_m)}{\prod_{i=1}^m f_{i|r}(y_i^0|y_1^0, \ldots, y_{i-1}^0, y_{i+1}, \ldots, y_m)} \cdot \frac{\prod_{i=1}^m f_{i|r}(z_i^0|z_1^0, \ldots, z_{i-1}^0, y_{i+1}, \ldots, y_m)}{\prod_{i=1}^m f_{i|r}(y_i|z_1^0, \ldots, z_{i-1}^0, y_{i+1}, \ldots, y_m)} \tag{9.4}$$

does not depend on $\mathbf{y}$ for $\mathbf{y}^0 \neq \mathbf{z}^0$.

Note that results symmetric to (9.3) and (9.4), with the indices of $\mathbf{y}$ permuted, also hold. This symmetry is useful for getting a general form in cases of specific conditional distributions.

More generally, one can consider other sets of conditional distributions. A result from Gelman and Speed (1993) is the following. Let $z_1, \ldots, z_m$ be dummy variables associated with random variables $Z_1, \ldots, Z_m$. If a set of conditional densities uniquely determines a joint density, then there is a permutation $(y_1, \ldots, y_m)$ of $(z_1, \ldots, z_m)$ such that the set is of the form $\{f_{i|A_i \cup \{k:k>i\}}(y_i|y_\ell, \ell \in A_i, y_{i+1}, \ldots, y_m) : i = 1, \ldots, m\}$, where A_i is a subset, possibly non-empty, of $\{1, \ldots, i-1\}$. Unless all of the sets A_i are empty, the conditional densities must be checked for consistency.

9.2 Examples

We illustrate the theory from the previous section with a few examples: exponential conditional densities, exponential family conditional densities, binary conditional densities that are logistic regressions, and more general binary conditional densities. The model with conditional logistic regressions is used in Section 11.1.

9.2.1 Conditional exponential density

Let $\mathbf{y}_{-i}$ denote $\mathbf{y}$ with the ith element y_i deleted. Suppose $\mathbf{Y}$ is an m-dimensional random vector of non-negative rvs, with conditional densities of the form:

$$f_{i|r}(y_i|\mathbf{y}_{-i}) = \lambda_i(\mathbf{y}_{-i})\, e^{-\lambda_i(\mathbf{y}_{-i})y_i},\ 1 \le i \le m,\ y_1, \dots, y_m \ge 0,$$

with the functions λ_i being positive and differentiable. The compatibility condition (9.4) simplifies to

$$\sum_{i=1}^{m}(y_i - y_i^0)\,\lambda_i(y_1^0, \dots, y_{i-1}^0, y_{i+1}, \dots, y_m)$$

$$+\sum_{i=1}^{m}(z_i^0 - y_i)\,\lambda_i(z_1^0, \dots, z_{i-1}^0, y_{i+1}, \dots, y_m) \qquad (9.5)$$

not depending on $\mathbf{y}$ for $\mathbf{y}^0 \ne \mathbf{z}^0$. Expression (9.5) actually does not depend on y_1 because terms with y_1 cancel out. The terms of (9.5) with y_2 are

$$(z_1^0 - y_1^0)\,\lambda_1(y_2, \dots, y_m) + (y_2 - y_2^0)\,\lambda_2(y_1^0, y_3, \dots, y_m)$$
$$+(z_2^0 - y_2)\,\lambda_2(z_1^0, y_3, \dots, y_m).$$

Hence differentiation of (9.5) with respect to y_2 followed by equating to 0 leads to:

$$\frac{\partial \lambda_1}{\partial y_2}(y_2, \dots, y_m) = \frac{\lambda_2(z_1^0, y_3, \dots, y_m) - \lambda_2(y_1^0, y_3, \dots, y_m)}{z_1^0 - y_1^0}.$$

Since the left-hand side does not depend on y_1^0, z_1^0, $\lambda_2(\mathbf{y}_{-2})$ must be linear in y_1, i.e., $\lambda_2(\mathbf{y}_{-2}) = \alpha_2(y_3, \dots, y_m) + \gamma_2(y_3, \dots, y_m)\, y_1$ for some functions α_2, γ_2. Hence by integration,

$$\lambda_1(\mathbf{y}_{-1}) = \alpha_1(y_3, \dots, y_m) + \gamma_2(y_3, \dots, y_m)\, y_2 \qquad (9.6)$$

for some function α_1. By symmetry, $\lambda_1(\mathbf{y}_{-1})$ must have the linear form of (9.6) when y_2 is exchanged with y_j, $j \ge 3$. Hence λ_1 is multilinear in $y_2, \dots, y_m$ with the form:

$$\lambda_1(\mathbf{y}_{-1}) = \alpha_1 + \sum_{S \subset \{2,\dots,m\}} \beta_{1S} \prod_{j \in S} y_j.$$

By symmetry, λ_i has the form:

$$\lambda_i(\mathbf{y}_{-i}) = \alpha_i + \sum_{S \subset \{1,\dots,m\}\setminus\{i\}} \beta_{iS} \prod_{j \in S} y_j.$$

Finally we can substitute back into (9.5) to check on the conditions for the α and β coefficients in order to achieve compatibility. From comparing coefficients of terms of the form

$$y_1^0 \cdots y_i^0 y_{i+1} \cdots y_k \quad \text{and} \quad z_1^0 \cdots z_i^0 y_{i+1} \cdots y_k,$$

one can determine that the constraints on the βs are that, for all i, S with $i \notin S$, the value of β_{iS} depends only on the indices in $S \cup \{i\}$. For example, if $m \geq 3$, the coefficient of $y_1^0 y_2 y_3$ is $\beta_{213} - \beta_{123}$ and this must be zero in order that (9.5) does not depend on $\mathbf{y}$ for any $\mathbf{y}^0 \neq \mathbf{z}^0$.

From (9.3), the multivariate density $f_{1\cdots m}(\mathbf{y})$ of $\mathbf{Y}$ has the form proportional to:

$$\exp\Big\{-\sum_{i=1}^{m}(y_i - y_i^0)\Big[\alpha_i + \sum_{S \subset \{1,\ldots,m\}\setminus\{i\}} \beta_{\{i\}\cup S} \prod_{j\in S, j<i} y_j^0 \prod_{j \in S, j>i} y_j\Big]\Big\}$$

$$\propto \exp\Big\{-\sum_{S \in \mathcal{S}_m} \gamma_S \prod_{j \in S} y_j\Big\} \tag{9.7}$$

for some constants γ_S, $S \in \mathcal{S}_m$. In order that (9.7) is a proper density (with finite integral), the parameters γ_S must be non-negative, and, for each i, there is a set S containing i such that $\gamma_S > 0$.

If $f_{1\cdots m}(\mathbf{y}) = A \exp\{-\sum_{S \in \mathcal{S}_m} \gamma_S \prod_{j \in S} y_j\}$, where A is a normalizing constant, then it is straightforward to verify that the conditional densities are exponential. That is, with a null product being equal to 1,

$$f_{i|r}(y_i|\mathbf{y}_{-i}) = \frac{f_{1\cdots m}(\mathbf{y})}{A(\sum_{S:i\in S} \gamma_S \prod_{j\in S, j\neq i} y_j)^{-1} \exp\{-\sum_{S:i\notin S} \gamma_S \prod_{j\in S} y_j\}}$$

$$= \Big(\sum_{S\in\mathcal{S}_m : i \in S} \gamma_S \prod_{j\in S, j\neq i} y_j\Big) \exp\Big\{-y_i \sum_{S\in\mathcal{S}_m : i\in S} \gamma_S \prod_{j \in S, j\neq i} y_j\Big\},$$

$i = 1, \ldots, m$.

We next show that (9.7) has negative dependence in the sense of RR_2 (see Section 2.1.5). For the bivariate case, (9.7) becomes

$$f_{12}(y_1, y_2) = A \exp\{-\gamma_1 y_1 - \gamma_2 y_2 - \gamma_{12} y_1 y_2\}, \quad y_1, y_2 > 0, \tag{9.8}$$

for $\gamma_1, \gamma_2, \gamma_{12} \geq 0$. For $\gamma_{12} > 0$, this density is RR_2 so that it has negative dependence. The univariate margins of (9.8) have the form $f_j(y_j) \propto (\gamma_{3-j} + \gamma_{12} y_j)^{-1} \exp\{-\gamma_j y_j\}$. More generally, for $m \geq 3$, the density is MRR_2 or RR_2 in any two variables with the remainder fixed, but this condition does not imply that all bivariate margins are RR_2. An analysis of the trivariate case shows that it is possible for one bivariate margin to be TP_2 (but not all three). For $m = 3$, (9.7) becomes

$$f_{123}(\mathbf{y}) = A \exp\{-\gamma_1 y_1 - \gamma_2 y_2 - \gamma_3 y_3 - \gamma_{12} y_1 y_2 - \gamma_{13} y_1 y_3 - \gamma_{23} y_2 y_3 - \gamma_{123} y_1 y_2 y_3\},$$

for $y_1, y_2, y_3 > 0$. The (1,2) bivariate margin has density

$$\begin{aligned} f_{12}(y_1, y_2) &= A\left(\gamma_3 + \gamma_{13}y_1 + \gamma_{23}y_2 + \gamma_{123}y_1y_2\right)^{-1} \\ &\quad \cdot \exp\{-\gamma_1 y_1 - \gamma_2 y_2 - \gamma_{12}y_1y_2\}. \end{aligned}$$

It is straightforward to show that $(\gamma_3 + \gamma_{13}y_1 + \gamma_{23}y_2 + \gamma_{123}y_1y_2)^{-1}$ is RR_2 if $\gamma_3\gamma_{123} \geq \gamma_{13}\gamma_{23}$ and TP_2 if $\gamma_3\gamma_{123} \leq \gamma_{13}\gamma_{23}$. Since the exponential term is RR_2, $\gamma_3\gamma_{123} \geq \gamma_{13}\gamma_{23}$ is a sufficient condition for f_{12} to be RR_2. If $\gamma_{12} = 0$ and $\gamma_3\gamma_{123} < \gamma_{13}\gamma_{23}$, then f_{12} is TP_2. Symmetric conditions (by interchanging subscripts) hold for the (1,3) and (2,3) bivariate marginal densities f_{13}, f_{23}. For example, if $\gamma_{12} = \gamma_3 = \gamma_{123} = 0$ and $\gamma_1 = \gamma_2 = \gamma_{13} = \gamma_{23} = 1$, then f_{12} is TP_2, and f_{13}, f_{23} are RR_2.

This example of negative dependence arising from conditional densities within a given family is not unusual. Another example is given in the next subsection.

9.2.2 Conditional exponential families

Let family j (for $j = 1, \ldots, m$) be the exponential family

$$g_j(y; \boldsymbol{\theta}_j) = r_j(y)\beta_j(\boldsymbol{\theta}_j)\exp\{\boldsymbol{\theta}_j^T \mathbf{q}_j(y)\},$$

where $\boldsymbol{\theta}_j$ and $\mathbf{q}_j$ are column vectors of length ℓ_j. Let $(Y_1, \ldots, Y_m)$ be our random vector. Suppose we want to consider the model where the density of Y_j given $Y_i = y_i$, $i \in \{1, \ldots, m\}\backslash\{j\}$, is in the family j with parameter $\boldsymbol{\theta}_j(\mathbf{y}_{-j})$, where $\mathbf{y}_{-j}$ is $\mathbf{y}$ with the jth element deleted. The joint density f then must necessarily be of the form

$$\begin{aligned} f(\mathbf{y}) = \prod_{i=1}^{m} r_i(y_i)\exp\Big\{B &+ \sum_i \mathbf{a}_i^T \mathbf{q}_i(y_i) + \sum_{i_1<i_2} \mathbf{q}_{i_1}^T(y_{i_1})M_{i_1i_2}\mathbf{q}_{i_2}(y_{i_2}) \\ &+ \sum_{i_1<i_2<i_3}\sum_{k_1=1}^{\ell_{i_1}}\sum_{k_2=1}^{\ell_{i_2}}\sum_{k_3=1}^{\ell_{i_3}} M_{i_1i_2i_3}(k_1,k_2,k_3)\, q_{i_1k_1}q_{i_2k_2}q_{i_3k_3} + \cdots \\ &+ \sum_{k_1=1}^{\ell_1}\cdots\sum_{k_m=1}^{\ell_m} M_{1\cdots m}(k_1,\ldots,k_m)\, q_{1k_1}\cdots q_{mk_m}\Big\}, \end{aligned} \tag{9.9}$$

for suitable choices of vectors $\mathbf{a}_i$, matrices $M_{i_1i_2}$, and higher-order arrays $M_{i_1\cdots i_j}$, $i_1 < \cdots < i_j$, $3 \leq j \leq m$; the term B is a normalizing constant.

The proof of the sufficiency of the form (9.9) is not difficult. If

(9.9) holds, then

$$f_{1|2,\ldots,m}(y_1|y_2,\ldots,y_m) = f(\mathbf{y})[r_2(y_2)\cdots r_m(y_m)]^{-1}$$

$$\cdot \exp\Big\{-B'(y_2,\ldots,y_m) - \sum_{i=2}^{m} \mathbf{a}_i^T \mathbf{q}_i(y_i)$$

$$- \sum_{2\le i_1<i_2} \mathbf{q}_{i_1}^T(y_{i_1}) M_{i_1 i_2} \mathbf{q}_{i_2}(y_{i_2}) - \cdots\Big\}$$

$$= r_1(y_1)\exp\{-B'(y_2,\ldots,y_m) + \boldsymbol{\theta}_1^T(y_2,\ldots,y_m)\mathbf{q}_1(y_1)\},$$

where

$$\boldsymbol{\theta}_1(y_2,\ldots,y_m) = \mathbf{a}_1 + \sum_{j=2}^{m} M_{1j}\mathbf{q}_j(y_j) + \cdots$$

$$+ \sum_{k_2=1}^{\ell_2} \cdots \sum_{k_m=1}^{\ell_m} M_{1\cdots m}(\cdot, k_2,\ldots,k_m)\, q_{2k_2}\cdots q_{mk_m}.$$

By symmetry, one can obtain the other conditional distributions.

We outline the proof of necessity for $m = 2$. From (9.2), we want $f_{1|2}(y_1^*|y_2)f_{2|1}(y_2|y_1)/[f_{1|2}(y_1|y_2)f_{2|1}(y_2|y_1^*)]$ to be independent of y_2 for all $y_1 \neq y_1^*$. This reduces to the condition of

$$\boldsymbol{\theta}_1^T(y_2)[\mathbf{q}_1(y_1^*) - \mathbf{q}_1(y_1)] + [\boldsymbol{\theta}_2(y_1) - \boldsymbol{\theta}_2(y_1^*)]^T\mathbf{q}_2(y_2) \qquad (9.10)$$

being independent of y_2. Hence if $\boldsymbol{\theta}_1$ and $\boldsymbol{\theta}_2$ are not constant functions, they must be of the form:

$$\boldsymbol{\theta}_1(y_2) = \mathbf{a}_1 + M_{12}\mathbf{q}_2(y_2), \quad \boldsymbol{\theta}_2(y_1) = \mathbf{a}_2 + M_{21}\mathbf{q}_1(y_1),$$

for some matrices M_{12}, M_{21} with respective dimensions $\ell_1 \times \ell_2$ and $\ell_2 \times \ell_1$. Furthermore, after substitution of these functions into (9.10), one must have $M_{12} = M_{21}^T$ in order for (9.10) to be independent of y_2. Now apply (9.1) to get

$$f(y_1, y_2) \propto f_{1|2}(y_1|y_2)f_{2|1}(y_2|y_1^*)/f_{1|2}(y_1^*|y_2)$$

$$= \frac{r_1(y_1)\beta_1(\boldsymbol{\theta}_1(y_2))\exp\{[\mathbf{a}_1^T + \mathbf{q}_2^T(y_2)M_{12}^T]\mathbf{q}_1(y_1)\}}{r_1(y_1^*)\beta_1(\boldsymbol{\theta}_1(y_2))\exp\{[\mathbf{a}_1^T + \mathbf{q}_2^T(y_2)M_{12}^T]\mathbf{q}_1(y_1^*)\}}$$

$$\cdot r_2(y_2)\beta_2(\boldsymbol{\theta}_2(y_1^*))\exp\{[\mathbf{a}_2^T + \mathbf{q}_1^T(y_1^*)M_{12}]\mathbf{q}_2(y_2)\}$$

$$\propto r_1(y_1)r_2(y_2)\exp\{\mathbf{a}_1^T\mathbf{q}_1(y_1) + \mathbf{a}_2^T\mathbf{q}_2(y_2) + \mathbf{q}_1^T(y_1)M_{12}\mathbf{q}_2(y_2)\}.$$

Some special cases are the following.

1. (Normal.) $m = 2$, $q_j(y) = (y, y^2)^T$, $r_j \equiv 1$, $j = 1, 2$. Then (9.9) has the form

$$f(y_1, y_2) = \exp\{B + \alpha_{11}y_1 + \alpha_{12}y_1^2 + \alpha_{21}y_2 + \alpha_{22}y_2^2 + \gamma_{11}y_1y_2 + \gamma_{12}y_1y_2^2 + \gamma_{21}y_1^2y_2 + \gamma_{22}y_1^2y_2^2\}. \tag{9.11}$$

   Some constraints are needed on the parameters in order to get a density, including $\gamma_{22} \le 0$. The univariate margins of this density do not have a simple form.

2. (Poisson.) $m = 2$, $q_j(y) = y$, $r_j(y) = (y!)^{-1}$, $j = 1, 2$. Then (9.9) has the form

$$f(y_1, y_2) = [y_1!y_2!]^{-1}\exp\{B + \alpha_1y_1 + \alpha_2y_2 + \gamma y_1y_2\}, \tag{9.12}$$

   for $y_1, y_2 = 0, 1, \ldots$, where γ must be non-positive. Again the density has the RR_2 property. The univariate margin has the form $f_1(y_1) = [y_1!]^{-1}\exp\{B + \alpha_1y_1 + e^{\alpha_2+\gamma y_1}\}$.

9.2.3 Conditional binary: logistic regressions

In this section, we suppose that there is a multivariate binary response vector $\mathbf{Y}$ and a covariate (column) vector $\mathbf{x}$ (say of dimension r). These are measured for each subject. We look at conditions for which $\Pr(Y_j = 1|Y_k = y_k, k \ne j,\ \mathbf{x}) = G(\alpha_j + \boldsymbol{\beta}_j\mathbf{x} + \sum_{k\ne j}\gamma_{jk}y_k)$, $j = 1, \ldots, m$, for some cdf G, where the $\boldsymbol{\beta}_j$ are row vectors of length r. The case of G being the logistic distribution, leading to conditional logistic regressions, is studied before the general case.

Consider first the case of two binary response variables Y_1, Y_2 and a covariate vector $\mathbf{x}$. Suppose that Y_1 conditional on $\mathbf{x}$ and $Y_2 = y_2$ is logit and that Y_2 conditional on $\mathbf{x}$ and $Y_1 = y_1$ is logit, i.e.,

$$\operatorname{logit}[\Pr(Y_1 = 1|Y_2 = y_2, \mathbf{x})] = \log\left[\frac{\Pr(Y_1 = 1|Y_2 = y_2, \mathbf{x})}{\Pr(Y_1 = 0|Y_2 = y_2, \mathbf{x})}\right] = \alpha_1 + \boldsymbol{\beta}_1\mathbf{x} + \gamma_{12}y_2, \tag{9.13}$$

$$\operatorname{logit}[\Pr(Y_2 = 1|Y_1 = y_1, \mathbf{x})] = \log\left[\frac{\Pr(Y_2 = 1|Y_1 = y_1, \mathbf{x})}{\Pr(Y_2 = 0|Y_1 = y_1, \mathbf{x})}\right] = \alpha_2 + \boldsymbol{\beta}_2\mathbf{x} + \gamma_{21}y_1. \tag{9.14}$$

What are necessary and sufficient conditions for (9.13) and (9.14) to be compatible conditional distributions? The answer, from (9.2), is that the conditional distributions are compatible if and only if $\gamma_{12} = \gamma_{21}$, in which case the joint distribution, from (9.1), is:

$$p_{12}(y_1, y_2|\mathbf{x}) = [c(\mathbf{x})]^{-1} \exp\{(\alpha_1+\boldsymbol{\beta}_1\mathbf{x})y_1+(\alpha_2+\boldsymbol{\beta}_2\mathbf{x})y_2+\gamma_{12}y_1y_2\}, \tag{9.15}$$

where $c(\mathbf{x}) = 1+\exp\{\alpha_1+\boldsymbol{\beta}_1\mathbf{x}\}+\exp\{\alpha_2+\boldsymbol{\beta}_2\mathbf{x}\}+\exp\{(\alpha_1+\alpha_2)+(\boldsymbol{\beta}_1+\boldsymbol{\beta}_2)\mathbf{x}+\gamma_{12}\}$.

The model as given in (9.13)–(9.15) generalizes for dimension m. For the general multivariate case with binary response variables $Y_1, \ldots, Y_m$, suppose that for $j = 1, \ldots, m$, Y_j conditional on $\mathbf{x}$ and $Y_k = y_i$, $k \neq j$, is logit with parameters $\alpha_j, \boldsymbol{\beta}_j, \gamma_{jk}$, $k \neq j$. That is, for $j = 1, \ldots, m$,

$$\text{logit}\,[\Pr(Y_j = 1|Y_k = y_k, k \neq j,\ \mathbf{x})] = \alpha_j+\boldsymbol{\beta}_j\mathbf{x}+\sum_{k\neq j}\gamma_{jk}y_k. \tag{9.16}$$

The necessary and sufficient conditions for compatibility of the conditional distributions are $\gamma_{ij} = \gamma_{ji}$, $i \neq j$. The resulting joint distribution is

$$p_{1\cdots m}(\mathbf{y}|\mathbf{x}) = [c(\mathbf{x})]^{-1} \exp\Big\{\sum_{i=1}^{m}(\alpha_i + \boldsymbol{\beta}_i\mathbf{x})y_i + \sum_{1\leq i<j\leq m}\gamma_{ij}y_iy_j\Big\}, \tag{9.17}$$

$y_j = 0, 1$, $j = 1, \ldots, m$, with normalizing constant

$$c(\mathbf{x}) = \sum_{y_1=0}^{1}\cdots\sum_{y_m=0}^{1}\exp\Big\{\sum_{i=1}^{m}(\alpha_i + \boldsymbol{\beta}_i\mathbf{x})y_i + \sum_{i<j}\gamma_{ij}y_iy_j\Big\}.$$

Proof. For $k = 1, \ldots, m$, let $p_{-k}(\cdot|\mathbf{x})$ be the marginal distribution of $(Y_1, \ldots, Y_{k-1}, Y_{k+1}, \ldots, Y_m)$. Given the model (9.17) with $\gamma_{ij} = \gamma_{ji}$, $i \neq j$, we have

$$\begin{aligned}
&p_{-k}(y_1, \ldots, y_{k-1}, y_{k+1}, \ldots, y_m|\mathbf{x}) \cdot c(\mathbf{x}) \\
&= \exp\Big\{(\alpha_k + \boldsymbol{\beta}_k\mathbf{x}) + \sum_{i\neq k}(\alpha_i + \boldsymbol{\beta}_i\mathbf{x})y_i + \sum_{j\neq k}\gamma_{kj}y_j \\
&\quad + \sum_{i<j,i,j\neq k}\gamma_{ij}y_iy_j\Big\} + \exp\Big\{\sum_{i\neq k}(\alpha_i + \boldsymbol{\beta}_i\mathbf{x})y_i + \sum_{i<j,i,j\neq k}\gamma_{ij}y_iy_j\Big\} \\
&= \exp\Big\{\sum_{i\neq k}(\alpha_i + \boldsymbol{\beta}_i\mathbf{x})y_i + \sum_{i<j,i,j\neq k}\gamma_{ij}y_iy_j\Big\} \\
&\quad \cdot\Big[1 + \exp\big\{\alpha_k + \boldsymbol{\beta}_k\mathbf{x} + \sum_{j\neq k}\gamma_{kj}y_j\big\}\Big].
\end{aligned}$$

Hence, we obtain

$$\Pr(Y_k = y_k|Y_j = y_j, j \neq k, \mathbf{x}) = \frac{p_{1\cdots m}(y_1, y_2, \ldots, y_m|\mathbf{x})}{p_{-k}(y_1, \ldots, y_{k-1}, y_{k+1}, \ldots, y_m|\mathbf{x})}$$

$$= \frac{\exp\{\sum_{i=1}^m (\alpha_i + \boldsymbol{\beta}_i \mathbf{x}) y_i + \sum_{1\le i<j\le m} \gamma_{ij} y_i y_j\}}{\exp\{\sum_{i\neq k} (\alpha_i + \boldsymbol{\beta}_i \mathbf{x}) y_i + \sum_{1\le i<j\le m, i,j\neq k} \gamma_{ij} y_i y_j\}} \cdot \frac{1}{1 + \exp\{\alpha_k + \boldsymbol{\beta}_k \mathbf{x} + \sum_{j\neq k} \gamma_{kj} y_j\}}$$

$$= \frac{\exp\{(\alpha_k + \boldsymbol{\beta}_k \mathbf{x} + \sum_{j\neq k} \gamma_{kj} y_j) y_k\}}{1 + \exp\{\alpha_k + \boldsymbol{\beta}_k \mathbf{x} + \sum_{j\neq k} \gamma_{kj} y_j\}}.$$

This equation indicates that the conditional probability distributions $\Pr(Y_k = y_k \mid Y_j = y_j, j \neq k, \mathbf{x})$ for $k = 1, \ldots, m$ are logistic regressions.

Next we turn to the proof of the necessity. The notation $f_{j|r}$ is used for the pmf of Y_j given $\mathbf{x}$ and the rest of the Ys. From (9.3)

$$\Pr(Y_1 = y_1, \ldots, Y_m = y_m \mid \mathbf{x})$$
$$\propto \frac{f_{1|r}(y_1|y_2, \ldots, y_m, \mathbf{x})}{f_{1|r}(y_1^0|y_2, \ldots, y_m, \mathbf{x})} \frac{f_{2|r}(y_2|y_1^0, y_3, \ldots, y_m, \mathbf{x})}{f_{2|r}(y_2^0|y_1^0, y_3, \ldots, y_m, \mathbf{x})} \cdot \frac{f_{3|r}(y_3|y_1^0, y_2^0, y_4, \ldots, y_m, \mathbf{x})}{f_{3|r}(y_3^0|y_1^0, y_2^0, y_4, \ldots, y_m, \mathbf{x})} \cdots \frac{f_{m|r}(y_m|y_1^0, \ldots, y_{m-1}^0, \mathbf{x})}{f_{m|r}(y_m^0|y_1^0, \ldots, y_{m-1}^0, \mathbf{x})}$$

for a fixed and arbitrary $\mathbf{y}^0$. This simplifies to

$$\Pr(Y_1 = y_1, \ldots, Y_m = y_m \mid \mathbf{x})$$
$$\propto \frac{\prod_{i=1}^m \exp\{(\alpha_i + \boldsymbol{\beta}_i \mathbf{x} + \sum_{j<i} \gamma_{ij} y_j^0 + \sum_{j>i} \gamma_{ij} y_j) y_i\}}{\prod_{i=1}^m \exp\{(\alpha_i + \boldsymbol{\beta}_i \mathbf{x} + \sum_{j<i} \gamma_{ij} y_j^0 + \sum_{j>i} \gamma_{ij} y_j) y_i^0\}}$$
$$\propto \exp\Big\{\sum_{i=1}^m \Big(\alpha_i + \boldsymbol{\beta}_i \mathbf{x} + \sum_{j>i} \gamma_{ij} y_j\Big) y_i + \sum_{i=1}^m \Big[\Big(\sum_{j<i} \gamma_{ij} y_j^0\Big) y_i - \Big(\sum_{j>i} \gamma_{ij} y_j\Big) y_i^0\Big]\Big\}. \tag{9.18}$$

Take $y_1^0 = \cdots = y_m^0 = 0$ to obtain

$$\Pr(Y_1 = y_1, \ldots, Y_m = y_m|\mathbf{x}) \propto \exp\Big\{\sum_{i=1}^m \Big(\alpha_i + \boldsymbol{\beta}_i \mathbf{x} + \sum_{j>i} \gamma_{ij} y_j\Big) y_i\Big\},$$

the form of the joint probability distribution.

For the condition for compatibility, take the ratio of (9.18) with $\mathbf{y}^0$ and $\mathbf{y}^*$; the ratio should not depend on $y_1, \ldots, y_m$. It is

$$R(\mathbf{y};\mathbf{y}^0,\mathbf{y}^*) = \frac{\exp\{\sum_{i=1}^m(\sum_{j<i}\gamma_{ij}y_j^0)y_i - \sum_{i=1}^m(\sum_{j>i}\gamma_{ij}y_j)y_i^0\}}{\exp\{\sum_{i=1}^m(\sum_{j<i}\gamma_{ij}y_j^*)y_i - \sum_{i=1}^m(\sum_{j>i}\gamma_{ij}y_j)y_i^*\}}$$

$$= \frac{\exp\{\sum_{i=1}^m(\sum_{j>i}\gamma_{ji}y_j)y_i^0 - \sum_{i=1}^m(\sum_{j>i}\gamma_{ij}y_j)y_i^0\}}{\exp\{\sum_{i=1}^m(\sum_{j>i}\gamma_{ji}y_j)y_i^* - \sum_{i=1}^m(\sum_{j>i}\gamma_{ij}y_j)y_i^*\}}$$

$$= \exp\Big\{\sum_{i=1}^m\Big[\sum_{j>i}(\gamma_{ji}-\gamma_{ij})y_j\Big](y_i^0 - y_i^*)\Big\}. \tag{9.19}$$

In (9.19), one can take appropriate values for y_i^0, y_i^*, e.g., $\mathbf{y}^* = (0,\ldots,0)$ and $\mathbf{y}^0 = (0,\ldots,0,1,1), (0,\ldots,0,1,1,1), \ldots, (1,\ldots,1)$. The first choice of $\mathbf{y}^0$ yields $\exp\{(\gamma_{m,m-1} - \gamma_{m-1,m})y_m\}$ and this is independent of $y_1,\ldots,y_m$ only if $\gamma_{m-1,m} = \gamma_{m,m-1}$. The other choices yield the other inequalities. Alternatively, the symmetry in the variables and parameters then implies that $\gamma_{ij} = \gamma_{ji}$ for all $i \neq j$. □

The parameters γ_{ij} have interpretations as conditional log-odds ratios since

$$\exp\{\gamma_{ij}\} = \frac{\Pr(Y_i=1, Y_j=1,\ Y_k=y_k, k\neq i,j \mid \mathbf{x})}{\Pr(Y_i=1, Y_j=0,\ Y_k=y_k, k\neq i,j \mid \mathbf{x})} \cdot \frac{\Pr(Y_i=0, Y_j=0,\ Y_k=y_k, k\neq i,j \mid \mathbf{x})}{\Pr(Y_i=0, Y_j=1,\ Y_k=y_k, k\neq i,j \mid \mathbf{x})}$$

$$= \frac{\Pr(Y_i=1, Y_j=1 \mid \mathbf{x},\ Y_k=y_k, k\neq i,j)}{\Pr(Y_i=1, Y_j=0 \mid \mathbf{x},\ Y_k=y_k, k\neq i,j)} \cdot \frac{\Pr(Y_i=0, Y_j=0 \mid \mathbf{x},\ Y_k=y_k, k\neq i,j)}{\Pr(Y_i=0, Y_j=1 \mid \mathbf{x},\ Y_k=y_k, k\neq i,j)}.$$

For $m = 2$, there are no Y_k so that γ_{12} is also the unconditional log-odds ratio of $p_{12}(\cdot|\mathbf{x})$ and it is constant over $\mathbf{x}$. For $m \geq 3$, let $\pi_{ij}(y_i, y_j|\mathbf{x})$ be the (i,j) bivariate marginal distribution of (9.17). The unconditional log-odds ratio of π_{ij} will depend on $\mathbf{x}$. For notational simplicity and without loss of generality, we write the log-odds ratio only for the case of $(i,j) = (1,2)$. Let

$$q_m(y_1,y_2,x) = \sum_{y_3=0}^1 \cdots \sum_{y_m=0}^1 \exp\Big\{\sum_{k=3}^m[\alpha_k + \boldsymbol{\beta}_k\mathbf{x} + \gamma_{1k}y_1 + \gamma_{2k}y_2]y_k + \sum_{3\leq k<k'}\gamma_{kk'}y_ky_{k'}\Big\}.$$

Then

$$\pi_{12}(y_1, y_2|\mathbf{x}) = \frac{q_m(y_1, y_2, \mathbf{x})}{c(\mathbf{x})} \exp\Big\{\sum_{j=1}^{2}(\alpha_j + \boldsymbol{\beta}_j\mathbf{x})y_j + \gamma_{12}y_1y_2\Big\}$$

and the log-odds ratio is

$$\gamma_{12} \log\{[q_m(1,1,\mathbf{x})q_m(0,0,\mathbf{x})]/[q_m(1,0,\mathbf{x})q_m(0,1,\mathbf{x})]\}.$$

If $\mathbf{x}$ has dimension r, then model (9.17) is an exponential family model $[c(\boldsymbol{\theta}, \mathbf{x})]^{-1} \exp\{\boldsymbol{\theta}^T\mathbf{s}\}$ with 'sufficient' statistic vector

$$\mathbf{s}^T = (y_1, \ldots, y_m, y_1y_2, \ldots, y_{m-1}y_m, x_ky_j, 1 \le k \le r, 1 \le j \le m).$$

Given data of the form $(y_{i1}, \ldots, y_{im}, x_{i1}, \ldots, x_{ir})$, $i = 1, \ldots, n$, and corresponding vectors $\mathbf{s}_i$, the sufficient statistic vector is $\sum_{i=1}^{n} \mathbf{s}_i$. The estimate of $\boldsymbol{\theta}$ can be obtained using the Newton–Raphson method for an exponential family log-likelihood.

Note that the exponential family model (9.17) is not closed under margins. This is typical of multivariate exponential family models that are not multivariate normal.

9.2.4 *Binary: other conditional models*

The use of conditional logistic regressions leads to simple compatibility conditions and a multivariate distribution in the exponential family. With other forms for the conditional binary distributions, the analysis is not simple. This is illustrated for the bivariate and trivariate cases.

For the bivariate case, without covariates, suppose

$$f_{1|2}(y_1|y_2) = [p_1(y_2)]^{y_1}[q_1(y_2)]^{1-y_1},$$

$$f_{2|1}(y_2|y_1) = [p_2(y_1)]^{y_2}[q_2(y_1)]^{1-y_2},$$

for $y_1, y_2 = 0, 1$, $q_j = 1 - p_j$, $j = 1, 2$. From (9.2), the compatibility condition simplifies to

$$[p_1(y_2)]^{z_1^0 - y_1^0}[q_1(y_2)]^{y_1^0 - z_1^0} \left[\frac{p_2(y_1^0)q_2(z_1^0)}{p_2(z_1^0)q_2(y_1^0)}\right]^{y_2}$$

independent of y_2 for all choice of y_1^0, z_1^0. Since y_2 takes only values of 0 and 1, this means

$$\left[\frac{p_1(0)}{q_1(0)}\right]^{z_1^0 - y_1^0} = \left[\frac{p_1(1)}{q_1(1)}\right]^{z_1^0 - y_1^0} \left[\frac{p_2(y_1^0)q_2(z_1^0)}{p_2(z_1^0)q_2(y_1^0)}\right].$$

This is an identity if $z_1^0 = y_1^0$ so the compatibility comes from the case in which z_1^0, y_1^0 are different, say $z_1^0 = 1$, $y_1^0 = 0$. This leads to the condition:

$$\frac{p_2(1)}{q_2(1)} = \frac{p_2(0)}{q_2(0)} \frac{p_1(1)}{q_1(1)} \frac{q_1(0)}{p_1(0)}, \tag{9.20}$$

i.e., there are three degrees of freedom in the choice of the parameters $p_1(0), p_1(1), p_2(0), p_2(1)$ of the conditional binary distributions.

If the p_j are written in the form $p_1(y_2) = G(\alpha_1 + \gamma_{12} y_2)$, $p_2(y_1) = G(\alpha_2 + \gamma_{21} y_1)$, where G need not be the logistic cdf, then the compatibility condition (9.20) becomes

$$\frac{G(\alpha_2 + \gamma_{21})}{1 - G(\alpha_2 + \gamma_{21})} = \frac{G(\alpha_2)}{1 - G(\alpha_2)} \frac{1 - G(\alpha_1)}{G(\alpha_1)} \frac{G(\alpha_1 + \gamma_{12})}{1 - G(\alpha_1 + \gamma_{12})}. \tag{9.21}$$

With G, $\alpha_1, \alpha_2, \gamma_{12}$ fixed, there is a single value of γ_{21} that solves (9.21) since the left-hand side of (9.21) is strictly increasing in γ_{21}. Unlike the case of the logistic distribution, the solution need not be $\gamma_{21} = \gamma_{12}$. With the addition of covariates to a model based on G, the analysis and compatibility conditions are left as a problem.

The trivariate case without covariates does not lend itself to a simple analysis. Suppose

$$f_{1|r}(y_1|y_2, y_3) = [p_1(y_2, y_3)]^{y_1}[q_1(y_2, y_3)]^{1-y_1},$$

$$f_{2|r}(y_2|y_1, y_3) = [p_2(y_1, y_3)]^{y_2}[q_2(y_1, y_3)]^{1-y_2},$$

$$f_{3|r}(y_3|y_1, y_2) = [p_3(y_1, y_2)]^{y_3}[q_3(y_1, y_2)]^{1-y_3}.$$

From (9.4), one gets

$$\left[\frac{p_2(y_1^0, y_3)}{q_2(y_1^0, y_3)}\right]^{y_2 - y_2^0} \left[\frac{p_3(y_1^0, y_2^0) q_3(z_1^0, z_2^0)}{p_3(z_1^0, z_2^0) q_3(y_1^0, y_2^0)}\right]^{y_3}$$

$$\cdot \left[\frac{p_1(y_2, y_3)}{q_1(y_2, y_3)}\right]^{z_1^0 - y_1^0} \left[\frac{p_2(z_1^0, y_3)}{q_2(z_1^0, y_3)}\right]^{z_2^0 - y_2}$$

independent of y_2, y_3 for all $\mathbf{y}^0, \mathbf{z}^0$. Hence the compatibility conditions come from substituting (y_2, y_3) equal to $(0,0)$, $(0,1)$, $(1,0)$ and $(1,1)$ and equating. The remaining analysis is left as an unsolved problem.

Enough evidence has been presented in this section to show that conditional logistic regressions are a very convenient choice for mathematical tractibility.

9.3 Bibliographic notes

Models which are based on compatible conditional distributions in given parametric families are the main topic of the book by Arnold, Castillo and Sarabia (1992). Papers on this topic include Arnold and Strauss (1988; 1991), Arnold and Press (1989) and Arnold (1990); the paper by Arnold and Strauss (1988) has more details on distributions with conditional exponential densities, and Arnold (1990) has more details on negatively dependent distributions. The approach in this chapter that is different is the use of the compatibility condition that generalizes that in Gelman and Speed (1993). Another reference following that of Gelman and Speed (1993) is Arnold, Castillo and Sarabia (1995). The model in Section 9.2.3 for multivariate binary data based on compatible conditionally specified logistic regressions is given in Joe and Liu (1996).

9.4 Exercises

9.1 Prove the multivariate extension in Section 9.1, given by (9.3) and (9.4).

9.2 What are the constraints on the parameters for (9.11) to be a density?

9.3 For the bivariate density in (9.12) with conditional Poisson margins, show that γ must be non-positive. [Hint: check on the convergence of $\sum_y f(y, y)$.]

9.4 Study other cases with conditionally specified densities in given parametric families.

(Arnold, Castillo and Sarabia 1992)

9.5 Generalize (9.17) when there are the additional interaction terms in the logistic regressions in (9.16): $\mathbf{x}\prod_{k\in S} y_k$, $|S|\geq 1$, and $\prod_{k\in S} y_k$, $|S|\geq 2$, where S is a non-empty subset of $\{1,\ldots,m\}\backslash\{j\}$.

9.6 Complete the details in Section 9.2.2.

9.5 Unsolved problems

9.1 Do a further analysis of conditionally specified binary regressions, when the latent distribution is not logistic.

CHAPTER 10

Statistical inference and computation

This chapter is devoted to statistical inference theory and data analysis methods for multivariate models, and numerical methods for the estimation of parameters for these models.

For almost all of the multivariate models in this book, much of the classical statistical inference theory is not applicable. This includes exponential family results, sufficient statistics, ancillary statistics, minimum variance unbiased estimators, etc. One should not expect estimators with closed forms, rather one should assume that numerical methods are needed to get estimates. Practically the only theory that can be applied is the asymptotic maximum likelihood (ML) theory. But this can be applied in a way that leads to simpler computations and greater robustness, especially for multivariate models where different parameters are associated with different marginal distributions. For these models, which include copula-based models in which univariate parameters are separated from multivariate parameters, rather than maximizing the multivariate log-likelihood in all of the parameters together, one can estimate different parameters from log-likelihoods associated with different marginal distributions of the multivariate distribution. This theory is developed in Section 10.1. It can be considered as part of estimating equation or inference function theory, with each inference function being a score function from a likelihood of a marginal distribution. This is the new statistical inference theory that comes from multivariate non-normal models. It has not been studied in the statistical literature, and it is not needed for the MVN distribution, because estimation for the MVN distribution has a closure property under the taking of margins that generally does not hold (see Exercise 10.1).

For multivariate data, initial data analysis often starts with univariate analyses, then bivariate analyses, and finally higher-dimensional multivariate analyses. Can multivariate modelling follow the same steps? The inference method in Section 10.1 is an attempt to follow a similar sequence for the multivariate models which have the property that all parameters are associated with marginal distributions — see Section 4.1.

Next we comment on statistical inference for models for longitudinal data which fit within the multivariate framework. Longitudinal or repeated measures data can appear in many different forms, with three cases being:

(a) short time series (of the same length) on many different subjects;

(b) a single long time series;

(c) moderate-length or long time series (with possibly different lengths) on many different subjects.

These cases will be used to illustrate the different methods, models and analyses that might be appropriate. For case (a), if the lengths of the time series are all the same and the observation times are all the same (either in an absolute or relative sense), then this can be treated as multivariate data where reasonable models to try would have dependence decreasing with lag. Such models are given in Chapter 8. For case (b), there is no replication over units so asymptotic theory depends on ergodicity-related results. Markov chain models might be appropriate, including the autoregressive models in Section 8.4. For case (c), a random effects type of model might be appropriate. If parameter estimates can be obtained for each subject, the estimation of the parameters for the random effects distribution might be done in a second stage. Of course, as mentioned in Section 1.7, appropriate models and methods depend on inferences of interest. Examples for all three types of longitudinal data are given in Chapter 11.

Subsections of Section 10.1 are devoted to the asymptotic covariance matrix, estimation of standard errors, asymptotic efficiency, and assessment of estimation consistency for the method of using inference functions that are derived from the log-likelihoods of marginal distributions. Also there are several examples illustrating this method. In Section 10.2, some results in Section 10.1 are extended to the inclusion of covariates in the model. Also there are extensions to situations in which there are parameters common to more than one margin. Section 10.3 is on choice and comparison of models through (penalized) log-likelihoods and predictive ability. Section 10.4 has some results on inference for Markov chains. Section 10.5 comments on research needed for Bayesian methods to be applied to multivariate models. Section 10.6 discusses numerical methods, especially numerical optimization.

10.1 Estimation from likelihoods of margins °

In this section, we study the estimation of parameters of a copula-based multivariate model, based on the likelihoods of marginal distributions of the model. Following terminology of McLeish and Small (1988) and Xu (1996), we call this the method of **inference functions for margins** or IFM method. The inference or estimating functions are score functions of likelihoods of marginal distributions. The method actually applies to a larger class of models that have certain closure properties for the parameters (see Section 4.1). Here we assume that we have iid observations. The extension to include covariates is given in Section 10.2.

Consider a copula-based parametric model for the random vector $\mathbf{Y}$, with cdf

$$F(\mathbf{y};\boldsymbol{\alpha}_1,\ldots,\boldsymbol{\alpha}_m,\boldsymbol{\theta}) = C(F_1(y_1;\boldsymbol{\alpha}_1),\ldots,F_m(y_m;\boldsymbol{\alpha}_m);\boldsymbol{\theta}), \quad (10.1)$$

where $F_1,\ldots,F_m$ are univariate cdfs with respective parameters $\boldsymbol{\alpha}_1,\ldots,\boldsymbol{\alpha}_m$, and C is a family of copulas parametrized by a parameter $\boldsymbol{\theta}$. We assume that C has a density c (mixed derivative of order m). The vector $\mathbf{Y}$ could be discrete or continuous. In the former case, the joint pmf $f(\cdot;\boldsymbol{\alpha}_1,\ldots,\boldsymbol{\alpha}_m,\boldsymbol{\theta})$ for $\mathbf{Y}$ can be derived from the cdf in (10.1), and we let the univariate marginal pmfs be denoted by $f_1,\ldots,f_m$; in the latter case, we assume that F_j has density f_j for $j=1,\ldots,m$, and that $\mathbf{Y}$ has density

$$f(\mathbf{y};\boldsymbol{\alpha}_1,\ldots,\boldsymbol{\alpha}_m,\boldsymbol{\theta})= c(F_1(y_1;\boldsymbol{\alpha}_1),\ldots,F_m(y_m;\boldsymbol{\alpha}_m);\boldsymbol{\theta})\prod_{j=1}^{m} f_j(y_j;\boldsymbol{\alpha}_j).$$

For a sample of size n, with observed random vectors $\mathbf{y}_1,\ldots,\mathbf{y}_n$, we can consider the m log-likelihood functions for the univariate margins,

$$L_j(\boldsymbol{\alpha}_j) = \sum_{i=1}^{n} \log f_j(y_{ij};\boldsymbol{\alpha}_j), \quad j=1,\ldots,m,$$

and the log-likelihood function for the joint distribution,

$$L(\boldsymbol{\theta},\boldsymbol{\alpha}_1,\ldots,\boldsymbol{\alpha}_m) = \sum_{i=1}^{n} \log f(\mathbf{y}_i;\boldsymbol{\alpha}_1,\ldots,\boldsymbol{\alpha}_m,\boldsymbol{\theta}). \quad (10.2)$$

A simple case of the IFM method consists of doing m separate optimizations of the univariate likelihoods, followed by an optimization of the multivariate likelihood as a function of the dependence parameter vector. More specifically,

(a) the log-likelihoods L_j of the m univariate margins are separately maximized to get estimates $\tilde{\boldsymbol{\alpha}}_1, \ldots, \tilde{\boldsymbol{\alpha}}_m$;

(b) the function $L(\boldsymbol{\theta}, \tilde{\boldsymbol{\alpha}}_1, \ldots, \tilde{\boldsymbol{\alpha}}_m)$ is maximized over $\boldsymbol{\theta}$ to get $\tilde{\boldsymbol{\theta}}$.

That is, under regularity conditions, $(\tilde{\boldsymbol{\alpha}}_1, \ldots, \tilde{\boldsymbol{\alpha}}_m, \tilde{\boldsymbol{\theta}})$ is the solution of

$$(\partial L_1/\partial \boldsymbol{\alpha}_1, \ldots, \partial L_m/\partial \boldsymbol{\alpha}_m, \partial L/\partial \boldsymbol{\theta}) = \mathbf{0}. \tag{10.3}$$

This procedure is computationally simpler than estimating all parameters $\boldsymbol{\alpha}_1, \ldots, \boldsymbol{\alpha}_m, \boldsymbol{\theta}$ simultaneously from L in (10.2). A numerical optimization with many parameters is much more time-consuming compared with several numerical optimizations, each with fewer parameters.

If the copula model (10.1) has further structure such as a parameter associated with each bivariate margin, simplifications of the second step (b) can be made, so that no numerical optimization with a large number of parameters is needed. For example, with a MVN latent distribution, there is a simplification shown in Section 10.1.4; the correlation parameters can be estimated from separate likelihoods of the bivariate margins.

If it is possible to maximize L to get estimates $\hat{\boldsymbol{\alpha}}_1, \ldots, \hat{\boldsymbol{\alpha}}_m, \hat{\boldsymbol{\theta}}$, then one could compare these with the estimates $\tilde{\boldsymbol{\alpha}}_1, \ldots, \tilde{\boldsymbol{\alpha}}_m, \tilde{\boldsymbol{\theta}}$ as an estimation consistency check to evaluate the adequacy of fit of the copula. For comparison with the IFM method, the MLE refers to $(\hat{\boldsymbol{\alpha}}_1, \ldots, \hat{\boldsymbol{\alpha}}_m, \hat{\boldsymbol{\theta}})$. Under regularity conditions, this comes from solving

$$(\partial L/\partial \boldsymbol{\alpha}_1, \ldots, \partial L/\partial \boldsymbol{\alpha}_m, \partial L/\partial \boldsymbol{\theta}) = \mathbf{0}; \tag{10.4}$$

contrast with (10.3). Note for MVN distributions, consisting of the MVN copula with correlation matrix $\boldsymbol{\theta} = R$ and $N(\mu_j, \sigma_j^2)$ univariate margins ($\boldsymbol{\alpha}_j = (\mu_j, \sigma_j^2)$), that $\hat{\boldsymbol{\alpha}}_j = \tilde{\boldsymbol{\alpha}}_j$, $j = 1, \ldots, m$, and $\hat{\boldsymbol{\theta}} = \tilde{\boldsymbol{\theta}}$. The equivalence of the estimators generally does not hold. Possibly because the MVN distribution is dominant in multivariate statistics, attention has not been given to variations of maximum likelihood estimation for multivariate models.

Since it is computationally easier to obtain $(\tilde{\boldsymbol{\alpha}}_1, \ldots, \tilde{\boldsymbol{\alpha}}_m, \tilde{\boldsymbol{\theta}})$, a natural question is its asymptotic relative efficiency compared with $(\hat{\boldsymbol{\alpha}}_1, \ldots, \hat{\boldsymbol{\alpha}}_m, \hat{\boldsymbol{\theta}})$. Apart from efficiency considerations, the former set of estimates provides a good starting point for the latter if one needs and can compute $(\hat{\boldsymbol{\alpha}}_1, \ldots, \hat{\boldsymbol{\alpha}}_m, \hat{\boldsymbol{\theta}})$. Approximations leading to the asymptotic distribution of $(\tilde{\boldsymbol{\alpha}}_1, \ldots, \tilde{\boldsymbol{\alpha}}_m, \tilde{\boldsymbol{\theta}})$ are given in Section 10.1.1. Then, one can (numerically) compare the asymptotic covariance matrices of $(\tilde{\boldsymbol{\alpha}}_1, \ldots, \tilde{\boldsymbol{\alpha}}_m, \tilde{\boldsymbol{\theta}})$ and $(\hat{\boldsymbol{\alpha}}_1, \ldots, \hat{\boldsymbol{\alpha}}_m, \hat{\boldsymbol{\theta}})$. Also an estimate of the covariance matrix of $(\tilde{\boldsymbol{\alpha}}_1, \ldots, \tilde{\boldsymbol{\alpha}}_m, \tilde{\boldsymbol{\theta}})$ can be obtained. The theory is a special case of using a set of estimating equations to estimate a vector of parameters.

10.1.1 Asymptotic covariance matrix

Throughout this subsection, we assume that the usual regularity conditions (see Serfling, 1980) for asymptotic maximum likelihood theory hold for the multivariate model as well as for all of its margins. For the IFM method, there is a set of inference or estimating functions in which each function is a score function or the (partial) derivative of a log-likelihood of some marginal density.

Let $\boldsymbol{\eta} = (\boldsymbol{\alpha}_1, \ldots, \boldsymbol{\alpha}_m, \boldsymbol{\theta})$ be the row vector of parameters and let $\mathbf{g}$ be a row vector of functions with the same dimension as $\boldsymbol{\eta}$. For example, $\mathbf{g}$ could be the vector in the summands on the left-hand side of (10.3). The vector of **inference functions** is $\sum_{i=1}^n \mathbf{g}(\mathbf{y}_i, \boldsymbol{\eta})$.

Let $\mathbf{Y}, \mathbf{Y}_1, \ldots, \mathbf{Y}_n$ be iid with the density $f(\cdot; \boldsymbol{\eta})$. Suppose the vector of **estimating equations** for the estimator $\tilde{\boldsymbol{\eta}}$ is

$$\sum_{i=1}^{n} \mathbf{g}(\mathbf{Y}_i, \tilde{\boldsymbol{\eta}}) = \mathbf{0}. \tag{10.5}$$

Let $\partial \mathbf{g}^T / \partial \boldsymbol{\eta}$ be the matrix with (j, k) component $\partial g_j(\mathbf{y}, \boldsymbol{\eta}) / \partial \eta_k$, where g_j is the jth component of $\mathbf{g}$ and η_k is the kth component of $\boldsymbol{\eta}$. From an expansion of (10.5) similar to the derivation of the asymptotic distribution of an MLE, under the regularity conditions, the asymptotic distribution of $n^{1/2}(\tilde{\boldsymbol{\eta}} - \boldsymbol{\eta})^T$ is equivalent to that of

$$\left\{-\mathrm{E}\left[\partial \mathbf{g}^T(\mathbf{Y}, \boldsymbol{\eta}) / \partial \boldsymbol{\eta}\right]\right\}^{-1} n^{-1/2} \sum_{i=1}^{n} \mathbf{g}^T(\mathbf{Y}_i, \boldsymbol{\eta}). \tag{10.6}$$

Hence it has the same asymptotic distribution as

$$\left\{-\mathrm{E}\left[\partial \mathbf{g}^T(\mathbf{Y}, \boldsymbol{\eta}) / \partial \boldsymbol{\eta}\right]\right\}^{-1} \mathbf{Z},$$

where $\mathbf{Z} \sim N(\mathbf{0}, \mathrm{Cov}(\mathbf{g}(\mathbf{Y}, \boldsymbol{\eta})))$. That is, the asymptotic covariance matrix of $n^{1/2}(\tilde{\boldsymbol{\eta}} - \boldsymbol{\eta})^T$, called the inverse Godambe information matrix, is

$$V = D_{\mathbf{g}}^{-1} M_{\mathbf{g}} (D_{\mathbf{g}}^{-1})^T,$$

where

$$D_{\mathbf{g}} = \mathrm{E}\left[\partial \mathbf{g}^T(\mathbf{Y},\boldsymbol{\eta})/\partial\boldsymbol{\eta}\right], \quad M_{\mathbf{g}} = \mathrm{E}\left[\mathbf{g}^T(\mathbf{Y},\boldsymbol{\eta})\mathbf{g}(\mathbf{Y},\boldsymbol{\eta})\right].$$

(If $\tilde{\boldsymbol{\eta}}$ is the MLE and $\ell = \log f$, then $\mathbf{g} = \partial\ell/\partial\boldsymbol{\eta}$, $\partial\mathbf{g}^T(\mathbf{Y},\boldsymbol{\eta})/\partial\boldsymbol{\eta} = \partial^2\ell/\partial\boldsymbol{\eta}^T\partial\boldsymbol{\eta}$, and this is the standard result for the MLE.)

Estimation of the asymptotic covariance matrix $n^{-1}V$ of $\tilde{\boldsymbol{\eta}}$ involves $-\mathrm{E}\left[\partial\mathbf{g}^T(\mathbf{Y},\boldsymbol{\eta})/\partial\boldsymbol{\eta}\right]$ and hence the need to compute many derivatives (which can be tedious both analytically and in the coding for a computer program if symbolic manipulation cannot be used). This can be avoided by making using of the jackknife to estimate $n^{-1}V$. Let $\tilde{\boldsymbol{\eta}}^{(i)}$ be the estimator of $\boldsymbol{\eta}$ with the ith observation $\mathbf{Y}_i$ deleted, $i = 1,\ldots,n$. Assuming $\tilde{\boldsymbol{\eta}}$ and the $\tilde{\boldsymbol{\eta}}^{(i)}$ are row vectors, the jackknife estimator of $n^{-1}V$ is

$$\sum_{i=1}^{n}(\tilde{\boldsymbol{\eta}}^{(i)} - \tilde{\boldsymbol{\eta}})^T(\tilde{\boldsymbol{\eta}}^{(i)} - \tilde{\boldsymbol{\eta}}).$$

Proof. The (non-rigourous) justification of this comes from substituting an analogy of (10.6) for $\tilde{\boldsymbol{\eta}}^{(i)}$ to get

$$\begin{aligned}
(\tilde{\boldsymbol{\eta}}^{(i)} - \boldsymbol{\eta})^T &\approx \left\{-\mathrm{E}\left[\partial\mathbf{g}^T(\mathbf{Y},\boldsymbol{\eta})/\partial\boldsymbol{\eta}\right]\right\}^{-1}(n-1)^{-1}\sum_{k\neq i}\mathbf{g}^T(\mathbf{Y}_k,\boldsymbol{\eta}) \\
&\approx \{-\mathrm{E}\left[\partial\mathbf{g}^T(\mathbf{Y},\boldsymbol{\eta})/\partial\boldsymbol{\eta}\right]\}^{-1}(n-1)^{-1} \\
&\quad \cdot\left[n\{-\mathrm{E}\left[\partial\mathbf{g}^T(\mathbf{Y},\boldsymbol{\eta})/\partial\boldsymbol{\eta}\right]\}(\tilde{\boldsymbol{\eta}} - \boldsymbol{\eta})^T - \mathbf{g}^T(\mathbf{Y}_i,\boldsymbol{\eta})\right] \\
&= n(n-1)^{-1}(\tilde{\boldsymbol{\eta}} - \boldsymbol{\eta})^T \\
&\quad -(n-1)^{-1}\{-\mathrm{E}\left[\partial\mathbf{g}^T(\mathbf{Y},\boldsymbol{\eta})/\partial\boldsymbol{\eta}\right]\}^{-1}\mathbf{g}^T(\mathbf{Y}_i,\boldsymbol{\eta}),
\end{aligned}$$

so that

$$(\tilde{\boldsymbol{\eta}}^{(i)} - \tilde{\boldsymbol{\eta}})^T \approx -n^{-1}\{-\mathrm{E}\left[\partial\mathbf{g}^T(\mathbf{Y},\boldsymbol{\eta})/\partial\boldsymbol{\eta}\right]\}^{-1}\mathbf{g}^T(\mathbf{Y}_i,\boldsymbol{\eta}) + O_p(n^{-3/2}).$$

Finally,

$$\begin{aligned}
\sum_{i=1}^{n}(\tilde{\boldsymbol{\eta}}^{(i)} - \tilde{\boldsymbol{\eta}})^T(\tilde{\boldsymbol{\eta}}^{(i)} - \tilde{\boldsymbol{\eta}}) &\approx n^{-2}\left\{-\mathrm{E}\left[\partial\mathbf{g}^T(\mathbf{Y},\boldsymbol{\eta})/\partial\boldsymbol{\eta}\right]\right\}^{-1} \\
&\quad \cdot\left[\sum_{i=1}^{n}\mathbf{g}^T(\mathbf{Y}_i,\boldsymbol{\eta})\,\mathbf{g}(\mathbf{Y}_i,\boldsymbol{\eta})\right]\cdot\left\{-\mathrm{E}\left[\partial\mathbf{g}^T(\mathbf{Y},\boldsymbol{\eta})/\partial\boldsymbol{\eta}\right]^T\right\}^{-1} \\
&\approx n^{-1}\left\{-\mathrm{E}\left[\partial\mathbf{g}^T(\mathbf{Y},\boldsymbol{\eta})/\partial\boldsymbol{\eta}\right]\right\}^{-1}\mathrm{Cov}\,(\mathbf{g}(\mathbf{Y},\boldsymbol{\eta})) \\
&\quad \cdot\left\{-\mathrm{E}\left[\partial\mathbf{g}^T(\mathbf{Y},\boldsymbol{\eta})/\partial\boldsymbol{\eta}\right]^T\right\}^{-1} + O_p(n^{-3/2}) \\
&= n^{-1}V + O_p(n^{-3/2}).
\end{aligned}$$

□

Actually, if each inference function is the partial derivative of a log-likelihood function of some marginal density, then the vector $\mathbf{g}$ need not be explicitly obtained for numerical computation of $\tilde{\boldsymbol{\eta}}$. From the log-likelihoods and a quasi-Newton routine (see Section 10.5) for maximizing the log-likelihoods, the components of $\tilde{\boldsymbol{\eta}}$ can be obtained sequentially, beginning with the components corresponding to univariate parameters. The jackknife method can then be used for the estimation of the asymptotic covariance matrix of the estimators. Hence, with the combination of the quasi-Newton routine and the jackknife method, one can avoid taking analytic derivatives of log-likelihoods.

For large samples, the jackknife can be modified into estimates from deletions of more than one observation at a time in order to reduce the total amount of computation time. Suppose $n = n_1 n_2$, with n_2 groups or blocks of n_1; n_2 estimators can be obtained with the kth estimate based on $n - n_1$ observations after deleting the n_1 observations in the kth block. (It is probably best to randomize the n observations into the n_2 blocks. The disadvantage is that the jackknife estimates depend on the randomization. However, reasonable estimates for SEs should obtain and there are approximations in estimating SEs for any method.) Note that the simple (delete-one) jackknife has $n_1 = 1$.

Let $\tilde{\boldsymbol{\eta}}^{(k)}$ be the estimator of $\boldsymbol{\eta}$ with the kth block deleted, $k = 1, \ldots, n_2$. For the asymptotic approximation, think of n_1 as fixed with $n_2 \to \infty$. Assuming $\tilde{\boldsymbol{\eta}}$ and the $\tilde{\boldsymbol{\eta}}^{(k)}$ are row vectors, the jackknife estimator of $n^{-1}V$ is

$$\sum_{k=1}^{n_2} (\tilde{\boldsymbol{\eta}}^{(k)} - \tilde{\boldsymbol{\eta}})^T (\tilde{\boldsymbol{\eta}}^{(k)} - \tilde{\boldsymbol{\eta}}).$$

Proof. Let B_k consist of the indices in the kth block of observations to be deleted. Substitute an analogy of (10.6) for $\tilde{\boldsymbol{\eta}}^{(k)}$ to get

$$\begin{aligned}
(\tilde{\boldsymbol{\eta}}^{(k)} - \boldsymbol{\eta})^T &\approx \left\{-\mathrm{E}\left[\partial \mathbf{g}^T(\mathbf{Y}, \boldsymbol{\eta})/\partial \boldsymbol{\eta}\right]\right\}^{-1} (n - n_1)^{-1} \sum_{i \notin B_k} \mathbf{g}^T(\mathbf{Y}_i, \boldsymbol{\eta}) \\
&\approx \{-\mathrm{E}\left[\partial \mathbf{g}^T(\mathbf{Y}, \boldsymbol{\eta})/\partial \boldsymbol{\eta}\right]\}^{-1} (n - n_1)^{-1} \\
&\quad \cdot \left[n\{-\mathrm{E}\left[\partial \mathbf{g}^T(\mathbf{Y}, \boldsymbol{\eta})/\partial \boldsymbol{\eta}\right]\} (\tilde{\boldsymbol{\eta}} - \boldsymbol{\eta})^T - \sum_{i \in B_k} \mathbf{g}^T(\mathbf{Y}_i, \boldsymbol{\eta}) \right]
\end{aligned}$$

$$= \frac{n}{n - n_1} (\tilde{\boldsymbol{\eta}} - \boldsymbol{\eta})^T - \frac{1}{n - n_1} \left\{-\mathrm{E}\left[\partial \mathbf{g}^T(\mathbf{Y}, \boldsymbol{\eta})/\partial \boldsymbol{\eta}\right]\right\}^{-1} \sum_{i \in B_k} \mathbf{g}^T(\mathbf{Y}_i, \boldsymbol{\eta}),$$

so that

$$(\tilde{\boldsymbol{\eta}}^{(k)} - \tilde{\boldsymbol{\eta}})^T \approx -(n-n_1)^{-1}\{-\mathrm{E}\,[\partial \mathbf{g}^T(\mathbf{Y},\boldsymbol{\eta})/\partial\boldsymbol{\eta}]\}^{-1} \sum_{i\in B_k} \mathbf{g}^T(\mathbf{Y}_i,\boldsymbol{\eta}) + O_p(n^{-3/2}).$$

Hence,

$$\sum_k (\tilde{\boldsymbol{\eta}}^{(k)} - \tilde{\boldsymbol{\eta}})^T(\tilde{\boldsymbol{\eta}}^{(k)} - \tilde{\boldsymbol{\eta}}) \approx (n-n_1)^{-2}\{-\mathrm{E}\,[\partial \mathbf{g}^T(\mathbf{Y},\boldsymbol{\eta})/\partial\boldsymbol{\eta}]\}^{-1} \cdot \Big[\sum_{i=1}^{n} \mathbf{g}^T(\mathbf{Y}_i,\boldsymbol{\eta})\,\mathbf{g}(\mathbf{Y}_i,\boldsymbol{\eta}) + \sum_k \sum_{i\neq i',i,i'\in B_k} \mathbf{g}^T(\mathbf{Y}_i,\boldsymbol{\eta})\,\mathbf{g}(\mathbf{Y}_{i'},\boldsymbol{\eta})\Big] \cdot \{-\mathrm{E}\,[\partial \mathbf{g}^T(\mathbf{Y},\boldsymbol{\eta})/\partial\boldsymbol{\eta}]^T\}^{-1}.$$

The term $(n-n_1)^{-2}\sum_{i=1}^{n} \mathbf{g}^T(\mathbf{Y}_i,\boldsymbol{\eta})\,\mathbf{g}(\mathbf{Y}_i,\boldsymbol{\eta})$ is $O_p(n^{-1})$. If $n_1 > 1$, it also dominates the other term

$$(n-n_1)^{-2}\sum_k \sum_{i\neq i',i,i'\in B_k} \mathbf{g}^T(\mathbf{Y}_i,\boldsymbol{\eta})\,\mathbf{g}(\mathbf{Y}_{i'},\boldsymbol{\eta}),$$

which asymptotically is

$$(n-n_1)^{-2}\Big\{n_2\mathrm{E}\Big[\sum_{i\neq i',1\le i,i'\le n_1} \mathbf{g}^T(\mathbf{Y}_i,\boldsymbol{\eta})\,\mathbf{g}(\mathbf{Y}_{i'},\boldsymbol{\eta})\Big] + O_p(n_2^{1/2})\Big\}$$

or $O_p(n^{-3/2})$ since the expectation is zero and $n_2, n \to \infty$. □

The jackknife method can also be used for estimates of functions of parameters (such as probabilities of being in some category or probabilities of exceedances). The delta or Taylor method requires partial derivatives (of the function with respect to the parameters) and the jackknife method eliminates the need for this. As before, let $\tilde{\boldsymbol{\eta}}^{(k)}$ be the estimator of $\boldsymbol{\eta}$ with the kth block deleted, $k = 1,\ldots,n_2$, and let $\tilde{\boldsymbol{\eta}}$ be the estimator based on the entire sample. Let $b(\boldsymbol{\eta})$ be a (real-valued) quantity of interest. In addition to $b(\tilde{\boldsymbol{\eta}})$, the estimates $b(\tilde{\boldsymbol{\eta}}^{(k)})$, $k = 1,\ldots,n_2$, from the subsamples are obtained. The jackknife estimate of the SE of $b(\tilde{\boldsymbol{\eta}})$ is

$$\Big\{\sum_{k=1}^{n_2} \big[b(\tilde{\boldsymbol{\eta}}^{(k)}) - b(\tilde{\boldsymbol{\eta}})\big]^2\Big\}^{1/2}.$$

What the jackknife approach means for the computation sequence is that one should maintain a table of the parameter estimates for the full sample and each jackknife subsample. Then one can use this table for computing estimates of one or more functions of the parameters, together with the corresponding SEs.

10.1.2 Efficiency

The efficiency of the IFM method relative to the ML method (e.g., comparison of (10.3) and (10.4)) for multivariate models can be assessed using various methods. One comparison of efficiency is through the asymptotic covariance matrices of the MLE and the estimator from the IFM method. Another comparison consists of Monte Carlo simulations to compare the two estimators (through mean squared errors, etc.). Both methods are difficult, because of the general intractability of the asymptotic covariance matrices, and the computation time needed to obtain the MLE based on the likelihood in all of the univariate and multivariate parameters. Nevertheless, comparisons of various types are made in Xu (1996) for a number of multivariate models. All of these comparisons suggest that the IFM method is highly efficient compared with maximum likelihood. Intuitively, we expect the IFM method to be quite efficient because it depends heavily on maximum likelihood, albeit from likelihoods of marginal distributions.

Examples of models for which asymptotic covariance matrices were compared are:

(a) the trivariate probit model for binary data with cutoff points of zero and unknown correlation parameters;

(b) the trivariate probit model for binary data with general known cutoff points and unknown correlation parameters;

(c) the latent variable model for binary data with a trivariate copula having bivariate margins in the family B10 in Section 5.1.

Examples of models for which Monte Carlo simulations were used for comparisons are:

(a) the multivariate probit model for binary data, up to $m = 4$ with covariates;

(b) the multivariate probit model for ordinal data, up to $m = 4$ without covariates;

(c) the multivariate Poisson model based on the MVN copula.

In almost all of the simulations, the relative efficiency, as measured by the ratio of the mean squared errors of the IFM estimator to the MLE, is close to 1.

10.1.3 Estimation consistency

There are several ways to assess the adequacy of fit of a multivariate copula-based model. One approach is comparison of estimates from higher-order margins and lower-order margins. This comparison will be called the **estimation consistency** check. A rough assessment could be based on differences of the estimators relative to the SEs or on a likelihood interval.

In general terms, the estimation consistency check is the following. Let

$$F(\mathbf{y};\boldsymbol{\alpha}_1,\ldots,\boldsymbol{\alpha}_m,\boldsymbol{\theta}) = C(F_1(y_1;\boldsymbol{\alpha}_1),\ldots,F_m(y_m;\boldsymbol{\alpha}_m);\boldsymbol{\theta})$$

be a family of m-variate distributions with margins $F_S(\mathbf{y}_S;\boldsymbol{\alpha}_j, j\in S,\boldsymbol{\theta}_S)$, $S\in\mathcal{S}_m$, $|S|\geq 2$, where $\boldsymbol{\theta}$ is the multivariate parameter and $\boldsymbol{\alpha}_j$ is the parameter for the jth univariate margin. Note that $\boldsymbol{\theta}_S$ is uniquely determined from $\boldsymbol{\theta}$ and S. Let S_1 and S_2 be two subsets with S_1 being a proper subset of S_2 and $|S_1|\geq 2$. Then $\boldsymbol{\theta}_{S_1} = \mathbf{a}(\boldsymbol{\theta}_{S_2})$ for a function $\mathbf{a}$. Let the data be $\mathbf{Y}_i$, $i=1,\ldots,n$. Assume that the univariate parameters are known, or are estimated based on individual univariate likelihoods and then assumed known. For $k=1,2$, let $\tilde{\boldsymbol{\theta}}_{S_k}$ be the estimator using the IFM method based on the $\mathbf{Y}_{i,S_k}=(Y_{ij}:j\in S_k)$, $i=1,\ldots,n$. Then in general, $\tilde{\boldsymbol{\theta}}_{S_1}$ and $\mathbf{a}(\tilde{\boldsymbol{\theta}}_{S_2})$ are different. The closeness of $\tilde{\boldsymbol{\theta}}_{S_1}$ and $\mathbf{a}(\tilde{\boldsymbol{\theta}}_{S_2})$ for all S_1,S_2 with $S_1\subset S_2$ is a necessary (but not sufficient) indication that the model $F(\mathbf{y};\boldsymbol{\alpha}_1,\ldots,\boldsymbol{\alpha}_m,\boldsymbol{\theta})$ is an adequate fit to the data. A rough SE for $\tilde{\boldsymbol{\theta}}_{S_1}-\mathbf{a}(\tilde{\boldsymbol{\theta}}_{S_2})$ could be obtained from the jackknife method if needed, or one could use SEs from the separate optimizations. Another method of assessing the closeness is to check if $\mathbf{a}(\tilde{\boldsymbol{\theta}}_{S_2})$ is in the likelihood-based confidence interval for $\boldsymbol{\theta}_{S_1}$; this interval is

$$\{\boldsymbol{\theta}: 2[L(\tilde{\boldsymbol{\theta}}_{S_1};S_1)-L(\boldsymbol{\theta};S_1)]\geq \chi^2_{p_{S_1},1-\alpha}\},$$

where $\chi^2_{p_S,1-\alpha}$ is the upper α quantile of a chi-square distribution with p_S degrees of freedom, p_S is the dimension of $\boldsymbol{\theta}_S$, and $L(\cdot;S_1)$ is the log-likelihood based on the density of $F(\mathbf{y}_{S_1};\boldsymbol{\alpha}_j, j\in S_1,\boldsymbol{\theta}_{S_1})$.

There are variations on the use of estimation consistency for comparisons of parameter estimates from likelihoods of different margins. Some of these are mentioned in the specific examples in the next subsection and in Chapter 11. See also the end of the next subsection for another consistency check that could be done.

10.1.4 Examples

This subsection consists of examples to illustrate the theory described earlier in this section.

Example 10.1 (Multivariate probit model with no covariates.) The model for the multivariate binary response vector $\mathbf{Y}$ is $Y_j = I(Z_j \le \alpha_j)$, $j = 1, \ldots, m$, where $\mathbf{Z} \sim N_m(\mathbf{0}, R)$, and $\boldsymbol{\theta} = R = (\rho_{jk})$ is a correlation matrix. Let the data be $\mathbf{y}_i = (y_{i1}, \ldots, y_{im})$, $i = 1, \ldots, n$. For $j = 1, \ldots, m$, let $N_j(0)$ and $N_j(1)$ be the number of 0s and 1s among $y_{1j}, \ldots, y_{nj}$. For $1 \le j < k \le m$, let $N_{jk}(i_1, i_2)$ be the frequency of (i_1, i_2) among the pairs $(y_{1j}, y_{1k}), \ldots, (y_{nj}, y_{nk})$, for (i_1, i_2) equal to (0,0), (0,1), (1,0) and (1,1). From maximizing the jth univariate likelihood,

$$L_j^*(\alpha_j) = [\Phi(\alpha_j)]^{N_j(1)}[1 - \Phi(\alpha_j)]^{N_j(0)},$$

$\tilde{\alpha}_j = \Phi^{-1}(N_j(1)/n)$, $j = 1, \ldots, m$. For this example, one can estimate the dependence parameters from maximizing separate bivariate likelihoods $L_{jk}^*(\rho_{jk}) = L_{jk}^*(\rho_{jk}, \alpha_j, \alpha_k)$ rather than using the m-variate log-likelihood L in (10.2). Let Φ_ρ be the BVSN cdf with correlation ρ. Then

$$\begin{aligned} L_{jk}^*(\rho_{jk}, \alpha_j, \alpha_k) &= [\Phi_{\rho_{jk}}(\alpha_j, \alpha_k)]^{N_{jk}(11)}[\Phi(\alpha_j) - \Phi_{\rho_{jk}}(\alpha_j, \alpha_k)]^{N_{jk}(10)} \\ &\quad \cdot [\Phi(\alpha_k) - \Phi_{\rho_{jk}}(\alpha_j, \alpha_k)]^{N_{jk}(01)} \\ &\quad \cdot [1 - \Phi(\alpha_j) - \Phi(\alpha_k) + \Phi_{\rho_{jk}}(\alpha_j, \alpha_k)]^{N_{jk}(00)}, \end{aligned}$$

and $\tilde{\rho}_{jk}$ is the root ρ of $N_{jk}(11)/n = \Phi_\rho(\tilde{\alpha}_j, \tilde{\alpha}_k)$.

The maximization of the bivariate likelihood L_{jk}^* jointly in the parameters $\alpha_j, \alpha_j, \rho_{jk}$ leads to the same estimates as above. An assessment of the estimation consistency of the multivariate probit model can be made from comparisons of estimators from trivariate likelihoods L_{jks}^* in the parameters $\alpha_j, \alpha_k, \alpha_s, \rho_{jk}, \rho_{js}, \rho_{ks}$.

Another check that could be done is the positive definiteness of the matrix $\tilde{R} = (\tilde{\rho}_{jk})$. It is possible that $\tilde{R}$ is not positive definite, especially if the true correlation matrix R is close to being singular.

Possibly one may be considering a simple correlation structure from the nature of the variables or perhaps the correlation estimates suggest a simple structure. Examples are an exchangeable correlation matrix with all correlations equal to ρ, and an AR(1) correlation matrix with the (j, k) component equal to $\rho^{|j-k|}$ for some ρ. For these cases, see Section 10.2.2. □

Example 10.2 (Multivariate probit model with covariates.) Let the data be $(\mathbf{Y}_i, \mathbf{x}_i)$, $i = 1, \ldots, n$, where $\mathbf{Y}_i$ is an m-variate response vector and $\mathbf{x}_i$ is a covariate (column) vector. The multivariate probit model has stochastic representation $Y_{ij} = I(Z_{ij} \le \beta_{j0} + \boldsymbol{\beta}_j \mathbf{x}_i)$, where $\mathbf{Z}_i$ are iid $N_m(\mathbf{0}, R)$, $R = (\rho_{jk})$ and the $\boldsymbol{\beta}_j$ are row vectors. With $\boldsymbol{\gamma}_j = (\beta_{j0}, \boldsymbol{\beta}_j)$, the jth univariate likelihood is

$$L_j^*(\boldsymbol{\gamma}_j) = \prod_{i=1}^{n} [\Phi(\beta_{j0} + \boldsymbol{\beta}_j \mathbf{x}_i)]^{y_{ij}} [1 - \Phi(\beta_{j0} + \boldsymbol{\beta}_j \mathbf{x}_i)]^{1-y_{ij}}.$$

The maximization of L_j^* leads to $\tilde{\boldsymbol{\gamma}}_j$. For $1 \le j < k \le m$, the (j,k) bivariate likelihood is:

$$\begin{aligned} L_{jk}^*(\rho_{jk}, \boldsymbol{\gamma}_j, \boldsymbol{\gamma}_k) = \prod_{i=1}^{n} \Big\{ & [\Phi_{\rho_{jk}}(\beta_{j0} + \boldsymbol{\beta}_j \mathbf{x}_i, \beta_{k0} + \boldsymbol{\beta}_k \mathbf{x}_i)]^{I(y_{ij}=y_{ik}=1)} \\ \cdot \quad & [\Phi(\beta_{j0} + \boldsymbol{\beta}_j \mathbf{x}_i) - \Phi_{\rho_{jk}}(\beta_{j0} + \boldsymbol{\beta}_j \mathbf{x}_i, \beta_{k0} + \boldsymbol{\beta}_k \mathbf{x}_i)]^{I(y_{ij}=1, y_{ik}=0)} \\ \cdot \quad & [\Phi(\beta_{k0} + \boldsymbol{\beta}_k \mathbf{x}_i) - \Phi_{\rho_{jk}}(\beta_{j0} + \boldsymbol{\beta}_j \mathbf{x}_i, \beta_{k0} + \boldsymbol{\beta}_k \mathbf{x}_i)]^{I(y_{ij}=0, y_{ik}=1)} \\ \cdot \quad & [1 - \Phi(\beta_{j0} + \boldsymbol{\beta}_j \mathbf{x}_i) - \Phi(\beta_{k0} + \boldsymbol{\beta}_k \mathbf{x}_i) \\ & + \Phi_{\rho_{jk}}(\beta_{j0} + \boldsymbol{\beta}_j \mathbf{x}_i, \beta_{k0} + \boldsymbol{\beta}_k \mathbf{x}_i)]^{I(y_{ij}=y_{ik}=0)} \Big\}. \end{aligned}$$

The maximization of $L_{jk}^*(\rho_{jk}, \tilde{\boldsymbol{\gamma}}_j, \tilde{\boldsymbol{\gamma}}_k)$ in ρ_{jk} leads to $\tilde{\rho}_{jk}$. An assessment of the estimation consistency can be made from a bivariate likelihood in the parameters $\beta_{j0}, \beta_{k0}, \boldsymbol{\beta}_j, \boldsymbol{\beta}_k$, ρ_{jk} or a trivariate likelihood in the parameters $\beta_{j0}, \beta_{k0}, \beta_{s0}, \boldsymbol{\beta}_j, \boldsymbol{\beta}_k, \boldsymbol{\beta}_s$, $\rho_{jk}, \rho_{js}, \rho_{ks}$. Again, see Section 10.2.2 if some of the parameters (regression or correlation) can be assumed to be the same. □

Example 10.3 (Multivariate Poisson-lognormal distribution.) Suppose we have a random sample of iid random vectors $\mathbf{y}_i$, $i = 1, \ldots, n$, from the density (7.28) in Section 7.2.3. Let $\boldsymbol{\alpha}_j = (\mu_j, \sigma_{jj})$, $j = 1, \ldots, m$. The jth univariate marginal density is

$$f_j(y_j; \boldsymbol{\alpha}_j) = \int_0^\infty \frac{e^{-\lambda_j} \lambda_j^{y_j}}{y_j! \sqrt{2\pi\sigma_{jj}}\, \lambda_j} \exp\{-\tfrac{1}{2}(\log \lambda_j - \mu_j)^2 / \sigma_{jj}\}\, d\lambda_j.$$

The (j,k) bivariate marginal density is

$$\begin{aligned} f_{jk}(y_j, y_k; \boldsymbol{\alpha}_j, \boldsymbol{\alpha}_k, \rho_{jk}) = & \int_0^\infty \int_0^\infty \frac{e^{-\lambda_j} \lambda_j^{y_j} e^{-\lambda_k} \lambda_k^{y_k}}{y_j! y_k!\, 2\pi \lambda_j \lambda_k \sqrt{\sigma_{jj}\sigma_{kk} - \sigma_{jk}^2}} \\ & \cdot \exp\Big\{-\tfrac{1}{2}(1 - \rho_{jk}^2)^{-1} \big[(\log \lambda_j - \mu_j)^2/\sigma_{jj} + (\log \lambda_k - \mu_k)^2/\sigma_{kk} \\ & \quad -2\rho_{jk}(\log \lambda_j - \mu_j)(\log \lambda_k - \mu_k)/\sqrt{\sigma_{jj}\sigma_{kk}}\big]\Big\}\, d\lambda_j d\lambda_k, \end{aligned}$$

where $\rho_{jk} = \sigma_{jk}/\sqrt{\sigma_{jj}\sigma_{kk}}$. Suppose $\tilde{\boldsymbol{\alpha}}_j = (\tilde{\mu}_j, \tilde{\sigma}_{jj})$ obtains from maximizing $L_j(\boldsymbol{\alpha}_j) = \sum_{i=1}^n \log f_j(y_{ij}; \boldsymbol{\alpha}_j)$, and given the estimates of the univariate parameters, $\tilde{\rho}_{jk}$ obtains from maximizing

$$L_{jk}(\rho_{jk}, \tilde{\boldsymbol{\alpha}}_j, \tilde{\boldsymbol{\alpha}}_k) = \sum_{i=1}^n \log f_{jk}(y_{ij}, y_{ik}; \tilde{\boldsymbol{\alpha}}_j, \tilde{\boldsymbol{\alpha}}_k, \rho_{jk})$$

as a function of ρ_{jk}. The log-likelihood maximizations are straightforward using a quasi-Newton routine, with good starting points, which in this case can come from method of moments estimates. Let $\overline{y}_j, s_j^2$ be the sample mean and sample variance for the jth univariate margin, and let r_{jk} be the (j,k) sample correlation. Based on the expected values given in Section 7.2.3, for L_j, the method of moments estimates are

$$\tilde{\sigma}_{jj}^0 = \log\bigl([s_j^2 - \overline{y}_j]/\overline{y}_j^2 + 1\bigr), \quad \tilde{\mu}_j^0 = \log \overline{y}_j - \tfrac{1}{2}\tilde{\sigma}_{jj}^0,$$

and, for L_{jk}, the method of moments estimate is

$$\tilde{\sigma}_{jk}^0 = \log\bigl(r_{jk} s_j s_k / [\overline{y}_j \overline{y}_k] + 1\bigr).$$

Assessment of estimation consistency can come from comparisons of bivariate and trivariate likelihoods with the univariate parameters already estimated. The trivariate log-likelihoods, with three-dimensional integrations, would be time-consuming to maximize but should be possible.

The comments for a special correlation structure from Example 10.1, and the extensions to include covariates in Example 10.2, apply here as well. □

Example 10.4 (MEV model with partially exchangeable dependence structure.) We illustrate some ideas for models from Section 4.2 using the family M6 in the trivariate case. In the MSMVE form, the exponent is

$$A(\mathbf{z}; \beta_{1,2}, \beta_{1,3}) = \bigl((z_1^{\beta_{1,2}} + z_2^{\beta_{1,2}})^{\beta_{1,3}/\beta_{1,2}} + z_3^{\beta_{1,3}}\bigr)^{1/\beta_{1,3}},$$

with $\beta_{1,2} \geq \beta_{1,3} \geq 1$. The bivariate margins have exponents in the family B6, $A(w_1, w_2) = (w_1^\delta + w_2^\delta)^{1/\delta}$, with parameters $\beta_{1,2}$, $\beta_{1,3}$, $\beta_{1,3}$ for the (1,2), (1,3), (2,3) bivariate margins, respectively. For trivariate extreme value data (maxima), (y_{i1}, y_{i2}, y_{i3}), $i = 1, \ldots, n$, one can maximize separate GEV univariate log-likelihoods L_j to get parameter estimates $\tilde{\mu}_j, \tilde{\sigma}_j, \tilde{\gamma}_j$, $j = 1, 2, 3$. Then one can transform the univariate margins to exponential survival functions with $z_{ij} = (1 + \gamma_j[y_{ij} - \mu_j]/\sigma_j)_+^{-1/\gamma_j}$, substituting in the estimated parameters. Next, $\beta_{1,2}, \beta_{1,3}$ can be estimated from the bivariate log-likelihoods and the model M6 may be plausible if the estimates from the (1,3) and (2,3) bivariate likelihoods are about the same and smaller than the estimate from the (1,2) bivariate likelihood. If the estimates from the bivariate likelihoods have the right pattern, one could go on to the trivariate likelihood (with given $\tilde{\mu}_j, \tilde{\sigma}_j, \tilde{\gamma}_j$).

The survival function of the family M6 has closed form so that the density can be obtained by symbolic manipulation software in the form of Fortran or C code for use with a quasi-Newton routine for numerical maximization of the log-likelihood. This comment applies to any MEV or MSMVE model with closed-form cdf or survival function. □

Example 10.5 (MEV model MM6 or MM7.) Consider the trivariate case, with GEV univariate margins. After the transform to z_{ij}, given in Example 10.4, one can estimate the multivariate parameters. Because of the asymmetry in the variables, one has to decide which variables go with which indices. Once this is decided, the parameters θ_{12}, θ_{23} can be estimated from the (1,2) and (2,3) bivariate log-likelihoods and then the parameter θ_{13} can be estimated from the (1,2,3) trivariate log-likelihood. If there is no natural order to the indices, one could compare the trivariate log-likelihood values, evaluated at estimated parameters, for the three distinct ways of assigning indices to variables. □

Example 10.6 (MEV model M8.) The univariate parameters can be estimated from the GEV log-likelihoods. After the transform to z_{ij}, given in Example 10.4, the parameter δ_{jk}, $1 \leq j < k \leq m$, can be estimated from the (j, k) bivariate log-likelihood to get $\tilde{\delta}_{jk}$. For estimation consistency, comparisons can be made with the estimates from the trivariate log-likelihoods (with the univariate parameters fixed). □

Example 10.7 (Molenberghs–Lesaffre construction of Section 4.8.) Assume that the bivariate margins all belong to a common parametric family of copulas, such as the family B2 or B3. Also assume that the parameters ψ_S, $S \in \mathcal{S}_m$, $|S| \geq 3$, are in a range such that the construction in Section 4.8 leads to proper multivariate distributions. Suppose that the multivariate copula is used as a latent variable model for multivariate binary or ordinal responses with a logistic distribution (compare the model in Section 7.1.7).

A sequence for estimation of parameters is: first, the univariate parameters for the separate univariate logistic or ordinal regressions; second, the bivariate parameters from separate bivariate log-likelihoods; and then the parameters ψ_S, for dimensions

$|S| \geq 3$. For each multivariate log-likelihood, estimates from lower-order margins are used, so that it involves only the maximization over one variable. For trivariate and higher-order margins, a reasonable starting point for the numerical optimization is 1, because of the maximum entropy interpretation in Section 4.8 and because this leads to reflection symmetry as discussed in Section 7.1.7. □

Example 10.8 (Families MM1–MM3 with νs equal to 0.) For the use of these multivariate families of copulas, the sequence of estimation of parameters is: first, the univariate parameters from univariate log-likelihoods; second, bivariate analysis to determine degrees of bivariate dependence; and then the multivariate parameters from the multivariate log-likelihood with the univariate parameters given. □

As noted in Example 10.1, with the same possibility for other multivariate models, when dependence parameters are estimated based on bivariate or lower-dimensional margins, then one should check if the set of estimators is compatible for the model. For example, for any model with the MVN as a latent distribution, one needs to check if the estimated correlations from the bivariate margins lead to a positive definite matrix. If the sample size is sufficiently large, then non-compatibility would not be expected to be a problem, unless the vector of dependence parameters is near the boundary of the parameter space.

10.2 Extensions

In this section, we outline extensions of the results in the preceding section to include covariates (e.g., the multivariate probit model with covariates) and to situations for which parameters are common to more than one margin (e.g., the multivariate probit model with AR(1) dependence structure).

10.2.1 Covariates

Under certain regularity conditions, the results of Section 10.1 extend to the inclusion of covariates. There are two ways of looking at the asymptotics: one is a more standard approach to extend from the iid case to the independent, non-identically distributed case and the other is to view the response vector and covariate vector pair $(\mathbf{Y}_i, \mathbf{x}_i)$ as iid. An outline of the asymptotic results is given below.

There are common ways of extending many univariate distributions to include covariates (often, transformations of parameters are linear functions of some functions of covariates). A difficult modelling question may be whether dependence parameters should be functions of covariates. If so, what are natural functions to choose? This is not even clear for the multivariate probit model. For data analysis, one could split the covariate space into several clusters or subgroups, and then do a separate estimation of dependence parameters by clusters. If there are only binary (or categorical) predictor variables, then one could do estimation for each combination if the resulting (sub)sample sizes are large enough.

Inclusion of covariates: approach 1

Assume that we have independent, non-identically distributed random vectors $\mathbf{Y}_i$, $i = 1, \ldots, n$, with $\mathbf{Y}_i$ having density $f(\cdot; \boldsymbol{\eta}_i)$, $\boldsymbol{\eta}_i = \boldsymbol{\eta}(\mathbf{x}_i, \boldsymbol{\gamma})$ for a function $\boldsymbol{\eta}$ and a parameter function $\boldsymbol{\gamma}$. The necessary conditions for the asymptotic results depend somewhat on the specific models. However, we indicate the general types of conditions that must hold.

For something like (10.1), we assume that each component of $\boldsymbol{\eta} = (\boldsymbol{\alpha}_1, \ldots, \boldsymbol{\alpha}_m, \boldsymbol{\theta})$ is a function of $\mathbf{x}$, more specifically, $\boldsymbol{\alpha}_j = \mathbf{a}_j(\mathbf{x}, \boldsymbol{\gamma}_j)$, $j = 1, \ldots, m$, and $\boldsymbol{\theta} = \mathbf{t}(\mathbf{x}, \boldsymbol{\gamma}_{m+1})$, with $\mathbf{a}_1, \ldots, \mathbf{a}_m, \mathbf{t}$ having components that are each functions of linear combinations of the functions of the components of $\mathbf{x}$. We assume that the inference function vector has a component for each parameter in $\boldsymbol{\gamma} = (\boldsymbol{\gamma}_1, \ldots, \boldsymbol{\gamma}_m, \boldsymbol{\gamma}_{m+1})$.

We explain the notation here for Examples 10.1 and 10.2. With no covariates, $\boldsymbol{\eta} = (\alpha_1, \ldots, \alpha_m, \boldsymbol{\theta})$, where α_j is the cutoff point for the jth univariate margin, and $\boldsymbol{\theta} = R = (\rho_{jk})$ is a correlation matrix. With covariates, $\alpha_j = a_j(\mathbf{x}, \beta_{j0}, \boldsymbol{\beta}_j) = \beta_{j0} + \boldsymbol{\beta}_j\mathbf{x}$, $j = 1, \ldots, m$, and $t(\mathbf{x}, \boldsymbol{\theta}) = \boldsymbol{\theta}$, so that $\boldsymbol{\gamma} = (\beta_{10}, \boldsymbol{\beta}_1, \ldots, \beta_{m0}, \boldsymbol{\beta}_m, \boldsymbol{\theta})$ with $\boldsymbol{\gamma}_j = (\beta_{j0}, \boldsymbol{\beta}_j)$ and $\boldsymbol{\gamma}_{m+1} = \boldsymbol{\theta}$.

For (10.1), in place of $f(\mathbf{y}; \boldsymbol{\alpha}_1, \ldots, \boldsymbol{\alpha}_m, \boldsymbol{\theta})$ and $f_j(y_j, \boldsymbol{\alpha}_j)$ in the case of no covariates, we now have the densities

$$f_{\mathbf{Y}|\mathbf{x}}(\mathbf{y}|\mathbf{x}; \boldsymbol{\gamma}) \stackrel{\text{def}}{=} f\big(\mathbf{y}; \mathbf{a}_1(\mathbf{x}, \boldsymbol{\gamma}_1), \ldots, \mathbf{a}_m(\mathbf{x}, \boldsymbol{\gamma}_m), \mathbf{t}(\mathbf{x}, \boldsymbol{\gamma}_{m+1})\big)$$

and

$$f_{Y_j|\mathbf{x}}(y_j|\mathbf{x}; \boldsymbol{\gamma}_j) \stackrel{\text{def}}{=} f_j\big(y_j; \mathbf{a}_j(\mathbf{x}, \boldsymbol{\gamma}_j)\big), \quad j = 1, \ldots, m.$$

In a simple case, the estimate $\tilde{\boldsymbol{\gamma}}$ from the IFM method has component $\tilde{\gamma}_j$ coming from the maximization of

$$L_j(\gamma_j) = \sum_{i=1}^{n} \log f_{Y_j|\mathbf{x}}(y_{ij}|\mathbf{x}_i;\gamma_j), \quad j = 1,\ldots,m, \tag{10.7}$$

and $\tilde{\gamma}_{m+1}$ comes from the maximization of $L(\tilde{\gamma}_1,\ldots,\tilde{\gamma}_m,\gamma_{m+1})$ in γ_{m+1}, where

$$L(\gamma) = \sum_{i=1}^{n} \log f_{\mathbf{Y}|\mathbf{x}}(\mathbf{y}_i|\mathbf{x}_i;\gamma). \tag{10.8}$$

Alternatively, the components of γ_{m+1} may be estimated from log-likelihoods of lower-dimensional margins such as in Example 10.2. In any case, let $\sum \mathbf{g}(\mathbf{y}_i,\mathbf{x}_i,\gamma)$ be the vector of inference functions from partial derivatives of log-likelihood functions of margins.

Conditions for the asymptotic results to hold have the following sense:

(a) mixed derivatives of $\mathbf{g}$ of first and second order are dominated by integrable functions;

(b) products of these derivatives are uniformly integrable;

(c) the functions $\mathbf{a}_1,\ldots,\mathbf{a}_m,\mathbf{t}$ are twice continuously differentiable with first- and second-order derivatives bounded away from zero;

(d) covariates are uniformly bounded, and the sample covariance matrix of the covariates $\mathbf{x}_i$ is strictly positive definite;

(e) a Lindeberg–Feller type condition holds.

References for these types of conditions and proofs of asymptotic normality are Bradley and Gart (1962) and Hoadley (1971).

Assuming that the conditions hold, then the asymptotic normality result has the form:

$$n^{-1/2} V_n^{-1/2} (\tilde{\gamma} - \gamma)^T \to_d N(\mathbf{0}, I),$$

where $V_n = D_n^{-1} M_n (D_n^{-1})^T$ with

$$D_n = n^{-1} \sum_{i=1}^{n} \mathrm{E}\left[\partial \mathbf{g}^T(\mathbf{Y}_i,\mathbf{x}_i,\gamma)/\partial\gamma\right]$$

and

$$M_n = n^{-1} \sum_{i=1}^{n} \mathrm{E}\left[\mathbf{g}^T(\mathbf{Y}_i,\mathbf{x}_i,\gamma)\,\mathbf{g}(\mathbf{Y}_i,\mathbf{x}_i,\gamma)\right].$$

Details for the case of multivariate discrete models are given in Xu (1996); the results generalize to the continuous case when the assumptions hold.

Inclusion of covariates: approach 2.

A second approach to asymptotics allows for parameters to be more general functions of the covariates, and treats the covariates as realizations of random vectors. This approach assumes a joint distribution for response vector and covariates, with the parameters for the marginal distribution of the covariate vector treated as nuisance parameters. This assumption might be reasonable for a random sample of subjects in which $\mathbf{x}_i$ and $\mathbf{Y}_i$ are observed together.

Similar to the preceding approach to the inclusion of covariates, we write the conditional density as

$$f_{\mathbf{Y}|\mathbf{x}}(\mathbf{y}|\mathbf{x};\boldsymbol{\gamma}) = f(\mathbf{y};\boldsymbol{\eta}(\mathbf{x},\boldsymbol{\gamma})).$$

Let $\mathbf{Z}_i = (\mathbf{Y}_i, \mathbf{x}_i)$, $i = 1, \ldots, n$. These are treated as iid random vectors from the density

$$f_{\mathbf{Z}}(\mathbf{z};\boldsymbol{\gamma}) = f_{\mathbf{Y}|\mathbf{x}}(\mathbf{y}|\mathbf{x};\boldsymbol{\gamma})\, f_{\mathbf{X}}(\mathbf{x};\boldsymbol{\omega}). \tag{10.9}$$

For inference, we are interested in $\boldsymbol{\gamma}$ and $\boldsymbol{\eta}(\mathbf{x},\boldsymbol{\gamma})$, and not in $\boldsymbol{\omega}$. Marginal distributions of (10.9) are:

$$f_{Y_j|\mathbf{x}}(y_j|\mathbf{x};\boldsymbol{\gamma}_j)\, f_{\mathbf{X}}(\mathbf{x};\boldsymbol{\omega}), \quad j = 1, \ldots, m.$$

If $\boldsymbol{\omega}$ is treated as a nuisance parameter, then the log-likelihood in $\boldsymbol{\gamma}$ from (10.10) below is essentially the same as that in the first approach.

Let $\boldsymbol{\gamma}, \boldsymbol{\alpha}_j, \mathbf{a}_j, \boldsymbol{\theta}, \mathbf{t}$ be the same as before, except that $\mathbf{a}_j$, $j = 1, \ldots, m$, and $\mathbf{t}$ could be more general functions of the covariate vector $\mathbf{x}$. The vector estimate from the IFM method has component $\tilde{\boldsymbol{\gamma}}_j$ coming from the maximization of

$$L_j(\boldsymbol{\gamma}_j) = \sum_{i=1}^{n} \log\left[f_{Y_j|\mathbf{x}}(y_{ij}|\mathbf{x}_i;\boldsymbol{\gamma}_j) f_{\mathbf{X}}(\mathbf{x}_i;\boldsymbol{\omega})\right], \quad j = 1, \ldots, m, \tag{10.10}$$

and $\tilde{\boldsymbol{\gamma}}_{m+1}$ coming from the maximization of $L(\tilde{\boldsymbol{\gamma}}_1, \ldots, \tilde{\boldsymbol{\gamma}}_m, \boldsymbol{\gamma}_{m+1})$ in $\boldsymbol{\gamma}_{m+1}$, where

$$L(\boldsymbol{\gamma}) = \sum_{i=1}^{n} \log\left[f_{\mathbf{Y}|\mathbf{x}}(\mathbf{y}_i|\mathbf{x}_i;\boldsymbol{\gamma}) f_{\mathbf{X}}(\mathbf{x}_i;\boldsymbol{\omega})\right]. \tag{10.11}$$

Note that optimization of (10.7) and (10.10), and of (10.8) and (10.11), is the same. Alternatively, the components of $\boldsymbol{\gamma}_{m+1}$ may be estimated from log-likelihoods of lower-dimensional margins. In any case, let $\sum \mathbf{g}(\mathbf{y}_i, \mathbf{x}_i, \boldsymbol{\gamma})$ be the vector of inference functions based on partial derivatives of log-likelihood functions of margins.

Assuming that the standard regularity conditions hold for $f_{\mathbf{Z}}$ and its margins, then the asymptotic theory for the iid case holds for the estimates using the IFM method in Section 10.1. The asymptotic normality result is

$$n^{-1/2}(\tilde{\gamma} - \gamma)^T \to_d N(\mathbf{0}, V),$$

where $V = D_{\mathbf{g}}^{-1} M_{\mathbf{g}} (D_{\mathbf{g}}^{-1})^T$, with $D_{\mathbf{g}} = \mathrm{E}\,[\partial \mathbf{g}^T(\mathbf{Y}, \mathbf{x}, \gamma)/\partial \gamma]$, $M_{\mathbf{g}} = \mathrm{E}\,[\mathbf{g}^T(\mathbf{Y}, \mathbf{x}, \gamma)\, \mathbf{g}(\mathbf{Y}, \mathbf{x}, \gamma)]$.

The asymptotic covariance matrices for $n^{-1/2}(\tilde{\gamma} - \gamma)^T$ may be different in the two asymptotic approaches. However, the inference functions are the same. The use of the empirical distribution function to estimate the inverse Godambe information matrix or the use of the jackknife would lead to the same standard error estimates in the two approaches.

10.2.2 Parameters common to more than one margin

There are situations in which a parameter can appear in more than one margin. Examples are special dependence structures; for the exchangeable dependence structure a parameter is common to all bivariate margins, and for the AR(1) dependence structure for a latent MVN distribution each bivariate margin has a parameter which is the power of the lag 1 correlation parameter. Other examples arise when different univariate margins have a common parameter; e.g., in a repeated measures study with short time series over a short period of time, it may be reasonable to assume common regression coefficients for the different time points.

There are several ways in which the theory of Section 10.1 can be extended:

(a) use higher-dimensional margins;

(b) average or weight the estimators from the log-likelihoods of the margins with the common parameter;

(c) create inference functions based on the sum of log-likelihoods of the margins that have the common parameter.

If a dependence parameter appears in more than one bivariate margin, one possibility is to go directly to the m-dimensional multivariate log-likelihood with the univariate parameters given. For the exchangeable dependence structure, this is usually easy to do computationally, but for the AR(1) dependence structure for the MVN copula it is harder, and one of the other two approaches may be computationally easier. For all of these methods, under regularity conditions, the use of the jackknife approach to obtain standard errors of estimates of parameters is still valid.

Details, including some comparisons of efficiency, are given in Xu (1996).

10.3 Choice and comparison of models

Choices of models might depend on what are the most useful summaries at the initial data analysis stage and what are the inferences or predictions of interest. It may be a good idea to try more than one statistical model in order to check on the sensitivity of inferences to model assumptions. Also one should do diagnostic checks on the adequacy of fit of models. The (internal) estimation consistency check is one method mentioned in Section 10.1.3. There are several other ways to compare the adequacy of models.

One method of comparing fits of different models consists of using log-likelihoods L or Akaike information criterion (AIC) values $L - n_p$, where n_p is the number of parameters in a model. Note that in this form (rather than $-2L + 2n_p$), we are using the AIC as a penalized log-likelihood, with the penalty being the number of parameters. The method is useful even if models are not nested within each other. AIC values can be evaluated from the multivariate log-likelihood with the estimates obtained using the IFM method. This means that one has the log-likelihoods or AIC values at points that should be near the MLE and one does not have to do time-consuming multi-parameter numerical optimizations.

A second, possibly more relevant, comparison is the predictive ability of the models. This would be based on some comparisons of 'observed' summaries from the data and 'predicted' summaries from the models. The best choice of 'observed' and 'predicted' summaries depends on the data, the study and the inferences of interest. In the case of a multivariate discrete response, the summaries could be frequencies, possibly collapsed over categories. See Chapter 11 for comparisons used in different examples.

If several different models lead to similar inferences and have similar predictive ability then this is reassuring. If there is much sensitivity to the different models, then there is further work to do as one must think more about the assumptions in the models, or check for influential observations, etc., that might be affecting the models.

10.4 Inference for Markov chains

This section concerns inference for long time series that may be modelled using a Markov chain.

For Markov chains of first order, results are given in Billingsley (1961). Essentially, under certain regularity conditions, the asymptotic likelihood theory and numerical ML from the iid case can be extended. An outline of the regularity conditions and asymptotic results is given below. The extension to the case where transition probabilities depend on covariates should also be possible, by making use of the first approach in Section 10.2.1.

Notation used in this section is as follows.

1. $\{Y_t : t = 1, 2, \ldots\}$ is a Markov chain of order 1 with state space $\mathcal{Y}$.
2. $h(y_t|y_{t-1};\boldsymbol{\theta})$ is a family of transition densities (with respect to a measure ν) with column vector parameter $\boldsymbol{\theta}$ of dimension r in the parameter space $\boldsymbol{\Theta}$.
3. There exists a stationary distribution with density $f(\cdot;\boldsymbol{\theta})$ (with respect to the measure ν).
4. $\ell(\boldsymbol{\theta}; y_{t-1}, y_t) = \log h(y_t|y_{t-1};\boldsymbol{\theta})$.
5. With the observed Markov chain $y_1, \ldots, y_n$, the log-likelihood function is taken to be $L_n(\boldsymbol{\theta}) = \sum_{t=2}^{n} \ell(\boldsymbol{\theta}; y_{t-1}, y_t)$ for asymptotic analysis, since asymptotically the likelihood contribution of the first observation y_1 does not matter (for a specific example, this could be included if relevant).
6. $\partial\ell/\partial\boldsymbol{\theta}$ is the (column) vector of partial derivatives $\partial\ell/\partial\theta_u$, $u = 1, \ldots, r$, and the components are denoted more simply by ℓ_u.
7. $\partial^2\ell/[\partial\boldsymbol{\theta}\partial\boldsymbol{\theta}^T]$ is the matrix of second-order partial derivatives, with components denoted by ℓ_{uv}.
8. Third-order (mixed) derivatives are denoted by ℓ_{uvw}.
9. Similar notation for derivatives is used for other functions of $\boldsymbol{\theta}$.
10. $\mathrm{E}_{\boldsymbol{\theta}}$ means expectation assuming that the true parameter value is $\boldsymbol{\theta}$ and Y_1 starts with a stationary distribution.

Note that for applications, the measure ν is usually taken to be Lebesgue measure on a Euclidean space or counting measure.

Regularity conditions are the following.

(a) The maximum $\hat{\boldsymbol{\theta}}$ of $L_n(\boldsymbol{\theta})$ is assumed to satisfy $\partial L_n/\partial\boldsymbol{\theta} = \mathbf{0}$.

(b) All states of the Markov chain communicate with each other (meaning that there are no transient states).

(c) The set of y for which $h(y|x;\boldsymbol{\theta})$ is positive does not depend on $\boldsymbol{\theta}$.

(d) h_u, h_{uv}, h_{uvw}, $u, v, w = 1, \ldots, r$, exist and are continuous in $\boldsymbol{\theta}$ (and hence the same is true for $\ell_u, \ell_{uv}, \ell_{uvw}$).

(e) For $\boldsymbol{\theta} \in \boldsymbol{\Theta}$, there exists a neighbourhood $N_{\boldsymbol{\theta}}$ of $\boldsymbol{\theta}$ such that for all u, v, w and x,

$$\int_{\mathcal{Y}} \Big\{ \sup_{\boldsymbol{\theta}' \in N_{\boldsymbol{\theta}}} \big|h_u(y|x;\boldsymbol{\theta}')\big| \Big\} \nu(dy) < \infty,$$

$$\int_{\mathcal{Y}} \Big\{ \sup_{\boldsymbol{\theta}' \in N_{\boldsymbol{\theta}}} \big|h_{uv}(y|x;\boldsymbol{\theta}')\big| \Big\} \nu(dy) < \infty,$$

$$\mathrm{E}_{\boldsymbol{\theta}}\big[\sup_{\boldsymbol{\theta}' \in N_{\boldsymbol{\theta}}} |\ell_{uvw}(\boldsymbol{\theta}', Y_1, Y_2)|\big] < \infty.$$

(f) For $u = 1, \ldots, r$, $\mathrm{E}_{\boldsymbol{\theta}}[|\ell_u(\boldsymbol{\theta}; Y_1, Y_2)|^2] < \infty$, and $\Sigma(\boldsymbol{\theta}) = (\sigma_{uv}(\boldsymbol{\theta}))$ is a non-singular $r \times r$ matrix with

$$\sigma_{uv}(\boldsymbol{\theta}) = \mathrm{E}_{\boldsymbol{\theta}}\big[\ell_u(\boldsymbol{\theta}; Y_1, Y_2)\, \ell_v(\boldsymbol{\theta}; Y_1, Y_2)\big].$$

(g) $h(\cdot|x;\boldsymbol{\theta})$ is absolutely continuous with respect to $f(\cdot;\boldsymbol{\theta})$.

With the given regularity conditions, asymptotic results are the following. There exists a root $\hat{\boldsymbol{\theta}}_n$ of $\partial L_n/\partial\boldsymbol{\theta} = 0$ such that $\hat{\boldsymbol{\theta}}_n$ converges in probability to the true $\boldsymbol{\theta}$ and the asymptotic distribution of $n^{-1/2}(\hat{\boldsymbol{\theta}}_n - \boldsymbol{\theta})$ is $N(\mathbf{0}, \Sigma^{-1}(\boldsymbol{\theta}))$. Also log-likelihood ratios for hypotheses involving nested models for the parameter $\boldsymbol{\theta}$ have asymptotic null chi-square distributions. The proof involves the use of ergodicity and a martingale central limit theorem.

A practical implication of the asymptotic result for numerical ML is that, as in the iid case, the negative inverse Hessian of $L_n(\boldsymbol{\theta})$ evaluated at the MLE $\hat{\boldsymbol{\theta}}$ can be used as an estimated covariance matrix of $\hat{\boldsymbol{\theta}}$. That is, for large n,

$$-\left[\frac{\partial^2 L_n(\boldsymbol{\theta})}{\partial\boldsymbol{\theta}\partial\boldsymbol{\theta}^T}\bigg|_{\hat{\boldsymbol{\theta}}}\right]^{-1} = -\left[\frac{\partial^2 \sum_t \ell(\boldsymbol{\theta}; y_{t-1}, y_t)}{\partial\boldsymbol{\theta}\partial\boldsymbol{\theta}^T}\bigg|_{\hat{\boldsymbol{\theta}}}\right]^{-1}$$

is an approximation to $n^{-1}\Sigma^{-1}(\hat{\boldsymbol{\theta}})$.

The theory in Billingsley (1961) should apply for higher-order Markov chains, assuming that the order of Markov chain is known. One can perhaps either extend the proof and conditions for a first-order Markov chain to a transition density $h(x_t|x_{t-k}, \ldots, x_{t-1})$, or change a kth-order Markov chain with state space $\mathcal{Y}$ to a first-order Markov chain with state space $\mathcal{Y}^k$.

For comparison of Markov chains with different orders the AIC can be used, but note that log-likelihood ratios do not have asymptotic null chi-square distributions in this case, because a lower-order Markov chain will have parameters at the boundary of the parameter space of a higher-order Markov chain.

10.5 Comments on Bayesian methods

This section briefly discusses how multivariate models could be used and compared using Bayesian methods.

For a given parametric multivariate model and a prior distribution on the vector of parameters, one could compute the posterior distribution of the parameter vector and then make inferences and predictions. The computations would be more difficult than the likelihood-based estimating equation approach of Sections 10.1, 10.2 and 10.4. Also there has been little research on how one might choose a prior to reflect prior opinions about the dependence structure and the strength of dependence.

Furthermore, in order to compare different multivariate models, more research is needed in Bayesian model comparisons. A consistent way would be needed to convert opinions about the dependence structure as well as the univariate margins into priors of parameter vectors of (non-nested) models. For example, for two substantially different multivariate models, each covering a wide dependence structure, to be consistent, one would want the prior distributions for the parameter vectors of the two models to be similar in the probabilistic assessment of $(\zeta_{12}(\mathbf{x}), \ldots, \zeta_{m-1,m}(\mathbf{x}))$, where $\zeta_{jk}(\mathbf{x})$, $j \neq k$, is a bivariate dependence measure for (Y_j, Y_k) as a function of the covariate $\mathbf{x}$.

10.6 Numerical methods

A traditional approach for numerical optimization is the Newton–Raphson method, which requires first- and second-order derivatives of the objective function (which for applications in this book is the log-likelihood function). This is the preferred method if the derivatives can easily be analytically obtained and coded in a program, as in the case of the log-likelihood of an exponential family model. Modern symbolic manipulation software, such as Maple and Mathematica, may be useful since they can output equations in the form of Fortran or C code.

Many multivariate models, such as those based on copulas, are specified via the cdf. The likelihood involves either a mixed derivative of the cdf to get a pdf or a discretization of the cdf to get a pmf if the cdf is used for a latent variable model. To get the derivatives of the parameters after this may be possible but tedious.

A numerical method that is useful for many multivariate models in this book is the quasi-Newton optimization (usually minimization) routine. This requires that only the objective function (say, negative log-likelihood) is coded; the gradients are computed numerically and the inverse Hessian matrix of second-order derivatives is updated after each iteration. An example of a quasi-Newton routine is that in Nash (1990); this is useful for statistical applications because it outputs the estimated inverse Hessian at the (local) optimum — this corresponds to the estimated asymptotic covariance matrix of the parameters for an objective function that is a log-likelihood.

For all numerical optimization methods, a good starting point is important. In general, an objective function may have more than one local optimum. Having a good starting point based on a simple method is better than trying many random starting points. Having a model with interpretable parameters makes it easier to have a good starting point. (This is a reason for the emphasis of multivariate models with interpretable parameters — numerical estimation is easier.) Note also that computational complexity is increasing linearly to quadratically in the number of parameters.

A quasi-Newton routine works fine if the objective function can be computed to arbitrary precision, say ϵ_0. The numerical gradients are then based on a step size $\epsilon > \epsilon_0$ (more specifically, one should have $\epsilon > 10\epsilon_0$). A precision of even three or four significant digits may be difficult to attain if the objective function involves a multidimensional integral; one-dimensional numerical integrals are usually no problem and even two-dimensional numerical integrals can be computed quite quickly to around six digits of precision, but there is a problem of computation time in trying to achieve many digits of precision for numerical integrals of dimension 3 or more. There is a need for research into numerical optimization methods for imprecisely computed objective functions.

For numerical integration, methods include Romberg and adaptive integration, and Monte Carlo simulation, with the latter being especially useful for high-dimensional integrals.

10.7 Bibliographic notes

Sections 10.1 and 10.2 consist of material from Xu's (1996) doctoral thesis; this has the rigourous details for the asymptotics of the estimation method of inference functions for margins (IFM) and the associated jackknife method for estimation of standard errors, as well as assessment of efficiency and simulation results for various discrete multivariate models. Also Joe and Xu (1996) has an introduction to the IFM method, with some simulation results to demonstrate efficiency and some examples with data sets. The concept of estimation consistency is from Joe (1994).

A reference for the jackknife is Miller (1974), and references for the theory of estimating and inference functions are Godambe (1991) and McLeish and Small (1988). A one-step jackknife with estimating equations is used in a context of clustered survival data in Lipsitz, Dear and Zhao (1994). Concerning existence and uniqueness of the MLE for the multivariate probit model, see Lesaffre and Kaufmann (1992).

In Section 10.3, a reference for the AIC is Sakamoto, Ishiguro and Kitagawa (1986), and initial data analysis is used in the sense of Chatfield (1995). The main reference for Section 10.4 is Billingsley (1961).

Quasi-Newton optimization methods are also known as variable metric methods. An example is the compact routine in Nash (1990). It is available in several programming languages. My experience with it is good. References for symbolic manipulation software are Char *et al.* (1991) and Abell and Braselton (1992).

For numerical integration, I have used the Romberg integration method in Davis and Rabinowitz (1984) (good for two to about four dimensions) and the adaptive integration method of Berntsen, Espelid and Genz (1991) (useable for up to nine to ten dimensions, but requires more and more memory as the dimension increases). A reference for approximating integrals in statistics is Evans and Swartz (1995). For computing the MVN cdf, a Fortran program is given in Schervish (1984); see Genz (1992) for other computational methods and Joe (1995) for good approximation methods for the MVN cdf and rectangle probabilities.

10.8 Exercises

10.1 Let $H(\mathbf{y};\boldsymbol{\theta})$ be a family of m-variate distribution functions, with $\boldsymbol{\theta}$ being a column vector parameter. Let $\mathbf{Y}_i$, $i=1,\ldots,n$, be a random sample from $H(\mathbf{y};\boldsymbol{\theta})$. Let $\{H_S(\mathbf{y}_S;\boldsymbol{\theta}_S) : S \in \mathcal{S}_m\}$ be the set of marginal distributions. Let S_1 and S_2 be two subsets, with S_1 being a proper subset of S_2. Then $\boldsymbol{\theta}_{S_1} = \mathbf{a}(\boldsymbol{\theta}_{S_2})$ for a function $\mathbf{a}$. For $k = 1, 2$, let $\hat{\boldsymbol{\theta}}_{S_k}$ be the MLE based on the $\mathbf{Y}_{i,S_k} = (Y_{ij} : j \in S_k)$, $i = 1, \ldots, n$. Then, in general, $\hat{\boldsymbol{\theta}}_{S_1}$ and $\mathbf{a}(\hat{\boldsymbol{\theta}}_{S_2})$ are different. Verify this by studying the likelihood equations for some multivariate models. Show that they are the same for the MVN family.

10.2 Show that the regularity conditions in Section 10.4 apply to the Markov chain copula models in Section 8.1 (provided univariate margins satisfy the usual regularity conditions) and to the AR(1) models in Section 8.4.1.

10.3 Extend the asymptotic results of Section 10.4 (and obtain some regularity conditions) to the case where Markov chain transition probabilities depend on covariates (which are possibly time-varying).

10.9 Unsolved problems

10.1 Study further examples in which analytic calculations to assess the efficiency of the IFM method can be performed.

10.2 Study further examples which permit analytic calculations to compare the estimation methods in Section 10.2.2.

CHAPTER 11

Data analysis and comparison of models

In this chapter, models are applied to and compared on some real data sets. We illustrate the estimation procedures of Chapter 10, as well as much of the theory and models in Chapters 4 to 9. The examples show the stages of initial data analysis, modelling, inference and diagnostic checking. Models are compared on the adequacy of predictions. See Section 1.7 on a view of statistical modelling.

A summary of the first six sections in this chapter is the following.

- Section 11.1 involves a cardiac surgery data set consisting of a multivariate binary response with covariates. Models that are compared are the multivariate probit model, various multivariate logit models, and the model with conditional logistic regressions. See Sections 7.1.7 and 9.2.3 for the models.
- Section 11.2 involves a data set from a stress study consisting of a multivariate/longitudinal ordinal reponse with a binary covariate. Models that are compared are the multivariate probit model and various multivariate logit models. See Sections 7.1.7 and 7.3 for the models.
- Section 11.3 involves an air quality data set consisting of multivariate extremes of ozone concentrations over time. MEV models and theory from Sections 6.1 to 6.3 are used.
- Section 11.4 involves a data set, arising from a medical study, with longitudinal binary time series. Markov models (Section 8.1) with and without random effects are compared.
- Section 11.5 involves a data set of longitudinal counts with covariates, arising from a study of the health effects of pollution. Models that are compared are the AR(1) and AR(2) Poisson and negative binomial time series models and Markov chain models

based on bivariate and trivariate copulas. See Sections 8.4 and 8.1 for the models and theory.

- Section 11.6 involves a time series data set of daily air quality measurements to show the effects of assuming iid data for inference when in fact there is serial dependence.

11.1 Example with multivariate binary response data

In this section, several models for multivariate binary response data with covariates are applied to a data set from cardiac surgery. MCR (Merged, Multi-Center, Multi-Specialty Clinical Registries) is a data base developed by Health Data Research Institute (Portland, Oregon) in which information of patients who had heart-related surgery was recorded. The data set is large so we use a random subset of the data (5000 subjects) for a detailed analysis; this is large enough to check predictions from the models, and computations can still be done in a reasonable time. The main response variable is an indicator of survival 30 days after surgery, but there are also other binary response variables that are known immediately after surgery. These dependent variables include for indicators of (mild or severe) renal complications (REC), pulmonary complications (PUC), neurological complications (NEC) and low-output syndrome (LOS). The complication variables are related to the quality of life after surgery. Zhang (1993) gives further documentation of the data set and some initial data analysis. We concentrate on multivariate models for the four response variables REC, PUC, NEC and LOS. The analyses in Zhang (1993) suggest that, among the many possible predictor variables, the more important predictors or risk factors for the complication variables are: AGE (in years), SEX (0 for male, 1 for female), and indicators of prior myocardial infarction (PMI), diabetes (DIA), renal disease (REN) and chronic obstructive pulmonary disease (COP).

In the models below, we use these six pre-operation variables as covariates for all four response variables, even though one could have simpler final models in which not all covariates are used for all responses.

Summaries from the initial data analysis are given in tabular form. By way of a univariate summary, Table 11.1 has the percentages of 1s for the binary response and predictor variables. The age variable ranges from below 20 to 90 with a mean of 63 and a standard deviation of 11. Table 11.2 has the frequencies of 4-vectors for (REC, PUC, NEC, LOS) when ignoring the effects of

Table 11.1. *Cardiac surgery data. Percentages for binary variables.*

Variable	1s	Percentage
REC	224	4.5
PUC	1394	27.9
NEC	350	7.0
LOS	472	9.4
SEX	1403	28.1
PMI	2086	41.7
DIA	637	12.7
REN	168	3.4
COP	364	7.3

the covariates. Table 11.3 has the pairwise log-odds ratio for the response variables, ignoring the covariates; it gives some indication of the dependence in the response variables, in addition to Table 11.2.

Multivariate binary response models that were used to model the data are the following.

1. The multivariate probit model from Section 7.1.7.
2. The multivariate logit model from Section 7.1.7:
   (a) with the construction in Section 4.8 using bivariate copulas from the family B2;
   (b) with the construction in Section 4.8 using bivariate copulas from the family B3;
   (c) with copulas from Section 4.3 that are mixtures of max-id distributions.
3. The conditionally specified logistic regression model from Section 9.2.3.
4. The multivariate logit model with the permutation-symmetric copula M3.

For the models in (2c), parametric families of copulas of the form (4.25) were tried with $\psi(\cdot;\theta)$ in one of the LT families LTA, LTB, LTC, LTD, and with $K_{ij}(\cdot) = K(\cdot;\delta_{ij})$ in one of the bivariate copula families B3 to B7. The summaries below are given only for a choice with a high AIC value. Model (4) is used to compare a

Table 11.2. *Cardiac surgery data. Frequencies of the response vector (REC, PUC, NEC, LOS).*

Vector	Frequency	Rel. freq.
0 0 0 0	3172	0.6344
0 0 0 1	163	0.0326
0 0 1 0	141	0.0282
0 0 1 1	19	0.0038
0 1 0 0	984	0.1968
0 1 0 1	164	0.0328
0 1 1 0	95	0.0190
0 1 1 1	38	0.0076
1 0 0 0	70	0.0140
1 0 0 1	19	0.0038
1 0 1 0	14	0.0028
1 0 1 1	8	0.0016
1 1 0 0	38	0.0076
1 1 0 1	40	0.0080
1 1 1 0	14	0.0028
1 1 1 1	21	0.0042

Table 11.3. *Cardiac surgery data. Pairwise log-odds ratios for REC, PUC, NEC, LOS.*

Pair	Odds	Log-odds (SE)
REC, PUC	2.78	1.02 (0.14)
REC, NEC	5.22	1.65 (0.17)
REC, LOS	7.40	2.00 (0.15)
PUC, NEC	2.58	0.95 (0.11)
PUC, LOS	3.78	1.33 (0.10)
NEC, LOS	3.60	1.28 (0.14)

simple permutation-symmetric copula model with the other models which all allow a general dependence structure.

For model (3), which is an exponential family model, all parameter estimates were obtained using the Newton–Raphson iterative method. For every other model referred to in the preceding paragraph, the IFM method from Section 10.1 was used; univariate (regression) parameters were estimated from separate univariate likelihoods (using the Newton–Raphson method), and bivariate and multivariate parameters were estimated from bivariate, trivariate, or 4-variate likelihoods, using a quasi-Newton routine, with univariate parameters fixed as estimated from the separate univariate likelihoods. That is, for models (1), (2a), (2b), (4), there are separate quasi-Newton optimizations of log-likelihoods for each parameter (see the examples in Section 10.1.4). Estimation for model (2c) involves a quasi-Newton optimization in up to seven dependence parameters simultaneously (if the ν_j are assumed zero or fixed, and there is a parameter associated with ψ and one for each K_{ij}), since its parameters cannot be assigned to lower-dimensional margins. (Note that the parameter of ψ for model (2c) represents a general minimum level of dependence and the remaining parameters, indexed by bivariate margins, represent bivariate dependence exceeding the minimum dependence). Model (2c) has the advantage of being a copula with a closed-form cdf; this leads to faster computation of probabilities of the form $\Pr(\mathbf{Y} = \mathbf{y}|\mathbf{x})$.

For standard errors (SEs) of parameter estimates and prediction probabilities, the delta method was used for model (3) (see Exercise 11.1) and otherwise the jackknife method from Section 10.1.1 was used with 50 groups of 100.

Summaries of the modelling process are given in several tables. Table 11.4 contains the estimates and SEs of the regression parameters for the univariate probit and logit models for the four binary responses. It is well known that the univariate probit and logit models are comparable; in Table 11.4, the ratios of estimates of a single regression parameter are roughly equal to the ratios of standard deviations of the standard normal and logistic distributions. Table 11.5 contains estimates and SEs of the bivariate dependence parameters for models in (1), (2a) and (2b); it also has the trivariate and 4-variate parameters for models (2a) and (2b). For models (2a) and (2b), the SEs as well as the comparisons of the trivariate and 4-variate log-likelihoods at a parameter value of 1 and at the estimate from the IFM method suggest that one could simplify to models with higher-order dependence parameters

Table 11.4. *Cardiac surgery data. Estimates of regression parameters for univariate marginal probit and logistic regressions.*

Response: covariate	Probit Estimate (SE)	Logit Estimate (SE)
1. REC:		
constant	−3.33 (0.24)	−6.61 (0.56)
AGE	0.021 (0.004)	0.048 (0.008)
SEX	0.135 (0.084)	0.294 (0.179)
PMI	0.085 (0.073)	0.170 (0.156)
DIA	0.133 (0.085)	0.277 (0.177)
REN	1.11 (0.12)	2.09 (0.22)
COP	0.228 (0.127)	0.448 (0.262)
2. PUC:		
constant	−1.29 (0.13)	−2.15 (0.23)
AGE	0.010 (0.002)	0.017 (0.004)
SEX	0.053 (0.058)	0.088 (0.098)
PMI	0.017 (0.049)	0.028 (0.081)
DIA	0.296 (0.062)	0.488 (0.099)
REN	−0.017 (0.139)	−0.028 (0.232)
COP	0.222 (0.088)	0.365 (0.145)
3. NEC:		
constant	−3.47 (0.28)	−6.90 (0.61)
AGE	0.030 (0.004)	0.064 (0.008)
SEX	−0.049 (0.059)	−0.115 (0.116)
PMI	0.049 (0.056)	0.120 (0.114)
DIA	0.132 (0.085)	0.248 (0.167)
REN	0.328 (0.107)	0.582 (0.187)
COP	0.138 (0.087)	0.271 (0.169)
4. LOS:		
constant	−2.78 (0.20)	−5.18 (0.44)
AGE	0.019 (0.003)	0.039 (0.006)
SEX	0.252 (0.057)	0.487 (0.107)
PMI	0.179 (0.060)	0.343 (0.115)
DIA	0.081 (0.073)	0.164 (0.138)
REN	0.417 (0.157)	0.748 (0.281)
COP	0.297 (0.078)	0.559 (0.143)

Table 11.5. *Cardiac surgery data. Estimates of dependence parameters in models (1), (2a), (2b);* $\mathbf{Y}$ *=(REC,PUC,NEC,LOS).*

Margin	(1) Estimate (SE)	(2a) Estimate (SE)	(2b) Estimate (SE)
12	0.269 (0.052)	2.56 (0.48)	1.99 (0.43)
13	0.366 (0.050)	3.91 (0.76)	3.20 (0.56)
14	0.470 (0.050)	5.78 (1.14)	4.31 (0.66)
23	0.265 (0.043)	2.34 (0.34)	1.78 (0.32)
24	0.387 (0.076)	3.52 (0.80)	2.74 (0.58)
34	0.296 (0.040)	2.90 (0.44)	2.41 (0.41)
123		0.84 (0.30)	0.85 (0.31)
124		1.19 (0.46)	1.19 (0.46)
134		0.80 (0.20)	0.82 (0.20)
234		1.03 (0.42)	1.03 (0.42)
1234		0.62 (0.48)	0.62 (0.48)

Table 11.6. *Cardiac surgery data. Log-likelihoods associated with dependence parameters in models (1), (2a), (2b);* $\mathbf{Y}$ *=(REC,PUC,NEC,LOS).*

Margin	(1)	(2a)	(2b)
12	−3730.6	−3732.2	−3732.3
13	−2005.9	−2008.8	−2009.6
14	−2268.0	−2272.3	−2274.5
23	−4093.7	−4094.2	−4094.4
24	−4337.0	−4337.0	−4337.1
34	−2671.9	−2670.3	−2670.4
123		−4885.9	−4886.8
124		−5105.6	−5107.8
134		−3431.3	−3434.0
234		−5494.4	−5494.6
1234		−6248.4	−6251.1

Table 11.7. *Cardiac surgery data. Estimates of dependence parameters in model (2c), ψ from family LTB, K_{ij} from family B6;* $\mathbf{Y}$ *=(REC,PUC,NEC,LOS).*

Parameter	Estimate (SE)
θ	0.21 (0.03)
δ_{12}	1 (—)
δ_{13}	1.07 (0.10)
δ_{14}	1.69 (0.38)
δ_{23}	1 (—)
δ_{24}	1.37 (0.18)
δ_{34}	1 (—)

Table 11.8. *Cardiac surgery data. Conditionally specified logistic regression model (3), regression parameter estimates and standard errors.*

$\mathbf{x} \setminus \mathbf{y}$	REC	PUC	NEC	LOS
const.	−6.01 (0.52)	−1.86 (0.16)	−6.70 (0.44)	−5.01 (0.36)
AGE	0.031 (0.008)	0.010 (0.003)	0.056 (0.007)	0.027 (0.005)
SEX	0.18 (0.16)	0.03 (0.07)	−0.21 (0.13)	0.48 (0.11)
PMI	0.07 (0.15)	−0.02 (0.07)	0.07 (0.12)	0.33 (0.10)
DIA	0.15 (0.20)	0.48 (0.09)	0.14 (0.16)	−0.01 (0.14)
REN	2.05 (0.21)	−0.35 (0.19)	0.16 (0.26)	0.28 (0.23)
COP	0.23 (0.23)	0.28 (0.12)	0.11 (0.20)	0.43 (0.16)

Table 11.9. *Cardiac surgery data. Conditionally specified logistic regression model (3), dependence parameter estimates and standard errors.*

Pair $(y_j, y_{j'})$	γ (SE)
REC, PUC	0.56 (0.15)
REC, NEC	1.03 (0.19)
REC, LOS	1.48 (0.17)
PUC, NEC	0.67 (0.12)
PUC, LOS	1.14 (0.10)
NEC, LOS	0.68 (0.15)

being 1 (see Section 4.8 on the maximum entropy interpretation in this case). Table 11.6 contains the log-likelihoods of the bivariate parameters of the models in (1), (2a) and (2b). This and Table 11.5 suggest that the models are comparable; their conclusions about which bivariate margins are more dependent or less dependent are the same. Table 11.7 contains the parameter estimates and SEs for a model of the form (2c) with a high log-likelihood value. For comparison, the estimate of the dependence parameter for model (4), with a permutation-symmetric copula, is 2.28. Tables 11.8 and 11.9 contain parameter estimates and their SEs for model (3).

For the 20 parametric families of the form (4.25) that were tried, generally some dependence parameter δ_{ij} (corresponding to K_{ij}) reached either the lower bound (independence copula) or the upper bound (Fréchet upper bound copula) in the estimation. This causes a problem with SEs of estimates but not for AIC values or prediction probabilities. (The asymptotic theory of Chapter 10 does not strictly apply to the SE calculations when parameter values are on the boundary, but using the jackknife with some dependence values fixed at boundary values, the resulting SEs for prediction probabilities are similar to those obtained from other models.) For model (2c) for this data set, the log-likelihood values were affected greatly by the choice of the LT family $\psi(\cdot;\theta)$, but not by the family of bivariate copulas $K(\cdot;\delta_{ij})$. The 'best' fit was with LT family LTB; for the summaries in the tables, we use the bivariate copula family B6 for $K(\cdot;\delta_{ij})$. Because some δ_{ij} were approaching the boundaries for the numerical optimization with seven parameters, we simplify the model and numerical computations with $\delta_{12} = \delta_{23} = \delta_{34} = 1$, $\nu_1 = \nu_4 = -1$ and $\nu_2 = \nu_3 = -2$. (That is, we assume a common level of dependence for the (1,2), (2,3) and (3,4) bivariate margins and a higher level of dependence for the remaining three bivariate margins — compare Tables 11.3 and 11.5.) The jackknife estimates of SEs for model (2c) used much more computer time than the other models because of a multi-parameter optimization (versus many one-dimensional optimizations in the other models).

Turning to inferences and predicted probabilities, Table 11.10 contains estimates of probabilities of the form

$$\Pr(Y_j = y_j\,, j = 1,2,3,4 \mid \mathbf{x})$$

for various $\mathbf{y}$ and $\mathbf{x}$, from most of the models listed earlier. (Comparisons with 'observed' frequencies are given in Table 11.12.) To save space, for each line the entries for model (2b), which are very close to those of (2a), are not given, and only the maximum estim-

Table 11.10. *Cardiac surgery data. Estimates of* Pr($\mathbf{Y}$ = $\mathbf{y}$ | $\mathbf{x}$) *from various models;* $\mathbf{Y}$ = *(REC, PUC, NEC, LOS),* $\mathbf{x}$ =*(AGE, SEX, PMI, DIA, REN, COP).*

x	y	Prob. (1)	(2a)	(2c)	(3)	(4)	ind.	Max SE
(80,0,0,0,0,0)	1111	0.006	0.007	0.006	0.009	0.002	0.000	0.002
(80,0,0,0,0,0)	0000	0.567	0.562	0.558	0.560	0.570	0.496	0.038
(80,1,1,1,1,1)	1111	0.118	0.162	0.116	0.249	0.145	0.061	0.053
(80,1,1,1,1,1)	0000	0.168	0.162	0.106	0.120	0.165	0.053	0.041
(50,0,0,0,0,0)	1111	0.000	0.000	0.001	0.000	0.000	0.000	0.000
(50,0,0,0,0,0)	0000	0.753	0.746	0.754	0.738	0.746	0.729	0.027
(50,1,1,1,1,1)	1111	0.020	0.020	0.015	0.022	0.012	0.002	0.009
(50,1,1,1,1,1)	0000	0.380	0.402	0.370	0.425	0.407	0.286	0.059
(63,0,0,0,0,0)	0000	0.685	0.680	0.687	0.679	0.681	0.648	0.032
(63,0,1,0,0,0)	0000	0.664	0.661	0.666	0.663	0.662	0.621	0.037
(63,0,0,1,0,0)	0000	0.584	0.581	0.584	0.581	0.585	0.541	0.049
(63,0,0,0,0,1)	0000	0.590	0.591	0.592	0.598	0.594	0.541	0.047
(63,0,1,1,0,0)	0000	0.564	0.562	0.563	0.566	0.566	0.513	0.053
(63,1,0,0,0,0)	0000	0.653	0.651	0.655	0.654	0.651	0.608	0.028
(63,1,1,0,0,0)	0000	0.627	0.626	0.628	0.630	0.627	0.574	0.038
(63,1,0,1,0,0)	0000	0.552	0.551	0.551	0.556	0.554	0.499	0.043
(75,0,0,0,0,0)	0000	0.605	0.601	0.602	0.602	0.606	0.547	0.037
(75,1,0,0,0,0)	0000	0.571	0.569	0.566	0.572	0.574	0.501	0.030
(63,0,0,0,0,0)	0100	0.195	0.197	0.195	0.197	0.192	0.218	0.027
(63,0,0,0,0,0)	0101	0.023	0.022	0.023	0.022	0.023	0.014	0.007
(63,0,0,0,0,0)	0001	0.022	0.025	0.021	0.025	0.026	0.041	0.004
(63,0,0,0,0,0)	1000	0.008	0.011	0.009	0.011	0.010	0.017	0.002
(63,0,0,0,0,0)	0010	0.026	0.027	0.025	0.028	0.023	0.037	0.004

ated SE over models (1), (2a), (2b), (2c), (3) is given. Actually the SEs are quite close to each other. The selected $\mathbf{x}$ and $\mathbf{y}$ values in Table 11.10 are extremes in the covariate space, or common values in the data set. Tables 11.10 and 11.12 suggest that the simple exchangeable dependence model is adequate for predictive purposes, since the prediction probabilities are comparable with those from the models with general dependence structure, when the SEs are considered, and they are comparable with the 'observed' frequencies. The large divergences in estimated probabilities occur only

Table 11.11. *Cardiac surgery data. Log-likelihoods and AIC values.*

Model	Log-lik	AIC
(1)	−6245.6	−6279.6
(2a)	−6248.3	−6287.3
(2b)	−6251.1	−6290.1
(2c)	−6242.7	−6277.7
(3)	−6250.3	−6284.3
(4)	−6286.0	−6315.0
indep.	−6448.9	−6476.9

in the cases in which the vector $\mathbf{x}$ is at the extreme of the covariate space. For another comparison, the prediction probabilities for the multivariate logit model with the independence copula are also given in Table 11.10. This shows that the assumption of independence leads to poor estimated probabilities in several cases.

Hence for this data set, a simple exchangeable dependence model appears adequate for prediction probabilities. There is no reason to expect this in general. An explanation may be that the dependences in the bivariate margins are different but not different enough to make a difference in prediction probabilities. Another possibility may be the dominance of the response vector (0,0,0,0). The comparison of exchangeable versus general dependence models can be investigated further through other examples.

Table 11.11 contains log-likelihoods and AIC values for all of the models; the AIC value for a model is the log-likelihood, evaluated at the IFM estimate (MLE for model (3)), minus the number of parameters in the model. For model (1), the second-order version of the approximation of Section 4.7.1 was used for faster computations. Note also that we do not do the fuller analysis of AIC values based on different sets of covariates. (Some analyses showed that the six covariates yield a wider range of univariate predictive probabilities, $\Pr(Y_j = y_j \,|\mathbf{x})$, $j = 1, 2, 3, 4$, than the best set of five covariates, and also that for model (3) the six-covariate case had the largest AIC value.) The AIC values for models (1), (2a), (2b), (2c) and (3) are comparable and that for model (4) is much smaller. A conclusion from Tables 11.10 and 11.11 is that the AIC values separate out the models more than the predictive probabilities.

Finally, we mention some diagnostic checks. For models (1), (2a),

Table 11.12. *Cardiac surgery data. Observed frequencies for comparison with Table 11.10 (match in* **x** *space for ages within 10 years);* **Y** *=(REC,PUC,NEC,LOS),* **x***=(AGE,SEX,PMI,DIA,REN,COP).*

x	n^*	**y**	Prop.	Freq. (probit)
(80,0,0,0,0,0)	410	(1,1,1,1)	0.012	0.006
(80,0,0,0,0,0)	410	(0,0,0,0)	0.583	0.567
(80,1,1,1,1,1)	0	(1,1,1,1)	—	0.118
(80,1,1,1,1,1)	0	(0,0,0,0)	—	0.168
(50,0,0,0,0,0)	636	(1,1,1,1)	0.000	0.000
(50,0,0,0,0,0)	636	(0,0,0,0)	0.761	0.753
(50,1,1,1,1,1)	0	(1,1,1,1)	—	0.020
(50,1,1,1,1,1)	0	(0,0,0,0)	—	0.380
(63,0,0,0,0,0)	738	(0,0,0,0)	0.690	0.685
(63,0,1,0,0,0)	555	(0,0,0,0)	0.669	0.664
(63,0,0,1,0,0)	114	(0,0,0,0)	0.570	0.584
(63,0,0,0,0,1)	89	(0,0,0,0)	0.584	0.590
(63,0,1,1,0,0)	129	(0,0,0,0)	0.519	0.564
(63,1,0,0,0,0)	456	(0,0,0,0)	0.664	0.653
(63,1,1,0,0,0)	208	(0,0,0,0)	0.620	0.627
(63,1,0,1,0,0)	75	(0,0,0,0)	0.520	0.552
(75,0,0,0,0,0)	732	(0,0,0,0)	0.619	0.605
(75,1,0,0,0,0)	407	(0,0,0,0)	0.602	0.571
(63,0,0,0,0,0)	1069	(0,1,0,0)	0.195	0.195
(63,0,0,0,0,0)	1069	(0,1,0,1)	0.021	0.023
(63,0,0,0,0,0)	1069	(0,0,0,1)	0.022	0.022
(63,0,0,0,0,0)	1069	(1,0,0,0)	0.011	0.008
(63,0,0,0,0,0)	1069	(0,0,1,0)	0.024	0.026

(2b), comparisons were made between the estimates of the dependence parameters based on one-parameter likelihoods and on the multivariate likelihoods with the univariate parameters fixed (this is an example of the estimation consistency check in Section 10.1.3). These were about the same, with the former likelihoods being much faster to compute. For model (3), the check of $\gamma_{jk} = \gamma_{kj}$ was done based on separate logistic regressions with dependent variables being covariates for other dependent variables; the estimates from the separate logistic regressions were quite close to the MLEs (and were good starting points for the ML estimation).

Table 11.13. *Cardiac surgery data. Dependence parameters for subsets, multivariate probit model;* $\mathbf{Y}$ *=(REC,PUC,NEC,LOS).*

Subset	n^*	Margin					
		1,2	1,3	1,4	2,3	2,4	3,4
all	5000	0.269	0.367	0.479	0.265	0.387	0.296
AGE> 63	2621	0.301	0.344	0.459	0.204	0.407	0.312
AGE≤ 63	2379	0.192	0.433	0.496	0.393	0.351	0.248
FEMALE	1403	0.266	0.273	0.429	0.268	0.364	0.300
PMI=1	2086	0.250	0.406	0.460	0.259	0.383	0.326
5 binary vars =0	1628	0.311	0.377	0.502	0.259	0.396	0.253

A rough assessment of observed versus predicted frequencies was done as follows. We can compute the sample relative frequency for a given $\mathbf{y}$ over the subset of subjects with a covariate vector near a given $\mathbf{x}$. Since all the covariates except for age are discrete, we consider 'near a given $\mathbf{x}$' as meaning 'age within 10 years' (the use of 7 in place of 10 led to similar results). Table 11.12 has the observed frequency and the subset size n^* for each $\mathbf{x}$; it also repeats the predicted frequencies from model (1) for easier comparisons with Table 11.10.

Table 11.13 contains dependence parameters for the multivariate probit model for several subsets, chosen by limiting the range of the covariate space; the estimates of univariate parameters from all 5000 subjects were used. The dependence parameters from the subsets are generally close to those from all 5000 subjects, with the largest divergences from the subset of AGE≤ 63. This type of analysis could also be done for the other models. It is done here to illustrate what might be done in general for the multivariate binary models used in this section, even though Table 11.10 suggests that this is not needed for this data set.

Conclusions: The models have similar predictive abilities and inferences for this data set. The IFM method, which allows reduction to one-dimensional numerical optimizations in models (1), (2a) and (2b), and the use of jackknife estimates of SEs were quite important in allowing estimation and inference to be done computationally quickly.

11.2 Example with multivariate ordinal response data

In this section, several models for multivariate ordinal response data are applied to a longitudinal data set in Fienberg *et al.* (1985) and Conaway (1989), which comes from a study on the psychological effects of the accident at the Three Mile Island nuclear power plant in 1979. We use multivariate probit and logit models (or, equivalently, copula models with univariate normal or logistic margins). These are different from the models used in the cited papers; they are used to highlight features in the data that are not as clear from the other models.

The study focuses on the changes in levels of stress of mothers of young children living within 10 miles of the plant. Four waves of interviews were conducted in 1979, 1980, 1981 and 1982, and one variable measured at each time point is the level of stress (categorized as low, medium, or high from a composite score of a 90-item checklist). Hence stress is treated as an ordinal response variable with three categories, now labelled as L, M, H. There were 268 mothers in the study, and they were stratified into two groups, those living within 5 miles of the plant, and those living between 5 and 10 miles from the plant. There were 115 mothers in the first group and 153 in the second group.

Over the four time points and three levels of the ordinal response variable, there are 81 possible 4-tuples of the form LLLL to HHHH. Table 11.14 lists the frequencies of the 4-tuples by group (based on distance); only the 35 four-tuples with non-zero frequency in at least one of the two groups are listed. The table shows that there is only one subject with a big change in the stress level (L to H or H to L) from one year to another; 42% of the subjects are categorized into the same stress level in all four years. The frequencies by univariate margin (or by year) are given in Table 11.15. The medium stress category predominates and there is a higher relative frequency of subjects in the high stress category for the group within 5 miles of the plant compared with the group exceeding 5 miles. Table 11.15 shows that there are no big changes over time, but there is a small trend towards lower stress levels for the group exceeding 5 miles.

Next we consider some multivariate models for the data. Because the single covariate, distance, is dichotomous, we fit latent multivariate distributions, separately by the value of the categorized distance. With this approach, one does not have (initially) to think about the how the univariate and dependence parameters

Table 11.14. *Stress data. Stress levels for 4 years following accident at Three Mile Island, four-tuples with non-zero frequencies.*

4-tuple	Distance (miles) < 5	> 5
L L L L	2	1
L L L M	0	2
L L M L	2	2
L L M M	3	0
L M L L	0	1
L M L M	1	0
L M M L	2	0
L M M M	4	3
M L L L	5	4
M L L M	1	4
M L M L	1	5
M L M M	4	15
M L M H	0	1
M M L L	3	2
M M L M	2	2
M M M L	2	6
M M M M	38	53
M M M H	4	6
M M H M	2	5
M M H H	3	1
M H M M	2	1
M H M H	0	1
M H H M	1	3
M H H H	1	1
H L L H	0	1
H M M L	0	1
H M M M	4	13
H M M H	3	0
H M H M	1	0
H M H H	4	0
H H M L	1	1
H H M M	2	7
H H M H	0	2
H H H M	5	2
H H H H	12	7

Table 11.15. *Stress data. Univariate marginal (and relative) frequencies.*

	Year			
Outcomes	1979	1980	1981	1982
	< 5 mi.			
L	14 (0.122)	18 (0.157)	14 (0.122)	18 (0.157)
M	69 (0.600)	73 (0.635)	72 (0.626)	70 (0.609)
H	32 (0.278)	24 (0.209)	29 (0.252)	27 (0.235)
	> 5 mi.			
L	9 (0.059)	35 (0.229)	17 (0.111)	23 (0.150)
M	110 (0.719)	93 (0.608)	117 (0.765)	110 (0.719)
H	34 (0.222)	25 (0.163)	19 (0.124)	20 (0.131)
	all			
L	23 (0.086)	53 (0.198)	31 (0.116)	41 (0.153)
M	179 (0.668)	166 (0.619)	189 (0.705)	180 (0.672)
H	66 (0.246)	49 (0.183)	48 (0.179)	47 (0.175)

are functions of the covariate. For continuous covariates, this would have to be done; for example, should the regression coefficients for different cutpoints of a single ordinal response variable be the same or different? The comparison of models is not as detailed as in the multivariate binary response example in Section 11.1. We leave as exercises other comparisons and checks of whether simpler models are adequate fits to the data, for example, one could consider Markov or exchangeable dependence structure, or common univariate parameters for different time points and/or groups.

Multivariate ordinal response models that were used to model the data are the following.

1. The multivariate probit model from Sections 7.1.7 and 7.3.
2. The multivariate logit model from Sections 7.1.7 and 7.3:
   (a) with the construction in Section 4.8 using bivariate copulas from the family B2;
   (b) with the construction in Section 4.8 using bivariate copulas from the family B3;

Table 11.16. *Stress data. Estimates of cutpoints for probit and logistic models, by distance category and year.*

Distance, year	Probit Estimate (SE)	Logit Estimate (SE)
<5 mi., 1979:		
α_{11}	−1.16 (0.16)	−1.98 (0.29)
α_{12}	0.59 (0.13)	0.95 (0.21)
1980:		
α_{21}	−1.01 (0.14)	−1.68 (0.26)
α_{22}	0.81 (0.13)	1.33 (0.23)
1981:		
α_{31}	−1.17 (0.16)	−1.98 (0.29)
α_{32}	0.67 (0.13)	1.09 (0.22)
1982:		
α_{41}	−1.01 (0.14)	−1.68 (0.26)
α_{42}	0.73 (0.13)	1.18 (0.22)
>5 mi., 1979:		
α_{11}	−1.57 (0.17)	−2.77 (0.36)
α_{12}	0.76 (0.11)	1.25 (0.20)
1980:		
α_{21}	−0.74 (0.11)	−1.22 (0.19)
α_{22}	0.98 (0.12)	1.63 (0.22)
1981:		
α_{31}	−1.22 (0.14)	−2.10 (0.26)
α_{32}	1.15 (0.13)	1.95 (0.25)
1982:		
α_{41}	−1.04 (0.13)	−1.73 (0.23)
α_{42}	1.12 (0.13)	1.89 (0.25)

(c) with copulas from Section 4.3 that are mixtures of max-id distributions.

That is, models of the form $F(\cdot;\delta) = C(F_0, F_0, F_0, F_0; \delta)$ for the latent vector $\mathbf{Z}$ with univariate margins F_0 wére used. For the jth ordinal variable, the category k obtains if $\alpha_{j,k-1} < Z_j \leq \alpha_{j,k}$. For model (1), F_0 is the standard normal distribution, and for model (2), F_0 is the standard logistic distribution. For the models in (2c), parametric families of copulas of the form (4.25) were tried with

Table 11.17. *Stress data. Estimates of dependence parameters in models (1), (2a), (2b), by distance category.*

Margin	(1) Estimate (SE)	(2a) Estimate (SE)	(2b) Estimate (SE)
< 5 mi.:			
12	0.785 (0.060)	17.9 (7.5)	8.4 (1.8)
13	0.696 (0.067)	10.4 (3.7)	6.4 (1.2)
14	0.653 (0.086)	9.6 (3.7)	5.5 (1.3)
23	0.806 (0.059)	20.8 (9.1)	9.1 (2.0)
24	0.636 (0.096)	9.1 (3.7)	5.3 (1.4)
34	0.844 (0.052)	27.0 (12.3)	10.2 (2.3)
123		0.67 (1.59)	0.61 (1.15)
124		1.27 (1.73)	1.25 (1.57)
134		1.03 (2.06)	0.80 (1.23)
234		0.60 (0.81)	0.51 (0.65)
> 5 mi.:			
12	0.678 (0.079)	13.8 (6.3)	6.4 (1.7)
13	0.463 (0.111)	5.0 (2.1)	3.6 (1.2)
14	0.436 (0.108)	4.6 (1.9)	3.3 (1.1)
23	0.750 (0.066)	16.7 (7.2)	8.2 (1.8)
24	0.510 (0.104)	6.6 (2.6)	3.9 (1.1)
34	0.562 (0.116)	8.5 (3.5)	5.1 (1.7)
123		0.10 (0.16)	0.14 (0.30)
124		1.36 (1.74)	1.66 (2.05)
134		0.30 (0.36)	0.36 (0.40)
234		0.16 (0.18)	0.17 (0.19)

$\psi(\cdot;\theta)$ in one of the LT families LTA, LTB, LTC, LTD, and with $K_{ij}(\cdot) = K(\cdot;\delta_{ij})$ in one of the bivariate copulas families B3 to B7. The summaries below are given only for the choice that led to the best AIC value for both groups (categorized by distance).

For the models referred to in the preceding paragraph, the IFM method in Section 10.1 was used; univariate parameters were estimated from separate univariate likelihoods (leading to $\hat{\alpha}_{j,k} = F_0^{-1}(n_{jk}/n)$, where n is the sample size and n_{jk} is the number in the jth univariate margin that are in the kth category or below),

Table 11.18. *Stress data. Estimates of dependence parameters in model (2c); ψ from family LTA, K_{ij} from family B7.*

Parameter	< 5 mi. Estimate (SE)	> 5 mi. Estimate (SE)
θ	1.93 (0.20)	1.47 (0.12)
δ_{12}	0.75 (0.45)	0.85 (0.29)
δ_{13}	0 (—)	0 (—)
δ_{14}	0 (—)	0 (—)
δ_{23}	1.21 (1.02)	1.62 (1.91)
δ_{24}	0 (—)	0 (—)
δ_{34}	1.21 (0.79)	0.22 (0.21)

Table 11.19. *Stress data. Comparisons of log-likelihood and AIC values, with given univariate parameters.*

Model	< 5 mi. Log-lik	AIC	Model	> 5 mi. Log-lik	AIC
(1)	−323.7	−337.7	(1)	−417.1	−431.1
(2a)	−325.6	−343.6	(2a)	−413.5	−431.5
(2b)	−323.1	−341.1	(2b)	−415.7	−433.7
(2c)	−323.8	−338.8	(2c)	−416.4	−431.4

and bivariate and multivariate parameters were estimated from bivariate, trivariate, or 4-variate likelihoods, using a quasi-Newton routine, with univariate parameters fixed as estimated from the separate univariate likelihoods. This is similar to the use of these models in Section 11.1. For SEs of parameter estimates, the (delete-one) jackknife method from Section 10.1.1 was used.

Summaries of the modelling process are given in several tables. Table 11.16 has estimates and SEs of univariate parameters. Tables 11.17 and 11.18 have estimates and SEs of dependence parameters. For models (2a) and (2b), the 4-variate parameter was taken to be 1 (for numerical stability in the estimation consistency check). The models (2a) and (2b) allow one to assess whether there is multivariate structure beyond that given in the set of bivariate margins. For both groups, two of the trivariate parameters are

Table 11.20. *Stress data. Observed versus expected frequencies for several models, $<$ 5 mi. group.*

4-tuple	Observed	Expected			
		(1)	(2a)	(2b)	(2c)
L L L L	2	4.9	4.9	4.4	3.3
L L M L	2	1.0	1.1	1.2	1.1
L L M M	3	2.3	2.0	2.1	2.4
L M L M	1	0.2	0.1	0.2	0.4
L M M L	2	1.0	0.7	0.7	1.5
L M M M	4	2.6	2.8	3.4	2.8
M L L L	5	2.2	2.1	2.0	2.4
M L L M	1	1.3	1.4	1.6	1.4
M L M L	1	0.8	0.6	0.7	1.2
M L M M	4	4.1	3.9	4.6	4.4
M M L L	3	2.2	2.0	2.3	2.3
M M L M	2	1.2	1.3	1.7	1.8
M M M L	2	4.7	4.5	5.4	4.5
M M M M	38	35.9	37.1	34.7	37.4
M M M H	4	3.8	3.2	3.8	3.0
M M H M	2	3.0	3.1	3.3	2.5
M M H H	3	3.7	3.4	3.4	3.2
M H M M	2	2.1	2.1	2.4	1.3
M H H M	1	1.5	1.2	1.0	1.7
M H H H	1	1.8	1.5	1.3	1.0
H M M M	4	6.9	6.7	7.1	6.7
H M M H	3	2.0	2.0	1.7	1.3
H M H M	1	1.2	0.8	0.4	1.1
H M H H	4	3.2	3.0	3.4	2.7
H H M L	1	0.1	0.1	0.1	0.1
H H M M	2	3.0	2.5	2.6	2.4
H H H M	5	3.3	3.0	3.6	1.9
H H H H	12	10.9	12.0	12.3	14.2
others	0	4.1	5.9	3.5	4.9

Table 11.21. *Stress data. Observed versus expected frequencies for several models, > 5 mi. group.*

4-tuple	Observed	Expected			
		(1)	(2a)	(2b)	(2c)
L L L L	1	2.0	1.6	1.4	1.0
L L L M	2	1.1	1.0	1.0	1.3
L L M L	2	1.4	1.8	1.8	1.2
L M L L	1	0.0	0.2	0.1	0.1
L M M M	3	1.2	1.2	1.6	1.7
M L L L	4	4.5	4.7	4.6	2.9
M L L M	4	5.2	4.8	5.8	5.1
M L M L	5	3.9	4.7	4.5	4.8
M L M M	15	13.1	11.9	12.4	14.6
M L M H	1	0.6	0.6	0.6	0.7
M M L L	2	1.2	1.5	1.6	1.5
M M L M	2	2.1	1.4	1.3	3.4
M M M L	6	7.8	6.1	7.2	8.2
M M M M	53	51.3	53.9	50.2	52.7
M M M H	6	5.7	6.0	6.5	4.8
M M H M	5	3.7	5.1	5.6	2.6
M M H H	1	1.5	0.7	0.7	0.6
M H M M	1	3.9	4.0	4.8	3.0
M H M H	1	0.8	0.9	1.0	0.5
M H H M	3	2.7	1.3	1.4	2.5
M H H H	1	1.6	1.2	1.2	0.6
H L L H	1	0.0	0.0	0.0	0.0
H M M L	1	0.9	0.7	0.7	0.8
H M M M	13	12.0	12.1	12.5	10.5
H H M L	1	0.2	0.3	0.2	0.3
H H M M	7	5.7	6.2	6.2	5.2
H H M H	2	2.1	0.8	0.6	2.0
H H H M	2	3.6	3.2	3.6	2.7
H H H H	7	4.0	6.4	5.6	7.3
others	0	9.2	8.7	8.3	10.3

significantly below 1. For model (2c), the 'best' fit of the form (4.25) was with the LT family LTA and the bivariate copula family B7, with $\delta_{13} = \delta_{14} = \delta_{24} = 0$, $\nu_1 = \nu_4 = -2$ and $\nu_2 = \nu_3 = -1$. For the 20 parametric families of the form (4.25) that were tried, generally some dependence parameter δ_{ij} (corresponding to K_{ij}) reached either the lower bound (independence copula) or the upper bound (Fréchet upper bound copula) in the estimation. This causes a problem with SEs of estimates but not for AIC values or prediction probabilities (see the comment about this in Section 11.1). For the SE estimates in Table 11.18, the jackknife subsample with the observation HLLH deleted led to substantially different estimates of the δ_{ij} so that this case is not included in the reported SE calculation.

As would be expected, the dependence parameters for consecutive years are larger. In comparisons of the two groups ($<$ 5 mi. and $>$ 5 mi.), the dependence parameters are larger for the first group. This means that the mothers in the first group are probably more consistent over time in the original 90-item checklist; there could be a number of reasons for this. The estimation consistency check from Section 10.1.3 was done; for example, in model (1), the estimates of the correlation parameters based on a 4-variate likelihood were very close to those in Table 11.17 (the maximum absolute difference was 0.018). Table 11.19 has the log-likelihoods and AIC values for four models, evaluated at the IFM estimates. Tables 11.20 and 11.21 contain expected frequencies from the models for comparisons with the observed frequencies. Tables 11.19, 11.20 and 11.21 show that the four models are comparable in fit.

11.3 Example with multivariate extremes

In this section, we describe modelling and inference with multivariate extremes for an air quality data set with ozone concentrations from a regional network of monitoring stations.

The data consist of daily maxima of hourly averages of ozone concentrations — in parts per billion (ppb) — collected over six years (1983–1988) from 24 air quality monitoring stations in the southern Ontario region. Some stations were eliminated because of excessive missing data. For ease of illustration of some multivariate ideas, we use nine stations that altogether have very few missing values over the six years. (See Figure 11.1(a) for a map showing the latitudes and longitudes of the nine monitoring stations, with the indexing used for the analysis.) We illustrate inferences concerning the probability that k or more stations of a group in a region will have annual maxima in exceedance of a certain threshold over a year. For these inferences, we convert the data to weekly maxima (consecutive weekly maxima are much less serially dependent than daily maxima). The sample size then is large enough for some asymptotic inference.

Ozone concentrations have two patterns, one diurnal and one annual. Over the course of a day, the largest values of hourly averages almost always occur in the afternoon to early evening. Hence a daily maximum ozone concentration is roughly the maximum of about six or seven hourly averages and a weekly maximum is the maximum of 40+ (weakly dependent) hourly averages. (Note that the asymptotic extreme value theory extends from maxima of independent observations to maxima of weakly dependent observations.) Daily maximum ozone concentrations have an annual cyclic pattern with higher values for the summer months and lower values for the winter months. An exploratory data analysis of the weekly maxima derived from the present data set shows that the 'high' period consists of weeks 23 to 33 inclusive of the year, and this extends to weeks 17 to 40 allowing for a slow drop-off at both ends of the period. This pattern can be seen in Table 11.22, in which averages are obtained over all years and stations for a given week (14 to 43) of the year, and standard deviations are over years for the averages of stations per year. See also Figure 11.1(b) for the time series of the weekly maxima, averaged over stations, for the six years; this suggests that there are no time trends over the six years.

For further analyses, we use data from weeks 17 to 40 inclusive of the year, and assume that annual maxima always occur in this period. For the univariate analysis, we assume a GEV model (see Section 6.1) for each station with the location parameter depending on the week of the year as a piecewise linear function. That is, for the jth station, the parameters of the GEV model are $(\gamma_j, \mu_j, \sigma_j)$ with

$$\mu_j = \mu_j(t) = \begin{cases} \mu_j^* - \beta_{1j}(23-t), & 17 \le t \le 22, \\ \mu_j^*, & 23 \le t \le 33, \\ \mu_j^* - \beta_{2j}(t-33), & 34 \le t \le 40, \end{cases}$$

where t is an index for the week of the year. (We had also tried having the scale parameter depend on the location parameter through a function $\sigma_j(t) = \nu_j[\mu_j(t)]^{\alpha_j}$ but the estimated α_j was usually 0 or very close to 0.) The parameters $(\gamma_j, \mu_j^*, \sigma_j, \beta_{1j}, \beta_{2j})$ were es-

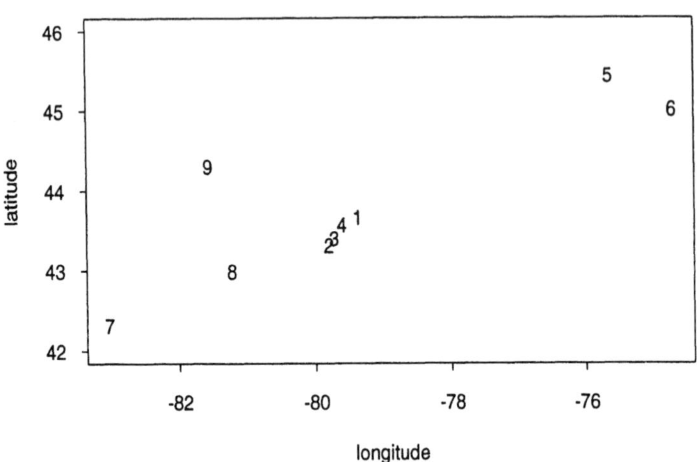

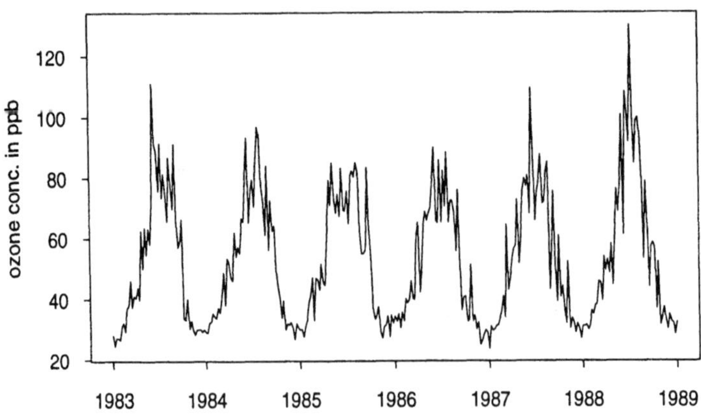

Figure 11.1. *Plots for ozone data*

Table 11.22. *Ozone data. Average of weekly maximum ozone concentrations by weeks over years and stations.*

Week	Average	SD	Week	Average	SD
14	51.0	8.4	29	81.7	10.6
15	48.0	5.7	30	86.9	5.6
16	50.4	7.6	31	80.6	13.2
17	66.4	7.9	32	78.7	12.2
18	56.8	10.6	33	81.5	8.3
19	65.0	10.3	34	72.2	9.7
20	67.2	9.9	35	60.2	14.4
21	69.5	5.6	36	68.7	15.5
22	72.2	8.5	37	64.7	8.7
23	81.6	12.3	38	55.8	16.4
24	80.5	19.2	39	54.0	11.2
25	84.8	21.0	40	55.1	8.9
26	84.9	11.6	41	40.8	9.0
27	75.6	9.8	42	36.8	4.7
28	85.7	23.7	43	41.4	8.4

timated separately for each j using numerical ML with a quasi-Newton routine, and then were fixed for estimating bivariate and multivariate parameters using the IFM method in Section 10.1. For multivariate extreme value data, GEV margins were transformed to exponential survival margins, so that MSMVE models could be used (see the examples in Section 10.1.4). If the ith observation vector is $(t_i, y_{i1}, \ldots, y_{i9})$, where t_i is an index for the week, the transformed vector is $\mathbf{z}_i$, with $z_{ij} = [1 + \hat{\gamma}_j(y_{ij} - \hat{\mu}_j(t_i))/\hat{\sigma}_j]^{-1/\hat{\gamma}_j}$.

In order to assess the dependence pattern among the stations, we initially fitted the bivariate family B8 separately to each of the 36 bivariate margins (since the family B8 extends to the multivariate family M8 which has all bivariate margins in the family B8 with possibly different parameter values). Applying the IFM method with bivariate margins, the bivariate parameters were estimated with a quasi-Newton routine, and are given in Table 11.23 (the (j, k) entry, with $j < k$, corresponds to the station pair (j, k)). The estimated SEs, with the univariate parameters assumed fixed, are in the range 0.10 to 0.14.

Table 11.23. *Ozone data. Bivariate dependence parameters for station pairs using model B8.*

Stn	1	2	3	4	5	6	7	8	9
1	–	1.33	1.48	1.71	0.81	0.94	1.13	1.07	1.02
2		–	1.38	1.18	0.78	0.81	0.90	0.86	0.78
3			–	1.85	0.74	0.76	0.80	1.04	1.00
4				–	0.77	0.78	0.88	1.16	1.04
5					–	1.49	0.68	0.82	0.85
6						–	0.67	0.94	0.90
7							–	0.86	0.82
8								–	1.05

In matching the values in Table 11.23 to the map in Figure 11.1(a), the larger dependence parameter values occur for pairs of stations that are closer to each other (a cluster of four in the middle and another cluster of two off to the side); otherwise there seems to be roughly a common (lower) dependence level for the remaining pairs. The overall dependence pattern suggests the use of the family MM1 in the form given next.

We use model MM1 with a parameter θ for the minimal dependence level (over pairs) and additional parameters δ_{jk} for the pairs of stations in the set $B = \{(1,2),(1,3),(1,4),(2,3),(3,4),(5,6)\}$; this set consists of the pairs with a higher level of dependence. With $\psi(s) = \exp\{-s^\theta\}$ in the family LTA, and K_{jk} being bivariate copulas in the family B6, then following (4.24) and (4.25) in Section 4.3, we write the copula in the form

$$\psi\Big(-\sum_{(j,k)\in B} \log K_{jk}(H_j, H_k) - \sum_{j=7}^{9} \log H_j\Big)$$

$$= \psi\Big(-\sum_{(j,k)\in B} \log K_{jk}\big(e^{-p_j\psi^{-1}(u_j)}, e^{-p_k\psi^{-1}(u_k)}\big) + \sum_{j=7}^{9} \psi^{-1}(u_j)\Big)$$

$$= \psi\Big(\sum_{(j,k)\in B} \big[(p_j z_j^\theta)^{\delta_{jk}} + (p_k z_k^\theta)^{\delta_{jk}}\big]^{1/\delta_{jk}} + z_7^\theta + z_8^\theta + z_9^\theta\Big), \quad (11.1)$$

where $z_j = -\log u_j$, $j = 1, \ldots, 9$, and $p_1 = p_3 = 1/3$, $p_2 = p_4 = 1/2$, $p_5 = p_6 = 1$. Model (11.1) is a copula as a function of $\mathbf{u}$ and a MSMVE survival function as a function of $\mathbf{z}$. It has seven dependence parameters. (The model with an additional parameter δ_{24} was not an improvement on this one.)

A simpler model than (11.1) that was also tried was the permutation-symmetric form of the family M6. In MSMVE form, this is

$$\exp\left\{-\left[\sum_{j=1}^{9} z_j^{\theta}\right]^{1/\theta}\right\}. \tag{11.2}$$

For likelihood inference with (11.1) and (11.2), the densities were obtained using symbolic manipulation software, and then applying the IFM method in Section 10.1, the dependence parameters were estimated with a quasi-Newton routine (much more computer time was spent in the symbolic differentiation). SEs were obtained using the jackknife approach of Section 10.1.1, with 36 subsamples (of 140) from dividing the 144 observation vectors into random blocks of four. Table 11.24 consists of the log-likelihoods at the estimated parameter vectors, and estimates and SEs for the dependence parameters and some of the univariate parameters.

The log-likelihoods or AIC values suggest that model MM1 is a much better fit to the data than model M6. Next we show how they compare for the inference of the probability that at least one (or two) of the nine stations exceeds a threshold in a year (assuming no change in the ozone levels due to stricter air quality requirements). The concentration of 120 ppb for ozone is used in some air quality standards for the annual maximum and a tolerable range for an hourly average is 80 to 150 ppb. In Table 11.25, the probability inferences with SEs are given for thresholds of 120, 130, 150 and 160 ppb. A reason for considering probabilities of 'at least k stations exceeding a threshold', with $k > 1$, is because this can assess if an exceedance is more local or global over a region. Let F_a be the cdf for a vector of annual maxima and let $F(\cdot;t)$ be the cdf for a vector of weekly maxima in week t. Based on the aforementioned assumptions, the probability that at least one station exceeds a threshold T in a year is

$$P_1 = 1 - F_a(T\mathbf{1}_9) = 1 - \prod_{t=17}^{40} F(T\mathbf{1}_9;t),$$

and the probability that at least two stations exceed T is

$$P_2 = 1 - \sum_{j=1}^{m} F_a(T\mathbf{i}_j) + (m-1)F_a(T\mathbf{1}_m),$$

Table 11.24. *Ozone data. Parameter estimates and log-likelihoods.*

Model	Estimate (SE)	Log-lik
M6	θ: 1.31 (0.04)	−1085.5
MM1	θ: 1.29 (0.04)	−1016.1
	δ_{12}: 1.41 (0.40)	
	δ_{13}: 1.43 (0.65)	
	δ_{14}: 2.59 (0.68)	
	δ_{23}: 1.66 (0.56)	
	δ_{34}: 2.48 (0.44)	
	δ_{56}: 1.39 (0.12)	
univariate	γ_1: −0.06 (0.07)	
	μ_1^*: 72.6 (2.5)	
	σ_1: 17.4 (1.6)	
	β_{11}: 4.49 (0.73)	
	β_{21}: 5.77 (0.87)	
univariate	γ_9: −.15 (0.06)	
	μ_9^*: 79.3 (3.2)	
	σ_9: 19.8 (1.4)	
	β_{19}: 2.57 (0.90)	
	β_{29}: 3.34 (1.01)	

with $m = 9$ and $\mathbf{i}_j$ being a vector of 1s except for an ∞ in the jth position. If either (11.1) or (11.2) is denoted as $\exp\{-A(\mathbf{z})\}$, then $F(\mathbf{y};t)$ is $\exp\{-A(z_1(y_1,t),\ldots,z_9(y_9,t))\}$, with $z_j(y_j,t) = [1+\hat{\gamma}_j(y_j-\hat{\mu}_j(t))/\hat{\sigma}_j]^{-1/\hat{\gamma}_j}$.

With the same value of θ, the cdf in (11.1) is more PLOD and higher in the $\prec_c^{\mathrm{pw}}$ ordering than the cdf in (11.2). This explains why the point estimate for P_1 is smaller for model MM1 than for model M6. However, the point estimates in Table 11.25 are quite close for the two models, when the SEs are considered. The SEs show that there is not a lot of precision in the probability estimates which are in the middle range; there is less uncertainty for thresholds that have probability near zero or near one of being exceeded. With weak positive serial dependence by week, the probabilities of exceedance in a year are a little smaller.

Table 11.25. *Ozone data. Probability of at least one (two) stations having an annual maximum exceeding* T.

T	M6 P_1 (SE)	MM1 P_1 (SE)	M6 P_2 (SE)	MM1 P_2 (SE)
120	0.979 (0.022)	0.974 (0.025)	0.894 (0.083)	0.879 (0.082)
130	0.84 (0.11)	0.81 (0.11)	0.58 (0.18)	0.57 (0.17)
150	0.34 (0.15)	0.31 (0.13)	0.11 (0.07)	0.14 (0.09)
160	0.19 (0.11)	0.17 (0.10)	0.050 (0.039)	0.068 (0.053)

Table 11.26. *Ozone data. Observed and expected cdfs (models M6 and MM1) for the weekly maxima over all nine stations.*

x	M6	MM1	Observed
70	0.059	0.064	0.104
80	0.159	0.169	0.236
85	0.236	0.248	0.313
90	0.329	0.344	0.382
95	0.433	0.448	0.486
100	0.539	0.554	0.569
105	0.638	0.652	0.667
110	0.725	0.737	0.736
120	0.854	0.862	0.833
130	0.928	0.933	0.903
170	0.995	0.996	0.993

Finally, we illustrate one diagnostic check of the MEV models, although many other diagnostic checks could be made. We compare in Table 11.26 the observed and expected relative frequencies of weeks for which the maximum weekly maxima over all nine stations is less than or equal to x for various x values; using the preceding notation, this is $(24)^{-1}\sum_{t=17}^{40} F(x\mathbf{1}_9;t)$. As a function of x, the curves of expected frequencies for models M6 and MM1 cross with the curve of observed frequencies, with the curve from MM1 being closer to that of the observed frequencies for a wider range (for the probability). For the inferences and diagnostics considered here, the simpler exchangeable dependence model does about as well, even though it is a worse fit using the AIC. However, for other inferences and other data, the more complex models may perform much better.

11.4 Example with longitudinal binary data

In this section, models are fitted to a data set with binary time series of length 4 to 16 for different subjects. Markov models with and without random effects are used, and model parameters are compared for two treatment groups. The models are used to summarize how the two groups are different in the response variable.

The data, consisting of a binary time series for each subject, are given in Table 11.27; there are two treatment groups labelled A and B, the time unit is a week and the binary response is an indicator of bacteriuria (bacteria in the urine) for the week. The subjects are acutely spinal cord injured patients with chronic urinary tract infections. Retaining only those patients with at least four weeks of observation, there were 36 subjects on each of two treatments for bacteriuria. Patients were in the study for at most 16 weeks. For treatment A, patients were treated for all episodes of bacteriuria. For treatment B, patients were treated for episodes of bacteriuria, only if they were accompanied by two specific symptoms. Patients were assigned randomly to the two treatment groups; patients entered the study with bacteriuria, so that the first response for each patient is 1. Note that having bacteriuria for a longer period of time does not necessarily mean that the patient is sicker. See Gribble, McCallum and Schechter (1988) for some background.

In Table 11.27, a few (11) missing values were imputed. For treatment A, the imputed value was 0 in five cases; for treatment B, in six cases, the imputed value was 1, if among a long string of 1s, and 0 otherwise. The inferences are not really affected by these few imputed values. The imputation is done for simplicity (so that missing data did not have to be considered in the computer programming).

From other information collected for the study, treatment B subjects would often get one 'bug' in their urinary tract and not get rid of it for some time, whereas for treatment A subjects bacteria were removed so did not stay around for a lengthy period.

Summary statistics, by subject, are given in Table 11.28 for the proportions of weeks without bacteriuria, not including the first week in the study. Over all subjects, the proportions of weeks without bacteriuria are 0.616 and 0.322 for treatments A and B,

Table 11.27. *Bacteriuria data. Binary weekly time series data; response is indicator of bacteriuria.*

Weeks	Treatment A Series	Weeks	Treatment B Series
13	1001010100101	9	111100011
5	10001	16	1111111111111111
16	1010010010101010	15	111111111111001
9	101000101	15	100011111110111
15	100011010111111	16	1111111111010000
16	1100100010101001	16	1111100111111111
9	100010101	16	1000011110111001
4	1000	12	111000111011
16	1001000000100010	16	1110111111111111
7	1010011	4	1110
10	1010110101	10	1011111111
10	1001010101	16	1011111111111111
6	100110	16	1111111100000110
14	10100011010100	6	100010
5	10010	16	1001111011011111
16	1000101100100100	5	10011
16	1010100110110100	12	111111100000
16	1000100010000000	16	1000111111111000
16	1010001010011001	16	1111001100011111
12	101000011011	16	1001111111111111
15	101000111011101	16	1000000000000000
10	1001000100	16	1100111111100111
16	1000101001110001	6	100100
16	1010001010010011	11	11111100000
13	1010101001001	12	100111110100
16	1000101010100010	10	1111111111
10	1010001001	16	1111111110111111
12	100010110101	16	1111101110111111
6	111000	6	110011
16	1001000010100010	16	1100000110010000
11	10000101001	15	100001111111111
9	110011011	15	100011110101001
15	100101100010001	16	1111110011111111
14	11010011001010	16	1111000101111111
8	10010010	10	1100000111
6	100100	7	1111000

Table 11.28. *Bacteriuria data. Summary statistics for proportions of weeks without bacteriuria.*

Treatment	Mean	SD	Min	Q1	Med	Q3	Max
A	0.628	0.132	0.357	0.551	0.600	0.707	1
B	0.349	0.236	0	0.133	0.333	0.483	1

respectively. Hence it appears that treatment A is effective in treating bacterial infection of the urinary tract.

From the initial analysis of the data, a Markov chain model of order 1 seems reasonable. Table 11.29 summarizes the transition frequencies to state 1 by subject; it seems to indicate that there is enough variability in the transition probabilities to consider a random effects model, e.g., transition probabilities p_{01}, p_{11} are random over subjects with Beta (α_k, β_k) distributions, and $\pi_k = \beta_k/(\alpha_k + \beta_k)$ and $\eta_k = (\alpha_k + \beta_k)^{-1}$, $k = 0, 1$. The shapes of the histograms of the four columns of proportions in Table 11.29 suggest that the beta distributions are a reasonable model to try, but of course other distributions could be used for a random effects model.

Other models that were tried have no random effects or different order for the Markov chain. The models are:

1. Markov chains of order 1 with random effects;
2. Markov chains of order 1 with no random effects;
3. Markov chains of order 2 with no random effects (parameters $p_{i_1 i_2 1}$, $i_1, i_2 = 0, 1$);
4. iid observations after first week, no random effects (parameter p_1 for probability of 1).

Note that for the Markov chain models with no random effects, the MLE of the transition probability p_{i1} or $p_{i_1 i_2 1}$ comes from $n_{i,1}/(n_{i,1} + n_{i,0})$ or $n_{i_1 i_2,1}/(n_{i_1 i_2,1} + n_{i_1 i_2,0})$, where n_{s_1,s_2} is the number of transitions (over all subjects in a treatment group) to state s_2 given the recent past s_1. For model (1), MLEs were estimated separately for two treatment groups, using a quasi-Newton routine. For model (3), the SEs also came from using a quasi-Newton routine. Parameter estimates, SEs and log-likelihoods are given in Table 11.30. The main interpretations of the results in Table 11.30 are the following.

Table 11.29. *Bacteriuria data. Transition proportions to the state of 1, by subject.*

Trt A $0 \to 1$	Trt A $1 \to 1$	Trt B $0 \to 1$	Trt B $1 \to 1$
0.714	0.000	0.333	0.800
0.333	0.000	0.500	1.000
0.750	0.000	0.500	0.917
0.600	0.000	0.500	0.800
0.600	0.667	0.250	0.818
0.556	0.167	0.500	0.923
0.600	0.000	0.429	0.625
0.000	0.000	0.500	0.714
0.273	0.000	1.000	0.929
0.667	0.333	0.500	0.667
1.000	0.200	1.000	0.875
0.800	0.000	1.000	0.929
0.500	0.333	0.200	0.800
0.571	0.167	0.333	0.000
0.500	0.000	0.750	0.727
0.444	0.167	0.500	0.500
0.714	0.250	0.000	0.857
0.167	0.000	0.200	0.800
0.556	0.167	0.400	0.800
0.500	0.400	0.500	0.923
0.667	0.500	0.000	0.000
0.333	0.000	0.500	0.818
0.444	0.333	0.333	0.000
0.556	0.167	0.000	0.833
0.714	0.000	0.500	0.571
0.556	0.000	0.500	1.000
0.500	0.000	1.000	0.929
0.667	0.200	1.000	0.846
0.000	0.667	0.500	0.667
0.400	0.000	0.200	0.400
0.429	0.000	0.250	0.900
0.667	0.600	0.571	0.429
0.444	0.200	0.500	0.923
0.667	0.286	0.500	0.818
0.500	0.000	0.200	0.750
0.333	0.000	0.000	0.750

Table 11.30. *Bacteriuria data. Estimates and log-likelihoods for the models.*

Model	Treatment A		Treatment B	
	Estimates (SEs)	Log-lik	Estimates (SEs)	Log-lik
(1)	π_0: 0.528 (0.035)	−235.5	π_0: 0.366 (0.055)	−226.4
	η_0: 0.000 (0.039)		η_0: 0.083 (0.079)	
	π_1: 0.208 (0.040)		π_1: 0.809 (0.028)	
	η_1: 0.063 (0.085)		η_1: 0.019 (0.039)	
(2)	p_{01}: 0.520 (0.034)	−237.0	p_{01}: 0.347 (0.042)	−227.4
	p_{11}: 0.200 (0.031)		p_{11}: 0.816 (0.022)	
(3)	p_{001}: 0.596 (0.049)	−233.3	p_{001}: 0.378 (0.056)	−221.9
	p_{011}: 0.208 (0.040)		p_{011}: 0.805 (0.062)	
	p_{101}: 0.460 (0.045)		p_{101}: 0.302 (0.063)	
	p_{111}: 0.286 (0.086)		p_{111}: 0.851 (0.024)	
(4)	p_1: 0.384 (0.025)	−347.4	p_1: 0.678 (0.022)	−271.4

(a) Although there is some variability in the transition proportions from state to state, the variability is not more than that expected from a model with no random effects. (However, with longer time series, we would expect that a random effects model would be needed to explain variability in such summaries as in Table 11.29.) The parameters, η_0, η_1, are not significantly different from 0 so that the degenerate model with p_{01}, p_{11} set at single values suffices to explain the variability in the data. (This conclusion of no random effects could be checked with a sensitivity analysis by using an alternative random effects model; e.g., two-component mixture models for the transition probabilities to state 1. This is left as an exercise.)

(b) Markov chains of order 1 are good fits to the data; the Markov chains of order 2 are slightly better fits and the iid model in (4) is a much worse fit to the data.

From model (2), treatment B patients are much more likely to remain in state 1 if they are already in this state: $p_{11}^{(B)} - p_{11}^{(A)} = 0.616$, with SE of 0.038. They are also less likely to go from state 0 to state 1: $p_{01}^{(B)} - p_{01}^{(A)} = -0.173$, with SE of 0.054. This latter summary may be an indication that treatment A is not necessarily better.

The analyses in this section are fairly straightforward. The ideas of Markov chains and random effects generalize to data which consist of equally spaced time series for many subjects, provided lengths of time series are sufficiently long.

11.5 Example with longitudinal count data

In this section, several models are applied to a data set of longitudinal counts, which comes from daily totals over individual binary time series for many subjects. A description of the study and data set is given next before the explanation for treating the data set as a single count time series rather than many binary time series.

There have been many recent studies with the aim of assessing pollution effects on human health. Such studies are not easy to do. For this particular study, details are given in Vedal *et al.* (1997). The population consists of elementary school children in Port Alberni, BC, Canada, a small town of about 30 000 with pulp and paper as the main industry causing ambient particulate pollution. The children were classified into asthmatic, non-asthmatic and slight abnormality. These groups constituted a census for the study, and a control group was randomly chosen to match them. After receiving instruction and feedback, the children filled in diaries at home. The analysis here combines all four groups in order to have a larger sample size for making initial inferences.

Binary variables to be measured daily were indicators of cough (S1), phlegm (S2), burning, aching or redness of the eyes (S3), runny or stuffed nose (S4), sore throat (S5), wheezing (S6), chest tightness (S7) and shortness of breath (S8). There were other continuous response variables which will not be considered here. Covariates for the study included daily temperature (TEMP), humidity, indicator of precipitation and concentration of PM10 from two different stations (labelled as PM10A and PM10B). PM10 is particulate matter of diameter 10 μm and below.

Altogether there were 208 subjects and the period of study was from September 1990 to March 1992. However, there was a break in July and August 1991 so that the number of days for the study was 493.

There were missing data for most subjects, i.e., only a few had the daily data for all 493 days. The proportions of days with symptoms varied greatly from subject to subject; see Figure 11.2 for a

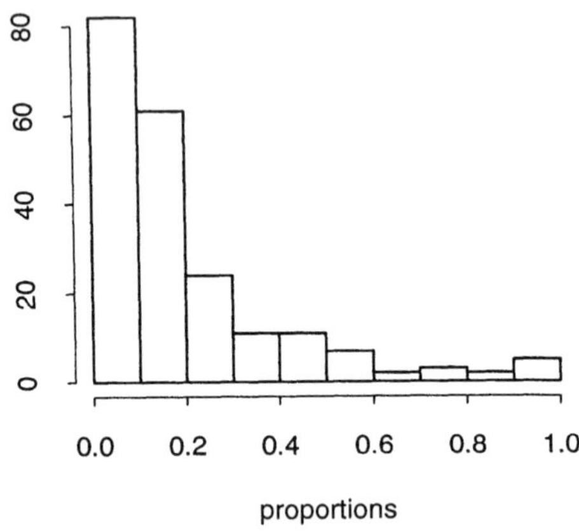

Figure 11.2. *Health effects data. Histogram of proportions of days by subjects with at least one symptom.*

histogram of the proportions of days by subject with at least one of the eight symptoms. These two features make (random effects) modelling of many binary series difficult.

The initial data analysis indicated that the assumption that data are missing at random is acceptable, i.e., there is no suggestion that subjects with a tendency to more symptoms had more days with missing data. This assessment is based on plots and correlations (a) by subject, of the proportions p_s of days with at least one symptom and the number of observed days, and (b) by day t, of the numbers n_t of subjects observed and the average $\sum_s p_s I(s \text{ obs. on day } t)/n_t$ of proportions among the observed subjects. The correlation in (a) is -0.058 and that in (b) is -0.013.

There appears not be enough power in the study to show effects of pollutants on specific individuals; hence we use totals by day, as explained next.

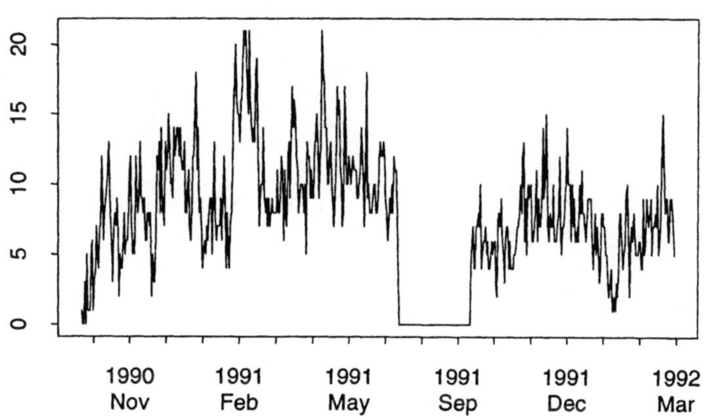

Figure 11.3. *Health effects data. Daily count time series plot for symptom 1 (cough).*

Because the proportions are widely varying by subject and because missing data appear to be missing at random, we can consider the Poisson distribution as a model for Y_t, the daily total number of subjects with a given symptom. (Essentially this model comes from the Poisson approximation to the sum of dependent Bernoulli rvs with different probability parameters.) The series Y_t is serially dependent with an expectation that is a function of n_t and covariates.

Summary statistics (means, standard deviations and quartiles) are given in the Table 11.31 for response variables and covariates that are used, including some grouped symptom variables G1, G2, G3 and G4. For a subject on a given day, G1 is the indicator of whether symptom 1 or 2 occurs, G2 that of whether symptom 3, 4 or 5 occurs, G3 that of whether symptom 6, 7 or 8 occurs, and G4 that of whether at least one of the eight symptoms occurs. Figure 11.3 shows the time series of total daily counts for symptom 1. In Table 11.31, temperature is measured in degrees Celsius. The variable humidity is not used for further analysis because it is moderately negatively correlated with temperature and it had missing values during weekends in 1992. Also a few values of PM10 were missing and were estimated from the expectation–maximization (EM) algorithm. For the models given below, a scaled temperature (SCTEMP) variable was used, with definition SCTEMP=(TEMP−10.3)/6.35 (based on mean and standard deviation over a longer period of time than that for this study). Finally, the variable names beginning with CUM are explained below.

Count time series models, with incorporation of covariates, that were used to model the data are:

1. the AR(1) Poisson time series model in Sections 8.4.1 and 8.4.4;
2. the AR(2) Poisson time series model in Sections 8.4.3 and 8.4.4;
3. the AR(1) negative binomial time series model in Sections 8.4.1 and 8.4.4;
4. Markov chain models based on bivariate and trivariate copula models (Sections 8.1 and 4.3).

For all these models, θ_t in Section 8.4.4 is taken as $\exp\{\beta_0 + \boldsymbol{\beta}\mathbf{x}_t\}$ for various choices of the covariate vector $\mathbf{x}_t$ (which is time-dependent). The first covariate is $\log n_t$ because of the tendency to have higher daily counts of occurrences of symptoms for days with more subjects reporting. The model in (1) was the first one tried, since the initial data analysis showed strong lag 1 correlation but not significant overdispersion relative to Poisson (as indicated, for example, by the ratios of sample variances to sample means for the response variables in Table 11.31). The model in (2) allows for stronger dependence than AR(1) for lags of order 2 or more. The model in (3) provides a model-based method to check on whether one needs to account for overdispersion relative to Poisson. The main focus will be in the models in items (1) to (3) because the AR models are more interpretable for Poisson and negative binomial margins, and conclusions are similar with the various models. However, we also want to illustrate the use of copula-based Markov chain time series models which may be more suitable for other ap-

Table 11.31. *Health effects data. Summary statistics (over days).*

Variable	Mean	SD	Min	Q1	Med	Q3	Max
n_t	105	25	59	83	107	122	157
S1	8.6	3.8	0	6	8	11	21
S2	4.9	2.5	0	3	5	6	13
S3	1.5	1.3	0	1	1	2	7
S4	10.6	3.5	1	8	10	13	23
S5	4.4	2.2	0	3	4	6	13
S6	2.4	1.5	0	1	2	3	7
S7	1.9	1.3	0	1	2	3	8
S8	3.2	1.7	0	2	3	4	10
G1	10.7	4.3	1	8	10	13	24
G2	13.3	4.2	2	10	13	16	27
G3	5.2	2.3	0	4	5	7	15
G4	19.8	6.3	6	15	19	24	40
TEMP	7.8	5.3	−8.4	4.4	7.2	11.6	20.2
SCTEMP	−0.40	0.84	−2.95	−0.93	−0.49	0.20	1.56
PM10A	17.4	9.5	0.5	11	16	22	67
PM10B	25.7	19.3	0.2	14	22	30	159
CUM4A	17.3	6.6	3.5	13.5	16.8	20.7	45.0
CUM5A	17.3	6.3	3.6	13.7	16.6	20.3	41.4
CUM6A	17.3	6.0	3.7	13.8	16.8	20.1	37.3
CUM7A	17.3	5.8	3.8	14.0	16.9	20.0	36.9
CUM4B	25.6	16.1	5.7	15.9	22.2	29.0	116
CUM5B	25.6	15.7	7.4	15.8	22.1	28.9	111
CUM6B	25.6	15.3	8.3	15.9	22.2	28.8	106
CUM7B	25.5	15.0	8.3	15.8	22.1	28.3	99.2

plications. The use of (4) and comparisons with the other models (1) to (3) are made at the end of this section.

MLEs were obtained using a quasi-Newton routine. (See Section 10.4 on the asymptotic properties of MLEs of parameters of Markov chains.) Good starting points for the regression coefficients come from the Poisson regression model, $Y_t \sim \text{Poisson}(\exp\{\beta_0 + \boldsymbol{\beta}\mathbf{x}_t\})$, $t = 1, 2, \ldots$, assuming independent rvs over days (because this is an exponential family model, numerical estimates are easily

Table 11.32. *Health effects data. Correlations of measurements at the two monitoring stations.*

Variable	Correlation
log PM10	0.576
log CUM2	0.716
log CUM3	0.720
log CUM4	0.725
log CUM5	0.729
log CUM6	0.731
log CUM7	0.732

obtained using a Newton–Raphson routine). For the autoregressive parameters, initial estimates can come from autocorrelations — the lag 1 correlation for the AR(1) model, and $\tilde{\alpha}_1 \approx \tilde{\rho}_1 - \tilde{\alpha}_3$, $\tilde{\alpha}_2 \approx \tilde{\rho}_2 - \tilde{\alpha}_3$ for the AR(2) model, with $\tilde{\rho}_1, \tilde{\rho}_2$ being the autocorrelations of lags 1 and 2, and $\tilde{\alpha}_3 > 0$ chosen so that all initial estimates are positive. SEs and the asymptotic covariance matrix of the parameters were obtained from the quasi-Newton routine.

The models were tried for several of the response variables, in particular the responses S1, S4, G3, G4, with the larger averaged counts. Different choices and transforms of the covariates were used, including averaged cumulative PM10 concentrations over the past k days. These variables are denoted as CUMkA and CUMkB, and some descriptive statistics for them are given in Table 11.31. For the PM10 concentrations and their cumulative lags, the log transform was used because of the skewness of their distributions. Table 11.32 shows the correlations between transformed PM10 concentrations from stations A and B.

The main conclusions about covariates and models are the following.

(a) The PM10 concentration seems to have a contribution towards increased counts for some of the symptom variables (S1, S4, G4) through the averaged cumulative concentration over about 4 to 7 days. For S1, the averaged cumulative concentration from station B is a better predictor, and for S4, that from station A is better. For G3, no additional covariate is very good as a predictor. For G4, the averaged cumulative concentrations from both stations have about the same predictive value, with that from station A being a little better; the lags from the two stations are highly correlated and there is no model improvement with two lags as covariates.

(b) Temperature as a single covariate (without pollutant variables) is significant only for some of the response variables. Another covariate that was considered was an indicator of regular school day (versus weekend or holiday). This indicated a tendency of marginally higher counts on school days but the variable was not significant except with the response variable G3 and it had no effect on the conclusion concerning PM10 referred to in (a).

(c) The AR(2) model is not a big improvement on AR(1); there is an increase in log-likelihood, but not much improvement in predictive ability (see below).

(d) Negative binomial response variable models which allow for overdispersion relative to Poisson are not an improvement in model fit (the overdispersion parameter $\nu = \sigma^2/\mu - 1$ is estimated at zero or near zero in all cases).

The conclusions in (a) and (b) correspond roughly to the correlations between daily proportions of a symptom response variable and the covariates listed in Table 11.31. The better predictors have a correlation of 0.20 to 0.35 with the daily proportions. The conclusions are similar to those of Vedal *et al.* (1997), based on other statistical methods.

Table 11.33 contains log-likelihoods and estimates for the four response variables S1, S4, G3, G4, with corresponding sets of covariates that are roughly the best fits found. For comparison, the summaries for the AR(1) models with only $\log n_t$ as the covariate are given for S1 and G4, and the summaries with AR(2) models are given for all four reponses.

The fitted parameters for the AR(2) models suggest some dependence at lag 2 beyond that expected from an AR(1) model; in other words, they show slightly more positive serial dependence than that for an AR(1) model. This is mainly through latent variables that are associated with every triple of consecutive observations since the parameter α_3 for this three-way dependence is quite positive, whereas the dependence parameter α_2 for pairs of observations separated by a lag of 2 is zero or near zero.

We give an intepretation of the rate of increased counts of symptoms as PM10 concentration increases. This is based on the AR(1)

Table 11.33. *Health effects data. Estimates and log-likelihoods for some AR(1) and AR(2) Poisson models.*

y	Mod.	x_k	β_k (SE)	Dep. par.	Log-lik
S1	AR1	const.	−2.79 (0.60)	$\alpha = 0.63$ (0.02)	−1118.6
		$\log n_t$	1.06 (0.13)		
S1	AR1	const.	−3.45 (0.61)	$\alpha = 0.62$ (0.02)	−1110.2
		$\log n_t$	1.07 (0.12)		
		SCTEMP	−0.024 (0.029)		
		log CUM4B	0.191 (0.055)		
S1	AR2	const.	−3.56 (0.80)	$\alpha_1 = 0.16$ (0.04)	−1104.9
		$\log n_t$	1.07 (0.12)	$\alpha_2 = 0.00$ (0.04)	
		SCTEMP	−0.023 (0.025)	$\alpha_3 = 0.49$ (0.06)	
		log CUM4B	0.190 (0.053)		
S4	AR1	const.	−2.27 (0.58)	$\alpha = 0.66$ (0.02)	−1114.7
		$\log n_t$	0.93 (0.11)		
		SCTEMP	0.024 (0.025)		
		log CUM7A	0.120 (0.069)		
S4	AR2	const.	−2.62 (0.67)	$\alpha_1 = 0.18$ (0.05)	−1109.2
		$\log n_t$	1.01 (0.13)	$\alpha_2 = 0.05$ (0.05)	
		SCTEMP	0.022 (0.025)	$\alpha_3 = 0.49$ (0.06)	
		log CUM7A	0.107 (0.077)		
G3	AR1	const.	−1.05 (0.74)	$\alpha = 0.45$ (0.04)	−1028.7
		$\log n_t$	0.57 (0.14)		
		SCTEMP	0.034 (0.038)		
		log CUM5A	0.031 (0.083)		
G3	AR2	const.	−1.81 (0.88)	$\alpha_1 = 0.15$ (0.07)	−1016.0
		$\log n_t$	0.75 (0.17)	$\alpha_2 = 0.10$ (0.07)	
		SCTEMP	0.032 (0.033)	$\alpha_3 = 0.32$ (0.08)	
		log CUM5A	−0.011 (0.083)		
G4	AR1	const.	−1.90 (0.38)	$\alpha = 0.62$ (0.03)	−1297.5
		$\log n_t$	1.05 (0.08)		
G4	AR1	const.	−2.39 (0.42)	$\alpha = 0.62$ (0.03)	−1293.7
		$\log n_t$	1.08 (0.08)		
		SCTEMP	−0.004 (0.020)		
		log CUM7A	0.126 (0.050)		
G4	AR2	const.	−2.60 (0.44)	$\alpha_1 = 0.24$ (0.06)	−1290.9
		$\log n_t$	1.13 (0.09)	$\alpha_2 = 0.08$ (0.06)	
		SCTEMP	−0.004 (0.019)	$\alpha_3 = 0.39$ (0.07)	
		log CUM7A	0.112 (0.050)		

Table 11.34. *Health effects data. Comparisons of observed versus predicted values by models.*

Response	Model	Cor(obs,pred)	RMS(obs-pred)
S1	AR0	0.584	3.06
S1	AR1	0.781	2.35
S1	AR2	0.786	2.33
S4	AR0	0.584	2.87
S4	AR1	0.748	2.35
S4	AR2	0.754	2.33
G3	AR0	0.283	2.19
G3	AR1	0.495	1.98
G3	AR2	0.531	1.94
G4	AR0	0.756	4.13
G4	AR1	0.846	3.37
G4	AR2	0.848	3.35

models; the result is similar for the AR(2) models because the regression coefficients for the averaged cumulative concentrations do not differ much. The approximate 95% confidence intervals for the ratio of the symptom count at the third quartile of averaged cumulative concentration to that at the median are the following:

(a) for S1, (1.02, 1.08);

(b) for S4, (1.00, 1.04);

(c) for G4, (1.00, 1.04).

That is, there is suggestion of a slight increase (point estimates of 2% for S4 and G4, 5% for S1) in expected counts of some symptoms as the (averaged cumulative) concentrations of PM10 increase from the median to the upper quartile.

The final assessment of the models is the predictive ability; this is mainly a comparison of the improvements of the AR(1) and AR(2) models over a Poisson regression model with the same covariates and no serial dependence. Let $\hat{y}_t$ be the predictive value. For the Poisson regression model, $\hat{y}_t = \exp\{\hat{\beta}_0 + \hat{\boldsymbol{\beta}}\mathbf{x}_t\}$. For the AR(1) model, $\hat{y}_t = \hat{\alpha}y_{t-1} + (\hat{\theta}_t - \hat{\alpha}\hat{\theta}_{t-1})$, with $\hat{\theta}_t = \exp\{\hat{\beta}_0 + \hat{\boldsymbol{\beta}}\mathbf{x}_t\}$. For the AR(2) model, $\hat{y}_t = \mathrm{E}\,[A_t(y_{t-2}, y_{t-1})] + (\hat{\theta}_t - \hat{\alpha}_1\hat{\theta}_{t-1} - \hat{\alpha}_2\hat{\theta}_{t-2} - \hat{\alpha}_3 \min\{\hat{\theta}_{t-1}, \hat{\theta}_{t-2}\})$; the expected value of the operator

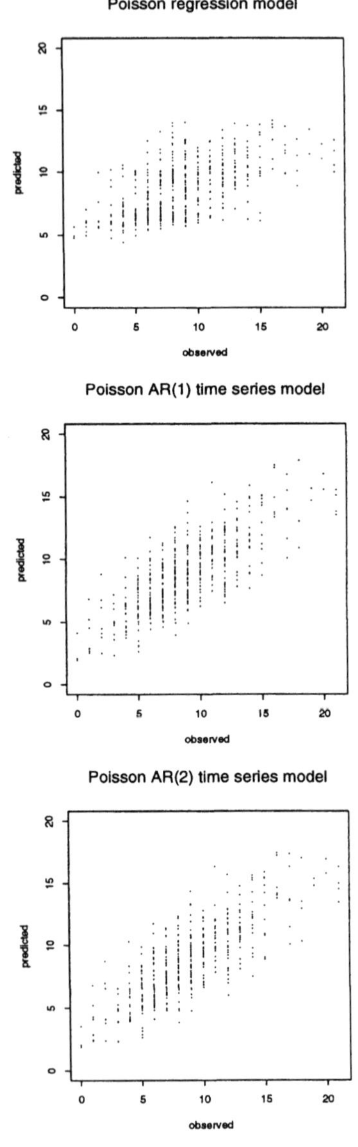

Figure 11.4. *Health effects data. Predicted versus observed values for different models for the S1 count data.*

A_t in (8.19) does not simplify analytically (as for the AR(1) model), but it can be computed numerically from the probability distribution associated with A_t.

Table 11.34 contains the correlations of y_t versus $\hat{y}_t$ for the four response variables in Table 11.33, and the root mean squared (RMS) error for prediction, $\{\sum_{t=1}^{T}(y_t - \hat{y}_t)^2/T\}^{1/2}$, with $T = 493$. Plots of predicted versus observed values for the three models for the S1 response are given in Figure 11.4. These show that the AR(1) is a big improvement on the Poisson regression model (which does not make use of the previous observations) based on the criterion of predictive ability, and that the AR(2) model does not improve much on the AR(1) model. An alternative to AR models are Poisson regression models with lagged counts as covariates; an advantage of the former over the latter is that predictions both with and without previous observed counts can be made more easily with the AR models.

To end this section, for illustration, we also apply some Markov chain copula models to the data set. However, for this data set, we think the AR models have more interpretability.

For a Markov chain of order 1, let $C(\cdot;\eta)$ be a bivariate copula. With $F(\cdot;\theta)$ being the Poisson cdf, a bivariate Poisson cdf obtains from $F_t^*(y_{t-1}, y_t) = C(F(y_{t-1};\theta_{t-1}), F(y_t;\theta_t);\eta)$. The transition density associated with this bivariate distribution is

$$h_t(y|x) = \Pr(Y_t = y \mid Y_{t-1} = x)$$

$$= \frac{F_t^*(x,y) - F_t^*(x,y-1) - F_t^*(x-1,y) + F_t^*(x-1,y-1)}{f(x;\theta_{t-1})},$$

where $f(\cdot;\theta)$ is the Poisson pmf. Note that θ_t is varying with time (because of time-varying covariates), and the transition density also depends on time.

For a Markov chain of order 2, we consider trivariate copulas $C(\cdot;\eta,\delta)$ of the form

$$C(\mathbf{u};\eta,\delta) = \psi_\eta\Big(\sum_{j=1,3}\Big\{-\log K\big(e^{-0.5\psi_\eta^{-1}(u_j)}, e^{-0.5\psi_\eta^{-1}(u_2)};\delta\big) + \tfrac{1}{2}\psi_\eta^{-1}(u_j)\Big\}\Big),$$

which is (4.29) in Section 4.3. A trivariate Poisson cdf obtains from

$$F_t^*(y_{t-2}, y_{t-1}, y_t) = C\big(F(y_{t-2};\theta_{t-2}), F(y_{t-1};\theta_{t-1}), F(y_t;\theta_t);\eta,\delta\big).$$

Table 11.35. *Health effects data. Estimates and log-likelihoods for Markov chain models with Poisson margins.*

y	Mod.	x_k	β_k (SE)	Dep. par.	Log-lik
S1	MC1/	const.	−2.90 (0.52)	$\eta = 1.65$ (0.07)	−1132.1
	B6	$\log n_t$	0.98 (0.10)		
		SCTEMP	−0.038 (0.028)		
		log CUM4B	0.159 (0.053)		
S1	MC2/	const.	−3.33 (0.43)	$\eta = 1.46$ (0.07)	−1127.6
	MM1	$\log n_t$	1.05 (0.09)	$\delta = 1.30$ (0.09)	
		SCTEMP	−0.006 (0.030)		
		log CUM4B	0.198 (.049)		
S4	MC1/	const.	−2.33 (0.49)	$\eta = 1.86$ (0.09)	−1115.8
	B6	$\log n_t$	0.96 (0.09)		
		SCTEMP	0.026 (0.023)		
		log CUM7A	0.095 (0.059)		
S4	MC2/	const.	−2.61 (0.35)	$\eta = 1.57$ (0.08)	−1109.0
	MM1	$\log n_t$	1.02 (0.07)	$\delta = 1.56$ (0.12)	
		SCTEMP	0.035 (0.023)		
		log CUM7A	0.097 (0.055)		
G3	MC1/	const.	−0.92 (0.63)	$\eta = 1.39$ (0.06)	−1035.1
	B6	$\log n_t$	0.53 (0.12)		
		SCTEMP	0.040 (0.031)		
		log CUM5A	0.046 (0.066)		
G3	MC2/	const.	−1.32 (0.51)	$\eta = 1.35$ (0.06)	−1020.2
	MM1	$\log n_t$	0.65 (0.10)	$\delta = 1.11$ (0.08)	
		SCTEMP	0.036 (0.032)		
		log CUM5A	0.001 (0.066)		
G4	MC1/	const.	−2.23 (0.38)	$\eta = 1.70$ (0.08)	−1298.1
	B6	$\log n_t$	1.07 (0.07)		
		SCTEMP	−0.014 (0.017)		
		log CUM7A	0.096 (0.043)		
G4	MC2/	const.	−2.45 (0.31)	$\eta = 1.42$ (0.06)	−1296.1
	MM1	$\log n_t$	1.11 (0.06)	$\delta = 1.54$ (0.07)	
		SCTEMP	−0.005 (0.018)		
		log CUM7A	0.111 (0.043)		

The transition density is

$$h_t(y|w,x) = \Pr(Y_t = y \mid Y_{t-2} = w, Y_{t-1} = x)$$

$$= \{F_t^*(w,x,y) - F_t^*(w,x,y-1) - F_t^*(w,x-1,y)$$
$$+F_t^*(w,x-1,y-1) - F_t^*(w-1,x,y) + F_t^*(w-1,x,y-1)$$
$$+F_t^*(w-1,x-1,y) - F_t^*(w-1,x-1,y-1)\}/\{F_t^*(w,x,\infty)$$
$$-F_t^*(w,x-1,\infty) - F_t^*(w-1,x,\infty) + F_t^*(w-1,x-1,\infty)\}.$$

Families of copulas that were tried were B6 for a first-order Markov chain and MM1 (K in the family B6 and ψ in the family LTA) for a second-order Markov chain. The family B6 was chosen because in a situation like for these data, one might have extreme value dependence (see Section 8.1.4). With θ_t depending on covariates as before, Table 11.35 has parameter estimates and log-likelihoods that can be compared with Table 11.33. Comparisons of log-likelihoods and AIC values suggest that the AR models are slightly better fits. From the regression coefficients, the conclusion about the effects of the pollutants is similar to before.

11.6 Example of inference for serially correlated data

In this section, we show through a simple example how the assumption of independent observations for serially correlated data may lead to SEs that are too small. The example illustrates the results at the end of Section 8.5.

We use a time series which consists of daily air quality measurements. Marginal distributions, such as lognormal, gamma, Weibull, or generalized gamma, etc., are commonly used (see Holland and Fitz-Simons 1982; Jakeman, Taylor and Simpson 1986; Marani, Lavagnini and Buttazzoni 1986; and references therein), and usually inferences are made assuming that the measurements are iid when in fact they are serially correlated. The SEs, computed assuming positive serial dependence, for parameters and quantiles of the marginal distribution, and for exceedances of thresholds, are usually larger than would be obtained with an assumption of iid observations. A Markov chain time series model based on a copula can be used as a means to get SEs that take into account the positive serial dependence. Different copulas can be used to check for sensitivity of the particular assumption for dependence.

We use the daily maxima of hourly averaged NO_2 concentrations (in ppm) from October 1984 to September 1986 at an air

Table 11.36. *NO_2 data. Estimates of parameters under various dependence models.*

Model	$\hat{\mu}$ (SE)	$\hat{\sigma}$ (SE)	Log-lik.	Median (SE)	UQ (SE)
indep.	−3.224	0.371	511.9	0.0398	0.0511
	(0.028)	(0.020)		(0.0011)	(0.0016)
B1	−3.223	0.372	530.0	0.0398	0.0512
	(0.044)	(0.024)		(0.0018)	(0.0024)
B2	−3.219	0.372	530.8	0.0400	0.0514
	(0.040)	(0.021)		(0.0016)	(0.0021)
B3	−3.222	0.373	531.3	0.0403	0.0518
	(0.041)	(0.021)		(0.0016)	(0.0022)
B4	−3.233	0.375	525.5	0.0395	0.0508
	(0.039)	(0.024)		(0.0016)	(0.0024)
B5	−3.222	0.384	522.9	0.0399	0.0516
	(0.044)	(0.026)		(0.0017)	(0.0027)
B6	−3.208	0.383	527.0	0.0404	0.0524
	(0.046)	(0.026)		(0.0018)	(0.0027)
B7	−3.211	0.382	527.2	0.0403	0.0522
	(0.043)	(0.026)		(0.0017)	(0.0026)

quality monitoring station in the Greater Vancouver Regional District. The range of the data is 0.014 to 0.145 ppm. The data were separated into four separate time periods so that checks could be made for trends and seasonal effects. Exploratory plots did not show any seasonal patterns or trends and the lognormal distribution appeared to be acceptable for the marginal distribution. The density of the two-parameter lognormal distribution is:

$$f(x;\mu,\sigma) = \frac{1}{(2\pi)^{1/2}\sigma x}\exp\{-\tfrac{1}{2}[(\log x - \mu)/\sigma]^2\}, \quad x > 0,$$

$-\infty < \mu < \infty, \sigma > 0$. For the (second) period from April to September 1985, the estimates of the parameters μ, σ (mean and standard deviation of logarithm of concentration) and their SEs are given in Table 11.36 for various models. The models are Markov chain stationary time series based on the families of copulas B1–B7 in Section 5.1, as well as the model of iid observations. The MLEs and corresponding SEs were computed using a quasi-Newton routine.

From Table 11.36, estimates of μ, σ are not sensitive to the copula used for the transition density, but the SEs for the MLEs are too small (especially for $\hat{\mu}$) from the likelihood assuming independence observations. The substantially larger log-likelihoods as well as the estimates of the dependence parameter δ for the copulas suggest that serial dependence is significant (for example, the correlation parameter in the BVN copula in the family B1 is 0.430). The estimates (and corresponding SEs) for the median and the upper quartile (UQ) of the concentration are also given in Table 11.36. Again the SEs based on the likelihood assuming independent observations are too small; compared with one of the likelihoods assuming dependence, the SEs are at least 25% too small and hence suggest more accuracy than really exists.

11.7 Discussion

In this chapter, different models are compared on several multivariate and longitudinal data sets. Different models seem to provide similar conclusions and predictions if they have the same qualitative features, such as the range of dependence covered. The fitting of different models also provides a sensitivity analysis. This is important because there is the common question of whether inferences (e.g., SEs) are valid after fitting many models and choosing one that is best under some criteria. If the inferences and predictions are not sensitive to the choice of models with similar qualitative features, then one could end up choosing the model that is in some ways more convenient, or one could report the inferences from more than one model to show the lack of sensitivity.

More experience and research are needed to assess whether dependence parameters should be functions of covariates, and if so, how to develop functional forms naturally. This is less of a problem for the conditional specified logistic regression model of Section 9.2.3, since diagnostic methods for logistic regression can be used. However, this latter model has the disadvantage of not being closed under the taking of margins.

In practice, missing data can occur. The examples here mainly illustrate ideas without the complication of missing data, although there were a few missing data in some of the examples. Provided missing data can be assumed to be missing at random, the IFM method can still be used. For example, a multivariate response vector with some components missing can be included in the likelihoods of margins corresponding to the non-missing components.

This, in fact, was the procedure used in the multivariate extremes example in Section 11.3.

11.8 Exercises

11.1 Write the conditionally specified logistic regression model in the form

$$\Pr(\mathbf{Y} = \mathbf{y} \mid \mathbf{x}; \boldsymbol{\theta}) = [c(\boldsymbol{\theta}, \mathbf{x})]^{-1} \exp\Big\{\sum_{\alpha} \theta_\alpha s_\alpha(\mathbf{y}, \mathbf{x})\Big\},$$

where $c(\boldsymbol{\theta}, \mathbf{x}) = \sum_{y'_j=0,1;j=1,\ldots,m} \exp\{\sum_\alpha \theta_\alpha s_\alpha(\mathbf{y}', \mathbf{x})\}$, and $\boldsymbol{\theta}$ has dimension $m(r+1)+m(m-1)/2$, r being the dimension of the covariate vector $\mathbf{x}$. Given $\mathbf{x}$ and $\mathbf{y}$, let $g(\boldsymbol{\theta}) = \Pr(\mathbf{Y} = \mathbf{y}|\mathbf{x}; \boldsymbol{\theta})$. For the delta method, the partial derivatives of g are needed to evaluate the SE of the probability. Show that

$$\frac{\partial g}{\partial \theta_\alpha} = g(\boldsymbol{\theta})\Big\{s_\alpha(\mathbf{y}, \mathbf{x}) - \sum_{\mathbf{y}'} s_\alpha(\mathbf{y}', \mathbf{x}) \Pr(\mathbf{Y} = \mathbf{y}' \mid \mathbf{x}; \boldsymbol{\theta})\Big\}.$$

11.2 Write a computer program for estimation with the IFM method for a multivariate logit model with some closed-form copula family.

11.3 For the multivariate ordinal data set in Section 11.2, do further analysis, e.g., fit models with fewer univariate parameters and possibly Markov dependence structure.

11.4 For the copulas in (11.1) and (11.2) with the same value of θ, show that (11.1) is more PLOD.

Appendix

Section A.1 presents properties of and results on Laplace transforms that are used in the construction of multivariate models, as well as a listing of parametric families of Laplace transforms useful for this purpose. Section A.2 consists of other definitions and background results that help to make the book more self-contained; topics include types of distribution functions and densities, convex functions and inequalities, and maximum entropy.

A.1 Laplace transforms

In this section, results on and properties of Laplace transforms (LTs) are summarized. A good reference is Feller (1971).

For a non-negative rv with cdf M, the **Laplace transform** $\phi = \phi_M$ is defined as

$$\phi(s) = \int_0^\infty e^{-sw} dM(w), \quad s \geq 0,$$

so that $\phi(-t)$ is the moment generating function of M. If π_0 is the mass of M at 0, then $\lim_{s\to\infty} \phi(s) = \phi(\infty) = \pi_0$. We assume throughout that LTs correspond to positive rvs, or that $\phi(\infty) = 0$. (This is because applications require that $\exp\{-\phi^{-1}(F(x))\}$ is a proper cdf when F is a univariate cdf.) Clearly ϕ is continuous and strictly decreasing, and $\phi(0) = 1$. Therefore the functional inverse ϕ^{-1} is strictly decreasing and satisfies $\phi^{-1}(0) = \infty$, $\phi^{-1}(1) = 0$. Furthermore, ϕ has continuous derviatives of all orders and the derivatives alternate in sign, i.e., $(-1)^i\phi^{(i)}(s) \geq 0$ for all $s \geq 0$, where $\phi^{(i)}$ is the ith derivative. The property of alternating signs in the derivatives is called the **completely monotone** property. LTs can be characterized as completely monotone functions on $[0, \infty)$ with a value of 1 at 0.

Other completely monotone functions are important in the construction of multivariate copulas based on one or more LTs. Let

$\mathcal{L}_\infty^*$ be the class of infinitely differentiable increasing functions of $[0,\infty)$ onto $[0,\infty)$, with alternating signs for the derivatives (see (1.2) in Section 1.3).

The property of $-\log\chi \in \mathcal{L}_\infty^*$ for a LT χ means that χ is the LT of an infinitely divisible rv. This is summarized in the theorem below.

Theorem A.1 *Let χ be a LT. Then χ^α is completely monotone for all $\alpha > 0$ if and only if $-\log\chi = \nu \in \mathcal{L}_\infty^*$.*

Proof. First suppose that $\nu \in \mathcal{L}_\infty^*$. Then it is easily checked that $d\chi^\alpha(s)/ds \le 0$ and that, by induction, the derivative of each summand of $(-1)^k(d^k\chi^\alpha(s)/ds^k)$ is opposite in sign. Hence sufficiency has been proved.

Next we show necessity. Let $\sigma = -\nu$. The first two derivatives of χ^α are $(\chi^\alpha)' = \alpha\sigma'\chi^\alpha$ and $(\chi^\alpha)^{(2)} = \alpha\sigma''\chi^\alpha + \alpha^2(\sigma')^2\chi^\alpha$, which has the form

$$\alpha\,\sigma^{(k)}\chi^\alpha + \alpha^2\chi^\alpha\sum_\ell c_{k\ell}(s)\,\alpha^{m_{k\ell}}$$

for $k = 2$, with $m_{k\ell}$ being non-negative integers for all ℓ. Suppose the kth derivative of χ^α has this form; then the $(k+1)$th derivative of χ^α is $\alpha\sigma^{(k+1)}\chi^\alpha + \alpha^2\sigma^{(k)}\sigma'\chi^\alpha + \alpha^3\sigma'\chi^\alpha\sum_\ell c_{k\ell}(s)\alpha^{m_{k\ell}} + \alpha^2\chi^\alpha\sum_\ell c'_{k\ell}(s)\alpha^{m_{k\ell}}$, which has the form

$$\alpha\,\sigma^{(k+1)}\chi^\alpha + \alpha^2\chi^\alpha\sum_\ell c_{k+1,\ell}(s)\,\alpha^{m_{k+1,\ell}},$$

where $m_{k+1,\ell} \ge 0$ for all ℓ. Therefore $\lim_{\alpha\to 0}(\chi^\alpha)^{(k)}/\alpha = \sigma^{(k)}$ for $k \ge 1$. Hence the complete monotonicity of $\sigma = -\nu$ is necessary for χ^α to be completely monotone for all α near 0. □

The importance of the above theorem is that it provides an indirect way to verify the complete monotonicity of χ^α, $\alpha > 0$. For some results (in Chapter 4) on multivariate distributions to hold, conditions of the form $\psi\circ\phi^{-1} \in \mathcal{L}_\infty^*$ or $\exp\{-\psi^{-1}\circ\phi\}$ being the LT of an infinitely divisible rv are needed.

Theorem A.2 *If ψ is a LT such that $-\log\psi \in \mathcal{L}_\infty^*$ and ϕ is another LT, then $\eta(s) = \phi(-\log\psi(s))$ is a LT.*

Proof. The proof can be obtained by straightforward differentiation. With the assumptions on ϕ and $-\log\psi$, the derivatives of each term of $\eta^{(k)}$ become opposite in sign. □

Some one-parameter families of LTs $\phi_\theta(s)$ that are used in Chapters 4, 5, 6 are:

LTA. (positive stable) $\exp\{-s^{1/\theta}\}$, $\theta \geq 1$;

LTB. (gamma) $(1+s)^{-1/\theta}$, $\theta \geq 0$;

LTC. (power series) $1-(1-e^{-s})^{1/\theta}$, $\theta \geq 1$;

LTD. (logarithmic series) $-\theta^{-1}\log[1-(1-e^{-\theta})e^{-s}]$, $\theta > 0$.

The corresponding functional inverses $\phi_\theta^{-1}(t)$ are:

LTA. $(-\log t)^\theta$;

LTB. $t^{-\theta}-1$;

LTC. $-\log[1-(1-t)^\theta]$;

LTD. $-\log[(1-e^{-\theta t})/(1-e^{-\theta})]$.

For the family LTD, the rv with the given LT has mass $(1-e^{-\theta})^i/(i\theta)$ on the integer i, $i=1,2,\ldots$. Similarly for the family LTC, the mass is θ^{-1} for $i=1$ and $\theta^{-1}\prod_{j=1}^{i-1}(j-\theta^{-1})$ for $i=2,3,\ldots$.

For the LT families LTA to LTD ϕ_θ, the condition of $-\log\phi_\theta \in \mathcal{L}_\infty^*$ is satisfied. The proof is direct for LTA and LTB. The proof for LTC and LTD comes from showing that ϕ_θ^α is a LT for all $\alpha > 0$, or that $e^s\phi_\theta(s)$ is the LT of an infinitely divisible distribution with support on the non-negative integers, and then applying Theorem A.1. For LTC, a Taylor expansion for $e^s\phi_\theta(s)$ yields $\sum_{i=0}^\infty p_i e^{-is}$, with $p_i = \prod_{k=1}^i (k\theta-1)/[(i+1)!\,\theta^{i+1}]$ (the null product is 1 for $i=0$). The ratio p_i/p_{i-1} for $i \geq 1$ is $\theta^{-1}(i\theta-1)/(i+1) = 1-(1+\theta^{-1})/(i+1)$ and this is increasing in i so, by the sufficient condition in Warde and Katti (1971), the infinite divisibility property follows. For LTD, a Taylor expansion for $e^s\phi_\theta(s)$ yields $\sum_{i=0}^\infty p_i e^{-is}$, with $p_i = c^{i+1}/[\theta(i+1)]$ with $c = 1-e^{-\theta}$. The ratio p_i/p_{i-1} for $i \geq 1$ is $ci/(i+1)$ which is increasing in i.

We next show that $\nu = \phi_{\theta_1}^{-1} \circ \phi_{\theta_2} \in \mathcal{L}_\infty^*$, $\theta_1 < \theta_2$, for the four families LTA to LTD. For LTA, $\nu(s) = s^\rho$, where $\rho = \theta_1/\theta_2$, so that it is easy to show that $\nu \in \mathcal{L}_\infty^*$. For LTB, $\nu(s) = (1+s)^\rho - 1$, where $\rho = \theta_1/\theta_2$, and again $\nu \in \mathcal{L}_\infty^*$ follows easily. For LTC, $\chi(s) = \exp\{-\nu(s)\} = 1-(1-e^{-s})^\rho$, where $\rho = \theta_1/\theta_2 \leq 1$, is within the family LTC. Hence, $\nu \in \mathcal{L}_\infty^*$ follows from the preceding paragraph. For LTD, $\chi(s) = \exp\{-\nu(s)\} = (1-[1-ce^{-s}]^\rho)/(1-e^{-\theta_1})$, where $\rho = \theta_1/\theta_2$ and $c = 1-e^{-\theta_2}$. This is similar to the family LTC and, using the approach of the preceding paragraph, χ^α is a LT for all $\alpha > 0$. Hence, by Theorem A.1, $\nu \in \mathcal{L}_\infty^*$.

The next theorem has results on cdfs deriving from LTs, when ψ, ϕ are LTs such that $\psi^{-1}\circ\phi \in \mathcal{L}_\infty^*$.

Theorem A.3 *Suppose ψ, ϕ are LTs such that*

$$\chi_\alpha = \exp\{-\alpha(\psi^{-1} \circ \phi)\}$$

is a LT for all $\alpha > 0$. Let $G_1(u) = \exp\{-\psi^{-1}(u)\}$, $G_2(u) = \exp\{-\phi^{-1}(u)\}$, $0 \le u \le 1$, let M_ψ be the distribution with LT ψ, and let $M_{\psi^{-1}\circ\phi}(\cdot;\alpha)$ be the distribution with LT χ_α. Then

$$G_1^\alpha(u) = \int_0^\infty G_2^\beta(u)\, dM_{\psi^{-1}\circ\phi}(\beta;\alpha). \tag{A.1}$$

Also, for $\alpha_2 > \alpha_1 > 0$,

$$M_{\psi^{-1}\circ\phi}(\cdot;\alpha_1) \prec^{st} M_{\psi^{-1}\circ\phi}(\cdot;\alpha_2)$$

Proof. By the definitions of $M_{\psi^{-1}\circ\phi}$, the right-hand side of (A.1) is $\chi_\alpha(-\log G_2(u)) = \chi_\alpha \circ \phi^{-1}(u) = \exp\{-\alpha\psi^{-1}(u)\} = G_1^\alpha(u)$. The second result follows immediately from the assumption that $\exp\{-\psi^{-1} \circ \phi\}$ is the LT of an infinitely divisible distribution with support on the positive real line. □

By making use of Theorem A.2, combinations of LTs of the form $\phi(-\log\psi(s))$, with ϕ from a one-parameter family of LTs and ψ from another, lead to two-parameter families of LTs. Some of these two-parameter families that appear in Section 5.2 are:

LTE. $(1 + s^{1/\delta})^{-1/\theta}$, $\delta \ge 1$, $\theta > 0$;

LTF. $[1 + \delta^{-1}\log(1 + s)]^{-1/\theta}$, $\delta, \theta > 0$;

LTG. $\exp\{-[\delta^{-1}\log(1 + s)]^{1/\theta}\}$, $\delta > 0$, $\theta \ge 1$;

LTH. $1 - [1 - \exp\{-s^{1/\delta}\}]^{1/\theta}$, $\delta, \theta \ge 1$;

LTI. $1 - [1 - (1 + s)^{-1/\delta}]^{1/\theta}$, $\delta > 0$, $\theta \ge 1$.

Other multi-parameter families of LTs that are used are:

LTJ. $\delta^{-1}\left[1 - \{1 - [1 - (1 - \delta)^\theta]e^{-s}\}^{1/\theta}\right]$, $\theta \ge 1$, $0 < \delta \le 1$ (LT of a discrete power series distribution on positive integers with mass $p_i = [1 - (1 - \delta)^\theta]/(\theta\delta)$ for $i = 1$ and $[1 - (1 - \delta)^\theta]^i[\prod_{j=1}^{i-1}(j - \theta^{-1})]/[\delta\theta i!]$ for $i > 1$);

LTK. $\beta^{-1}\left\{1 - [1 - (1 + \beta)^{-\zeta}]e^{-s}\right\}^{-1/\zeta} - \beta^{-1}$, $\zeta \ge 0$, $\beta > 0$ (LT of a discrete power series distribution on positive integers with mass $[1 - (1 + \beta)^{-\zeta}]^i[\prod_{j=0}^{i-1}(1 + j\zeta)]/[\beta\zeta^i i!]$ on the integer $i \ge 1$);

LTL. $\exp\{-(\alpha^\theta + s)^{1/\theta} + \alpha\}$, $\alpha \ge 0$, $\theta \ge 1$ (two-parameter family that includes LTA);

LTM. $[(1 - \theta)e^{-s}/(1 - \theta e^{-s})]^\alpha = [(1 - \theta)/(e^s - \theta)]^\alpha$, $0 \le \theta < 1$, $\alpha > 0$ (LT of a negative binomial distribution).

The corresponding functional inverses are:

LTE. $(t^{-\theta} - 1)^{\delta}$;

LTF. $\exp\{\delta(t^{-\theta} - 1)\} - 1$;

LTG. $\exp\{\delta(-\log t)^{\theta}\} - 1$;

LTH. $\{-\log[1 - (1 - t)^{\theta}]\}^{\delta}$;

LTI. $[1 - (1 - t)^{\theta}]^{-\delta} - 1$;

LTJ. $-\log\{[1 - (1 - \delta t)^{\theta}]/[1 - (1 - \delta)^{\theta}]\}$;

LTK. $-\log\{[1 - (1 + \beta t)^{-\zeta}]/[1 - (1 + \beta)^{-\zeta}]\}$;

LTL. $(\alpha - \log t)^{\theta} - \alpha^{\theta}$;

LTM. $\log[(1 - \theta)t^{-1/\alpha} + \theta]$.

Note that the family LTJ is a two-parameter generalization of LTC; LTC obtains when $\delta = 1$. Also note that LTK has the same form as LTJ with $\beta = -\delta$, $\zeta = -\theta$.

A.2 Other background results

This section consists of background definitions and results that are used in a few places in the book. The subsection topics are types of distribution functions and densities, convex functions and inequalities, and maximum entropy.

A.2.1 Types of distribution functions and densities

This subsection contains the key result concerning the three components of a distribution function (see, for example, Chung 1974), and explains the relevance to the multivariate models in this book. Also explained is the usage of the word 'density'. Some examples are used to illustrate the concepts.

Let F be a cdf on $\Re^m$. Then F can be written as a mixture with three components:

$$F = p_d F_d + p_s F_s + p_a F_a, \tag{A.2}$$

where F_d, F_s, F_a are cdfs from respectively the **discrete, singular and absolutely continuous components**, with corresponding probabilities p_d, p_s, p_a such that $p_d + p_s + p_a = 1$. The first component $p_d F_d$ comes from the point masses of F and the third component comes from integrating the mixed mth-order right derivative

f^* of F (this exists because a cdf F is right continuous and increasing). Hence,

$$p_a F_a(\mathbf{x}) = \int_{-\infty}^{x_1} \cdots \int_{-\infty}^{x_m} f^*(\mathbf{z})\, d\mathbf{z}.$$

For statistical modelling, the singular part has no practical importance for univariate distributions; however, it has importance for multivariate distributions since it exists when there are functional relationships among rvs. For example, if $\mathbf{X}$ has a singular MVN distribution, then there is a vector $\mathbf{a}$ and a constant b such that $\mathbf{a}\mathbf{X}^T = b$ (with probability 1); in this case, the covariance matrix of $\mathbf{X}$ is not positive definite, i.e., it has at least one zero eigenvalue. Another simple example of a random vector that has a distribution with a singular component is when there is a positive probability of ties among subsets of its variables.

A simple bivariate example to illustrate the decomposition (A.2) is as follows. Let X_1, X_2 be respectively the cumulative amount of summer rainfall in a certain location after the end of the first and second weeks of the summer. Suppose that the amount of rainfall in a summer week is 0 with probability 0.05 and an exponential rv with a mean of 1 cm if there is rainfall. Also suppose that the amount of rainfall is independent for different weeks. Then

- $\Pr(X_1 = X_2 = 0) = 0.0025$;
- $\Pr(X_1 = X_2 > 0) = 0.0475$;
- $\Pr(X_2 > X_1 = 0) = 0.0475$;
- $\Pr(X_2 > X_1 > 0) = 0.9025$.

For (A.2), $p_d = 0.0025$, $F_d(x_1, x_2)$ equals 1 if $x_1, x_2 \ge 0$ and is 0 otherwise. Also $p_s = 0.095$ with $F_s(x_1, x_2) = \frac{1}{2}(1 - e^{-(x_1 \wedge x_2)}) + \frac{1}{2}(1 - e^{-x_2})$, $x_1, x_2 \ge 0$. Finally, $p_a = 0.9025$, $F_a(x_1, x_2) = 1 - e^{-x_1} - x_1 e^{-x_2}$ for $x_2 > x_1 > 0$, and $F_a(x_1, x_2) = 1 - e^{-x_2} - x_2 e^{-x_2}$ for $x_1 \ge x_2 > 0$, with density $f_a(x_1, x_2) = e^{-x_2}$ for $x_2 > x_1 > 0$ and $f_a(x_1, x_2) = 0$ for $x_1 \ge x_2 > 0$.

Next we go on to the types of densities. Let ν be a measure on $\Re^m$. In this book, a measure is usually either counting measure for a discrete random vector or Lebesgue measure for a continuous random vector. The **density function** f with respect to ν is either the probability mass function (pmf) for a discrete random vector or the probability density function (pdf) for a continuous random vector (if F is absolutely continuous with respect to Lebesgue measure). *The reader who is unfamiliar with measures and measure spaces* *can take this to be the definition of a density.* If F is continuous but not absolutely continuous, then F does not have a density with respect to Lebesgue measure, but its absolutely continuous component F_a does have a density. Let F be the cdf of a random vector $\mathbf{X}$, and let h be a real-valued function. The **notation for the expected value** that covers random vectors of all types is:

$$\mathrm{E}\left[h(\mathbf{X})\right] = \int h\, dF = \int h\, f\, d\nu.$$

The middle integral can also be considered as a Riemann–Stieltjes integral. In the discrete case, $\int h\, dF = \sum_{\mathbf{x}} h(\mathbf{x}) f(\mathbf{x})$, where the summation is over the points of mass of $\mathbf{X}$. In the absolutely continuous case, $\int h\, dF = \int h(\mathbf{x}) f(\mathbf{x})\, d\mathbf{x}$.

A.2.2 Convex functions and inequalities

This subsection has results on convexity and convex functions. Related topics like star-shaped sets and functions, majorization and inequalities are also covered. References are Roberts and Varberg (1973), Rockafellar (1970) and Marshall and Olkin (1979).

We start with some basic definitions and results.

A set A in $\Re^d$ is **convex** if line segments joining points in A are also in A, or equivalently, $\mathbf{x}, \mathbf{y} \in A$ implies $\lambda \mathbf{x} + (1 - \lambda)\mathbf{y} \in A$ for all $0 < \lambda < 1$. A set A in $\Re^d$ is **star-shaped** with respect to a point $\mathbf{x}_0 \in A$ if, for other points $\mathbf{y} \in A$, the line segment joining $\mathbf{x}_0$ and $\mathbf{y}$ is in A, or equivalently, $\mathbf{y} \in A$ implies $\lambda \mathbf{x}_0 + (1 - \lambda)\mathbf{y} \in A$ for all $0 < \lambda < 1$.

Let $g : A \to \Re$ be a real-valued function with domain A being an open convex subset of $\Re^d$, $d \geq 1$. Then g is a **convex** function if

$$g(\lambda \mathbf{x} + (1 - \lambda)\mathbf{y}) \leq \lambda g(\mathbf{x}) + (1 - \lambda) g(\mathbf{y}), \quad \forall 0 < \lambda < 1, \mathbf{x}, \mathbf{y} \in A, \tag{A.3}$$

or if the set $\{(\mathbf{x}, z) : z \geq g(\mathbf{x}), \mathbf{x} \in A\}$ is a convex set, or equivalently, the region above the surface of g is a convex set. g is **strictly convex** if the inequality in (A.3) is strict (or one can replace $\leq$ with $<$). g is a **concave** function if $-g$ is convex. g is **star-shaped** if the region above the surface of g is star-shaped with respect to an appropriate point. For example, a real-valued function g defined on $[0, \infty)$ and satisfying $g(0) = 0$ is usually said to be star-shaped if the region above the curve of g is star-shaped with respect to the origin. This is equivalent to $g(x)/x$ increasing in $x > 0$.

If $d = 1$, in which case A is an (open) interval of the real line, then a convex function g has the following properties:

(a) g is continuous and has right and left derivatives (say, g'_+, g'_-) at each point;

(b) g'_+, g'_- are increasing and $g'_+(x) \geq g'_-(x)$ for all $x \in A$;

(c) if $g'_+ = g'_-$ and the second derivative g'' exists at x, then $g''(x) \geq 0$.

If $d \geq 2$, a convex function $g(\mathbf{x})$ on A is a convex function of one parameter on each line segment within A. From this, it follows that if g has derivatives of second order, then the Hessian matrix $H(\mathbf{x}) = (\partial^2 g/\partial x_i \partial x_j)$ of second-order derivatives is non-negative definite for all $\mathbf{x} \in A$ (i.e., $\mathbf{z}^T H(\mathbf{x}) \mathbf{z} \geq 0$ for all $\mathbf{z} \in \Re^d$).

We end this subsection with some inequalities for convex functions and give a definition of the majorization of vectors. Let g be a convex function on (a, b). Let $a < y_1 \leq x_1 \leq x_2 \leq y_2 < b$ be such that $x_1 + x_2 = y_1 + y_2$. Then

$$g(x_1) + g(x_2) \leq g(y_1) + g(y_2).$$

More generally,

$$\sum_{i=1}^{n} g(x_i) \leq \sum_{i=1}^{n} g(y_i), \tag{A.4}$$

if $\mathbf{x}$ is **majorized** by $\mathbf{y}$ or $(x_1, \ldots, x_n) \prec_m (y_1, \ldots, y_n)$ (i.e.,

$$\sum_{i=1}^{n} x_i = \sum_{i=1}^{n} y_i \quad \text{and} \quad \sum_{i=1}^{k} x_{[i]} \leq \sum_{i=1}^{k} y_{[i]} \ (k = 1, \ldots, n-1),$$

where $x_{[1]} \geq \cdots \geq x_{[n]}$ and $y_{[1]} \geq \cdots \geq y_{[n]}$ are the ordered x_i and y_i, respectively). An alternative definition of the **majorization** of vectors $\mathbf{x}, \mathbf{y}$ is that the inequality (A.4) holds for all convex continuous functions g.

A.2.3 Maximum entropy

This subsection contains some results on and examples of the concept of maximum entropy. References are Section 13.2 of Kagan, Linnik and Rao (1973) and Soofi (1994).

Maximum entropy is an approach to obtaining a density given partial information on a random variable or random vector, such as region of support, moments, expected values of certain functions, or marginal densities. The approach uses the information in a minimal sense. Densities that are obtained are 'smoothest' or closest to uniform given the constraints. If the measure is ν and $\mathbf{X}$ is a random vector with support on $R \subset \Re^m$, then the **maximum entropy** density f maximizes

$$-\int_R f(\mathbf{x}) \log f(\mathbf{x})\, d\nu,$$

subject to the constraints.

The most common usage of maximum entropy densities is when the constraints are in the form of expected values or moments. For example, if $\mathbf{X} = (X_1, \ldots, X_m)$ is a random vector with support on $R = \times_{j=1}^m (a_j, b_j)$, and $\mathrm{E}[h_k(\mathbf{X})] = \mu_k$, $k = 1, \ldots, K$, are compatible constraints, then the maximum entropy density has the form:

$$f(\mathbf{x}) = A \exp\Big\{\sum_{k=1}^{K} \lambda_k h_k(\mathbf{x})\Big\}, \tag{A.5}$$

where A is a normalizing constant and $\lambda_1, \ldots, \lambda_K$ are chosen so that the constraints are satisfied. The derivation of (A.5) can be obtained by the method of calculus of variations (or the method of Lagrange multipliers in the discrete case), or by using the inequality $\int_R g \log[g/f]\, d\nu \geq 0$ for densities f, g on R.

Special cases are the following.

(a) $b_j - a_j$ is finite, $j = 1, \ldots, m$, $K = 0$: (A.5) is the uniform density on R.

(b) $m = 1$, $R = [0, \infty)$, $K = 1$, $h_1(x) = x$: (A.5) is a one-parameter exponential density.

(c) $m = 1$, $R = (-\infty, \infty)$, $K = 2$, $h_1(x) = x$, $h_2(x) = x^2$: (A.5) is a two-parameter normal density.

(d) $R = \Re^m$, $K = m + m(m+1)/2$, $h_j(\mathbf{x}) = x_j$, $h_{m+j}(\mathbf{x}) = x_j^2$, $j = 1, \ldots, m$, $h_{2m+1}(\mathbf{x}) = x_1 x_2$, $\ldots$, $h_K = x_{m-1} x_m$: (A.5) is a multivariate normal density.

(e) $m = 1$, $R = [0, \infty)$, $K = 2$, $h_1(x) = x$, $h_2(x) = \log x$: (A.5) is a two-parameter gamma density.

(f) $m = 1$, $R = [0, 1]$, $K = 2$, $h_1(x) = \log x$, $h_2(x) = \log(1 - x)$: (A.5) is a two-parameter beta density.

Another use of maximum entropy is when the constraints are of the form of marginal densities. The maximum entropy distribution in the Fréchet class $\mathcal{F}(F_1, \ldots, F_m)$ of m-variate distributions with univariate cdfs $F_1, \ldots, F_m$ and corresponding pdfs $f_1, \ldots, f_m$ is $\prod_{j=1}^m F_j$, and this has density $\prod_{j=1}^m f_j$. An interpretation is that the distribution constructed from independence of the univariate margins is the most 'random' and uses the information of the univariate margins in a minimal sense. For another similar example, consider the Fréchet class $\mathcal{F}(F_{12}, F_{13})$ of trivariate distributions with given bivariate margins F_{12}, F_{13} and densities $f_{12}, f_{13}, f_1, f_2, f_3$. The maximum entropy density is $f_{12}f_{13}/f_1 = f_{2|1}f_{3|1}f_1$, which represents conditional independence of the second and third variables given the first variable. Again, the derivation of these maximum entropy densities can be obtained by the method of calculus of variations or Lagrange multipliers, or by using inequalities.

A.3 Bibliographic notes

The results and most of the LT families are given in Joe (1993) and Joe and Hu (1996). The extension of the LT family LTJ to LTK is due to T. Hu. On links between entropy, convexity and majorization, see Joe (1987; 1990b). The results in the last paragraph of Section A.2.3 can be found in Joe (1987).

References

Abell, M.L. and Braselton, J.P. (1992). *The Mathematica Handbook.* Academic Press, Boston.

Aitchison, J. and Ho, C.H. (1989). The multivariate Poisson-log normal distribution. *Biometrika*, **76**, 643–653.

Alam, K. and Saxena, K.M.L. (1981). Positive dependence in multivariate distributions. *Comm. Statist. A*, **10**, 1183–1196.

Albert, P.S. (1994). A Markov model for sequences of ordinal data from a relapsing-remitting disease. *Biometrics*, **50**, 51-60.

Al-Osh, M.A. and Alzaid, A.A. (1993). Some gamma processes based on the Dirichlet-gamma transform. *Comm. Statist. Stochastic Models*, **9**, 123–143.

Al-Osh, M.A. and Alzaid, A.A.(1994). A class of non-negative time series processes with ARMA structure. *J. Appl. Statist. Sci.*, **1**, 313–321.

Alzaid, A.A. and Al-Osh, M.A. (1993). Some autoregressive moving average processes with generalized Poisson marginal distributions. *Ann. Inst. Statist. Math.*, **45**, 223–232.

Alzaid, A.A. and Proschan, F. (1994). Max-infinite divisibility and multivariate total positivity. *J. Appl. Probab.*, **31**, 721–730.

Anderson, J.A. and Pemberton, J.D. (1985). The grouped continuous model for multivariate ordered categorical variables and covariate adjustment. *Biometrics*, **41**, 875–885.

Andrews, D.F. and Mallows, C.L. (1974). Scale mixtures of normal distributions. *J. Roy. Statist. Soc. B*, **36**, 99–102.

Arnold, B.C. (1967). A note on multivariate distributions with specified marginals. *J. Amer. Statist. Assoc.*, **62**, 1460–1461.

Arnold, B.C. (1990). Dependence in conditionally specified distributions. In *Topics in Statistical Dependence* (eds H.W. Block, A.R. Sampson and T.H. Savits), IMS Lecture Notes Monograph Series, **16**. Institute of Mathematical Statistics, Hayward, CA, pp. 13–18.

Arnold, B.C. (1991). Multivariate logistic distributions. In *Handbook of the Logistic Distribution* (ed. N. Balakrishnan). Marcel Dekker, New York, pp. 237–261.

Arnold, B.C. (1996). Distributions with logistic marginals and/or conditionals. In *Distributions with Fixed Marginals and Related Topics* (eds

L. Rüschendorf, B. Schweizer and M.D. Taylor), IMS Lecture Notes Monograph Series, **28**. Institute of Mathematical Statistics, Hayward, CA, pp. 15–32.

Arnold, B.C., Castillo, E. and Sarabia, J.M. (1992). *Conditionally Specified Distributions*, Lecture Notes in Statistics, **73**. Springer-Verlag, New York.

Arnold, B.C., Castillo, E. and Sarabia, J.M. (1995). General conditional specification models. *Comm. Statist. A*, **24**, 1-11.

Arnold, B.C. and Press, S.J. (1989). Compatible conditional distributions. *J. Amer. Statist. Assoc.*, **84**, 152–156.

Arnold, B.C. and Strauss, D. (1988). Bivariate distributions with exponential conditionals. *J. Amer. Statist. Assoc.*, **83**, 522–527.

Arnold, B.C. and Strauss, D. (1991). Bivariate distributions with conditionals in prescribed exponential families. *J. Roy. Statist. Soc. B*, **53**, 365–375.

Ashford, J.R. and Sowden, R.R. (1970). Multivariate probit analysis. *Biometrics*, **26**, 535–546.

Bahadur, R.R. (1961). A representation of the joint distribution of responses to n dichotomous items. In *Studies in Item Analysis and Prediction* (ed. H. Solomon), Stanford Mathematical Studies in the Social Sciences VI. Stanford University Press, Stanford, CA, pp. 158–168.

Balkema, A.A. and Resnick, S.I. (1977). Max-infinite divisibility. *J. Appl. Probab.*, **14**, 309–319.

Barlow, R.E. and Proschan, F. (1981). *Statistical Theory of Reliability and Life Testing.* To Begin With, Silver Spring, MD. (First published in 1975 by Holt, Rinehart and Winston)

Bernier, J. (1970). Inventaire des modèles de processus stochastiques applicables à la description des débits journaliers des rivières. *Rev. Inst. Internat. Statist.*, **38**, 49–61.

Berntsen, J., Espelid, T.O. and Genz, A. (1991). Algorithm 698: DCUHRE: An adaptive multidimensional integration routine for a vector of integrals. *Assoc. Comput. Machinery Trans. Math. Software*, **17**, 452–456.

Billingsley, P. (1961). *Statistical Inference for Markov Processes.* University of Chicago Press, Chicago.

Block, H.W. and Ting, M.L. (1981). Some concepts of multivariate dependence. *Comm. Statist. A*, **10**, 749–762.

Block, H.B., Savits, T.H. and Shaked, M. (1985). A concept of negative dependence using stochastic ordering. *Statist. Probab. Lett.*, **3**, 81–86.

Bluman, G.W. and Cole, J.D. (1974). *Similarity Methods for Differential Equations.* Springer-Verlag, New York.

Boland, P.J., Hollander, M., Joag-dev, K. and Kochar, S. (1996). Bivariate dependence properties of order statistics. *J. Multivariate Anal.*, **56**, 75–89.

Bonney, G.E. (1987). Logistic regression for dependent binary observations. *Biometrics*, **43**, 951–973.

Bradley, R.A. and Gart, J.J. (1962). The asymptotic properties of ML estimators when sampling from associated populations. *Biometrika*, **49**, 205–214.

Cambanis, S., Simons, G. and Stout, W. (1976). Inequalities for $\mathcal{E}k(X, Y)$ when the marginals are fixed. *Z. Wahrscheinlichkeitstheorie verw. Geb.*, **36**, 285–294.

Capéraà, P. and Genest, C. (1993). Spearman's ρ is larger than Kendall's τ for positively dependent random variables. *Nonparametric Statist.*, **2**, 183–194.

Char, B.W., Geddes, K.O., Gonnet, G.H., Leong, B.L., Monagan, M.B. and Watt, S.M. (1991). *MAPLE V Language Reference Manual.* Springer-Verlag, New York.

Chatfield, C. (1995). *Problem Solving: A Statistician's Guide*, second edition. Chapman & Hall, London.

Chung, K.L. (1974). *A Course in Probability Theory*, second edition. Academic Press, New York.

Clayton, D.G. (1978). A model for association in bivariate life tables and its application in epidemiological studies of familial tendency in chronic disease incidence. *Biometrika*, **65**, 141–151.

Coles, S.G. and Tawn, J.A. (1991). Modelling extreme multivariate events. *J. Roy. Statist. Soc. B*, **53**, 377–392.

Conaway, M.R. (1989). Analysis of repeated categorical measurements with conditional likelihood methods. *J. Amer. Statist. Assoc.*, **84**, 53–62.

Conaway, M.R. (1990). A random effects model for binary data. *Biometrics*, **46**, 317–328.

Connolly, M.A. and Liang, K.-Y. (1988). Conditional logistic regression models for correlated binary data. *Biometrika*, **75**, 501–506.

Consul, P.C. (1989). *Generalized Poisson Distribution: Properties and Applications.* Marcel Dekker, New York.

Cook, R.D. and Johnson, M.E. (1981). A family of distributions for modelling non-elliptically symmetric multivariate data. *J. Roy. Statist. Soc. B*, **43**, 210–218.

Cox, D.R. (1972). The analysis of multivariate binary data. *Appl. Statist.*, **21**, 113–120.

Cox, D.R. and Snell, E.J. (1989). *Analysis of Binary Data*, second edition. Chapman & Hall, London.

Cuadras, C.M. and Augé, J. (1981). A continuous general multivariate distribution and its properties. *Comm. Statist. A*, **10**, 339–353.

Dagsvik, J.K. (1995). How large is the class of generalized extreme value random utility models? *J. Math. Psych.*, **39**, 90–98.

Dall'Aglio, G. (1972). Fréchet classes and compatibility of distribution functions. *Sympos. Math.*, **IX**, 131–150.

Darlington, G.A. and Farewell, V.T. (1992). Binary longitudinal data analysis with correlation a function of explanatory variables. *Biometrical J.*, **34**, 899–910.

Davis, P.J. and Rabinowitz, P. (1984). *Methods of Numerical Integration*, second edition. Academic Press, Orlando.

Deheuvels, P. (1978). Caractérisation complète des lois extrêmes multivariées et de la convergence des types extrêmes. *Publ. Inst. Statist. Univ. Paris*, **XXIII** (3), 1–36.

Deheuvels, P. (1983). Point processes and multivariate extreme values. *J. Multivariate Anal.*, **13**, 257–272.

Diggle, P.J., Liang, K.-Y., and Zeger, S.L. (1994). *Analysis of Longitudinal Data.* Oxford University Press, Oxford.

Doss, D.C. (1979). Definition and characterization of multivariate negative binomial distributions. *J. Multivariate Anal.*, **9**, 460–464.

Esary, J.D. and Proschan, F. (1972). Relationships among some concepts of bivariate dependence. *Ann. Math. Statist.*, **43**, 651–655.

Esary, J.D., Proschan, F. and Walkup, D. (1967). Association of random variables, with applications. *Ann. Math. Statist.*, **38**, 1466–1474.

Evans, M. and Swartz, T. (1995). Methods for approximating integrals in statistics with special emphasis on Bayesian integration problems. *Statist. Sci.*, **10**, 254–272.

Fahrmeir, L. and Kaufman, H. (1987). Regression models for nonstationary categorical time series. *J. Time Series Anal.*, **8**, 147–160.

Fang, K.-T., Kotz, S. and Ng, K.-W. (1990). *Symmetric Multivariate and Related Distributions.* Chapman & Hall, London.

Fang, Z., Hu, T. and Joe, H. (1994). On the decrease in dependence with lag for stationary Markov chains. *Probab. Engrg. Inform. Sci.*, **8**, 385–401.

Fang, Z. and Joe, H. (1992). Further developments on some dependence ordering for continuous bivariate distributions. *Ann. Inst. Statist. Math.*, **44**, 501–517.

Feller, W. (1971). *An Introduction to Probability Theory and Its Applications*, second edition, Vol. II. Wiley, New York.

Fienberg, S.E., Bromet, E.J., Follman, D., Lambert, D. and May, S.M. (1985). Longitudinal analysis of categorical epidemiological data: a study of Three Mile Island. *Environ. Health Perspect.*, **63**, 241–248.

Follmann, D. (1994). Modelling transitional and joint marginal distributions in repeated categorical data. *Statist. in Medicine*, **13**, 467–477.

Frank, M.J. (1979). On the simultaneous associativity of $F(x, y)$ and $x + y - F(x, y)$. *Aequationes Math.*, **19**, 194–226.

Fréchet, M. (1935). Généralisations du théorème des probabilités totales. *Fund. Math.*, **25**, 379–387.

Fréchet, M. (1951). Sur les tableaux de corrélation dont les marges sont données. *Ann. Univ. Lyon, Sci.*, **14**, 53–77.

Fréchet, M. (1957). Les tableaux de corrélation dont les marges et des bornes sont données. *Ann. Univ. Lyon, Sci.*, **20**, 13–31.

Galambos, J. (1975). Order statistics of samples from multivariate distributions *J. Amer. Statist. Assoc.*, **70**, 674–680.

Galambos, J. (1987). *The Asymptotic Theory of Extreme Order Statistics*, second edition. Kreiger Publishing Co., Malabar, FL.

Gardner, W. (1990). Analyzing sequential categorical data: individual variation in Markov chains. *Psychometrika*, **55**, 263–275.

Gaver, D.P. and Lewis, P.A.W. (1980). First-order autoregressive gamma sequences and point processes. *Adv. Appl. Probab.*, **12**, 727–745.

Gelman, A. and Speed, T.P. (1993). Characterizing a joint probability distribution by conditionals. *J. Roy. Statist. Soc. B*, **55**, 185–188.

Genest, C. (1987). Frank's family of bivariate distributions. *Biometrika*, **74**, 549–555.

Genest, C. and MacKay, R.J. (1986). Copules archimédiennes et familles de lois bidimensionelles dont les marges sont données. *Canad. J. Statist.*, **14**, 145–159.

Genest, C., Molina, J.J.Q. and Lallena, J.A.R. (1995). De l'impossibilité de construire des lois à marges multidimensionnelles données à partir de copules. *C. R. Acad. Sci. Paris*, **320**, 723–726.

Genz, A. (1992). Numerical computation of multivariate normal probabilities. *J. Comp. Graph. Statist.*, **1**, 141–149.

Ghurye, S.G. and Marshall, A.W. (1984). Shock processes with aftereffects and multivariate lack of memory. *J. Appl. Probab.*, **21**, 786–801.

Glaz, J. and Johnson, B.M. (1984). Probability inequalities for multivariate distributions with dependence structures. *J. Amer. Statist. Assoc.*, **79**, 436–440.

Gleser, L.J. and Moore, D.S. (1983). The effect of dependence on chi-squared and empiric distribution tests of fit. *Ann. Statist.*, **11**, 1100–1108.

Glonek, G.F.V. and McCullagh, P. (1995). Multivariate logistic models. *J. Roy. Statist. Soc. B*, **57**, 533–546.

Godambe, V.P. (ed.) (1991). *Estimating Functions*. Oxford University Press, Oxford.

Gottschau, A. (1994). Markov chain models for multivariate binary panel data. *Scand. J. Statist.*, **21**, 57-71.

Gribble, M.J., McCallum, N.M. and Schechter, M.T. (1988). Evaluation of diagnostic criteria for bacteriuria in acutely spinal cord injured patients undergoing intermittent catheterization. *Diagnostic Microbiology and Infectious Diseases*, **9**, 197–206.

Gumbel, E.J. (1960a). Distributions des valeurs extrêmes en plusieurs dimensions. *Publ. Inst. Statist. Univ. Paris*, **9**, 171–173.

Gumbel, E.J. (1960b). Bivariate exponential distributions. *J. Amer. Statist. Assoc.*, **55**, 698–707.

Haan, L. de (1984). A spectral representation for max-stable processes. *Ann. Probab.*, **12**, 1194–1204.

Harris, R. (1970). A multivariate definition for increasing hazard rate distribution functions. *Ann. Math. Statist.*, **41**, 713–717.

Heckman, J.J. (1981). Statistical models for discrete panel data. In *Structural Analysis of Discrete Data with Econometric Applications* (eds C.F. Manski and D. McFadden). MIT Press, Cambridge, MA, pp. 114–178.

Hoadley, B. (1971). Asymptotic properties of maximum likelihood estimators for the independent not identically distributed case. *Ann. Math. Statist.*, **42**, 1977–1991.

Hoeffding, W. (1940). Maßstabinvariante Korrelationstheorie. *Schriften Math. Inst. Univ. Berlin*, **5**, 181–233.

Holland, D.M. and Fitz-Simons, T. (1982). Fitting statistical distributions to air quality data by the maximum likelihood method. *Atmospheric Environment*, **16**, 1071–1076.

Hougaard, P. (1986). A class of multivariate failure time distributions. *Biometrika*, **73**, 671–678.

Hougaard, P., Harvald, B. and Holm, N.V. (1992). Measuring the similarities between the lifetimes of adult Danish twins born between 1881–1930. *J. Amer. Statist. Assoc.*, **87**, 17–24.

Hu, T. and Joe, H. (1995). Monotonicity of positive dependence with time for stationary reversible Markov chains. *Probab. Engrg. Inform. Sci.*, **9**, 227–237.

Hüsler, J. and Reiss, R.-D. (1989). Maxima of normal random vectors: between independence and complete dependence. *Statist. Probab. Lett.*, **7**, 283–286.

Hutchinson, T.P. and Lai, C.D. (1990). *Continuous Bivariate Distributions, Emphasising Applications.* Rumsby, Sydney, Australia.

Jakeman, A.J., Taylor, J.A. and Simpson, R.W. (1986). Modeling distributions of air pollutant concentrations — II. Estimation of one or two parameter statistical distributions. *Atmospheric Environment*, **20**, 2435–2447.

Joag-dev, K., Perlman, M.D. and Pitt, L.D. (1983). Association of normal random variables and Slepian's inequality. *Ann. Probab.*, **11**, 451–455.

Joag-dev, K. and Proschan, F. (1983). Negative association of random variables, with applications. *Ann. Statist.*, **11**, 286–295.

Joe H. (1987). Majorization, randomness and dependence for multivariate distributions. *Ann. Probab.*, **15**, 1217–1225.

Joe H. (1990a). Families of min-stable multivariate exponential and multivariate extreme value distributions. *Statist. Probab. Lett.*, **9**, 75–81.

Joe, H. (1990b). Majorization and divergence. *J. Math. Anal. Appl.*, **148**, 287–305.

Joe, H. (1990c). Multivariate concordance. *J Multivariate Anal.*, **35**, 12–30.

Joe, H. (1993). Parametric family of multivariate distributions with given margins. *J. Multivariate Anal.*, **46**, 262–282.

Joe, H. (1994). Multivariate extreme value distributions with applications to environmental data. *Canad. J. Statist.*, **22**, 47–64.

Joe, H. (1995). Approximations to multivariate normal rectangle probabilities based on conditional expectations. *J. Amer. Statist. Assoc.*, **90**, 957–964.

Joe, H. (1996a). Families of m-variate distributions with given margins and $m(m-1)/2$ bivariate dependence parameters. In *Distributions with Fixed Marginals and Related Topics* (eds L. Rüschendorf, B. Schweizer and M.D. Taylor), IMS Lecture Notes Monograph Series, **28**. Institute of Mathematical Statistics, Hayward, CA, pp. 120–141.

Joe, H. (1996b). Time series models with univariate margins in the convolution-closed infinitely divisible class. *J. Appl. Probab.*, **33**, 664–677.

Joe, H. and Hu, T. (1996). Multivariate distributions from mixtures of max-infinitely divisible distributions. *J. Multivariate Anal.*, **57**, 240–265.

Joe, H. and Liu, Y. (1996). A model for a multivariate binary response with covariates based on compatible conditionally specified logistic regressions. *Statist. Probab. Lett.*, **31**, 113–120.

Joe, H., Smith, R.L. and Weissman, I. (1992). Bivariate threshold methods for extremes. *J. Roy. Statist. Soc. B*, **54**, 171–183.

Joe, H. and Xu, J.J. (1996). The estimation method of inference functions for margins for multivariate models. Technical Report no. 166, Department of Statistics, University of British Columbia.

Johnson, N.L. and Kotz, S. (1972). *Continuous Multivariate Distributions.* Wiley, New York.

Johnson, N.L. and Kotz, S. (1975). On some generalized Farlie–Gumbel–Morgenstern distributions. *Comm. Statist.*, **4**, 415–427.

Johnson, N.L. and Kotz, S. (1977). *Urn Models and Their Application.* Wiley, New York.

Kagan, A.M., Linnik, Yu.V. and Rao, C.R. (1973). *Characterization Problems in Mathematical Statistics.* Wiley, New York.

Karlin, S. (1968). *Total Positivity,* Vol. I. Stanford University Press, Stanford, CA.

Karlin, S. and Rinott, Y. (1980a). Classes of orderings of measures and related correlation inequalities I. Multivariate totally positive distributions. *J. Multivariate Anal.*, **10**, 467–498.

Karlin, S. and Rinott, Y. (1980b). Classes of orderings of measures and related correlation inequalities – II. Multivariate reverse rule distributions. *J. Multivariate Anal.*, **10**, 499–516.

Kelker, D. (1970). Distribution theory of spherical distributions and a location–scale parameter generalization. *Sankhyā A*, **32**, 419–430.

Kellerer, H.G. (1964). Verteilungsfunktionen mit gegebenen Marginalverteilungen. *Z. Wahrscheinlichkeitstheorie verw. Geb.*, **3**, 247–270.

Kendall, M.G. (1938). A new measure of rank correlation. *Biometrika*, **30**, 81–93.

Kepner, J.L., Harper, J.D. and Keith, S.Z. (1989). A note on evaluating a certain orthant probability. *Amer. Statist.*, **43**, 48–49.

Kibble, W.F. (1941). A two-variate gamma type distribution. *Sankhyā*, **5**, 137–150.

Kimeldorf, G. and Sampson, A.R. (1975). Uniform representations of bivariate distributions. *Comm. Statist.*, **4**, 617–627.

Kimeldorf, G. and Sampson, A.R. (1987). Positive dependence orderings. *Ann. Inst. Statist. Math.*, **39**, 113–128.

Kimeldorf, G. and Sampson, A.R. (1989). A framework for positive dependence. *Ann. Inst. Statist. Math.*, **41**, 31–45.

Kocherlakota, S. (1988). On the compounded bivariate Poisson distribution: a unified treatment. *Ann. Inst. Statist. Math.*, **40**, 61–76.

Kocherlakota, S. and Kocherlakota, K. (1992). *Bivariate Discrete Distributions.* Marcel Dekker, New York.

Krishnaiah, P.R. (1985). Multivariate gamma distributions. In *Encyclopedia of Statistical Science* (eds S. Kotz, N.L. Johnson and C.B. Read), vol. 6. Wiley, New York, pp. 63–66.

Lancaster, H. O. (1969). *The Chi-squared Distribution.* Wiley, New York.

Lawless, J.F. (1987). Negative binomial and mixed Poisson regression. *Canad. J. Statist.*, **15**, 209–225.

Leadbetter, M.R. (1983). Extremes and local dependence in stationary sequences. *Z. Wahrscheinlichkeitstheorie verw. Geb.*, **65**, 291–306.

Leadbetter, M.R., Lindgren, G. and Rootzén, H. (1983). *Extremes and Related Properties of Random Sequences and Processes.* Springer-Verlag, New York.

Lehmann, E.L. (1966). Some concepts of dependence. *Ann. Math. Statist.*, **37**, 1137–1153.

Lesaffre, E. and Kaufmann, H.K. (1992). Existence and uniqueness of the maximum likelihood estimator for a multivariate probit model. *J. Amer. Statist. Assoc.*, **87**, 805–811.

Lesaffre, E. and Molenberghs, G. (1991). Multivariate probit analysis: a neglected procedure in medical statistics. *Statist. in Medicine*, **10**, 1391–1403.

Lewis, P.A.W., McKenzie, E. and Hugus, D.K. (1989). Gamma processes. *Comm. Statist. Stochastic Models*, **5**, 1–30.

Li, H., Scarsini, M. and Shaked, M. (1996). Linkages: a tool for the construction of multivariate distributions with given nonoverlapping multivariate marginals. *J. Multivariate Anal.*, **56**, 20–41.

Lipsitz, S.R., Dear, K.B.G. and Zhao, L. (1994). Jackknife estimators of variance for parameter estimates from estimating equations with applications to clustered survival data. *Biometrics*, **50**, 842–846.

Marani, A., Lavagnini, I. and Buttazzoni, C. (1986). Statistical study of air pollutant concentrations via generalized gamma distributions. *J. Air Pollution Control Assoc.*, **36**, 1250–1254.

Marco, J.M. and Ruiz-Rivas, C. (1992). On the construction of multivariate distributions with given nonoverlapping multivariate marginals. *Statist. Probab. Lett.*, **15**, 259–265.

Mardia, K.V. (1970). *Families of Bivariate Distributions.* Griffin, London.

Marshall, A.W. and Olkin, I. (1967). A multivariate exponential distribution. *J. Amer. Statist. Assoc.*, **62**, 30–44.

Marshall, A.W. and Olkin, I. (1979). *Inequalities: Theory of Majorization and Its Applications.* Academic Press, New York.

Marshall, A.W. and Olkin, I. (1983). Domains of attraction of multivariate extreme value distributions. *Ann. Probab.*, **11**, 168–177.

Marshall, A.W. and Olkin, I. (1985). A family of bivariate distributions generated by the bivariate Bernoulli distribution. *J. Amer. Statist. Assoc.*, **80**, 332–338.

Marshall, A.W. and Olkin, I. (1988). Families of multivariate distributions. *J. Amer. Statist. Assoc.*, **83**, 834–841.

Marshall, A.W. and Olkin, I. (1990). Multivariate distributions generated from mixtures of convolutions and product families. In *Topics in Statistical Dependence* (eds H.W. Block, A.R. Sampson and T.H. Savits), IMS Lecture Notes Monograph Series, **16**. Institute of Mathematical Statistics, Hayward, CA, pp. 371–393.

Marshall, A.W. and Olkin, I. (1991). Functional equations for multivariate exponential distributions. *J. Multivariate Anal.*, **39**, 209–215.

McCloud, P.I. and Darroch, J.N. (1995). An analysis of categorical repeated measurements in the presence of an external factor. *Biometrika*, **82**, 47–55.

McCullagh, P. and Nelder, J.A. (1989). *Generalized Linear Models*, second edition. Chapman & Hall, London.

McFadden, D. (1974). Conditional logit analysis of qualitative choice behavior. In *Frontiers in Econometrics* (ed. P. Zarembka). Academic Press, New York, pp. 105–142.

McFadden, D. (1981). Economic models of probabilistic choice. In *Structural Analysis of Discrete Data with Econometric Applications* (eds C.F. Manski and D. McFadden). MIT Press, Cambridge, MA, pp. 198–272.

McKenzie, E. (1986). Autoregressive moving-average processes with negative-binomial and geometric marginal distributions. *Adv. Appl. Probab.*, **18**, 679–705.

McKenzie, E. (1988). Some ARMA models for dependent sequences of Poisson counts. *Adv. Appl. Probab.*, **20**, 822–835.

McLeish, D.L. and Small, C.G. (1988). *The Theory and Applications of Statistical Inference Functions*, Lecture Notes in Statistics, **44**. Springer-Verlag, New York.

Meester, S.G. and MacKay, J. (1994). A parametric model for cluster correlated categorical data. *Biometrics*, **50**, 954–963.

Metry, M.H. and Sampson, A.R. (1991). A family of partial orderings for positive dependence among fixed marginal bivariate distributions. In *Advances in Probability Distributions with Given Marginals* (eds G. Dall'Aglio, S. Kotz and G. Salinetti). Kluwer, Dordrecht, Netherlands, pp. 129–138.

Miller, R.G. (1974). The jackknife – a review. *Biometrika*, **61**, 1–15.

Molenberghs, G. and Lesaffre, E. (1994). Marginal modelling of correlated ordinal data using a multivariate Plackett distribution. *J. Amer. Statist. Assoc.*, **89**, 633–644.

Morgenstern, D. (1956). Einfache Beispiele zweidimensionaler Verteilungen. *Mitt. Math. Statist.*, **8**, 234–235.

Muenz, L.R. and Rubinstein, L.V. (1985). Markov models for covariate dependence of binary sequences. *Biometrics*, **41**, 91–101.

Nash, J.C. (1990). *Compact Numerical Methods for Computers: Linear Algebra and Function Minimisation*, second edition. Hilger, New York.

Nelsen, R.B. (1986). Properties of a one-parameter family of bivariate distributions with specified marginals. *Comm. Statist. A*, **15**, 3277–3285.

Nelsen, R.B. (1991). Copulas and association. In *Advances in Probability Distributions with Given Marginals* (eds G. Dall'Aglio, S. Kotz and G. Salinetti). Kluwer, Dordrecht, Netherlands, pp. 51–74.

Oakes, D. (1982). A model for association in bivariate survival data. *J. Roy. Statist. Soc. B*, **44**, 414–422.

Oakes, D. (1989). Bivariate survival models induced by frailties. *J. Amer. Statist. Assoc.*, **84**, 487–493.

O'Brien, G.L. (1987). Extreme values for stationary and Markov sequences. *Ann. Probab.*, **15**, 281–291.

Pickands, J. (1981). Multivariate extreme value distributions. *Proc. 43rd Session I.S.I.* (Buenos Aires), 859–878.

Plackett, R.L. (1965). A class of bivariate distributions. *J. Amer. Statist. Assoc.*, **60**, 516–522.

Prentice, R.L. (1986). Binary regression using an extended beta-binomial distribution, with discussion of correlation induced by covariate measurement error. *J. Amer. Statist. Assoc.*, **81**, 321–327.

Press, S.J. (1972). Multivariate stable distributions. *J. Multivariate Anal.*, **2**, 444–462.

Rachev, S.T. and Resnick, S. (1991). Max-geometric infinite divisibility and stability. *Comm. Statist. Stochastic Models* , **7**, 191–218.

Raftery, A.E. (1984). A continuous multivariate exponential distribution. *Comm. Statist. A*, **13**, 947–965.

Raftery, A.E. (1985). Some properties of a new continuous bivariate exponential distribution. *Statist. Decisions*, Supp. Issue 2, 53–58.

Resnick, S.I. (1987). *Extreme Values, Regular Variation, and Point Processes.* Springer-Verlag, New York.

Rinott, Y. and Pollak, M. (1980). A stochastic ordering induced by a concept of positive dependence and monotonicity of asymptotic test sizes. *Ann. Statist.*, **8**, 190–198.

Roberts, A.W. and Varberg, D.E. (1973). *Convex Functions.* Academic Press, New York.

Robertson, C.A. and Strauss, D.J. (1981). A characterization theorem for random utility variables. *J. Math. Psych.*, **23**, 184–189.

Rockafellar, R.T. (1970). *Convex Analysis.* Princeton University Press, Princeton, NJ.

Ruiz-Rivas, C. (1981). Un nuevo sistema bivariante y sus propiedades. *Estadíst. Española*, **87**, 47–54.

Sakamoto, Y., Ishiguro, M. and Kitagawa, G. (1986). *Akaike Information Criterion Statistics.* KTK Scientific Publishers, Tokyo.

Scarsini, M. (1985). Lower bounds for the distribution function of a k-dimensional n-extendible exchangeable process. *Statist. Probab. Lett.*, **3**, 57–62.

Scarsini, M. (1989). Copulae of probability measures on product spaces. *J. Multivariate Anal.*, **31**, 201–219.

Schervish, M.J. (1984). Multivariate normal probabilities with error bound. *Appl. Statist.*, **33**, 81–94.

Schriever, B.F. (1986). *Order Dependence*, CWI Tract 20. Centre for Mathematics and Computer Sciences, Amsterdam.

Schriever, B.F. (1987). An ordering of positive dependence. *Ann. Statist.*, **15**, 1208–1214.

Schweizer, B. (1991). Thirty years of copulas. In *Advances in Probability Distributions with Given Marginals* (eds G. Dall'Aglio, S. Kotz and G. Salinetti). Kluwer, Dordrecht, Netherlands, pp. 13–50.

Schweizer, B. and Wolff, E.F. (1981). On nonparametric measures of dependence for random variables. *Ann. Statist.*, **9**, 879–885.

Serfling, R.J. (1980). *Approximation Theorems of Mathematical Statistics.* Wiley, New York.

Shaked, M. (1975). A note on the exchangeable generalized Farlie–Gumbel–Morgenstern distributions. *Comm. Statist.*, **4**, 711–721.

Shaked, M. (1977a). A family of concepts of dependence for bivariate distributions. *J. Amer. Statist. Assoc.*, **72**, 642–650.

Shaked, M. (1977b). A concept of positive dependence for exchangeable random variables. *Ann. Statist.*, **5**, 505–515.

Shaked, M. (1979). Some concepts of positive dependence for bivariate interchangeable distributions. *Ann. Inst. Statist. Math.*, **31**, 67–84.

Shaked, M. and Shanthikumar, J.G. (1993). Dynamic conditional marginal distributions in reliability theory. *J. Appl. Probab.*, **30**, 421–428.

Shaked, M. and Shanthikumar, J.G. (eds) (1994). *Stochastic Orders and Their Applications.* Academic Press, New York.

Shea, G.A. (1983). Hoeffding's lemma. In *Encyclopedia of Statististical Science* (eds S. Kotz, N.L. Johnson and C.B. Read), vol. 3. Wiley, New York, pp. 648–649.

Sklar, A. (1959). Fonctions de répartition à n dimensions et leurs marges.

Publ. Inst. Statist. Univ. Paris, **8**, 229–231.

Smith, R.L. (1990). Extreme value theory. In *Handbook of Applicable Mathematics* (ed. W. Lederman), vol. 7. Wiley, Chichester, pp. 437–472.

Smith, R.L. (1992). The extremal index of a Markov chain. *J. Appl. Probab.*, **29**, 37–45.

Smith, R.L. and Weissman, I. (1994). Estimating the extremal index. *J. Roy. Statist. Soc. B*, **56**, 515–528.

Soofi, E.S. (1994). Capturing the intangible concept of information. *J. Amer. Statist. Assoc.*, **89**, 1243–1254 .

Spearman, C. (1904). The proof and measurement of association between two things. *Amer. J. Psych.*, **15**, 72–101.

Stefanski, L.A. (1991). A normal scale mixture representation of the logistic distribution. *Statist. Probab. Lett.*, **11**, 69–70.

Stiratelli, R., Laird, N. and Ware, J.H. (1984). Random-effects models for serial observations with binary response. *Biometrics*, **40**, 961–971.

Tawn J.A. (1988). Bivariate extreme value theory: models and estimation. *Biometrika*, **75**, 397–415.

Tawn, J.A. (1990). Modelling multivariate extreme value distributions. *Biometrika*, **77**, 245–253.

Tchen, A. (1980). Inequalities for distributions with given marginals. *Ann. Probab.*, **8**, 814–827.

Teicher, H. (1954). On the multivariate Poisson distribution. *Skandinavisk Aktuarietidskrift*, **37**, 1–9.

Thorin, O. (1977). On the infinite divisibility of the lognormal distribution. *Scand. Actuarial J.*, 121–148.

Tiago de Oliveira, J. (1980). Bivariate extremes: foundations and statistics. In *Multivariate Analysis V* (ed. P.R. Krishnaiah). North-Holland, Amsterdam, pp. 349–366.

Tong, Y.L. (1989). Inequalities for a class of positively dependent random variables with a common marginal. *Ann. Statist.*, **17**, 429–435.

Vedal, S., Petkau, A.J., White, R.A. and Blair, J. (1997). Acute effects of ambient inhalable particles in asthmatic and non-asthmatic children. *Amer. J. Respir. Crit. Care Med.*, **156**.

Warde, W.D. and Katti, S.K. (1971). Infinite divisibility of discrete distributions, II. *Ann. Math. Statist.*, **42**, 1088–1090.

Xu, J.J. (1996). Statistical Modelling and Inference for Multivariate and Longitudinal Discrete Response Data. Ph.D. thesis, Department of Statistics, University of British Columbia.

Yanagimoto, T. and Okamoto, M. (1969). Partial orderings of permutations and monotonicity of a rank correlation statistic. *Ann. Inst. Statist. Math.*, **21**, 489–506.

Zhang, H. (1993). Risk Prediction Models for Binary Response Variable for the Coronary Bypass Operation. M.Sc. thesis, Department of Statistics, University of British Columbia.

Index